W9-DDJ-066

Climate Change and Global Crop Productivity

Contents

Contributors

Göran I. Ågren, *Department of Ecology and Environmental Research, Swedish University of Agricultural Sciences, PO Box 7072, SE-750 Uppsala, Sweden*

Leon Hartwell Allen, Jr, *USDA-Agricultural Research Service, Institute of Food and Agricultural Sciences, S.W. 23rd St., University of Florida, PO Box 110965, Gainesville, FL 32611-0965, USA*

Jeffery T. Baker, *USDA-ARS/NRI/RSML, 10300 Baltimore Avenue, Beltsville, MD 20705-2350, USA*

Marco Bindi, *Department of Agronomy and Land Management, University of Firenze (DISAT), P. le delle Cascine 18, 50144 Firenze, Italy*

Herbert Blum, *Institute of Plant Sciences, Swiss Federal Institute of Technology (ETH), 8092 Zurich, Switzerland*

Kenneth J. Boote, *Agronomy Department, Institute of Food and Agricultural Sciences, University of Florida, Gainesville, FL 32611-0500, USA*

James A. Bunce, *Climate Stress Laboratory, USDA-ARS, Beltsville Agricultural Research Center, 10300 Baltimore Avenue, Beltsville, MD 20705-2350, USA*

Ruth E. Butterfield, *Environmental Change Unit (ECU), University of Oxford, 1a Mansfield Road, Oxford OX1 3TB, UK*

Bruce D. Campbell, *AgResearch, Grasslands Research Centre, Private Bag 11008, Palmerston North, New Zealand*

Reinhart Ceulemans, *Department of Biology, University of Antwerp, Universiteitsplein 1, B-2610 Wilrijk, Belgium*

Nordine Cheikh, *Monsanto Company, 1920 Fifth St, Davis, CA 95616, USA*

Jürg Fuhrer, *Institute of Plant Sciences, Swiss Federal Institute of Technology (ETH), 8092 Zurich, Switzerland*

James V. Groth, *Department of Plant Pathology, University of Minnesota, 495 Borlaug Hall, 1991 Upper Buford Circle, St Paul, MN 55108-6030, USA*

Andrew Paul Gutierrez, *Division of Ecosystem Science, University of California, 151 Hilgard Hall, Berkeley, CA 94720, USA*

Anthony E. Hall, *Department of Botany and Plant Sciences, University of California, Riverside, CA 92521-0124, USA*

Ann-Charlotte Hansson, *Biology and Crop Production Science, Swedish University of Agricultural Sciences, PO Box 7043, SE-755007 Uppsala, Sweden*

Jerry L. Hatfield, *USDA-Agricultural Research Service, National Soil Tilth Laboratory, 2150 Pammel Drive, Ames, IA 50011, USA*

Harry F. Hodges, *Department of Plant and Soil Sciences, 117 Dorman Hall, Box 9555, Mississippi State University, Mississippi State, MS 39762, USA*

Takeshi Horie, *Laboratory of Crop Science, Division of Agronomy, Graduate School of Agriculture, Kyoto University, Kitashirakawa Oiwake-cho, Sakyo-ku, Kyoto 606-8502, Japan*

Ivan A. Janssens, *Department of Biology, University of Antwerp, Universiteitsplein 1, B-2610 Wilrijk, Belgium*

Han Yong Kim, *Tohoku National Agriculture Experimental Station, Morioka, Iwate 020-0198, Japan*

Bruce A. Kimball, *USDA-ARS, Environmental and Plant Dynamics Research Unit, US Water Conservation Laboratory, 4331 East Broadway, Phoenix, AZ 85040, USA*

Ganesh Kishore, *Monsanto Company, 700 Chesterfield Parkway North, Chesterfield, MO 63198, USA*

Sagar V. Krupa, *Department of Plant Pathology, University of Minnesota, 495 Borlaug Hall, 1991 Upper Buford Circle, St Paul, MN 55108-6030, USA*

David W. Lawlor, *Biochemistry and Physiology Department, IACR-Rothamsted, Harpenden, Herts AL5 2JQ, UK*

Jan Lewandrowski, *US Department of Agriculture, Economic Research Service, 1800 M Street, Washington, DC 20036, USA*

Stephen P. Long, *Departments of Crop Sciences and Plant Biology, University of Illinois at Urbana-Champaign, 190 Edward R. Madigan Laboratories 1201, W. Gregory Drive, Urbana, IL 61801, USA*

Tsutomu Matsui, *Laboratory of Crop Science, Division of Agronomy, Graduate School of Agriculture, Kyoto University, Kitashirakawa Oiwake-cho, Sakyo-ku, Kyoto 606-01, Japan*

Linda O. Mearns, *Environmental and Societal Impacts Group, National Center for Atmospheric Research, PO Box 3000, Boulder, CO 80307-3000, USA*

Franco Miglietta, *Instituto Nazionale Analisi e Protezione Agro-ecosistemi, IATA-CNR, P. le delle Cascine 18, 50144 Firenze, Italy*

Philip W. Miller, *Monsanto Company, 700 Chesterfield Parkway North, Chesterfield, MO 63198, USA*

Rowan A.C. Mitchell, *Biochemistry and Physiology Department, IACR-Rothamsted, Harpenden, Herts AL5 2JQ, UK*

Jack A. Morgan, *USDA-ARS, Crops Research Laboratory, Fort Collins, CO 80526, USA*

Marianne Mousseau, *Laboratoire d'Ecologie Végétale, Université Paris-Sud XI, F-91405 Orsay Cedex, France*

Hiroshi Nakagawa, *Department of Agronomy, Graduate School of Agriculture, Kyoto University, Kitashirakawa Oiwake-cho, Sakyo-ku, Kyoto 606-01, Japan*

Park S. Nobel, *Department of Biology-OBEE, University of California, Los Angeles, CA 90095-1606, USA*

Josef Nösberger, *Institute of Plant Sciences, Swiss Federal Institute of Technology (ETH), 8092 Zurich, Switzerland*

Mary M. Peet, *Department of Horticultural Science, North Carolina State University, Raleigh, NC 27695-7609, USA*

H. Wayne Polley, *USDA-ARS, Grassland, Soil and Water Research Laboratory, 808 E. Blackland Rd, Temple, TX 76502, USA*

K. Raja Reddy, *Department of Plant and Soil Sciences, 117 Dorman Hall, Box 9555, Mississippi State University, Mississippi State, MS 39762, USA*

Donald C. Reicosky, *North Central Soil Conservation Laboratory, 803 Iowa Avenue, Morris, MN 56267, USA*

John Reilly, *MIT Joint Program on the Science and Policy of Global Change, Building E40-263, 77 Massachusetts Avenue, Cambridge, MA 02139, USA*

Ronald L. Sass, *Department of Ecology and Evolutionary Biology, Rice University, 6100 Main, Houston, TX 77005-1892, USA*

Ad. H.C.M. Schapendonk, *Research Institute for Agrobiology and Soil Fertility (AB-DLO), PO Box 14, 6700 AA Wageningen, The Netherlands*

David Schimmelpfennig, *US Department of Agriculture, Economic Research Service, 1800 M Street, Washington, DC 20036, USA*

Mark Stafford Smith, *National Rangelands Program, CSIRO Division of Wildlife and Ecology, PO Box 2111, Alice Springs, NT 0871, Australia*

Francesco P. Vaccari, *Instituto Nazionale Analisi e Protezione Agro-ecosistemi, IATA-CNR, P. le delle Cascina 18, 50144 Firenze, Italy*

David W. Wolf, *Cornell University, 134-A Plant Science Building, Tower Road, Ithaca, NY 14853-5908, USA*

Justus Wolf, *Department of Theoretical Production Ecology, Wageningen Agricultural University, PO Box 430, 6700 AK Wageningen, The Netherlands*

Kevin J. Young, *Departments of Crop Sciences and Plant Biology, University of Illinois at Urbana-Champaign, 190 Edward R. Madigan Laboratories, 1210 W. Gregory Drive, Urbana, IL 61801, USA*

Lewis H. Ziska, *Climate Stress Laboratory, USDA-ARS-BARC, B-046A, 10300 Baltimore Avenue, Beltsville, MD 20705-2350, USA*

Preface

Human activities are creating changes in our earth ecosystem. Emissions of carbon dioxide and other greenhouse gases are increasing. The evidence, using state-of-the-art computer models incorporating as much of the theoretical understanding of the earth's weather as possible, suggests that global warming is occurring along with shifting patterns of rainfall and incidences of extreme weather events. The rate of global climate change and warming expected over the next century is more than has occurred during the past 10,000 years. Changes in global environment will have profound effects and consequences for natural and agricultural ecosystems and for society as a whole. These changes could alter the location of the major crop production regions on the earth. Agricultural productivity is particularly vulnerable to disruption by weather. In the coming years, we have to produce more food, fibre and other commodities to cope with increasing population under diminishing per capita arable land and water and degrading soil resources and expanding biotic stresses. In addition to the above stringent constraints, shifting from 'normal weather', with its associated extreme events, zones of crop adaptation and cultural practices required for successful crop production will also surely change. Also, plant responses to climatic changes are not uniform and thus there will be winners and losers. Climate and weather-induced instability in food and fibre supplies will alter social and economic stability and regional competitiveness.

This book describes normal historical shifts in the earth's temperature and weighs the evidence concerning anthropogenically induced changes in temperature. It discusses the methods of predicting climatic changes and the role of today's agriculture in the production and release of greenhouse gases. The major aim of the text is to quantify the impact of altered climatic factors on different crops. The major food and fibre crops are evaluated, and crop responses to water and nutrient deficiencies in high CO_2 environments are predicted.

Chapters dealing with crop and weed interactions and insect population dynamics in an altered climatic setting are written by the world's most knowledgeable authorities. Topics include crop responses of crassulacean acid metabolism (CAM) plants, plants with C_3 and C_4 type photosynthesis, how and why respiration is depressed in high CO_2 environments, and how and why fruit and grain production is depressed at high temperatures. How growers and production specialists may cope with such environments and the need for improving cultivars with both conventional breeding methods and transgenic techniques are explored. Many other topics are discussed to illustrate the variabilities and similarities of the major economically important crop species.

Agronomists, horticulturalists, crop production specialists, environmental physiologists and others interested in global environmental issues will be interested in reading this book. The impact of climatic change is not only on the major food producing species (wheat, rice, soybean, maize, vegetable crops and root and tuberous crops), but also on productive grassland crops, rangelands, tree farming, and crops produced in deserts. The responses of cotton, the major textile-fibre crop, to environmental factors are quantified. The impact of climatic change on weed and insect pests of crops, as well as on economic, social and trade aspects, is also presented.

Acknowledgements

The editors wish to express their profound appreciation and gratitude to all authors of individual chapters for sharing their expertise expressed in this work. The helpful suggestions of many reviewers are acknowledged and thanks offered. The reviewers include: Drs Wayne Cole, Jann Conroy, Jerry Eastin, Otto Doering, Alex Friend, Elizebeth Jackson Heatherly, Mary Beth Kirkham, Rattan Lal, W.J. Manning, Gary Paulson, C. Potvin, John Read, Ray Wheeler, John Schneider, P.S. Verma, Tim Wheeler, Jeff Willers, Stan Wullschleger and Lewis Ziska. We especially acknowledge Dr Larry Heatherly for carefully reading all chapters and offering numerous suggestions. We also express our appreciation to Mississippi State University, Mississippi Agriculture and Forestry Experiment Station, the National Institute for Global Environmental Change and the US Department of Energy for supporting our research and encouraging the assembly and publication of this information.

1 Climate Change and Global Crop Productivity: an Overview

K. RAJA REDDY AND HARRY F. HODGES

Department of Plant and Soil Sciences, 117 Dorman Hall, Box 9555, Mississippi State University, Mississippi State, MS 39762, USA

Weather is the most important cause of year-to-year variability in crop production, even in high-yield and high-technology environments. There has been considerable concern in recent years about the possibility of climatic changes caused by human activities, because any change in weather will increase uncertainty regarding food production. Since the beginning of the industrial revolution, earth's population has increased dramatically, with accompanying large-scale burning of fossil fuels, the manufacture of cement, and intensive cultivation of lands not previously used for crops or livestock production. The largest population in human history will occur during the 21st century and thus dictate greater pertinence of climatic changes because the consequences may be so great and drastic.

Any observed or predicted changes in the global climate are of fundamental concern to man, because the present climate of the earth is so well-suited to support life. Alterations in our climate are governed by a complex system of atmospheric and oceanic processes and their interactions. In the context of crop production, relevant atmospheric processes consist of losses in beneficial stratospheric ozone concentration ($[O_3]$) and increasing concentrations of the surface-layer trace gases, including atmospheric carbon dioxide ($[CO_2]$), methane ($[CH_4]$), nitrous oxide ($[N_2O]$) and sulphur dioxide ($[SO_2]$). Surface level $[O_3]$, $[SO_2]$ and $[CO_2]$ have direct impacts on crops, while $[CO_2]$, $[CH_4]$ and $[N_2O]$ are critical in altering air temperature. Products of atmospheric processes also result in increases in surface-level ultraviolet radiation and changes in temperature and precipitation patterns.

There is considerable uncertainty associated with determining the current average temperature of the earth and determining historical temperatures is an even more daunting task. Obtaining global or even regional averages is difficult, because both diurnal and seasonal temperatures vary considerably from place to place. For this reason, the possibility of climatic change is somewhat controversial in the public view. However, the increasing atmospheric CO_2 and other radiative gases in recent years are well documented and theoretical reasons for higher concentrations of these gases to cause global warming are not disputed.

Changes in temperature during geological and even recent historical times have occurred as evidenced by ice ages and the cool period (sometimes called the Little Ice Age) during the 16th and 17th centuries. Unfortunately, there is not a clear explanation for these shifts in temperature, but it seems unlikely that pre-industrial variations were related in any way to human activity.

Much of our knowledge of future climatic change comes from studies using climate models. Climate models are complex mathematical representations of many of the processes known to be responsible for the climate. The processes include interactions between atmosphere and land surfaces to attain topographical effects, ocean currents and sea ice. The models simulate global distributions of variables such as temperature, wind, cloudiness and rain. These models have evolved in complexity over time as knowledge of atmospheric physics has increased and computational technology has improved. Today's three-dimensional atmospheric models are linked with mixed-layer ocean models that allow the annual seasonal cycle of solar radiation to be included. As understanding and computational power have increased, more detailed topography of land surfaces, vegetation and atmospheric interactions have been described more completely and smaller grids are being used. As such mechanistic details have increased, the simulation of current climates has improved both seasonally and spatially; however, relatively large discrepancies sometimes still occur.

Two climate models were run to the year 2100 for the National Assessment Program in the USA. One showed regional temperature increases of 5°C in winter and 3°C in summer by the year 2060, while the other predicted even greater increases. The models did not agree on specific regional climate changes, e.g. one had precipitation increases and the other decreases in the southeastern USA during the summer. It appears reasonable to conclude that since the concentration of radiative gases is clearly increasing in the atmosphere and the theoretical reasons for causing warming are not disputed, the conditions to induce warming are in place. Many atmospheric scientists agree with this assessment.

Agriculture provides a sizable contribution to the radiative gases that appear to be the driving forces in climatic change. The primary sources of these gases are the fossil fuel used in agricultural activities, soil carbon (C) loss because of tillage operations associated with crop culture, burning crop and forest residues, raising livestock and consequent manure-handling operations, manufacture and utilization of N fertilizer, and growing of flooded rice. Rice production in flooded paddies and lagoon storage of barnyard manure cause the production of relatively large quantities of CH_4, while various aspects of fertilization result in the release of N_2O. Methane and N_2O cause considerably more radiative forcing (21 and 310 times, respectively, per unit mass of gas) than does atmospheric [CO_2].

A large amount of C is stored in the soil and is relatively labile. It is subject to management as agricultural practices may result in the gain or loss of C from the soil. Less tillage usually results in more soil C accumulation and the resulting desirable attributes associated with soil conservation and sustainable crop production. Recent advances in herbicide technology make less tillage more economically feasible, because the primary reason for tillage in crop production is to control weeds. Crop production that utilizes improved herbicides allows reduced cultivation and also results in the use of less fossil fuel than do soil tillage operations that require high energy. In addition, less tillage usually results in secondary benefits such as better water infiltration, greater soil aggregate stability, lower susceptibility to erosion, and improved plant water relations with a lower incidence of crop drought injury. Organic C compounds bind soil aggregates and cause them to resist the breakdown but repeated tillage induces their degradation.

Animal manure represents a relatively small overall contribution of N_2O, CH_4 and CO_2 to the atmosphere. However, modern methods of livestock production that result in concentrations of animals have caused considerable concern among the public for finding ways to minimize undesirable odours, contamination of water sources, and atmospheric pollution by subsequent radiative gases from manure. Even though production of livestock in large concentrations accentuates the problem in a local community, the overall effect of producing a similar number of livestock animals that are evenly distributed over a wide area probably would not result in a very different overall impact, except for the production of methane in lagoons.

Methane production in flooded rice is correlated with biomass production during vegetative growth, but in areas where two crops per year are grown some management practices can be used to reduce CH_4 production and emission without yield loss. Methane emission is highly sensitive to water management; however N_2O emissions may result from practices that minimize CH_4 production. Additional information is needed to find ways to minimize both CH_4 and N_2O emissions during flooded rice production.

On a global scale, the C cycle consists of large C reservoirs having flows from one reservoir to another. The largest of these reservoirs is the ocean, followed by soil, atmosphere and, finally, living organic matter. Natural processes in the oceans and plant biomass are responsible for most CO_2 absorption and emission. The most important natural processes are the release of CO_2 from oceans, aerobic decay of plant materials, and plant and animal respiration. Green plants, through photosynthesis, sequester a great deal of C and at the same time return about 50% of that sequestered C to the atmosphere through respiration. The remaining 50% becomes biomass that is eventually oxidized slowly through microbial decomposition and also released to the atmosphere.

Soil is the major repository site of C where organic matter decomposition takes place. Organic matter in the soil consists of both living and non-living components. The living component is composed of plant roots, soil micro-organisms and animals, while the non-living component includes remnants of microorganisms and plant and animal materials. New non-living organic matter typically decomposes rapidly, leaving a residual quantity that becomes more recalcitrant with time so that some organic C may remain in soil for thousands of years. Plant materials produced in a C-rich environment may have a higher percentage of lignin and therefore be more resistant to decomposition. Little is known about the effects of high CO_2 environments on the rate of decomposition of the materials produced and therefore on litter and soil organic matter accumulation. The rate of decomposition depends on temperature, moisture, chemical composition of the decomposing material, the soil chemical environment and land use. Thus, the amount of C in soil depends on the balance between the input of photosynthetically fixed C and the loss of C through biomass decomposition. Agricultural practices modify both of these processes.

Substantial quantities of C are temporarily stored in plant materials. Natural ecosystems, particularly forests, store C for relatively long periods in tree trunks. The percentage of the earth's terrestrial surfaces covered with natural forests has decreased from 46% in pre-industrial times to only about 27% today. However, tree plantations are becoming more important in today's society and provide C storage for intermediate periods in vegetative structures. Fast-growing trees are managed to provide useful pulp and structural fibres and are grown on approximately 130 Mha.

The major changes in the earth's atmosphere are the concentrations of CO_2, which have increased by about 25% since the beginning of the industrial revolution. Carbon dioxide enhances photosynthesis and depresses plant respiration; these effects are expected to increase plant growth as well as affecting various other processes. However, a number of plant physiological processes are also affected by changes in temperature, ozone, ultraviolet radiation, nutrients and water, all of which are variable factors often associated with climatic change. This book addresses the way the most important food and fibre crops react to these physical and chemical changes and how we might expect the hypothesized changes to impact humanity's ability to live in the changing environment.

This book examines the case for man-induced climatic changes, the role of agriculture in these apparent changes, and the impact of those changes on agriculture. As a consequence of changes in food and fibre production, society will surely be affected. Some regions will likely be affected negatively, while other regions may benefit. Since trade and commerce are important today, the causes of economic shifts may not be transparent to the public but, nevertheless, the changes will occur. Most of the crops of major economic consequence are considered and the impacts of environmental changes are reviewed. Crops that are important for human food (rice, wheat, soybean, vegetable crops and root and tuberous crops) are examined. Crops that are used primarily as animal foods (maize, sorghum, productive forage crops and rangelands) are also reviewed. Crops grown primarily for fibres, such as cotton and tree crops, are included as well as some desert-grown species harvested for making beverages and other uses. The impact of environmental factors on various physiological processes and on the yield of the harvestable components are evaluated. Plants with C_4 and C_3 type photosynthesis and plants with crassulacean acid metabolism (CAM) are compared for their sensitivities to likely environmental changes.

Crop/weed and crop/insect pest interactions and the relative importance of pests on crop production in a changing climatic environment are discussed. The role that conventional breeding procedures have played in the past and this role in a changing environment, along with recent innovations in transgenic techniques, are contemplated. Transgenic techniques have only recently been applied to the solution of abiotic problems. Introductory studies suggest that dramatic and often unexpected responses to environmental stresses may be obtained by altering certain genes. Tolerance to more than one environmental stress may be increased as a result of changing a single gene. The importance of developing cultivars tolerant of various environmental stresses will increase and both transgenic and conventional breeding procedures will be essential.

Global warming will probably have a negative impact on tropical regions and any other areas where high temperature or inadequate rain often limits crop productivity. Regions where cold temperatures are the primary factor limiting crop production will probably benefit most from warming. Farmers in temperate regions, where most food is produced, will find ways of altering production practices to avoid the occurrence of particularly temperature-sensitive crop growth stages during brief periods of extremely high temperatures. The relative importance of particular crops in certain areas will likely change due to global warming and economic factors will determine these changes.

The importance of altering cultural practices and engineering techniques will also continue to be urgent to a thriving human culture. In temperate regions, the greatest risk to crop production associated with global climatic change is being caused by changes in frequency of extreme events. Unexpected early or late frosts can destroy the production of a crop in an otherwise favourable season. Increased incidences of drought or floods can likewise drastically alter crop production potential. These considerations and conditions will favour the industrialized regions of the world and place people in regions less able to change at an even greater disadvantage than they are today. Ways of extending the advantages of science and technology to disadvantaged people remain one of society's most daunting challenges.

2 Climatic Change and Variability

LINDA O. MEARNS

Environmental and Societal Impacts Group, National Center for Atmospheric Research, PO Box 3000, Boulder, CO 80307, USA*

2.1 Introduction

With the acceptance of the Kyoto Protocol to the United Nations Framework Convention on Climate Change (UNFCCC) in December 1997, possible climatic change due to anthropogenic pollution of the atmosphere in the 21st century became a higher profile global issue than ever before. Fears, uncertainties and confusion regarding necessary adjustments in environmental and economic policy in order to reduce emissions of greenhouse gases and aerosols have intensified. These fears and uncertainties have complicated scientific discussion on the effect of increased greenhouse gases on the climate and on how to reduce greenhouse gas emissions in a globally effective and equitable way. Moreover, the political posturing and strategizing of special interest groups in both developed and developing countries has sometimes led to the obfuscation of the issue. Hence, misinformation about the science of climatic change is often presented. Also, the media has a tendency to produce so-called 'balanced' news reports using representatives of two different sides of a question, with the comparative scientific merits of each side's arguments being largely unknown. This leads to much confusion for those interested and engaged in research into global climatic change.

These confusions and debates are particularly compelling in the arena of global climatic change and its relation to agriculture, because agricultural activities relate to and interact with so many different aspects of the climatic change issue (Rosenzweig and Hillel, 1998). Agricultural activities contribute significantly to the production of greenhouse gases, such as methane (CH_4) and nitrous oxide (N_2O) (see Chapter 3, this volume). In fact, agricultural activities account for 20% of the current increase in radiative forcing of the

* The National Center for Atmospheric Research is sponsored by the National Science Foundation.

climate system (Rosenberg *et al.*, 1998). Agricultural enterprise is a human-managed system with a fundamental biophysical base, and this complexity makes the task of discerning effects of climatic change difficult.

As a biophysical system dependent on climatic resources, agriculture can be affected through changes in crop yields and production. The positive fertilization effect of increased CO_2 on agricultural crops could have a beneficial effect on some crops, particularly the C_3 species (see Chapter 4, this volume). Since application of technology (e.g. new cultivars and pesticides, irrigation, and herbicide application) is such an important factor in agriculture, there is much discussion and debate about how successfully agriculture (regionally and globally) can adapt to whatever climatic changes may occur (Easterling, 1996; Reilly, 1996; Rosenzweig and Hillel, 1998).

This chapter reviews the science regarding climatic change induced by greenhouse gases. The fundamental physical climatic processes of such change, the current trends in the climate record, the climate-modelling tools used to study climatic change and results of relevant climate modelling experiments are reviewed. The most up-to-date information on what is now known and understood about climatic change is provided. Several of the more controversial issues in the climatic change debate are also discussed. The chapter begins with a brief review of past climatic changes to put the present climate in a broad perspective.

2.2 Past Climate Change

2.2.1 Distant past – millions to thousands of years before present (BP)

Climatic change occurs on diverse scales of time and space. The largest changes have occurred on the same time scale as that of drifting continents. However, large variations, such as interglacial periods that have marked the climate record during the past 3 million years or so, occurred in cycles that lasted tens to hundreds of thousands of years (Fig. 2.1a) (Schneider *et al.*, 1990). Thus, climatic change is the normal state of affairs for the earth/atmosphere system. This suggests that the notion of a stable, stationary climate is an erroneous concept, while that of ‘unceasing climatic change’ may be a more useful mental model by which to analyse the earth’s climate resources through time.

Climates of the past billion years have been about 13°C warmer to 5°C cooler than the current climate (Schneider *et al.*, 1990). Prominent in earth’s recent history have been the 100,000-year Pleistocene glacial/interglacial cycles when climate was cooler than at present (Fig. 2.1a). Global temperature varied by about 5°C through the ice age cycles. Some local temperature changes through these cycles were as great as 10–15°C in high latitude regions. During the last major glaciation, ice sheets covered much of North America and northern Europe, and sea level averaged 120 m below current values.

Since the last glaciation, there have been relatively small changes of probably less than 2°C (compared with the current global mean temperature) in

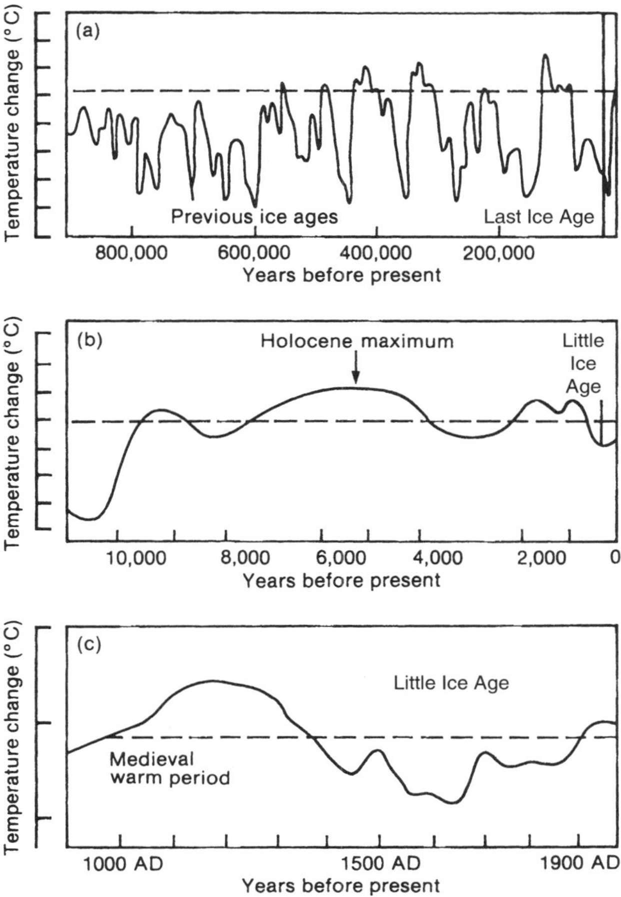

Fig. 2.1. Schematic diagrams of global temperature variations since the Pleistocene on three time-scales: (a) the last million years; (b) the last 10,000 years; and (c) the last 1000 years. The dotted line nominally represents conditions near the beginning of the 20th century. Each unit on the *x*-axis of all three panels represents 1°C. (Source: Folland *et al.*, 1990. Reprinted with permission of Hadley Centre for Climate Prediction and Research, Meteorological Office, Bracknell, UK.)

global average temperatures. During the mid-Holocene epoch between 4000 and 6000 years BP, however, temperatures rose significantly, particularly during the summer in the northern hemisphere (Fig. 2.1b) (Folland *et al.*, 1990).

2.2.2 Recent past – medieval optimum and the Little Ice Age

Fluctuations in the distant past are important for analysing the various causes of climatic fluctuations, but substantial fluctuations within recent human history (Fig. 2.1c) are more compelling for humans. For example, the so-called medieval optimum occurred from about the 10th to early 12th centuries. There is evidence that western Europe, Iceland and Greenland were exceptionally warm, with mean summer temperatures that were more than 1°C higher than current ones. In western and central Europe, vineyards extended as much as 5 degrees latitude farther north than today (Gribbin and Lamb, 1978). Not all regions experienced greater warmth; China, for example, was considerably colder in winter.

The most notable fluctuation in historical times was that known as the Little Ice Age, which lasted roughly from 1450 to the mid-19th century. earth's average temperature at one point was 1°C less than that of today (Fig. 2.1c) (Lamb, 1982). The effects of the Little Ice Age on everyday life are well documented: the freezing over of the river Thames in London; the freezing of New York harbour, which allowed citizens of New York to walk to Staten Island; abandonment of settlements in Iceland and Greenland; and crop failures in Scotland (Parry, 1978).

While there has been wide speculation on the cause or causes of the Little Ice Age (e.g. increased volcanism, reduced solar activity), there is no definitive explanation. This fluctuation is significant because it ended before the heavy industrialization of the late 19th century began, and is thus believed to be largely free of human causes. Some have argued that the increased global temperature that has been observed in the 20th century represents a 'recovery' from the Little Ice Age. However, without a definitive cause for the coolness of the Little Ice Age, the concept of a recovery from that anomalous cold period remains dubious.

2.3 20th Century Climate

Considerable effort has been expended to analyse the current historical climate record over the past 150 years in order to establish whether there are any trends that could be attributed to greenhouse gas warming. For example, a number of global data sets of near-surface temperature have been developed and analysed (Nicholls *et al.*, 1996). These data sets have been carefully corrected for errors or bias due to urban heat-island effects, non-homogeneities (such as instrument or location change) and changes in bucket types used to measure sea surface temperatures. One of the most carefully constructed data sets (Jones *et al.*, 1994, 1999) indicates there has been a 0.3°C to 0.6°C warming of the earth's surface since the late 19th century (Fig. 2.2). This trend continued through 1998. The global average temperature from January to December 1998, was the warmest on record for the period 1880–1998 (National Climate Data Center web site, www.ncdc.noaa.gov). Specifically, the global average temperature for January until June 1998, was 0.6°C higher than the 1961–1990 global mean temperature. However, the distinct warming has not been regionally homogeneous, since some regions have experienced cooling during the 20th century.

The diurnal temperature range has primarily decreased in most regions, indicating that minimum temperatures have warmed more than maximum

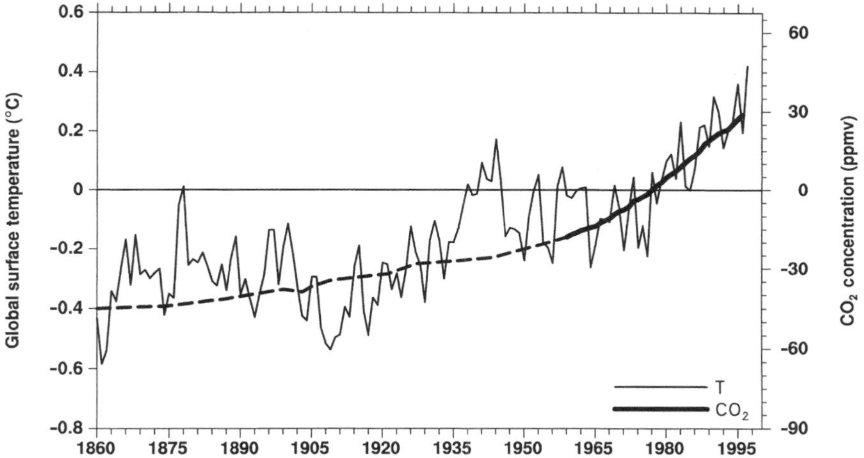

Fig. 2.2. Estimated changes in annual global-mean temperatures (thin line) and carbon dioxide (thick line) over the past 138 years relative to the 1961–1990 average (horizontal solid line). Earlier values for carbon dioxide are from ice cores (dashed line), and for 1959–1996 from direct measurements made at Mauna Loa, Hawaii. The scale for carbon dioxide is in parts per million (ppmv) relative to a mean of 333.7 ppmv. (Source: Hurrell, 1998.)

temperatures, or that cloudiness has increased in these areas (Karl *et al.*, 1993, 1996). A cooling of the lower stratosphere by 0.6°C has also occurred since 1979.

Sea level has increased on average between 10 and 25 cm over the past 100 years (Nicholls *et al.*, 1996). This rise is related to the increase in near-surface temperatures, which has caused thermal expansion of the oceans and melting of glaciers and ice caps. Thermal expansion of the oceans has contributed between 2 and 7 cm to the total increase in sea level.

There has also been a small increase in global average precipitation over land during the 20th century (Dai *et al.*, 1997). Recent investigations indicate that this mean increase has mainly influenced heavy precipitation rates (Groisman *et al.*, 1999). Since the late 1970s, there have been increases in the percentage of the globe experiencing extreme drought or severe moisture surplus (Dai *et al.*, 1998).

Many of the global climatic changes are analysed from the point of view of expected combinations of changes to different variables known as fingerprints. For example, the combination of cooling of the stratosphere, warming of the surface temperature and increased global mean precipitation is expected from increased greenhouse gas-induced climate change. These anticipated combinations of changes and their spatial patterns are based on our understanding of the physics of the earth/ocean/atmosphere system and results from climate models (Santer *et al.*, 1996). Fingerprinting will be discussed further in section 2.6.4 on climate models.

In the 1990s, remotely sensed temperature data of the lower troposphere (700 millibar [mbar] height, about 2.5 km) have been analysed to compare trends with those from surface observation stations. This has led to a debate regarding the robustness of results from surface observations compared with remotely sensed data (Christy and Spencer, 1995; Hurrell and Trenberth, 1997, 1998). For example, Christy and Spencer used data from the microwave sounding units (MSU) on board the National Oceanic Atmospheric Administration (NOAA) polar orbiting satellites and investigated the time series from 1979 to 1995. They found a slight global cooling of –0.04°C over this time period. According to surface observations, the temperature has increased since 1979 by 0.14°C. This putative discrepancy has been used to question the surface temperature record and to provide evidence that global warming is not occurring.

The intense political debates concerning global warming tend to polarize results from scientific research on the subject and result in oversimplification of the methods/techniques and research results. In fact, the differences in these global mean trends can largely be explained based on the differences in the physical quantities being measured (Hurrell, 1998). The MSU measures lower tropospheric temperatures at an altitude of about 2.5 km, whereas the surface measurements are made on land at a screen height of about 2 m. Differences in the thermal characteristics of land and ocean affect the relative correlations of temperature with altitude. Over the ocean, the sea surface temperatures will not necessarily be highly correlated with temperatures at 700 mbar heights.

These and other issues, such as the myriad technical difficulties associated with using remotely sensed data to establish trends (Hurrell, 1998; Hurrell and Trenberth, 1998; Kerr, 1998), suggest that the intense debate is more political than scientific.

2.4 Role of Greenhouse Gases

2.4.1 The greenhouse effect

The greenhouse effect is a natural feature of the climate system. In fact, without the atmosphere (and hence the greenhouse effect), the earth's average temperature would be approximately 33°C colder than it is currently. The earth/atmosphere system balances absorption of solar radiation with emission of longwave (infrared) radiation to space. The earth's surface primarily absorbs most of the shortwave solar radiation from the sun, but it also reradiates some of this radiation as longwave radiation (Fig. 2.3). Energy is lost before reaching the surface of the earth through reflection from clouds and aerosols in the atmosphere. Little is directly absorbed by the atmosphere, which is relatively transparent to shortwave radiation. Also, an average of about 30% is reflected off the earth's surface.

The atmosphere is more efficient at absorbing longwave radiation, which is then both emitted upward toward space and downward toward the earth. This downward emission serves to heat the earth further. This further warming by reradiated longwave radiation from the atmosphere is known as the greenhouse effect. The amount of longwave radiation that is absorbed and

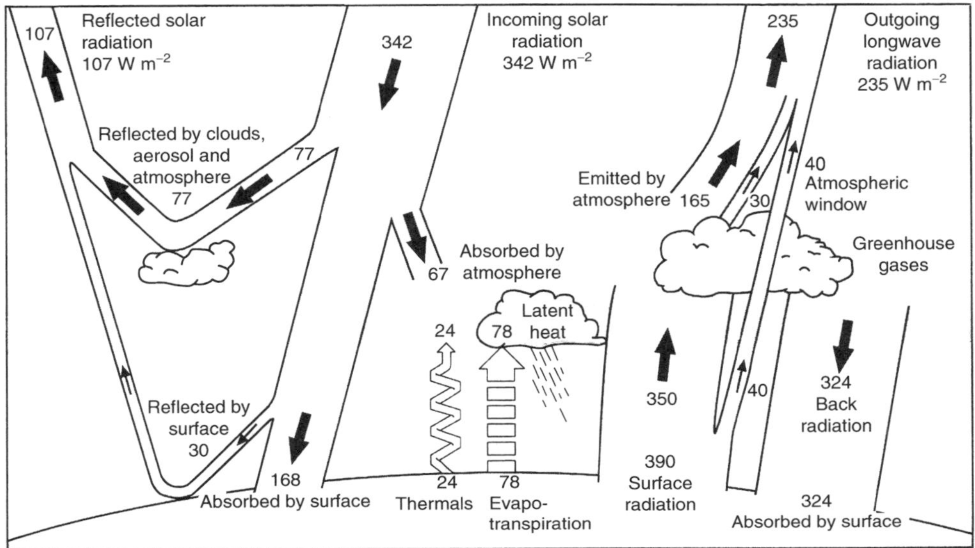

Fig. 2.3. The earth's radiation and energy balance. The net incoming radiation of 342 W m^{-2} is partially reflected by clouds and the atmosphere, or at the surface. Some of the heat absorbed at the earth's surface is returned to the atmosphere as sensible and latent heat. The remainder is radiated as thermal infrared radiation and most of that is absorbed by the atmosphere, which in turn emits radiation both up and down; this produces the greenhouse effect. (Source: Kiehl and Trenberth, 1997.)

then reradiated downward is a function of the constituents of the atmosphere. Certain gases in the atmosphere are particularly good at absorbing longwave radiation and are known as the greenhouse gases. These include water vapour, carbon dioxide (CO_2), methane (CH_4), some chlorofluorocarbons (CFCs) and nitrous oxide (N_2O) (see Chapter 3, this volume).

If the make-up of the atmosphere changes and the result is an increase in concentrations of the greenhouse gases, then more of the infrared radiation from earth will be absorbed by the atmosphere and then reradiated back to earth. This changes the radiative forcing of the climate system and results in increased temperature of the earth's surface. Such perturbations in the radiation balance of the earth system are known as changes in the radiative forcing, and the factors that affect this balance are known as radiative forcing agents (Shine *et al.*, 1990). One of the ways in which the effect of greenhouse gases is measured is by determining their radiative forcing. This can be viewed as a measure of their relative ability to alter the climate.

2.4.2 Current concentrations of greenhouse gases

This section discusses the major greenhouse gases and their relative contribution to the current radiative forcing of the atmosphere (Fig. 2.4), compared

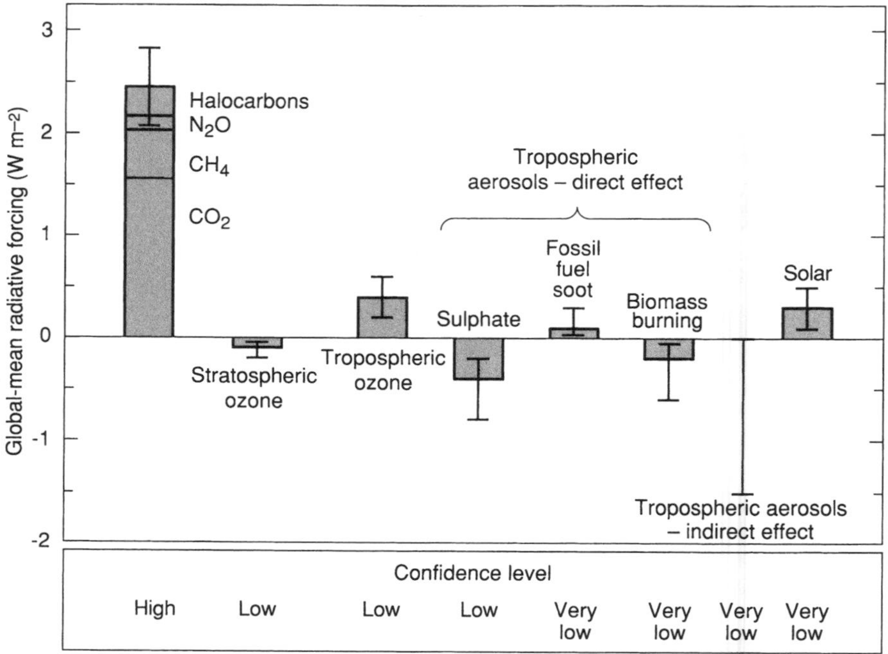

Fig. 2.4. Estimates of global and annual average radiative forcing (W m^{-2}) attributable to changes in greenhouse gases and aerosols, and the solar flux from 1850–1990. The height of a rectangular bar indicates either the best or mid-range estimate of the forcing; the vertical lines in the bars represent the uncertainty range. (Source: Schimel *et al.*, 1996.)

with pre-industrial times (1850). Because of human activities, atmospheric concentrations of greenhouse gases (except water vapour) have increased considerably since the beginning of the industrial revolution. Sources of these gases include fossil fuel burning, tropical deforestation, biomass burning, chemical industrial activities, and agricultural activities.

Carbon dioxide is the must abundant of the greenhouse gases. Concentrations of CO_2 [CO_2] have increased from about 280 parts per million by volume (ppmv) in the pre-industrial period to 358 ppmv in 1995, with a rate of increase of 1.6 ppmv per year (Fig. 2.2). The concentration by mid-1998 was 368 ppmv. Major sources of CO_2 emissions include burning of fossil fuels and production of cement. Tropical deforestation contributes to [CO_2] increase by removing vegetation, which is one of the major sinks of CO_2 (Weubbles and Rosenberg, 1998).

On a per molecule basis, methane is a more effective greenhouse gas than CO_2. The current globally averaged concentration is 1.72 ppmv and is increasing at the rate of 1% per year. For reasons that are not completely understood, the rate of increase in methane concentration decreased in the early 1990s, but has returned to a higher level of increase in the past few years. Methane is produced in rice culture, ruminant fermentation, landfills, and through losses during gas production and distribution and coal mining.

Nitrous oxide is even more efficient than methane in absorbing longwave radiation. Its mean concentration in 1990 was about 311 parts per billion by volume (ppbv) and has been increasing by about 0.2–0.3% per year. Its pre-industrial level was about 275 ppbv. Sources of the increase include fertilized soils that are used for crop production, biomass burning, industrial processes and feed lots. Hence, this is the greenhouse gas, along with methane, that is most strongly associated with agricultural activities (Schimel *et al.*, 1996; see also Chapter 3, this volume).

Chlorofluorocarbons (CFCs) are inordinately efficient greenhouse gases with a relatively long lifetime of about 100 years. Used primarily as propellants and refrigerants, they are perhaps better known for their role in the destruction of ozone in the stratosphere rather than as greenhouse gases. The CFCs 11 and 12 currently have the largest concentrations in the atmosphere (0.27 and 0.50 ppbv, respectively). Their concentrations have increased, but at a diminishing rate in the later 1990s. However, these and other CFCs accounted for 15% of the increase in radiative forcing since 1900 and contributed nearly 25% of the increased forcing in the 1980s (Houghton *et al.*, 1990). The continued phasing out of CFCs, as per the Montreal Protocol, indicates that these will become less significant greenhouse gases over time. They are being replaced by halogenated hydrocarbons (HCFCs), which, although still having some capacity as greenhouse gases, have less capacity than the CFCs they are replacing. Whether or not these changes are significant in the future depends on how large their emissions become.

The depletion of stratospheric ozone has been a problem because it has allowed increased ultraviolet radiation to reach the earth's surface. However, the increase in tropospheric ozone is also problematic, since tropospheric ozone acts as a greenhouse gas and is also a pollutant that affects humans, plants and animals. Tropospheric ozone contributed more than N_2O to the positive forcing of the climate system during the 20th century.

The relative contribution of these different gases to the change in radiative forcing from pre-industrial times to the present, as well as that of some other external forcing agents, such as variations in solar activity, is presented in Fig. 2.4. Clearly, the largest contributor to the positive forcing has been the increase in greenhouse gases. In order of importance, they are CO_2, CH_4, the halocarbons and N_2O (see Chapters 12 and 18, this volume).

2.4.3 Feedback to natural greenhouse gases (water vapour)

The most abundant greenhouse gas in the atmosphere is water vapour, but humans are not directly increasing its amount. However, it is anticipated that atmospheric increases in other greenhouse gases will lead to global warming, which will in turn lead to increased water vapour in the atmosphere because of increased evaporative capacity. Therefore, increase in water vapour is viewed as a feedback from the increases in the anthropogenically produced greenhouse gases CO_2, CH_4, N_2O and CFCs, rather than as an anthropogenically generated greenhouse gas.

2.4.4 Future increases in greenhouse gases

There is no doubt that there have been significant and even alarming increases in greenhouse gas concentrations. How they may change in the future is highly uncertain, and this is one of the most difficult problems in studying possible future climatic change, because changes in these gases greatly depend upon changes in the future economic and political activities of all nations. This is particularly true of those countries or regions, such as the USA, western Europe and China, that are or will be largely responsible for most of the future emissions of these gases. Predicting economic and environmental policy development on a global scale is a most daunting task.

Numerous scenarios of possible future increases of greenhouse gases have been constructed. They are based on different assumptions of future human activities, such as economic growth, technological advances, and human responses to environmental or socioeconomic constraints (Jager, 1988; Houghton *et al.*, 1990, 1992, 1996; German Bundestag Enquête Commission, 1991). Thus, it should be understood that these scenarios of future greenhouse gas emissions are highly uncertain and become more so as the length of time of the projection increases.

Four scenarios were developed for the first Report of the Intergovernmental Panel on Climate Change (IPCC) (Houghton *et al.*, 1990). These scenarios assumed identical population increases and economic development scenarios, but different technological development and environmental controls. Individual scenarios using changes in concentrations of CO_2, CH_4, N_2O and CFCs were developed. The scenarios varied from Business as Usual (BAU) with very little environmental control to scenario C with high levels of controls. The rate of increase of greenhouse gases decreased with increasing environmental control.

The IPCC 1992 Supplement (Houghton *et al.*, 1992) provided a more detailed set of scenarios that have been frequently used in a number of global change contexts. Six alternative scenarios (IS92a–f) to the year 2100 were constructed based on different quantitative assumptions about population growth, economic growth and energy supplies within different world sectors, i.e. developing and developed countries (Fig. 2.5a). The IS92a and IS92b scenarios were more or less updates of the scenarios presented in the 1990 IPCC report (Houghton *et al.*, 1990) and form 'middle-of-the-road' projections (Leggett *et al.*, 1992). The other scenarios assumed rates of change in emissions of greenhouse gases that encompassed a large total range. For example, gigatons (10^9 tons) of carbon (GtC) emitted by 2100 ranged from 4.6 (IS92c) to 35.8 (IS92e) and significantly departed from an actual 1990 value of 7.4 GtC $year^{-1}$. The IS92a 'middle-of-the-road' scenario assumed an emission of 20.3 GtC $year^{-1}$ (Fig. 2.5a).

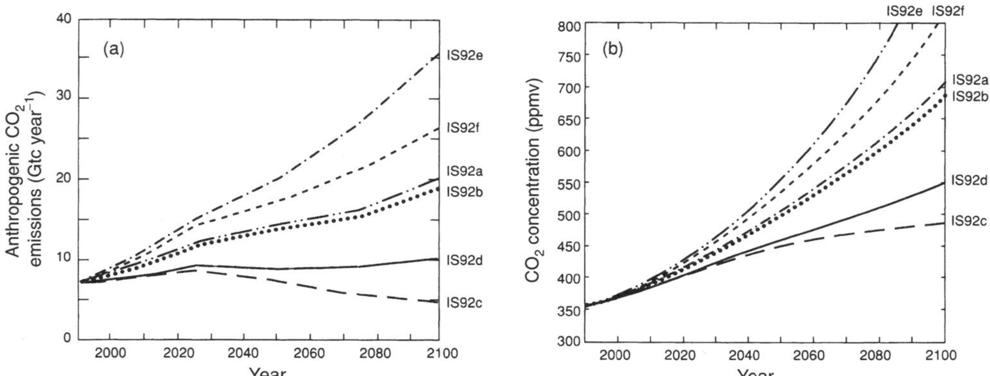

Fig. 2.5. (a) Total anthropogenic CO_2 emissions under the IS92 emission scenarios; (b) the resulting atmospheric CO_2 concentrations using a carbon cycle model. (Source: Houghton *et al.*, 1996.)

When given a particular quantity of emissions, the calculation of future concentrations of greenhouse gases in the atmosphere is determined by modelling the processes that transform and remove the relevant gases from the atmosphere. Future concentrations of CO_2, for example, are quantified using carbon cycle models that simulate exchange of CO_2 among atmosphere, oceans and biosphere. The CO_2 concentrations corresponding to the emissions scenarios listed above range from about 480 ppmv to well over 1000 ppmv, with about 710 ppmv for IS92a (Fig. 2.5b).

2.5 Anthropogenic Aerosol Effects

2.5.1 Physical processes and current distributions

In recent years the potentially countervailing influence of aerosol forcing to increased greenhouse gas forcing on the atmosphere has become a new focus of attention (Wigley, 1989; Houghton *et al.*, 1990). Aerosols are solid or liquid particles in the size range of 0.001 to 10 μm. Aerosols in the atmosphere influence the radiation balance of the earth directly, through scattering and absorption, and indirectly, by altering cloud properties. They affect the size, number and chemical composition of cloud droplets. When the number of aerosol particles increases, the number of cloud droplets increases, resulting in a higher cloud albedo and subsequent greater reflectivity.

Tropospheric aerosols result from combustion of fossil fuels, biomass burning and other activities and have led to a globally averaged direct negative radiative forcing (i.e. cooling effect) of about -0.5 W m^{-2} (Fig. 2.4) (Schimel *et al.*, 1996). The indirect effect of aerosols is more difficult to quantify. Note, however, that there is some positive forcing from black soot aerosols that directly absorb solar radiation (Fig. 2.4). An important difference between aerosol and greenhouse gas effects is that the former is heterogeneously distributed through the atmosphere, and thus has important local or regional effects. Aerosols are also much more short-lived in the atmosphere; thus, their concentrations respond relatively quickly to changes in emissions of these substances. They are predominantly found in the lower 2 km of the atmosphere. The major concentrations of sulphate aerosols are found over the eastern half of the USA, Europe and eastern China. These correspond to regions of intense industrial activity (Houghton *et al.*, 1995).

2.5.2 Future concentrations

Again, it is not easy to predict with any certainty what the level of aerosol emissions will be in the 21st century. To make any predictions, many assumptions must be made about future activity of regional economies and the availability and attractiveness of alternatives to burning fossil fuels, etc. The IPCC 1992 scenarios (Houghton *et al.*, 1992) included emission levels of aerosols for the future that were based on a number of economic, political and resource availability assumptions, as discussed above.

There has been recent new thinking on future levels of aerosol emissions. Now it is anticipated that aerosol emissions worldwide will continue to increase into the early part of the 21st century, but there will be regional shifts in the emissions. The largest current emissions are centred over the USA and Europe, but this distribution will shift because of larger emissions from China. However, by the middle of the 21st century it is widely assumed that aerosol emissions will decrease. Given the short lifetime of aerosols in the atmosphere, they may be much less significant in a longer time frame of climate change than was thought 5 years ago (S. Smith, National Center for Atmospheric Research, 1998, personal communication). More details on new emissions scenarios are provided in section 2.7.

2.6 Future Climate Change

2.6.1 Description of climate models – general circulation models (GCMs)

Much of our knowledge of future climate change comes from climate model experiments. Climate models are complex three-dimensional mathematical representations of the processes responsible for climate. These processes include complex interactions among atmosphere, land surface, oceans and sea ice. Climate models simulate the global distributions of variables such as temperature, wind, cloudiness and rainfall. Major climate processes represented in most state-of-the-art climate models are shown in Fig. 2.6. The equations describing the behaviour of the atmosphere are solved on a three-dimensional grid representing the surface of the earth and the vertical height of the atmosphere. The spatial resolution at which a model is configured is an important aspect of how well the model can reproduce the actual climate of the earth.

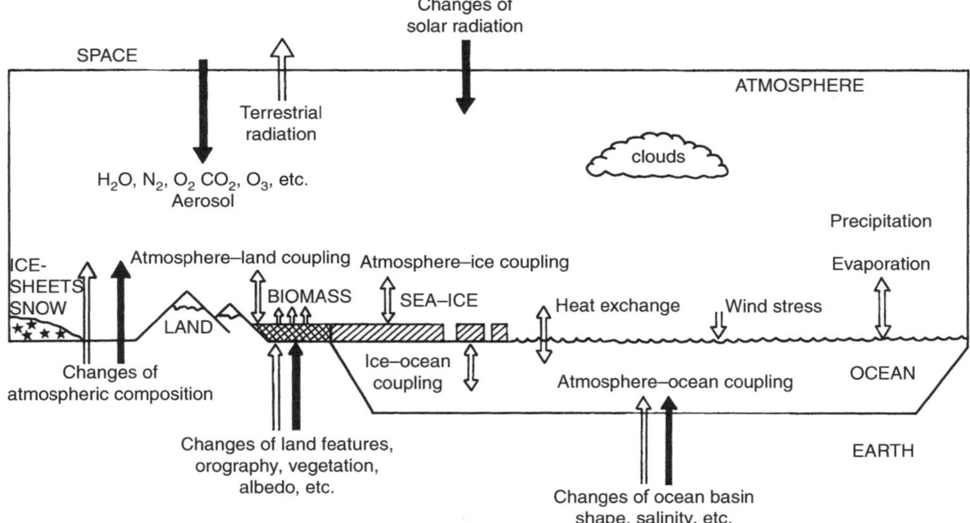

Fig. 2.6. Schematic illustration of the components of the coupled atmosphere/earth/ocean system. (Source: Cubasch and Cess, 1990.)

2.6.2 Climate model development

(a) History of GCM development

Climate models have developed considerably over the past few decades (Mearns, 1990). The earliest experiments that evaluated effects of increased greenhouse gases using climate models were performed in the 1960s and 1970s. General circulation models (GCMs) at that time were simple. They used very rudimentary geometric sectors to represent land masses and simple oceans. The oceans, which were referred to as swamp oceans, effectively consisted of a wet surface with zero heat capacity and essentially acted only as an evaporating surface (Manabe and Wetherald, 1975, 1980).

In the early and mid-1980s, climate models that included more realistic geography were developed. Mixed-layer oceans, which were usually about 50 m deep and included estimates of evaporation from their surface and heat diffusion throughout their depth, became the standard ocean model coupled to the atmospheric models (Washington and Meehl, 1984; Schlesinger and Mitchell, 1987). The inclusion of a mixed-layer ocean also allowed the annual seasonal cycle of solar radiation to be included. Climate modellers continued to develop better parameterizations for atmospheric processes, such as cloud formation and precipitation. The land surface, however, still was crudely represented. Soil moisture dynamics were handled via the simplified bucket approach in which the soil is assigned a certain field capacity (i.e. size of the bucket). When field capacity was exceeded, runoff would occur. Evaporation of soil water (water in the bucket) occurred at diminishing rates as the amount of water in the bucket decreased. Horizontal resolutions of these climate models were typically between 5 and 8 degrees (Schlesinger and Mitchell, 1987).

By the mid 1980s, more attention was directed toward improving the overly simplified surface of the earth. Several research groups (Dickinson *et al.*, 1986; Sellers *et al.*, 1986) developed sophisticated surface packages that included vegetation/atmosphere interactions and more realistic soil moisture representation. At the same time, the spatial resolution of the atmospheric modelling further increased, and other parameterizations for processes such as cloud formation and precipitation were also further improved (Mitchell *et al.*, 1990).

Since the late 1980s, atmospheric models have been coupled with three-dimensional dynamic ocean models, which allows for much more realistic modelling of interannual variability and longer-term variability of the coupled system. The ocean models allow for detailed modelling of horizontal and vertical heat transport within the ocean (Stouffer *et al.*, 1989; Washington and Meehl, 1989).

In the 1990s, the spatial resolutions of the atmospheric and oceanic components of models have been greatly improved. A relatively standard resolution of about 250–300 km (2.8 degrees) for the atmosphere and of about 100–200 km (1 or 2 degrees) for the ocean is used (Johns *et al.*, 1997; Boville and Gent, 1998). Another important new improvement is the coupling of atmospheric and oceanic models without using flux adjustment. Previously this adjustment was necessary to avoid coupled models drifting away from the observed climate. However, the result of the flux adjustment was that the models were less physically based, and their responses under perturbed conditions were somewhat constrained (Gates *et al.*, 1992). The most recent coupled models have largely resolved the problem, partially by increasing their resolution, and no flux adjustment is necessary (Gregory and Mitchell, 1997; Boville and Gent, 1998).

It is important to note that climate models are computationally quite expensive, i.e. they require large amounts of computer time. Computer power has increased tremendously over the past few years; however, the computer time required by the models has also increased because of the increasing sophistication of the modelling of various aspects of the climate system and the increased spatial resolution. Giorgi and Mearns (1991), for example, calculated the amount of computer time required to run the National Center for Atmospheric Research (NCAR) community climate model (CCM1) on the Cray X-MP, a state-of-the-art computer at the time. At a resolution of 4.5 degrees by 7.5 degrees, 1 cpu (central processing unit) minute was required to simulate 1 day of the global model run. To run the same model at a resolution of 0.3 degrees by 0.3 degrees would have required 3000 cpu minutes. This meant that it would have taken 2 days of computing time to simulate one actual day of climate with the CCM1. Computer power increases have kept pace with climate model developments throughout the 1990s, but power still remains a limitation for performing multi-ensemble, transient runs with fully coupled atmosphere/ocean models.

As climate models have improved with more detailed modelling of important processes and increasing spatial resolution, their ability to reproduce faithfully the current climate has improved significantly. The models can now represent most of the features of the current climate on a large regional or continental scale. The distributions of pressure, temperature, wind, precipitation and ocean currents are well represented in time (seasonally) and space. However, at spatial scales of less than several hundred kilometres, the models still can produce errors as large as 4 or 5°C in monthly average temperature and as large as 150% in precipitation (Risbey and Stone, 1996; Kittel *et al.*, 1998; Doherty and Mearns, 1999).

(b) Higher resolution models – regional climate models

Over the past 10 years, the technique of nesting higher resolution regional climate models within GCMs has evolved to increase the spatial resolution of the models over a region of interest (Giorgi and Mearns, 1991; McGregor, 1997). The basic strategy is to rely on the GCM to simulate the large-scale atmospheric circulation and the regional model to simulate sub-GCM-scale distributions of climatic factors such as precipitation, temperature and winds. The GCM provides the initial and lateral boundary conditions for driving the regional climate model. In numerous experiments, models for such regions as the continental USA, Europe, Australia and China have been driven by ambient and doubled CO_2 output from GCMs. The spatial pattern of changed climate, particularly changes in precipitation, simulated by these regional models often departs significantly from the more general pattern over the same region simulated by the GCM (Giorgi *et al.*, 1994; Jones *et al.*, 1997; Laprise *et al.*, 1998; Machenhauer *et al.*, 2000). The regional model is able to provide more detailed results because the spatial resolution is in the order of tens of kilometres, whereas the GCM scale is an order of magnitude coarser. This method, while often producing better simulations of the regional climate, is still dependent on the quality of the information provided by the GCM.

It is likely that the best regional climate simulations eventually will be performed by global models run at high spatial resolutions (tens of kilometres). In the meantime, the regional modelling approach affords climate scientists the opportunity to obtain greater insight into possible details of climatic change on a regional scale. It also provides researchers assessing the impact of climate with high-resolution scenarios of climatic change to use as input in models of climate impact, such as crop models. For example, Mearns *et al.* (1999, 2000) have used results from recent regional climate simulations over the USA to study the effect of the scale of climatic change scenarios on crop production in the Great Plains.

2.6.3 Climatic change experiments with climate models

To simulate possible future climatic change, climate models are run using changes in the concentrations of greenhouse gases (and aerosols) which then affect the radiative forcing within the models. In the early to mid-1980s, experiments were primarily conducted using doubled [CO_2] with climate models that possessed relatively simple mixed-layer oceans (described above). In general, control runs of 10–20 years duration were produced. In the climate change experiments, the amount of CO_2 was instantaneously doubled, and the climate model was run until it reached equilibrium in relation to the new forcing (Schlesinger and Mitchell, 1987).

In the late 1980s, coupled atmosphere/ocean general circulation models (AOGCMs) that used evolving changes in atmospheric CO_2 concentrations were used to simulate the response of the earth/atmosphere system over time. In these experiments, time-varying forcing by CO_2 and other greenhouse gases on a yearly basis was used, and the transient response of the climate was analysed. These first-generation experiments were still run at relatively coarse spatial resolutions of about 5 degrees latitude and longitude (Stouffer *et al.*, 1989; Washington and Meehl, 1989; Cubasch *et al.*, 1992; Manabe *et al.*, 1992). Results from these early runs indicated that the time evolving response could result in some patterns of climatic change different from those which resulted from equilibrium experiments. For example, initial cooling in the North Atlantic and off the coast of Antarctica was a common feature of these experiments (Stouffer *et al.*, 1989). These differences were a direct effect of the dynamic response of the ocean model to the change in radiative forcing.

More recent experiments have included detailed radiative models for each greenhouse gas and the effects of sulphate aerosols. These simulations with coupled models have been run at much higher spatial resolutions (e.g. 2.8 degrees) and have incorporated effects of increases in greenhouse gases including direct (and sometimes indirect) aerosol effects (Bengtsson, 1997; Johns *et al.*, 1997; Meehl *et al.*, 1996; Boer *et al.*, 2000a). However, in this generation of runs, the aerosol effect was highly parameterized: surface albedo was changed to simulate the direct effect, and cloud albedo was altered to simulate the indirect effect (Meehl *et al.*, 1996). By including the effects of aerosols, patterns distinct from those produced by greenhouse gases alone emerge.

2.6.4 Most recent results of climate models

A great deal of progress in the development of knowledge of climate and in the ability to model the climate system has occurred in the past 10 years. However, some of the fundamental statements made in the first IPCC report (Houghton *et al.*, 1990) still hold today. It is still considered likely that the doubled greenhouse gas equilibrium response of global surface temperature ranges from 1.5 to 4.5°C. It is also now inevitable that [CO_2] doubling will be surpassed.

(a) Summary results from IPCC 1995

To analyse climatic responses to a range of different scenarios of concentrations of emissions of greenhouse gas and aerosol amounts, simple (and comparatively less expensive) upwelling diffusion energy balance climate models are often used (Wigley and Raper, 1992). These models provide the mean global temperature response to the transient greenhouse gas and aerosol scenarios. Another factor considered is a range of model sensitivity. Climate model sensitivity is the equilibrium global mean warming per unit radiation forcing, usually expressed as the global mean warming simulated by the model for a doubling of [CO_2]. For the 1995 IPCC report (Houghton *et al.*, 1996), model sensitivities of 1.5, 2.5 and 3.5°C were used in conjunction with the emission scenarios from IPCC 1992 to provide a range of estimated global mean temperature for 2100. Across the three model sensitivities and the range of 1992 emission scenarios, the projected increase in global mean temperature by 2100 ranged from 0.9 to 3.5°C (Kattenberg *et al.*, 1996). Results based on a climate sensitivity of 2.5°C (medium range value) for all the scenarios are presented in Fig. 2.7. Any of these estimated rates of warming would be the greatest to occur in the past 10,000 years. Note also that all the scenarios showed warming even though the cooling effects of aerosols were accounted for in the simulations. Changes in sea level, based on the full range of climate models (energy balance and AOGCMs) and full range of 1992 scenarios, ranged between 13 and 94 cm (Warrick *et al.*, 1996). For the 'middle-of-the-road' IS92a scenario, the increase is projected to be about 50 cm, with a range from 20 to 86 cm.

There were regional differences in the various climate model runs, but some points of similarity occurred in all transient coupled model simulations, with and without aerosol effects. These included: (i) greater surface warming of land than of oceans; (ii) minimum warming around Antarctica and the northern Atlantic; (iii) maximum warming in high northern latitudes in late

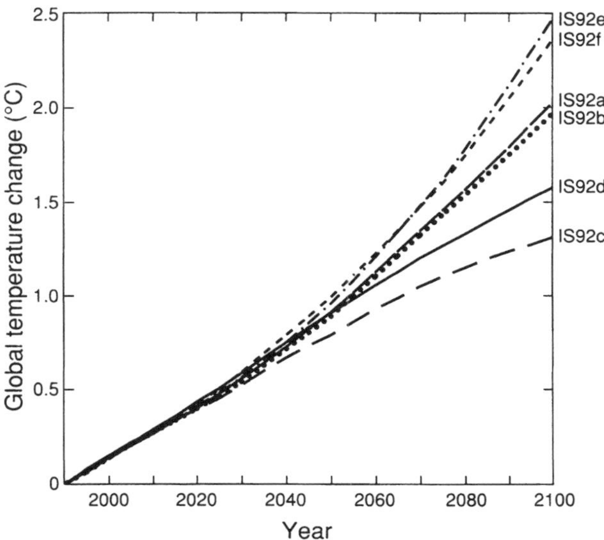

Fig. 2.7. Projected global mean surface temperature changes from 1990 to 2100 for the full set of IS92 emission scenarios. A climate sensitivity of 2.5°C is assumed. (Source: Houghton *et al.*, 1996.)

autumn and winter associated with reduced sea ice and snow cover; (iv) decreased diurnal temperature range over land in most seasons and most regions; (v) enhanced global mean hydrological cycle; and (vi) increased precipitation in high latitudes (Kattenberg *et al.*, 1996).

There were some important regional differences between the transient CO_2-only runs and those using increased CO_2 plus aerosols. For example, Mitchell *et al.* (1995) found that Asian monsoon rainfall increased in the CO_2-only run, but decreased in the CO_2 plus aerosols run. Also, precipitation decreased in southern Europe in the elevated CO_2-only case, but increased in the CO_2 plus aerosols run. This particular climate model did not include the indirect aerosols effect.

(b) Detection and attribution – the IPCC 1995 debate

Important issues in the effort to understand present and future climate are those of attribution and detection. The chapter dedicated to this topic in the IPCC 1995 volume (Santer *et al.*, 1996) inspired intense debate between the 'naysayers' on the issue of climate change and those scientists who participated in the IPCC process (Moss and Schneider, 1996). Detection, in the present context, refers to the detection of statistically significant changes in the global climate system. Attribution refers to determining the cause for these changes as at least partially anthropogenic. Thus, the combined goal of these two endeavours is to determine if there have been significant human-caused changes in the climate system, particularly during the 20th century. However, it is important to understand that both these concepts are inherently probabilistic in nature; i.e. there are no clearcut yes or no answers (Santer *et al.*, 1996).

Detection initially involved looking at one time series of one variable (e.g. global mean temperature). The purpose was to detect a signal of global warming by statistically separating out the natural variability in the time series from the possible anthropogenically generated variability. This is a difficult problem, given the high natural variability (on various time scales) of the climate system. Often this separation is aided by the use of climate model results in which only the natural variability is being modelled.

More recent work in detection and attribution has focused on examining patterns of change in temperature across many points on the earth's surface, through the various vertical heights of the atmosphere, or using a combination of both (three-dimensional analysis). The most sophisticated detection/attribution investigation involves patterns of multiple variables (e.g. temperature and precipitation). Santer *et al.* (1996) provides an excellent review of the work performed in this important and complex research area. Establishing the detectability of human-induced global climatic change significantly changes the perception of the climatic change problem, particularly by policy makers. In the IPCC 1995 chapter, a statement was made for the first time that global warming due to anthropogenic pollution of the atmosphere is most likely occurring.

Some of the strongest evidence for this in IPCC 1995 was from multivariate detection studies. These compared observed three-dimensional temperature patterns with the patterns found in AOGCM model runs that took into account 20th century historical changes in both CO_2 and sulphate aerosols. Often, comparisons of the relative strength of statistical agreement between observations and climate model runs with CO_2-only forcing were made against those with both CO_2 and aerosol forcing. In general, agreement was strongest between observations and climate model results with both forcings. The chapter concludes with the statement, 'The body of statistical evidence when examined in the context of our physical understanding of the climate system, now points towards a discernible human influence on global climate' (Santer *et al.*, 1996, p. 439).

When the IPCC 1995 report appeared, numerous editorials and opinion pieces alleged that more conservative statements in the chapter had been inappropriately changed by the lead authors after the final wording had been approved (e.g. F. Seitz, *Wall Street Journal*, 12 June 1996). These accusations were made without full knowledge of the carefully constructed approval procedures for the final IPCC document. A flurry of counter-editorials appeared (e.g. Bolin *et al.*, *Wall Street Journal*, 25 June 1996) defending the authors of the chapter, and the debate eventually subsided by autumn. However, by this time numerous scientists, policy makers, national politicians and world leaders had become involved.

Work in detection/attribution has continued to move forward since the IPCC 1995 report (Santer *et al.*, 1997). For example, Wigley *et al.* (1998) demonstrated that the serial correlation structure of observed temperature data was much stronger than that in two state-of-the-art climate models that did not account for increases in CO_2 or changes in aerosols. As climate models and observations continue to improve, and understanding of the external radiative forcing of the climate system increases, higher levels of detection and attribution will be obtained.

(c) Since IPCC 1995

Since the IPCC 1995 report, more coupled-climate runs have been performed with higher resolution models, and with both direct and indirect aerosol effects. Moreover, more physically based modelling of aerosol effects in the atmosphere is under way (Feichter *et al.*, 1997; Qian and Giorgi, 1999; J. Kiehl, 1998, personal communication).

In the current National Assessment Program in the USA, two of the most recent modelling transient experiments to the year 2100 are being used – that of the Canadian model CGCMI (Reader and Boer, 1998; Boer *et al.*, 2000b) and the British HADCM2 (Johns *et al.*, 1997; Mitchell and Johns, 1997). In the HADCM2 run, regional climate changes over the USA show that temperature increases in the order of 5°C in winter and 3°C in summer are expected to occur by 2060. The CGCMI model projects a 4–7°C greater warming over North America by 2060 than does the HADCM2. The problem still remains that the models do not always agree on the specific regional climate changes. For example, the CGCMI model predicts precipitation decreases in the south-eastern USA in the summer, while the HADCM2 model predicts increases (Doherty and Mearns, 1999).

2.6.5 Changes in climate variability

The discussions above have focused on changes in average climate conditions. However, under some new mean climate state, there will continue to be variability on various time scales around that mean. It is likely that the variability will also change as the mean climate state changes. The issue of change in variability in concert with change in mean conditions is a particularly important one from the perspective of impacts. It is well known that changes in variability can be equal to or more important than changes in the mean to many resource systems, such as agriculture. Semenov and Barrow (1997) and Mearns *et al.* (1997) have demonstrated that crop models are affected by changes in both the mean and variance of climate time series.

One of the important specific ways that changes in both climate mean and variability affect resource systems is the effect they have on changes in the frequency of extreme events (Mearns *et al.*, 1984). The effect of temperature and precipitation extremes on various agricultural crops has been investigated (Raper and Kramer, 1983; Acock and Acock, 1993). In essence, changes in variability have a greater effect on changes in the frequency of climatic extremes than do changes in the mean (Katz and Brown, 1992). However, it is the combination of the two types of changes that is most important. Hence, we need to understand how climatic variability (on various time scales) might change in a world warmed by greenhouse gases.

(a) Interannual time scale

El Niño Southern Oscillation (ENSO), which can roughly be characterized as an irregular oscillation of the coupled atmosphere/ocean system in the tropical Pacific, is associated with significant precipitation anomalies in the tropics and beyond (Kiladis and Diaz, 1989). These anomalous patterns have had serious impacts on human society worldwide (Glantz, 1996). Because ENSO events are the most important source of interannual climatic variability, the accuracy with which climate models reproduce the intensity and occurrence of such events and how well they 'predict' changes in them under conditions of global warming is of great importance.

Climate models are now becoming sophisticated enough to simulate correctly processes responsible for climatic variability on interannual to decadal time scales. Most state-of-the-art AOGCMs simulate ENSO-like behaviour to some degree, but the simulation is often weaker than the observed events (Knutson and Manabe, 1994; Roeckner *et al.*, 1996; Meehl and Arblaster, 1998).

ENSO events continue to occur under increased CO_2 conditions. Tett (1995) found no major changes in the frequency of El Niño events in his examination of the Hadley Centre (HADCM) coupled-transient run. Knutson and Manabe (1994), analysing the Geophysical Fluid Dynamics Laboratory (GFDL) AOGCM, found decreased amplitude in the Southern Oscillation Index (SOI) in a transient $4 \times [CO_2]$ experiment. They emphasized that their results were from a coupled model, the ocean component of which had a rather coarse resolution. Thus, this model may not realistically simulate coupled ocean/atmosphere responses in the tropical Pacific Ocean. Knutson *et al.* (1997) also found decreased amplitude of ENSO, but found no change in the frequency of El Niño/La Niña events with increasing concentrations of atmospheric CO_2. Newer coupled model runs with higher resolution provide better simulations of ENSO events (Roeckner *et al.*, 1996; Meehl and Arblaster, 1998), but they provide no definitive picture of how ENSO may change. Thus, predicting change in ENSO events in a world warmed by greenhouse gases is still open to speculation.

(b) Higher frequency variability changes

Compared with change in ENSO events, there have been clearer signals in climate models relating to how certain aspects of daily climate may change (for reviews of earlier results, see Mearns, 1992, 1993). The clearest indications from climate model experiments using increased greenhouse gas are decreased daily variability of winter temperature in northern mid-latitude climates and increased variability of daily precipitation (Gregory and Mitchell, 1995; Mearns *et al.*, 1995a,b; Kattenberg *et al.*, 1996). Changes in frequency and intensity of precipitation have been highly variable in climate models and are closely associated with the change in mean daily precipitation. However, Hennessy *et al.* (1997) found that greater intensity of precipitation occurred in many areas in two equilibrium runs of models with mixed-layer oceans. Often this was associated with an increased contribution of convective rainfall vs. non-convective rainfall. Zwiers and Kharin (1998) found precipitation extremes increased almost everywhere over the globe.

Other types of extremes have also been investigated. For example, Knutson *et al.* (1998) found intensification of hurricanes and tropical storms in the GFDL climate model. There are indications that increases in extremes of such variables as temperature and precipitation are likely. However, because climate models still cannot resolve all processes that are responsible for extreme events, all the results reported above should be interpreted cautiously.

2.7 The IPCC Third Assessment Report

New scenarios of future emissions of the various greenhouse gases and aerosols are currently being developed by Working Group III for the IPCC Third Assessment Report to be completed in 2001. Preliminary estimates that will be used in the newest climate model simulations have been made (H. Pitcher, Energy Modeling Forum Workshop, August 1998, Snowmass, Colorado; Nakicenovic, 2000).

These scenarios are formed differently from those of 1992. Basically, narrative story lines that include key scenario characteristics have been developed to describe a short history of a possible future development. The narratives explore what might happen if political, economic, technical and social development took particular alternative directions. From these narratives, quantifications are developed for the major factors that will influence emissions. These factors include population growth, gross domestic product and cumulative resource use. Four different macro-regions are considered: countries of the Organization for Economic Cooperation and Development (OECD), Eastern Europe and the former Soviet Union (EEFSU), Asia, and the rest of the world (ROW). The first two macro-regions represent developed countries, and the latter two represent developing countries. Interactions among industrialized and developing countries are particularly considered. There is neither a 'best guess' scenario (i.e. no probabilities are given for the four scenarios) nor an extreme 'disaster' one.

Scenario A1 describes a world with very rapid economic growth, low population growth, and rapid introduction of new and efficient technology. Scenario A2 describes a highly heterogeneous world with high population growth and less concern for rapid economic growth. Scenario B1 characterizes a convergent world with rapid change in economic structures, introduction of clean technologies, rapid technological development, and concern for global equity and environmental and social sustainability. Scenario B2 describes another heterogeneous world with less rapid and more diverse technological change. Scenario B1 has the lowest increase in total emissions of carbon by 2100; Scenario A1 has the highest. Scenarios A2 and B2 attain similar mid-range levels of carbon emissions by 2100. In all of these scenarios, sulphate aerosols are assumed to start decreasing before 2040, and in most of them, to start decreasing before 2030. All aerosols emission scenarios are considerably lower than the IS92a one. Aerosols essentially become relatively minor forcing factors by the end of the 21st century. These changes suggest less extreme scenarios than those of IS92. There still remains a set of scenarios based on the underlying economic and political assumptions, but the outer boundaries are narrower than those in earlier reports. Within the coming year, major climate modelling groups will use these emissions scenarios (once they are converted into concentrations) to generate new projections of climatic change to the year 2100.

2.8 Concluding Remarks

What do we need to know? When will we know more? When will we know enough? These are the kinds of questions often posed by the media to scientists who study future climatic change, and which often receive less than definitive answers.

It is probably disappointing to many scientists in agricultural and climatic change research fields that we still know so little about future climate. However, uncertainties about future climate are associated with uncertainties regarding emissions of greenhouse gases and aerosols, and these are in turn strongly associated with uncertainties regarding the economics and politics of most countries in the world. Thus, it is easy to see why there is such lack of predictability. Perhaps more discouraging is that there is still so much uncertainty about regional changes of climate and a lack of agreement among the results from climate models on regional scales when given a particular scenario of future emissions.

Even this latter uncertainty need not have such a daunting effect on our ability to cope with climatic change. We are learning more and more every day. With each new generation of climate model, new insights are gained into how the physical processes in the climate system work. Moreover, the agreement between observations and climate models improves as the climate models become better able to account for major forcings that have driven the climate system in the 20th century (Mitchell *et al.*, 1995). We may never know enough to determine 100 years in advance precisely how precipitation will change in eastern Kansas or southern Italy, but many scientists and policy makers feel we know enough now to implement global policies to limit the production of greenhouse gases. Indeed, the acceptance of the Kyoto Accords is a most compelling indication that imperfect knowledge of the future does not preclude taking judicious action to responsibly control the future of our environment.

References

Acock, B. and Acock, M.C. (1993) Modeling approaches for predicting crop ecosystem responses to climate change. In: Buxton, D.R., Shibles, R., Forsberg, R.A., Blad, B.L., Asay, K.H., Paulsen, G.M. and Wilson, R.F. (eds) *International Crop Science I.* Crop Science Society of America, Madison, Wisconsin, pp. 299–306.

Bengtsson, L. (1997) A numerical simulation of anthropogenic climate change. *Ambio* 26, 58–65.

Boer, G.J., Flato, G., Reader, M.C. and Ramsden, D. (2000a) A transient climate change simulation with greenhouse gas and aerosol forcing: experimental design and comparison with the instrumental record for the 20th century. *Climate Dynamics* (in press).

Boer, G.J., Flato, G., Reader, M.C. and Ramsden, D. (2000b) A transient climate change simulation with greenhouse gas and aerosol forcing: projected climate for the 21st century. *Climate Dynamics* (in press).

Boville, B.A. and Gent, P.R. (1998) The NCAR climate system model, version one. *Journal of Climate* 11, 1115–1130.

Christy, J.R. and Spencer, R.W. (1995) Assessment of precision in temperatures from the microwave sounding units. *Climatic Change* 30, 97–105.

Cubasch, U. and Cess, R.D. (1990) Processes and modeling. In: Houghton, J.T., Jenkins, G.J. and Ephraums, J.J. (eds) *Climate Change: the IPCC Scientific Assessment.* Cambridge University Press, Cambridge, UK, pp. 69–91.

Cubasch, U., Hasselmann, K., Hock, H., Maier Reimer, E., Mikolajewicz, U., Santer, B.D. and Sausen, R. (1992) Time-dependent greenhouse warming computations with a coupled ocean–atmosphere model. *Climate Dynamics* 8, 55–69.

Dai, A., Fung, I.Y. and Del Genio, A.D. (1997) Surface observed global land precipitation variations during 1900–88. *Journal of Climate* 10, 2943–2962.

Dai, A., Trenberth, K.E. and Karl, T.R. (1998) Global variations in droughts and wet spells: 1900–1995. *Geophysical Research Letters* 25, 3367–3370.

Dickinson, R.E., Henderson-Sellers, A., Kennedy, P.J. and Wilson, M.F. (1986) *Biosphere–Atmosphere Transfer Scheme (BATS) for the NCAR Community Climate Model.* NCAR Technical Note 275. National Center for Atmospheric Research, Boulder, Colorado.

Doherty, R. and Mearns, L.O. (1999) *A Comparison of Simulations of Current Climate from Two Coupled Atmosphere–Ocean GCMs Against Observations and Evaluation of Their Future Climates.* Report to the NIGEC National Office. National Center for Atmospheric Research, Boulder, Colorado, 47pp.

Easterling, W.E. (1996) Adapting North American agriculture to climate change. *Agricultural and Forest Meteorology* 80, ix–xi.

Feichter, J., Lohmann, U. and Schult, I. (1997) The atmospheric sulphur cycle in ECHAM-4 and its impact on the shortwave radiation. *Climate Dynamics* 13, 235–246.

Folland, C.K., Karl, T. and Vinnikov, K.Ya. (1990) Observed climate variations and change. In: Houghton, J.T., Jenkins, G.J. and Ephraums, J.J. (eds) *Climate Change: The IPCC Scientific Assessment.* Cambridge University Press, Cambridge, UK, pp. 195–238.

Gates, W.L., Mitchell, J.F.B., Boer, G.J., Cubasch, U. and Meleshko, V.P. (1992) Climate modeling, climate prediction and model validation. In: Houghton, J.T., Callander, B.A. and Varney, S.K. (eds) *Climate Change 1992: The Supplementary Report to the IPCC Scientific Assessment.* Cambridge University Press, Cambridge, UK, pp. 97–134.

German Bundestag Enquête Commission (1991) *Protecting the Earth: a Status Report with Recommendations for a New Energy Policy,* Vol. 1. Bonn University, Bonn, 672 pp.

Giorgi, F. and Mearns, L.O. (1991) Approaches to regional climate change simulation: a review. *Reviews of Geophysics* 29, 191–216.

Giorgi, F., Shields Brodeur, C. and Bates, G.T. (1994) Regional climate change scenarios over the United States produced with a nested regional climate model. *Journal of Climate* 7, 375–399.

Glantz, M.H. (1996) *Currents of Change: El Niño's Impact on Climate and Society.* Cambridge University Press, Cambridge, UK, 194 pp.

Gregory, J.M. and Mitchell, J.F.B. (1995) Simulation of daily variability of surface temperature and precipitation over Europe in the current $2 \times CO_2$ climates using the UKMO climate model. *Quarterly Journal of the Royal Meteorological Society* 121, 1451–1476.

Gregory, J.M. and Mitchell, J.F.B. (1997) The climate response to CO_2 of the Hadley Centre coupled AOGCM with and without flux adjustment. *Geophysical Research Letters* 24, 1943–1946.

Gribbin, J.R. and Lamb, H.H. (1978) Climatic change in historical times. In: Gribbin, J.R. (ed.) *Climatic Change.* Cambridge University Press, Cambridge, UK, pp. 68–82.

Groisman, Y.P., Karl, T.R., Easterling, D.R., Knight, R.W., Jamason, P.F., Hennessy, K.J., Suppiah, R., Page, C.M., Wibig, J., Fortuniak, K., Razuvaev, V.N., Douglas, A., Førland, E. and Zhai, P. (1999) Changes in the probability of heavy precipitation: important indicators of climatic change. *Climatic Change* 42, 246–283.

Hennessy, K.J., Gregory, J.M. and Mitchell, J.F.B. (1997) Changes in daily precipitation under enhanced greenhouse conditions. *Climate Dynamics* 13, 667–680.

Houghton, J.T., Jenkins, G.J. and Ephraums, J.J. (eds) (1990) *Climate Change: The IPCC Scientific Assessment.* Report prepared for IPCC by Working Group I. Cambridge University Press, Cambridge, UK, 364 pp.

Houghton, J.T., Callander, B.A. and Varney, S.K. (eds) (1992) *Climate Change 1992: The Supplementary Report to the IPCC Scientific Assessment.* Report prepared for IPCC by Working Group I. Cambridge University Press, Cambridge, UK, 200 pp.

Houghton, J.T., Meira Filho, L.G., Bruce, J., Lee, H., Callander, B.A., Haites, E., Harris, N. and Maskell, K. (eds) (1995) *Climate Change 1994: Radiative Forcing of Climate*

Change and an Evaluation of the IPCC I992 Emission Scenarios. Cambridge University Press, Cambridge, UK, 339 pp.

Houghton, J.T., Meira Filho, L.G., Callander, B.A., Harris, N., Kattenberg, A. and Maskell, K. (1996) *Climate Change 1995: the Science of Climate Change.* Cambridge University Press, Cambridge, UK, 572 pp.

Hurrell, J.W. (1998) Relationships among recent atmospheric circulation changes, global warming, and satellite temperatures. *Science Progress* 81, 205–224.

Hurrell, J.W. and Trenberth, K.E. (1997) Spurious trends in satellite MSU temperatures from merging different satellite records. *Nature* 386, 164–167.

Hurrell, J.W. and Trenberth, K.E. (1998) Difficulties in obtaining reliable temperature trends: reconciling the surface and satellite microwave sounding unit records. *Journal of Climate* 11, 945–967.

Jager, J. (1988) *Developing Policies for Responding to Climatic Change. A summary of the discussion and recommendations of the workshops held in Villach, 28 September to 2 October 1987.* World Meteorological Organization/TD-No. 225.

Johns, T.C., Carnell, R.E., Crossley, J.F., Gregory, J.M., Mitchell, J.F.B., Senior, C.A., Tett, S.F.B. and Wood, R.A. (1997) The second Hadley Centre coupled ocean–atmosphere GCM: model description, spinup and validation. *Climate Dynamics* 13, 103–134.

Jones, P.D., Wigley, T.M.L. and Briffa, K.R. (1994) Global and hemispheric anomalies: land and marine instrumental records. In: Boden, T.A., Kaiser, D.P., Sepanski, R.J. and Stoss, F.W. (eds) *Trends '93: a Compendium of Data on Global Change.* Carbon Dioxide Analysis Center, Oak Ridge National Laboratory, Oak Ridge, Tennessee, pp. 603–608.

Jones, P.D., Parker, D.E., Osborn, T.J. and Briffa, K.R. (1999) Global and hemispheric temperature anomalies — land and marine instrumental records. In: *Trends: a Compendium of Data on Global Change.* Carbon Dioxide Information Analysis Center, Oak Ridge National Laboratory, Oak Ridge, Tennessee (http://cdiac.esd.ornl.gov/trends/temp/jonescru/jones.html).

Jones, R.G., Murphy, J.M., Noguer M. and Keen, M. (1997) Simulation of climate change over Europe using a nested regional climate model. I: Comparison of driving and regional model responses to a doubling of carbon dioxide. *Quarterly Journal of the Royal Meteorological Society* 123, 265–292.

Karl, T.R., Jones, P.D., Knight, R.W., Kukla, G., Plummer, N., Razuvayev, V., Gallo, K.P., Lindseay, J., Charlson, R.J. and Peterson, T.C. (1993) Asymmetric trends of daily maximum and minimum temperature. *Bulletin of the American Meteteorological Society* 74, 1007–1023.

Karl, T.R., Knight, R.W., Easterling, D.R. and Quayle, R.G. (1996) Indices of climatic change for the USA. *Bulletin of the American Meteteorological Society* 77, 279–292.

Kattenberg, A., Giorgi, F., Grassl, H., Meehl, G.A., Mitchell, J.F.B., Stouffer, R.J., Tokioka, T., Weaver, A.J. and Wigley, T.M.L. (1996) Climate models – projections of future climate. In: Houghton, J.T., Meira Filho, L.G., Callander, B.A., Harris, N., Kattenberg, A. and Maskell, K. (eds) *Climate Change 1995: the Science of Climate Change.* Cambridge University Press, Cambridge, UK, pp. 285–357.

Katz, R.W. and Brown, B.G. (1992) Extreme events in a changing climate: variability is more important than averages. *Climatic Change* 21, 289–302.

Kerr, R.A. (1998) Among global thermometers, warming still wins out. *Science* 281, 1948–1949.

Kiehl, J.T. and Trenberth, K.E. (1997) Earth's annual global mean energy budget. *Bulletin of the American Meterological Society* 78, 197–208.

Kiladis, G.N. and Diaz, H.F. (1989) Global climatic extremes associated with extremes of the Southern Oscillation. *Journal of Climate* 2, 1069–1090.

Kittell, T.G.F., Giorgi, F. and Meehl, G.A. (1998) Intercomparison of regional biases and doubled CO_2 sensitivity of coupled atmosphere–ocean general circulation model experiments. *Climate Dynamics* 14, 1–15.

Knutson, T.R. and Manabe, S. (1994) Impact of increased CO_2 on simulated ENSO-like phenomena. *Geophysical Research Letters* 21, 2295–2298.

Knutson, T.R., Manabe, S. and Gu, D. (1997) Simulated ENSO in a global coupled ocean–atmospheric model: multidecadal amplitude modulation and CO_2 sensitivity. *Journal of Climate* 10, 138–161.

Knutson, T.R., Tuleya, R.E. and Kurihara, Y. (1998) Simulated increase in hurricane intensities in a CO_2-warmed climate. *Science* 279, 1018–1020.

Lamb, H.H. (1982) *Climate, History, and the Modern World.* Cambridge University Press, Cambridge, UK, 384 pp.

Laprise, R., Caya, D., Giguere, M., Gergeron, G., Cote, H., Blanchet, J.P., Boer, G.J. and McFarlane, N.A. (1998) Climate and climate change in western Canada as simulated by the Canadian Regional Climate Model. *Atmosphere–Ocean* 36, 119–167.

Leggett, J., Pepper, W.J. and Swart, R.J. (1992) Emissions scenarios for the IPCC: an update. In: Houghton, J.T., Callander, B.A. and Varney, S.K. (eds) *Climate Change 1992: the Supplementary Report to the IPCC Scientific Assessment.* Cambridge University Press, Cambridge, UK, pp. 69–96.

Machenhauer, B., Windelband, M., Botzet, M., Christensen, J.G., Deque, M., Jones, R.G., Ruti, P.M. and Visconti, G. (2000) Validation and analysis of regional present-day climate and climate change simulations over Europe. *Quarterly Journal of the Royal Meteorological Society* (in press).

Manabe, S. and Wetherald, R.T. (1975) The effects of doubling CO_2 concentration on the climate of a general circulation model. *Journal of the Atmospheric Sciences* 32, 3–15.

Manabe, S. and Wetherald, R.T. (1980) On the distribution of climate change resulting from an increase of CO_2 content of the atmosphere. *Journal of the Atmospheric Sciences* 37, 99–118.

Manabe, S., Spelman, M.J. and Stouffer, R.J. (1992) Transient responses of a coupled ocean–atmosphere model to gradual changes of atmospheric CO_2. Part II: Seasonal response. *Journal of Climate* 5, 105–126.

McGregor, J.J. (1997) Regional climate modeling. *Meteorological Atmospheric Physics* 63, 105–117.

Mearns, L.O. (1990) Future directions in climate modeling: a climate impacts perspective. In: Wall, J. and Sanderson, M. (eds) *Climate Change: Implications for Water and Ecological Resources.* Proceedings of an international symposium/workshop. Department of Geography, University of Waterloo, Waterloo, Canada, pp. 51–58.

Mearns, L.O. (1992) Changes in climate variability with climate change. In: Majumdar, S.K., Kalkstein, L.S., Yarnal, B., Miller, E.W. and Rosenfeld, L.M. (eds) *Global Climate Change: Implications, Challenges, and Mitigation Measures.* Pennsylvania Academy of Science, Easton, Pennsylvania, pp. 209–226.

Mearns, L.O. (1993) Implications of global warming on climate variability and the occurrence of extreme climatic events. In: Wilhite, D.A. (ed.) *Drought Assessment, Management, and Planning: Theory and Case Studies.* Kluwer Publishers, Boston, Massachusetts, pp. 109–130.

Mearns, L.O., Katz, R.W. and Schneider, S.H. (1984) Extreme high temperature events: changes in their probabilities with changes in mean temperature. *Journal of Climate and Applied Meteorology* 23, 1601–1613.

Mearns, L.O., Giorgi, F., McDaniel, L. and Shields, C. (1995a) Analysis of the diurnal range and variability of daily temperature in a nested modeling experiment: comparison with observations and 2 × CO_2 results. *Climate Dynamics* 11, 193–209.

Mearns, L.O., Giorgi, F., McDaniel, L. and Shields, C. (1995b) Analysis of the variability of daily precipitation in a nested modeling experiment: comparison with observations and 2 × CO_2 results. *Global and Planetary Change* 10, 55–78.

Mearns, L.O., Rosenzweig, C. and Goldberg, R. (1997) Mean and variance change in climate scenarios: methods, agricultural applications, and measures of uncertainty. *Climatic Change* 35, 367–396.

Mearns, L.O., Mavromatis, T., Tsvetsinskaya, E., Hays, C. and Easterling, W. (1999) Comparative responses of EPIC and CERES crop models to high and low resolution climate change scenarios. Special issue on new developments and applications with the NCAR Regional Climate Model (RegCM). *Journal of Geophysical Research* 104, 6623–6646.

Mearns, L.O., Easterling, W. and Hays, C. (2000) Comparison of agricultural impact of climate change calculated from high and low resolution climate model scenarios. *Climate Change* (in press).

Meehl, G.A. and Arblaster, J.M. (1998) Asian–Australian monsoon and El Niño–Southern Oscillation in the NCAR climate system model. *Journal of Climate* 11, 1356–1385.

Meehl, G.A., Washington, W.M., Erickson, J. III, Briegleb, B.P. and Jaumann, P.J. (1996) Climate change from increased CO_2 and direct and indirect effects of sulphate aerosols. *Geophysical Research Letters* 23, 3755–3758.

Mitchell, J.F.B. and Johns, T.C. (1997) On modification of global warming by sulfate aerosols. *Journal of Climate* 10, 245–267.

Mitchell, J.F.B., Manabe, S., Tokioka, T. and Meleshko, V. (1990) Equilibrium climate change. In: Houghton, J.T., Jenkins, G.J. and Ephraums, J.J. (eds) *Climate Change: the IPCC Scientific Assessment.* Cambridge University Press, Cambridge, UK, pp. 131–172.

Nakicenovic, N. (2000) Greenhouse gas emission scenarios. *Technological Forecasting and Social Change* (in press).

Mitchell, J.F.B., Johns, T.C., Gregory, J.M. and Tett, S.F.B. (1995) Climate response to increasing levels of greenhouse gases and sulphate aerosols. *Nature* 376, 501–504.

Moss, R. and Schneider, S. (1996) Characterizing and communicating scientific uncertainty: building on the IPCC second assessment. In: Hassol, S.J. and Katzenberger, J. (eds) *Elements of Change.* Aspen Global Change Institute, Aspen, Colorado, pp. 90–135.

Nicholls, N., Gruza, G.V., Jouzel, J., Karl T.R., Ogallo, L.A. and Parker, D.E. (1996) Observed climate variability and change. In: Houghton, J.T., Meira Filho, L.G., Callander, B.A., Harris, N., Kattenberg, A. and Maskell, K. (eds) *Climate Change 1995: The Science of Climate Change.* Cambridge University Press, Cambridge, UK, pp. 133–192.

Parry, M.L. (1978) *Climatic Change, Agriculture, and Settlement.* Dawson and Sons, Folkestone, UK, 214 pp.

Qian, Y. and Giorgi, F. (1999) Interactive coupling of regional climate and sulfate aerosol models over East Asia. *Journal of Geophysical Research* 104, 6477–6499.

Raper, C.D. and Kramer, P.J. (eds) (1983) *Crop Reactions to Water and Temperature and Stresses in Humid, Temperate Climates.* Westview Press, Boulder, Colorado, 373 pp.

Reader, M.C. and Boer, G.J. (1998) The modification of greenhouse gas warming by the direct effect of sulphate aerosols. *Climate Dynamics* 14, 593–607.

Reilly, J. (1996) Agriculture in a changing climate: impacts and adaptation. In: Watson, R.T., Zinyowera, M.C., Moss, R.H. and Dokken, D.J. (eds) *Climate Change 1995: Impacts, Adaptations and Mitigation of Climate Change Scientific–Technical Analyses.* Cambridge University Press, Cambridge, UK, pp. 427–467.

Risbey, J. and Stone, P. (1996) A case study of the adequacy of GCM simulations for input to regional climate change. *Journal of Climate* 9, 1441–1446.

Roeckner, E., Oberhuber, J.M., Bacher, A., Christoph, M. and Kirchner, I. (1996) ENSO variability and atmospheric response in a global coupled atmosphere–ocean GCM. *Climate Dynamics* 12, 737–754.

Rosenberg, N.J., Cole, C.V. and Paustian, K. (1998) Mitigation of greenhouse gas emissions by the agriculture sector. Special issue, *Climatic Change* 40, 1–5.

Rosenzweig, C. and Hillel, D. (1998) *Climate Change and the Global Harvest.* Oxford University Press, Oxford, UK, 324 pp.

Santer, B.D., Wigley, T.M.L., Barnett, T.P. and Anyamba, E. (1996) Detection of climate change and attribution of causes. In: Houghton, J.T., Meira Filho, L.G., Callander, B.A., Harris, N., Kattenberg, A. and Maskell, K. (eds) *Climate Change 1995: the Science of Climate Change.* Cambridge University Press, Cambridge, UK, pp. 407–443.

Santer, B.D., Taylor, K.E., Wigley, T.M.L., Johns, T.C., Jones, P.D., Karoly, D.J., Mitchell, J.F.B., Oort, A.H., Penner, J.E., Ramaswamy, V., Schwarzkopf, M.D., Stouffer, R.J. and Tett, S. (1997) A search for human influences on the thermal structure of the atmosphere. *Nature* 382, 39–46.

Schimel, D., Alves, D., Enting, I., Heimann, M., Joos, F., Raynaud, D., Wigley, T., Prather, M., DerWent, R., Ehhalt, D., Fraser P., Sanhueza, E., Zhou, X., Jonas, P., Charlson, R., Rodhe, H., Sadasivan, S., Shine, K.P., Fouquart, Y., Ramaswamy, V., Solomon, S., Srinivasan, J., Albritton, D., DerWent, R., Isaksen, I., Lal, M. and Wuebbles, D. (1996) Radiative forcing of climate change. In: Houghton, J.T., Meira Filho, L.G., Callander, B.A., Harris, N., Kattenberg, A. and Maskell, K. (eds) *Climate Change 1995: The Science of Climate Change.* Cambridge University Press, Cambridge, UK, pp. 65–131.

Schlesinger, M.E. and Mitchell, J.F.B. (1987) Climate simulation of the equilibrium climatic response to increased carbon dioxide. *Reviews of Geophysics* 25, 760–798.

Schneider, S.H., Gleick, P.H. and Mearns, L.O. (1990) Prospects for climate change. In: American Association for Advancement of Science *Climate and Water: Climate Change, Climatic Variability, and the Planning and Management of US Water Resources.* John Wiley & Sons, New York, pp. 41–73.

Sellers, P.J., Mintz, Y., Sud, Y.C. and Dalcher, A. (1986) A simple biosphere model (SiB) for use within general circulation models. *Journal of the Atmospheric Sciences* 43, 515–531.

Semenov, M.A. and Barrow, E. (1997) Use of a stochastic weather generator in the development of climate change scenarios. *Climatic Change* 35, 397–414.

Shine, K.P., DerWent, R.G., Wuebbles, D.J. and Morcrette, J.J. (1990) Radiative forcing of climate. In: Houghton, J.T., Jenkins, G.J. and Ephraums, J.J. (eds) *Climate Change: the IPCC Scientific Assessment.* Cambridge University Press, Cambridge, UK, pp. 41–68.

Stouffer, R.J., Manabe, S. and Bryan, K. (1989) Interhemispheric asymmetry in climate response to a gradual increase of atmospheric CO_2. *Nature* 342, 660–662.

Tett, S. (1995) Simulation of El Niño/Southern Oscillation-like variability in a global AOGCM and its response to CO_2 increase. *Journal of Climate* 8, 1473–1502.

Warrick, R.A., LeProvost, C., Meier, M.F., Oerlemans, J. and Woodworth, P.L. (1996) Changes in sea level. In: Houghton, J.T., Meira Filho, L.G., Callander, B.A., Harris, N., Kattenberg, A. and Maskell, K. (eds) *Climate Change 1995: the Science of Climate Change.* Cambridge University Press, Cambridge, UK, pp. 359–405.

Washington, W.M. and Meehl, G.A. (1984) Seasonal cycle experiment on the climate sensitivity due to a doubling of CO_2 with an atmospheric general circulation model coupled to a simple mixed layer ocean. *Journal of Geophysical Research* 89, 9475–9503.

Washington, W.M. and Meehl, G.A. (1989) Climate sensitivity due to increased CO_2: experiments with a coupled atmosphere and ocean general circulation model. *Climate Dynamics* 4, 1–38.

Weubbles, D.J. and Rosenberg, N.J. (1998) The natural science of global climate change. In: Rayner, S. and Malone, E. (eds) *Resources and Technology.* Battelle Press, Columbia, Ohio, pp. 1–78.

Wigley, T.M.L. (1989) Possible climate change due to SO_2-derived cloud condensation nuclei. *Nature* 339, 365–367.

Wigley, T.M.L. and Raper, S.C.B. (1992) Implications for climate and sea level of revised IPCC emission scenarios. *Nature* 357, 293–300.

Wigley, T.M.L., Smith, R.I. and Santer, B. (1998) Anthropogenic influence on the autocorrelation structure of hemispheric-mean temperatures. *Science* 282, 1676–1679.

Zwiers, F.W. and Kharin, V.V. (1998) Changes in the extremes of the climate simulated by CCC GCM2 under CO_2 doubling. *Journal of Climate* 11, 2200–2222.

3 Agricultural Contributions to Greenhouse Gas Emissions

DONALD C. REICOSKY[1], JERRY L. HATFIELD[2] AND RONALD L. SASS[3]

[1]*USDA – Agricultural Research Service, North Central Soil Conservation Laboratory, 803 Iowa Avenue, Morris, MN 56267, USA;* [2]*USDA-Agricultural Research Service, National Soil Tilth Laboratory, 2150 Pammel Drive, Ames, IA 50011, USA;* [3]*Department of Ecology and Evolutionary Biology, Rice University, 6100 Main, Houston, TX 77005-1892, USA*

3.1 Introduction

Agriculture provides both sources and sinks of greenhouse gases (GHGs). The global intensification of food and fibre production is an important factor influencing GHG emission. More than 97% of the world's food supply is produced on land that emits GHGs when intensively tilled and fertilized, and/or grazed by animals. While US agriculture is generally thought of as a minor source of GHGs, the increasing world population dictates a challenge to increase agricultural production without increasing the risks of GHG emissions and degrading environmental consequences. This review will attempt to put GHGs from agriculture in perspective, and briefly address fossil fuel in agriculture, soil carbon (C) loss from intensive tillage, emissions associated with fertilizers, emissions from animal production and manure management, and emissions associated with rice production. It has been estimated that 20% of the greenhouse effect (radiative forcing) is related to agricultural activities (Cole *et al.*, 1996). Other recent reviews on agriculture's contribution to GHGs and global change were presented by Houghton *et al.*, 1996; Cole *et al.*, 1997; Lal, 1997; Lal *et al.*, 1997a,b, 1998; Paul *et al.*, 1997; Paustian *et al.*, 1997a, 1998; and Rosenzweig and Hillel, 1998. Since the industrial revolution, the inflow and outflow of carbon dioxide have been disturbed by humans; atmospheric CO_2 concentrations ([CO_2]) have risen about 28% – principally because of fossil fuel combustion, which accounts for 99% of the total US CO_2 emissions (Houghton *et al.*, 1996). Agricultural activity, such as clearing forest for fields and pastures, transforming virgin soil into cultivated land, growing flooded rice, producing sugarcane, burning crop residues, raising cattle, and utilizing N fertilizers, are all implicated in the release of GHG into the atmosphere. The radiative forcing of GHGs and their relative amounts are shown in Fig. 3.1. Although CO_2, methane (CH_4), and nitrous oxide (N_2O) occur naturally in the atmosphere, their recent build-up is largely a result of human activities. Since

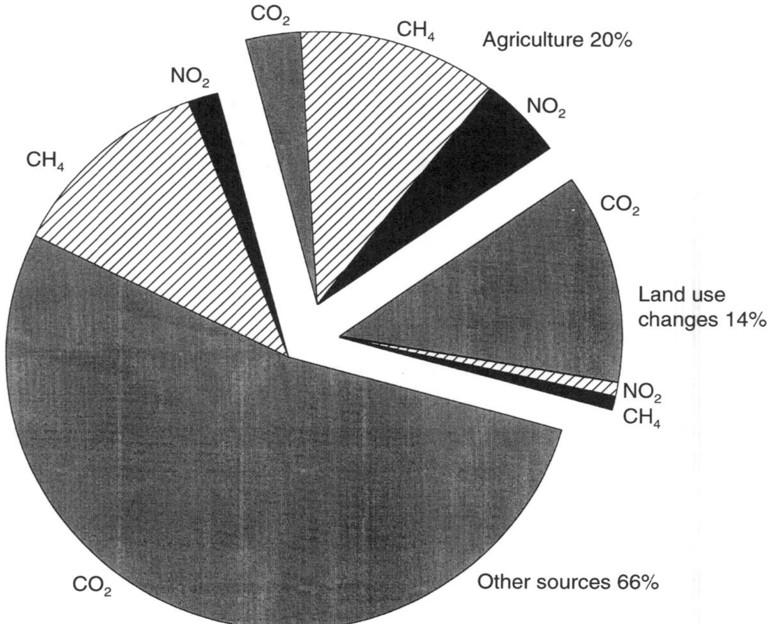

Fig. 3.1. Relative proportions of annual increase in global radiative forcing attributed to agriculture and land use change. (After Cole *et al.*, 1996.)

the 19th century, the atmospheric concentration of these greenhouse gases has increased by 30% for CO_2, 145% for CH_4 and 15% for N_2O (Houghton *et al.*, 1996).

The concept of global warming potential (GWP) has been developed by the Intergovernmental Panel on Climatic Change (IPCC) (Houghton *et al.*, 1996) to compare the ability of GHGs to trap heat in the atmosphere relative to CO_2. The GWP of a greenhouse gas is the ratio of radiative forcing from a unit mass of the gas to a unit mass of CO_2 over a 100-year period. The GWP for $CO_2 = 1$, for $CH_4 = 21$, and for $N_2O = 310$ (see Chapter 2, this volume). Man-made gases such as hydrofluorocarbons, perfluorocarbons and sulphur hexafluoride have significantly higher GWP, but are not from agricultural sources. To quantify the relative amounts of GHGs, IPCC (Houghton *et al.*, 1996) has chosen to express the GHGs in units of million metric tons of carbon equivalents (MMTCE), calculated as the products of the mass of gas (in teragrams Tg) $\times$ GWP $\times$ 12/44. The value of 12/44 is the ratio of the mass of C to the mass of CO_2. For consistency throughout this chapter, GHG units will be expressed as MMTCE. For the amount of CO_2 equivalent, the quantities can be multiplied by 3.67.

The global C cycle is made up of large C reservoirs (or pools) and flows (or fluxes) important to agriculture. The reservoirs of C are interconnected by pathways of exchange through various physical, geological and biological processes. Hundreds of billions of tons of C, in the form of CO_2, are absorbed by the oceans and by living biomass (through plant photosynthesis), considered to be C sinks. Comparable amounts are emitted to the atmosphere through natural and man-made processes, considered to be C sources. When the system is at equilibrium, the C fluxes or flows among the various pools are roughly balanced. Estimates of these pools are provided in Table 3.1. The largest pool is the oceans, which contain 38 million MMTCE. A large amount of soil C is stored as soil organic matter (SOM) in agricultural production systems. Over most of the earth's land surface, the quantity of C as SOM ranges from 1.4 million to 1.5 million MMTCE and exceeds, by a factor of two or three, the amount of C stored in living vegetation, estimated to be 560,000 MMTCE (Schlesinger, 1990; Eswaran *et al.*, 1993). The contribution of CO_2 released to the atmosphere from agricultural land represents 20–25% of the total amount released due to human activity (Duxbury, *et al.*, 1993). The amount of organic C contained in soils depends on the balance between the inputs of photosynthetically fixed C that go into plant biomass and the loss of C through microbial decomposition. Agricultural practices can modify the organic matter inputs from crop residues and their decomposition, thereby resulting in a net change in the flux of CO_2 to or from soils.

3.2 Fossil Fuel Use in Agriculture

Energy is required for all agricultural operations. Modern intensive agriculture requires much more energy input than did traditional farming methods, since it relies on the use of fossil fuels for tillage, transportation and grain drying, for the manufacture of fertilizers, pesticides and equipment used as agricultural inputs, and for generating electricity used on farms (Frye, 1984). Early estimates suggested that fossil fuel usage by agriculture, primarily of liquid fuels and electricity, constitute only 3–4% of the total consumption in developed countries (CAST, 1992; Enquête Commission, 1995). To provide a reference for agriculture's contribution, C emissions from fossil fuel use in the USA in 1996 were reported to be 286.7, 229.9, 477.5 and 445.5 MMTCE for residential, commercial, industrial and transportation sectors, respectively (EPA, 1998). The total amount of C emitted as CO_2 in the USA in 1996 from

Table 3.1. Estimates of global carbon pools.

	Total C content (MMTCE)[a]	
Pool	Bouwman, 1990	Eswaran *et al.*, 1993
Atmosphere	720,000	750,000
Biomass	560,000–835,000	550,000
Soil organic matter	1,400,000–2,070,000	1,500,000
Caliche[b]	780,000–930,000	–
Oceans	38,000,000	38,000,000

[a]Million metric tonnes of carbon equivalent.
[b]Petrocalcic horizons in arid and semi-arid regions.

fossil fuels was 1450.3 MMTCE, a value that has steadily increased with time. Revised estimates by Lal *et al.* (1998) showed that US agriculture has contributed 116 of the total 1596 MMTCE (i.e. 7.3%) of US emissions. These agricultural emissions include an additional 15 MMTCE due to soil erosion (not included in earlier estimates) and 27.9 MMTCE due to direct on-farm energy use and indirect fertilizer and pesticide production.

Tillage and harvest operations account for the greatest proportion of fuel consumption within intensive agricultural systems (Frye, 1984). Fuel requirements using no-till or reduced tillage systems were 55 and 78%, respectively, of that used for conventional systems that included mouldboard ploughing. On an aerial basis, savings of 23 kg C ha^{-1} per year in energy costs resulted from the conversion of conventional till to no-till. For the 186 Mha of cropland in the USA, this translates into potential C savings of 4.3 MMTCE per year. Kern and Johnson (1993) calculated that conversion of 76% of the cropland planted in the USA to conservation tillage could sequester as much as 286–468 MMTCE over 30 years and concluded that US agriculture could become a net sink for C. Lal (1997) provided a global estimate for C sequestration from conversion of conventional tillage to conservation tillage that was as high as 4900 MMTCE by 2020. Combining economics of fuel cost reductions and environmental benefits of conversion to conservation tillage is a positive first step for agriculture toward decreasing C emissions into the atmosphere.

A summary of the fossil fuels used in US agriculture in 1996 is presented in Table 3.2. The three major fuels used released more than 19 MMTCE directly in 1995, with diesel fuel being the largest contributor. Nitrogen fertilizers, which require the greatest amount of energy to produce, are used in larger amounts than any other fertilizer. Net energy use in fertilizer manufacture has declined up to 40% recently, due to substantial improvements in plant efficiencies and use of natural gas. Estimates of energy required are 45.5, 10.8 and 5.0 Btu g^{-1} of product for N, P_2O_5 and K_2O, respectively (Shapouri *et al.*, 1995). For example, converting from Btu to joules (J) (1 Btu = 1055.06 J) and using the C content for natural gas (13.6 kg C 10^{-9} J) yields 0.66 tons of C released per ton of N produced. Therefore, the annual global consumption of about 80 Tg of fertilizer N corresponds to the consumption of about 53 MMTCE released as CO_2. Fertilizer and chemical production in the USA has increased steadily since the 1940s, and contributed 8.3 MMTCE to the atmosphere in 1996. These combined estimates of fossil fuels used in US agriculture represent about 2% of the total US C emissions. However, this table does not include energy for electricity used for heating and cooling or energy for equipment manufacture. Pimentel and Heichel (1991) estimated that the energy required for making agricultural machinery is equal to the fuel used to grow the crop.

Pimentel (1984) indicated that 17% of the total energy used in the US economy is consumed in food systems, with about 6% for agricultural production, 6% for processing and packaging and 5% for distribution and preparation. This 17% of the total US energy use represents an annual per capita use of about 1500 litres of fuel just for food. Taking the C content of fuel oil as 0.73 kg C l^{-1} and multiplying by an estimated US population of 270 million in mid-1998 suggests that food production, processing and

Table 3.2. US agricultural emissions of carbon from use of fossil fuels directly in 1995 and indirectly for chemical inputs in 1996.

Fuels[a]	Volume (10^6 l)	Carbon (MMTCE)
Diesel	13,626	11.02
Gasoline	5,626	3.32
Propane	3,028	4.97
Sum		19.31
Chemicals[b]	Weight (kg × 10^6)	Carbon[c]
Nitrogen (N)	6,916.9	6.30
Phosphorus (P_2O_5)	2,904.0	0.61
Potash (K_2O)	3,181.0	0.31
Herbicide	155.1	0.68
Insecticide	18.8	0.08
Fungicide	3.32	0.01
Other chemicals	59.1	0.26
Sum		8.25

[a]Source: Agricultural Resources and Environmental Indicators #16, Dec. 1996; USDA-ERS, Office of Energy, based on data gathered by NASS.
[b] Source: USDA-NASS, Agricultural Statistics Board. 1996 Field Crops Summary, Agricultural Chemical Usage.
[c]Assuming fuel was natural gas at 14 mg C $(Btu)^{-1}$. Represents 100% of land farmed (123,968,000 ha), which was extrapolated from a survey that covered seven major crops and 71.5% of land.

preparation would emit about 296 MMTCE per year in the USA, or about 20% of the US total (Houghton *et al*., 1996). Fossil fuel requirements by the food sector as a whole (which includes processing, preservation, storage and distribution) account for 10–20% of the total fossil energy consumption (Pimentel *et al*., 1990; CAST, 1992). Thus, mitigation of energy use by agriculture should consider the 11% in non-production areas when considering solutions for decreasing the amount of CO_2 emitted by agriculture.

3.3 Management of Soil Carbon

Conversion of forest land to agricultural land or urban use can result in changes in emissions of soil C as CO_2. Conversely, net additions of forest and crop biomass can result in soil acting as a sink for CO_2 (Raich and Potter, 1995). Agriculture and intensive tillage have caused a decrease of between 30 and 50% in soil C since many soils were brought into cultivation more than 100 years ago (Schlesinger, 1986; Houghton, 1995). There needs to be a better understanding of tillage processes, the mechanisms leading to C loss and how this C loss can be linked to soil productivity, soil quality, C sequestration, and ultimately to crop production. Long-term studies of soil C point to the role of intensive tillage and residue management in soil C losses (Lal, 1997; Paul *et al.*, 1997; Paustian *et al.*, 1997b); however, extrapolation of these data to a global value is complicated by uncertainties in soil C quantities and distribution across the landscape. Paustian *et al.* (1998) estimated that better global management of agricultural soils, restoring degraded soils, permanent set-aside of surplus land and restoration of some wetlands now used for agriculture could sequester between 400 and 900 MMTCE per year in the soil. They caution that soils have a finite capacity to store additional C which likely will be realized within 50–100 years. The potential for improved management offers hope that agriculture can decrease GHG emissions.

Mineral soils generally have fairly shallow organic layers and, therefore, have low organic C content relative to organic soils (Lal *et al.*, 1997a; Paustian *et al.*, 1997b). Consequently, it is possible to deplete the C stock of a mineral soil within the first 10–20 years of tillage, depending on type of disturbance, climate and soil type. Once the majority of native C stocks have been depleted, an equilibrium is reached that reflects a balance between accumulation from plant residues and loss of C through decomposition. Lal (1997) calculates that if 15% of the C in crop residues is converted to passive soil organic C (SOC), it may lead to C sequestration at the rate of 200 MMTCE year^{-1} when used with less intensive tillage. If the current changes in improved residue management and conversion from conventional tillage to conservation tillage in mineral soils continue as they have in the recent decade, these changes may lead to cumulative global C sequestration that ranges from 1500 to 4900 MMTCE by the year 2020 (Lal, 1997). In addition to increasing SOM, combined ecological and economic benefits of conservation tillage also accrue from decreased soil erosion, lower energy costs, water conservation and quality improvements, soil temperature regulation and improved soil structure. These all contribute to enhanced environmental quality and increased crop production.

One example of what intensive tillage in agricultural production systems has done to soil organic C is illustrated in Fig. 3.2. These data illustrate the long-term trends in soil C at the Morrow plots in Champaign, Illinois (Peck, 1989), and Sanborn Field at the University of Missouri, Columbia, Missouri. (Wagner, 1989). Both locations show similar decreases in SOC over the last 100 years. The only experimental parameter or factor common to the two locations was use of a mouldboard plough to till the experimental plots. Different cropping systems or rotations yielded a difference in soil C, which shows that management options exist for controlling SOM and improving soil C levels. The large decline in soil C was a result of tillage-induced soil C losses caused by use of the mouldboard plough and disk harrow, and a change to annual species. Other work around the world shows similar trends (Lal, 1997; Paul *et al.*, 1997; Paustian *et al.*, 1997b) and supports the need for conservation tillage with improved residue management. The significant 'flush' of CO_2 immediately after tillage reported by Reicosky and Lindstrom (1993, 1995) partially explains the long-term role of tillage in affecting C flow within agricultural production systems. Tillage, particularly mouldboard ploughing, resulted in a loss of CO_2 within minutes of tillage. Nineteen days after mouldboard ploughing, C lost as CO_2 accounted for 134% of the C in the

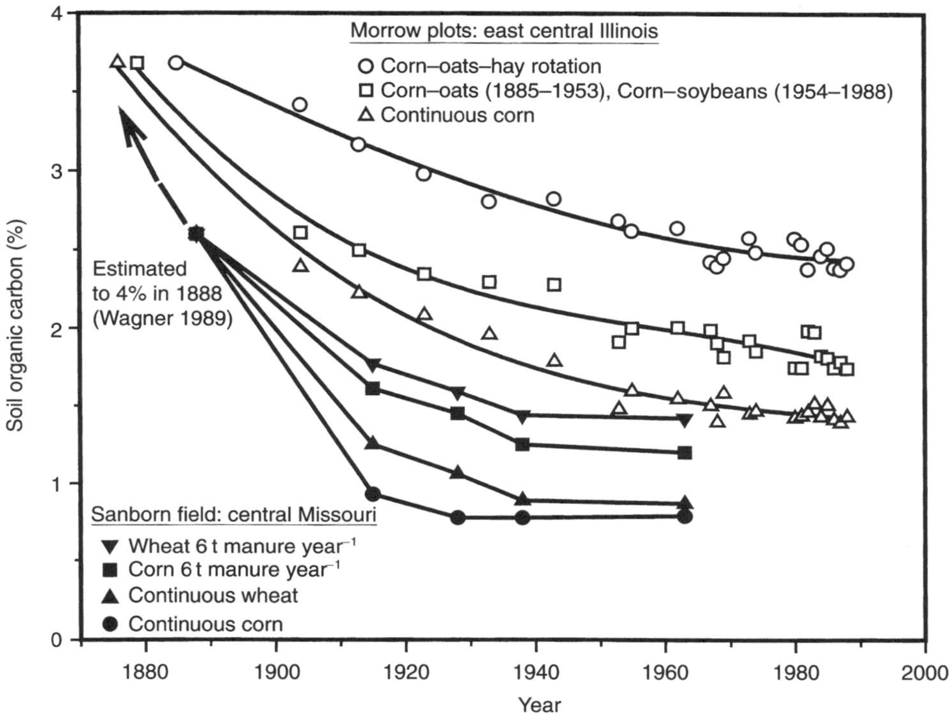

Fig. 3.2. Long-term effects of tillage and crop rotations on soil carbon in Midwest USA.

previous wheat residue. Mouldboard ploughing, one of the most disruptive types of tillage, appears to have two major effects: (i) to loosen and invert the soil, allowing rapid CO_2 loss and O_2 entry into the soil; and (ii) to incorporate/mix the crop residues, thus enhancing microbial attack. Tillage perturbs the soil system and causes a shift in the gaseous equilibrium by releasing CO_2 that enhances oxidation of soil C and organic matter loss. Conservation tillage, or any form of less intensive tillage, can minimize this tillage-induced C loss (Lal, 1997; Paustian *et al.*, 1997b).

Sustainable agriculture requires new technologies for efficient biomass C utilization. Crop stover or residue is an important and renewable resource that is manageable and serves as the primary input for soil C sequestration. Lal (1997) has estimated that the global arable land mass of about 1.4×10^9 ha annually produces 3.44×10^9 Mg of crop residue. At mean C content of 45 g kg^{-1} residue, the total global C assimilation is about 1500 MMTCE $year^{-1}$. While a large portion of crop residue C is recycled to CO_2 through microbial decomposition when the residue is mixed with soil by tillage, a small portion remains as humus that contributes to long-term sequestration in soil. The C from agricultural crop residues is only a small fraction (1%) of the estimated total global C fixed in photosynthesis; however, it is one amenable to management.

3.4 Nitrous Oxide and Methane Emissions from Animal Wastes and Lagoons

Nitrous oxide is produced from a wide variety of biological sources in soil, water, and animal wastes. During the last two centuries, human activities have increased N_2O concentration by 13% (EPA, 1998). The main activities producing N_2O are fossil fuel combustion, agricultural soil management and industrial sources. Use of large amounts of N fertilizer creates secondary problems associated with N_2O released in anaerobic conditions (Mosier *et al.*, 1998a). Agricultural soil management activities such as fertilizer application and cropping practices were the largest source of N_2O emission (56.5 MMTCE), accounting for 43% of the US total (EPA, 1998). Manure management in feedlots (3.7 MMTCE) and agricultural residue burning (0.1 MMTCE) are small sources of N_2O emissions.

Methane is second only to CO_2 in contributing to GHG emissions. Landfills are the largest contributor to CH_4 emissions in the USA, while the agricultural sector is responsible for 30% of US emissions. Of the total 176.7 MMTCE emitted in the USA in 1996 (EPA, 1998), agricultural emissions of CH_4 were: ruminant livestock fermentation, 34.5 MMTCE; agricultural manure management, 16.6 MMTCE; rice cultivation, 2.5 MMTCE; and biomass burning (Mosier *et al.*, 1998b).

Greenhouse gases are associated with storage and application of animal manure. Of these GHGs, the greatest attention has been given to CH_4 emissions generated by animals. There has been very little attention given to CO_2 production by manure storage systems. Among the agricultural sector's potential CH_4 emission sources, manure appears to contribute approximately 5% of the total (Table 3.3). Within the manure portion of CH_4 emissions, swine production constitutes the largest amount due to the type of manure handling and storage (Table 3.4). Nitrous oxide generation within manure is a result of the nitrification/denitrification process that occurs in manure storage and application. After field application, it would be difficult to separate the N_2O from manure sources from that of commercial fertilizer sources in the soil. Methane has been the gas most often measured in various studies; however, data comparing different production practices are sparse.

Table 3.3. Agricultural sources of atmospheric methane emissions in the US (EPA, 1994a).

Source	Methane emission	
	(Tg year^{-1})	(MMTCE)
Rice	65	372.3
Livestock	80	458.2
Manure	10	57.3
Biomass burning	30	171.8
Sum	185	1059.6

Table 3.4. Methane emissions from livestock manure in the USA and the World (EPA, 1994a,b).

	Methane emissions			
Species	USA (Tg year^{-1})	World (Tg year^{-1})	USA (MMTCE)	World (MMTCE)
Dairy	0.71	2.89	4.07	16.55
Beef	0.19	3.16	1.09	18.09
Swine	1.11	5.29	6.36	30.16
Sheep and goat	–	0.71	–	4.07
Poultry	0.23	1.28	1.32	7.33
Other	0.24	0.51	1.37	2.92
Sum	2.48	13.84	14.21	79.12

Production of GHG from manure storage systems has not been sufficiently measured over a large number of units and over a wide range of climatic conditions. Methane production within lagoons and earthen storage systems comes from the solid/liquid interface, with the CH_4-producing bacteria present at this interface. Anaerobic digestion of manure leads to the production of CH_4. Hill and Bolte (1989) described the anaerobic manure storage system as a complex set of interdependent biological systems. Methane production is part of the biological complex, and they proposed that loading rate, pH and temperature were factors causing shifts in the balance among the organisms. An illustration of these interactions is given by Burton (1992), who found that shifting the anaerobic manure storage to an aerobic storage reduced the potential NH_3 loss to the atmosphere. Unfortunately, this shift can lead to production of N_2O. However, he did not quantify the expected release of these gases. Safley *et al.* (1992) characterized the emission of CH_4 from different livestock systems and concluded that anaerobic manure storage systems would convert non-lignin organic matter into CH_4 under warm, moist, anaerobic conditions. Parsons and Williams (1987) developed a mathematical model for anaerobic storage systems based on these factors that could be adapted for prediction of GHG.

The annual release of CH_4 from different manure storage systems associated with swine vary from 10 kg per animal for subconfinement pits within buildings to about 90 kg per animal in a lagoon system. This variation in CH_4 production can be attributed to the amount of solids in the different manure systems and the bacterial populations present in the manure storage. Groenestein and Faassen (1996) found that deep-litter systems for swine reduced N_2O emissions because of changes in the manure digestion systems within the manure. Changes in manure management have had a positive impact on emission rates. Prueger and Hatfield (unpublished data, 1997) positioned a trace gas analyser over a lagoon and found that there was variation in the CH_4 fluxes throughout the day in response to diurnal changes in temperature. (Similar variation has been observed in rice fields; Sass *et al.*, 1991b; Satpathy *et al.*, 1997; Wang *et al.*, 1997a.) Prueger and Hatfield also documented variation in the exchange coefficient between the lagoon surface and the atmosphere. However, these data were not collected for a sufficient length of time to quantify seasonal changes in CH_4 production in response to a wide range of atmospheric conditions. Kinsman *et al.* (1995) measured CH_4 and CO_2 production from lactating dairy cows and found that stored manure contributed 5.8 and 6.1%, respectively, to CH_4 and CO_2 emissions under conditions of their experiment. Manure storage, particularly for ruminant animals, represents a small fraction of the total GHG load to the atmosphere.

Kuroda *et al.* (1996) measured the emissions of GHG emitted during composting of swine faeces under continuous aeration using laboratory-scale composting apparatus. Methane emission was observed within only 1 day from starting the composting, while N_2O and NH_3 repeatedly rose and fell after every turning. Of the total N loss during composting, the total amount of N_2O emission was a small fraction of NH_3 emissions. Lessard *et al.* (1996) measured N_2O emissions from agricultural soils after application of dairy cattle manure to cultivated land planted to maize (*Zea mays* L.). The manure application rates were 0, 170 and 339 kg N ha^{-1}, respectively. On the manured plots, 67% of the total N_2O emitted during the growing season occurred during the first 7 weeks following manure application. High N_2O fluxes coincided with periods when NO_3-N levels and soil water contents were relatively high. Fluxes were highest the first day after manure application, but returned to near pre-application levels 7 days later. There were short-lived peaks of N_2O flux, usually following rain. Only 1% of the manure N, which accumulated as N_2O, was potentially mineralizable over the snow-free season. In a similar study, Wassman *et al.* (1996) evaluated the effect of fertilizers and manure on CH_4 emission rates using an automated, closed-chamber system in Chinese rice *(Oryza sativa)* fields. The rate of increase in CH_4 emission was dependent on the total amount of organic manure applied. A single application of organic manure increased the relative short-term CH_4 emission rates by 2.7–4.1 times compared with fields without organic manure.

Reports on the literature indicate that there is a large amount of variation in the fluxes of GHG from animal manure storage and handling. These differences could be attributed to variations in species, diet, loading rates into the storage, type of storage and environmental conditions within the manure storage. Further studies giving greater attention to the physical and biological parameters affecting microbial production and emission of GHG are needed. These data will have to be coupled with dietary models for different species and a complete understanding of the chemical factors within manure storage systems in order to quantify the dynamics of GHG production and emission. This type of information will be essential in developing realistic mitigation scenarios.

3.5 Rice and Methane Production

Agricultural sources of CH_4 account for as much as one-third of the total atmospheric pool, with a significant portion contributed by rice cultivation.

A recent estimate suggests that CH_4 emitted from global rice paddies is 60 ± 40 Tg year^{-1} (344 ± 229 MMTCE) (Houghton *et al.*, 1992). Methane emission from rice fields is the result of bacterial processes – production in flooded anaerobic microsites and consumption (oxidation) in aerobic microsites). Flooding of rice fields promotes anaerobic fermentation of C sources supplied by the rice plants and other incorporated organics, resulting in the formation of CH_4. The process is governed by a complex set of parameters linking the physical and biological characteristics of flooded soil environments with specific agricultural management practices.

Rice is grown under a variety of climatic, soil and hydrological conditions in nearly 90 countries, and rice production can conveniently be divided into four categories based on water availability and CH_4 emission. The relative source strengths of CH_4 from these four rice production systems are: irrigated rice and favourable rainfed rice > flood-prone rainfed rice and deep-water rice > drought-prone rainfed rice > tidal wetland rice. Upland rice is not a source of CH_4 since it is grown on aerated soils (Neüe, 1997). Several other reviews on CH_4 emissions from rice fields have been published (Cicerone and Oremland, 1988; Neüe, 1993, 1997; Neüe and Sass, 1994; Sass and Fisher, 1996).

Conditions for CH_4 production in wetland rice soils have been categorized into six areas: water regime; Eh (redox potential)/pH buffer; carbon supply; temperature; texture and mineralogy; and salinity (Neüe, 1997). Methane production was influenced both by the reduction characteristics of the soils and by labile organic substrates (Gaunt *et al.*, 1997) and texture (Parashar *et al.*, 1991; Chen *et al.*, 1993). Sass *et al.* (1994) found a strong linear correlation between seasonal CH_4 emission and the percentage of sand in a sand : clay : silt gradient among three soils in Texas. In general, sandy soils high in organic C produce more CH_4 than clay soils with similar or lower C content (Neüe and Sass, 1994). Significantly decreased CH_4 emissions have been observed in soils with high percolation rates (Inubushi *et al.*, 1992). Increased percolation may transport sufficient dissolved oxygen in the soil to raise the Eh sufficiently either to inhibit CH_4 production or to increase CH_4 oxidation.

Variations in seasonal CH_4 emission from rice paddies are complex. A correlation with soil temperature has been reported in some studies, but not in others (Wang *et al.*, 1990; Neüe *et al.*, 1994; Neüe and Sass, 1994). Seasonal CH_4 fluxes observed in temperate rice fields show a general seasonal trend related to plant development (Sass *et al.*, 1991a,b, 1992). Methane emissions show a gradual rise during the vegetative phase that correlates with increasing plant biomass and peaks near panicle differentiation. This peak in emission may be attributed to a stabilization of soil pH and redox potential, root porosity, and an increasing amount of C substrate (Neüe and Sass, 1994; Kludze *et al.*, 1993). Prior to the end of the season, a second emission peak is sometimes observed which may be attributed to an increase in soil C due to leaf and root senescence (Neüe and Sass, 1994).

In irrigated double-cropped (two crops in one year) tropical rice paddies, both CH_4 emission and grain yield are consistently higher from the dry season crop than from the wet season crop (Neüe *et al.,* 1994). These results suggest that higher photosynthetic rates during the sunnier days of the dry season lead to larger amounts of C available to methanogenic bacteria and, consequently, to greater CH_4 emission rates. The addition of readily degradable C in sources such as rice straw before planting results in an additional early-season peak in CH_4 emission as the straw rapidly decomposes (Lindau *et al.,* 1991; Neüe and Sass, 1994). Other forms of C added by farmers, either for fertilization or to dispose of non-grain biomass, tend to increase both CH_4 production and emission (Sass *et al.,* 1991a,b; Neüe *et al.,* 1994; Minami, 1995). The incorporation of green manure leads to even higher emission levels (Denier van der Gon and Neüe, 1995).

Fertilizer is necessary to ensure adequate rice growth and root development. Wassmann *et al.* (1996) investigated the effect of fertilizers on CH_4 emission rates in Chinese rice fields, and found the rate of increase in CH_4 emission depended on the amount and timing of organic manure application. A potential mitigation technique involving double cropping was observed by these authors. Organic amendments are applied to the first rice crop (low CH_4 emission rates), and exclusively mineral fertilizers are applied to the second crop (high CH_4 emission rates). This fertilization distribution pattern does not reduce yields and results in a combined annual CH_4 emission that is only 56% of that emitted from fields treated with only blended mineral fertilizers over both seasons. Lindau *et al.* (1991) measured increased CH_4 emission with increased urea application in flooded rice fields of Louisiana, USA, where application of 200 kg urea-N ha^{-1} is typical. Similar emissions were measured with applications of either 200 or 300 kg urea-N ha^{-1}. However, emissions were lower when less than 200 kg urea-N fertilizer ha^{-1} was used. A reduction in CH_4 emission when ammonium sulphate fertilizer was used may be due to substrate competition by sulphate-reducing bacteria or to hydrogen sulphide toxicity (Neüe and Sass, 1994).

Methane emission rates are highly sensitive to water management. Periodic drainage of irrigated rice paddies, a common management practice in Japan, results in a significant decrease in CH_4 emissions (Yagi *et al.,* 1996; Cai *et al.,* 1997). Intermittent irrigation reduced CH_4 emissions by 36% compared with that from constant submergence of soil (Shin *et al.,* 1996). In the Philippines, draining for a period of 2 weeks at mid-tillering stage or at panicle initiation successfully suppressed CH_4 flux by up to 60%. However, N_2O flux increased sharply during the drainage period (Bronson *et al.,* 1997). Sass *et al.* (1992) found that a single mid-season drain reduced seasonal emission rates of CH_4 by 50%, and multiple short periods of drainage (2–3 days) reduced CH_4 emissions to an insignificant amount.

As shown in China (Yue *et al.*, 1997), an important contributor to variation in measured CH_4 emissions may be the use of different rice cultivars. Semi-dwarf varieties emit significantly less CH_4 than do tall varieties (Lindau *et al.,* 1995). In the Philippines (Neüe *et al.,* 1996; Wang *et al.,* 1997b), CH_4 emission rates from different cultivars showed a high correlation with root dry weight and total C released from roots. Cultivar-dependent variation in seasonal CH_4 emissions ranged from 18 to 41 g m^{-2} (Sass and Fisher, 1996;

Huang *et al.*, 1997a). Emission from a newly developed high-yielding, low tillering cultivar (IR65598) was very low. These differences in CH_4 emissions are attributed to differences in gas transport capacity among cultivars (Butterbachbahl *et al.*, 1997). Farmers' choice of the appropriate rice cultivar can therefore influence regional and global emissions of CH_4 without adversely affecting grain yields.

Because rice is an important crop globally, GHG mitigation efforts suggested by Yagi *et al.* (1997) must be based on sound agricultural practices and good science. Estimates of CH_4 emissions have been made in the following ways: by extrapolating field measurements to a regional or global scale (Wang *et al.*, 1994) assuming CH_4 emission as a constant fraction of rice net primary productivity (Bachelet and Neüe, 1993; Bachelet *et al.*, 1995); or by correlating CH_4 emissions with production (Anastasi *et al.*, 1992) or with organic matter inputs (Kern *et al.*, 1995). A field trial suggested that CH_4 emission can be predicted by a model utilizing environmental variables particular to a given region (Cao *et al.*, 1995). A semi-empirical model which predicts daily CH_4 emission from flooded rice fields (Huang *et al.*, 1998) is based on studies in Texas (Sass *et al.*, 1991a,b, 1992, 1994; Sass and Fisher, 1995; Huang *et al.*, 1997a,b; Sigren *et al.*, 1997a,b). Future research for mitigation will be directed toward using models, along with ground-truth, to interpret satellite-based sensor data for accurate assessments of regional, national and global trace-gas emissions from rice agriculture.

3.6 Mitigation Options for Agriculture

What agriculture can do to mitigate GHG emissions has been estimated by Cole *et al.* (1997), whose estimates of potential reduction of radiative forcing by the agricultural sector range from 1150 to 3300 MMTCE year^{-1}. Of the total potential global reduction in GHG emissions, approximately 32% could result from reduction in CO_2 emissions, 42% of the C offsets from biofuel production on 15% of the existing croplands, 16% from reduced CH_4 emissions and 10% from reduced emissions of N_2O.

Agriculture can contribute to mitigation of climatic change by adopting practices that promote stashing CO_2 as C in soil, crop biomass and trees, and by displacing the use of fossil fuels required for tillage, chemical manufacture equipment manufacture, and grain handling operations (Cole *et al.*, 1996; Paustian *et al.*, 1998). For the farm sector, the GHG mitigation potential through reduced fuel consumption is relatively small when compared with the rest of society; however, further reductions can be achieved. By combining appropriate land with best management practices to increase US crop production, Lal *et al.* (1998) suggest a soil C sequestration potential of 126 MMTCE year^{-1}. Much of the potential C sequestration (43% of US potential) comes from conservation tillage and crop residue management. Other strategies include eliminating fallow by using cover crops, improved irrigation scheduling, solar drying of crops, improved soil fertility, improved manure management, and producing more food using less land. Optimizing

N fertilizer efficiency, achieving higher yield per unit land area and using conservation tillage hold the most promise for indirectly mitigating N_2O and CO_2 emissions. Mitigation of CH_4 emissions from agriculture will require improved diets and rations for animals, aerobic conditions in manure management and improved rice production. Practices that will have the most impact on GHGs from rice production are water and carbon management, soil and variety selection, fertilizer type and amount, and soil preparation. Global understanding of these critical management practices will lead to enhanced soil and plant management and the development of new technologies that result in increased food production efficiency with minimum impact on environmental quality and GHGs. Acceptance of mitigation options will depend on the extent to which sustainable agricultural production can be achieved and the combined social, economic and environmental benefits.

References

Anastasi, C., Dowding, M. and Simpson, V.J. (1992) Future CH_4 emission from rice production. *Journal of Geophysical Research* 97, 7521–7525.

Bachelet, D. and Neüe, H.U. (1993) Methane emissions from wetland rice areas of Asia. *Chemosphere* 26, 219–237.

Bachelet, D., Kern, J. and Tolg, M. (1995) Balancing the rice carbon budget in China using a spatially-distributed data. *Ecological Modeling* 79, 167–177.

Bouwman, A.F. (1990) Global distributions of the major soils and land cover types. In: Bouwman, A.F. (ed.) *Soil and the Greenhouse Effect.* John Wiley & Sons, Chichester, UK, pp. 32–59.

Bronson, K.F., Neüe, H.U., Singh, U. and Abao, E.B. (1997) Automated chamber measurements of methane and nitrous oxide flux in a flooded rice soil. 1. Residue, nitrogen, and water management. *Soil Science Society of America Journal* 61, 981–987.

Burton, C.H. (1992) A review of the strategies in the aerobic treatment of pig slurry: purpose, theory and method. *Journal of Agricultural Engineering Research* 53, 249–272.

Butterbachbahl, K., Papen, H. and Rennenberg, H. (1997) Impact of gas transport through rice cultivars on methane emission from rice paddy fields. *Plant, Cell and Environment* 20, 1175–1183.

Cai, Z.C., Xing, G.X., Yan, X.Y., Xu, H., Tsuruta, H., Yagi, K. and Minami, K. (1997) Methane and nitrous oxide emissions from rice paddy fields as affected by nitrogen fertilizers and water management. *Plant and Soil* 296, 7–14.

Cao, M., Dent, J.B. and Heal, O.W. (1995) Modeling methane emission from rice paddies. *Global Biogeochemical Cycles* 9, 183–195.

CAST (1992) Preparing US Agriculture for Global Climatic Change. Task Force Report No. 119. Council for Agricultural Science and Technology, Ames, Iowa, 96 pp.

Chen, Z., Debo, L., Kesheng, S. and Bujun, W. (1993) Features of CH_4 emission from rice paddy fields in Beijing and Nanjing. *Chemosphere* 26, 239–245.

Cicerone, R.J. and Oremland, R.S. (1988) Biogeochemical aspects of atmospheric methane. *Global Biogeochemical Cycles* 2, 299–327.

Cole, C.V., Cerri, C., Minami, K., Mosier, A., Rosenberg, N., Sauerbeck, D. *et al.*, (1996) Agricultural options for mitigation of greenhouse gas emissions. In: Watson, R.,

Zinyowera, M.C. and Moss, R. (eds) *Climate Change 1995. Impacts, Adaptions, and Mitigation of Climate Change: Scientific–Technical Analyses.* IPCC Working Group II. Cambridge University Press, Cambridge, UK, pp 745–771.

Cole, C.V., Duxbury, J., Freney, J., Heinemeyer, O., Minami, K., Mosier, A., Paustian, K., Rosenberg, N., Sampson, N., Sauerbeck, D. and Zhao, Q. (1997) Global estimates of potential mitigation of greenhouse gas emissions by agriculture. *Nutrient Cycling in Agroecosystems* 49, 221–228.

Denier van der Gon, H.A.C. and Neüe, H.U. (1995) Influence of organic matter incorporation on the methane emission from a wetland rice field. *Global Biogeochemical Cycles* 9, 11–23.

Duxbury, J.M., Harper, L.A. and Mosier, A.R. (1993) Contributions of agroecosystems to global climate change. In: Harper, L., Duxbury, J.M., Mosier, A.R. and Rolston, D.S. (eds) *Agroecosystem Effects on Radiatively Important Trace Gases and Global Climate Change.* ASA Publication No. 55. American Society of Agronomy, Madison, Wisconsin, pp. 1–18.

EPA (1994a) *International Anthropogenic Methane Emissions: Estimates for 1990.* EPA-230-R-93–010. US Environmental Protection Agency, Washington, DC.

EPA (1994b) *Inventory of US Greenhouse Emissions and Sinks: 1990–1993.* EPA-230-R-94–014. US Environmental Protection Agency, Washington, DC.

EPA (1998) *Inventory of US Greenhouse Gas Emissions and Sinks 1990–1996.* US Environmental Protection Agency, Office of Policy, Planning and Evaluation. Washington, DC, 191 pp.

Enquête Commission (1995) *Protecting our Green Earth. How to manage global warming through environmentally sound farming and preservation of the world's forest.* Economica Verlag, Bonn, Germany, 683 pp.

Eswaran, H., Van Den Berg, E. and Reich, P. (1993) Organic carbon in soils of the world. *Soil Science Society of America Journal* 57, 192–194.

Frye, W.W. (1984) Energy requirements in no tillage. In: Phillips, R.E. and Phillips, S.H. (eds) *No-tillage Agricultural Principles and Practices.* Van Nostrand Reinhold, New York, pp. 127–151.

Gaunt, J.L., Neüe, H.U., Bragais, J., Grant, I.F. and Giller, K.E. (1997) Soil characteristics that regulate soil reduction and methane production in wetland rice soils. *Soil Science Society of America Journal* 61, 1526–1531.

Groenestein, C.M. and Faassen, H.G. van (1996) Volatilization of ammonia, nitrous oxide and nitric oxide in deep-litter systems for fattening pigs. *Journal of Agricultural Engineering Research* 65, 269–274.

Hill, D.T. and Bolte, J.P. (1989) Digestor stress as related to isobutyric and isovaleric acids. *Biological Wastes* 28, 33–37.

Houghton, J.T., Callander, B.A. and Varney, S.K. (eds) (1992) *Climate Change 1992: The Supplementary Report to the IPCC Scientific Assessment.* International Panel on Climate Change. Cambridge University Press, Cambridge, UK, 200 pp.

Houghton, J.T., Meira Filho, L.G., Callander, B.A., Harris, N., Kattenberg, A. and Maskell, K. (eds) (1996) *Climate change 1995: The Science of Climate Change.* International Governmental Panel on Climate Change. Cambridge University Press, Cambridge, UK.

Houghton, R.A. (1995) Changes in the storage of terrestrial carbon since 1850. In: Lal, R., Kimball, J., Levine, E. and Stewart, B.A. (eds) *Soils and Global Change.* CRC Press, Boca Raton, Florida, pp. 45–66.

Huang, Y., Sass, R.L. and Fisher, F.M. (1997a) Methane emission from Texas rice paddy soils. 1. Quantitative multi-year dependence of CH_4 emission on soil, cultivar and grain yield. *Global Change Biology* 3, 479–489.

Huang, Y., Sass, R.L. and Fisher, F.M. (1997b) Methane emission from Texas rice paddy soils. 2. Seasonal contribution of rice biomass production to CH_4 emission. *Global Change Biology* 3, 491–500.

Huang, Y., Sass, R.L. and Fisher, F.M. (1998) A semi-empirical model of methane emission from flooded rice paddy soils. *Global Change Biology* 4, 247–268.

Inubushi, K., Muramatsu, Y. and Umebayashi, M. (1992) Influence of percolation on methane emission from flooded paddy soil. *Japan Journal of Soil Science and Plant Nutrition* 63, 184–189.

Kern, J.S. and Johnson, M.G. (1993) Conservation tillage impacts on national soil and atmospheric carbon levels. *Soil Science Society of America Journal* 57, 200–210.

Kern, J.S., Bachelet, D. and Tölg, M. (1995) Organic matter inputs and methane emissions from soils in major rice growing regions of China. In: Lal, R., Kimble, J., Levine, E. and Stewart, B.A. (eds) *Soils and Global Change*. CRC Press, Boca Raton, Florida, pp. 189–198.

Kinsman, R., Sauer, F.D., Jackson, H.A. and Wolynetz, M.S. (1995) Methane and carbon dioxide emissions from dairy cows in full lactation monitored over a six-month period. *Journal of Dairy Science* 78, 2760–2766.

Kludze, H.K., DeLaune, R.D. and Patrick, W.H. Jr (1993) Aerenchyma formation and methane oxygen exchange in rice. *Soil Science Society of America Journal* 57, 386–391.

Kuroda, K., Osada, T., Yonaga, M., Kanematu, A., Nitta, T., Mouri, S. and Kojima, T. (1996) Emissions of malodorous compounds and greenhouse gases from composting swine feces. *Bioresources Technology* 56, 265–271.

Lal, R. (1997) Residue management conservation tillage and soil restoration for mitigating greenhouse effect by CO_2 enrichment. *Soil Tillage Research* 43, 81–107.

Lal, R., Kimble, J.M., Follett, R.F. and Stewart, B.A. (eds) (1997a) *Management of Carbon Sequestration in Soil*. CRC Press, Boca Raton, Florida. 480 pp.

Lal, R., Kimble, J.M., Follett, R.F. and Stewart, B.A. (eds) (1997b) *Soil Processes and the Carbon Cycle*. CRC Press, Boca Raton, Florida, 608 pp.

Lal, R., Kimble, J.M., Follet, R.F. and Cole, V. (1998) *Potential of US Cropland for Carbon Sequestration and Greenhouse Effect Mitigation*. USDA–NRCS, Washington, DC. Ann Arbor Press, Chelsea, Michigan, 128 pp.

Lessard, R., Rochette, P., Gregorich, E.G., Pattey, E. and Desjardins, R.L. (1996) Nitrous oxide fluxes from manure-amended soil under maize. *Journal of Environmental Quality* 25, 1371–1377.

Lindau, C.W., Bollich, P.K., DeLaune, R.D., Patrick, W.H. Jr and Law, V.J. (1991) Effect of urea fertilizer and environmental factors on methane emissions from a Louisiana, USA rice field. *Plant and Soil* 136, 195–203.

Lindau, C.W., Bollich, P.K. and Delaune, R.D. (1995) Effect of rice variety on methane emission from Louisiana rice. *Agriculture, Ecosystems and Environment* 54, 109–114.

Minami, K. (1995) The effect of nitrogen fertilizer use and other practices on methane emission from flooded rice. *Fertilizer Research* 40, 71–84.

Mosier, A.R., Duxbury, J.M., Freney, J.R., Heinemeyer, O. and Minami, K. (1998a). Assessing and mitigating N_2O emissions from agricultural soils. *Climate Change* 40, 7–38.

Mosier, A.R., Duxbury, J.M., Freney, J.R., Heinemeyer,O., Minami, K. and Johnson, D.J. (1998b) Mitigating agricultural emissions of methane. *Climate Change* 40, 39–80.

Neüe, H.U. (1993) Methane emission from rice fields. *BioScience* 43, 466–475.

Neüe, H.U. (1997) Fluxes of methane from rice fields and potential for mitigation. *Soil Use and Management* 13, 258–267.

Neüe, H.U. and Sass, R.L. (1994) Rice cultivation and trace gas exchange. In: Prinn, R.G. (ed.) *Global Atmospheric–Biospheric Chemistry*. Plenum Press, New York, pp. 119–147.

Neüe, H.U., Lantin, R.S., Wassmann, R., Aduna, J.B., Alberto, M.C.R. and Andales, M.J.F. (1994) Methane emission from rice soils of the Philippines. In: Minami, K., Mosier, A. and Sass R.L. (eds) *CH_4 and N_2O Global Emissions and Controls from Rice Fields and Other Agricultural and Industrial Sources.* Yokendo Publishers, Tokyo, pp. 55–63.

Neüe, H.U., Wassmann, R., Lantin, R.S., Alberto, M.A.C.R., Aduna, J.B. and Javellana, A.M. (1996) Factors affecting methane emission from rice fields. *Atmospheric Environment* 30, 1751–1754.

Parashar, D.C., Rai, J., Gupta, P.K. and Singh, N. (1991) Parameters affecting methane emission from paddy fields. *Indian Journal of Radio and Space Physics* 20, 12–17.

Parsons, D.J. and Williams, A.G. (1987) A mathematical model of bacterial growth in aerated pig slurry during anaerobic storage. *Journal of Agricultural Engineering Research* 38, 173–181.

Paul, E., Paustian, K., Elliott, E.T. and Cole, C.V. (eds) (1997) *Soil Organic Matter in Temperate Agroecosystems: Long-term Experiments in North America*. CRC Press, Boca Raton, Florida, 414 pp.

Paustian, K., Andren, O., Janzen, H.H., Lal, R., Smith, P., Tian, G., Tiessen, H., VanNoordwijk, M. and Woomer, P.L. (1997a) Agricultural soils as a sink to mitigate CO_2 emissions. *Soil Use and Management* 13, 230–244.

Paustian, K., Collins, H.P. and Paul, E.A. (1997b) Management controls on soil carbon. In: Paul, E.A, Paustian, K., Elliott, E.T. and Cole, C.V. (eds) *Soil Organic Matter in Temperate Agroecosystems; Long-term Experiments in North America*. CRC Press, Boca Raton, Florida, pp. 15–49.

Paustian, K., Cole, C.V., Sauerbeck, D. and Sampson, N. (1998) CO_2 mitigation by agriculture: an overview. *Climate Change* 40, 135–162.

Peck, T.R. (1989) Morrow Plots: long-term University of Illinois Field Research Plots. 1876 to present. In: Brown, J.R. (ed) *Proceedings of the Sanborn Centennial: a Celebration of 100 years of agricultural research*. Publication No. SR-415. University of Missouri, Columbia, pp. 49–52.

Pimentel, D. (1984) Energy flow in the food system. In: Pimentel, D. and Hall, C.W. (eds) *Food and Energy Resources.* Academic Press, New York.

Pimentel, D. and Heichel, G.H. (1991) Energy efficiency and sustainability of farming systems. In: Lal, R. and Pierce, F.J. (eds) *Soil Management for Sustainability.* Soil and Water Conservation Society, Ankeny, Iowa, pp. 113–122.

Pimentel, D., Dazhong, W. and Giampietro, M. (1990) Technological changes in energy use in US agricultural production. In: Gliessman, S.R. (ed.) *Agroecology: Researching the Ecological Basis for Sustainable Agriculture. Ecological studies: Analysis and Synthesis.* Springer, New York, pp. 305–321.

Raich, J.W. and Potter, C.S. (1995) Global patterns of carbon dioxide emissions from soils. *Global Geochemical Cycles* 9, 23–36.

Reicosky, D.C. and Lindstrom, M.J. (1993) Fall tillage method: effect on short-term carbon dioxide flux from soil. *Agronomy Journal.* 85, 1237–1243.

Reicosky, D.C. and Lindstrom, M.J. (1995) Impact of fall tillage on short-term carbon dioxide flux. In: Lal, R., Kimball, J., Levine, E. and Stewart, B.A (eds) *Soils and Global Change.* CRC Press, Boca Raton, Florida, pp. 177–187.

Rosenzwieg, C. and Hillel, D. (1998) Agricultural emissions of greenhouse gases. In: *Climate Change and the Global Harvest. Potential Impacts of the Greenhouse Effect on Agriculture.* Oxford University Press, New York, pp. 38–100.

Safley, L.M. Jr, Casada, M.E., Woodbury, J.W. and Roos, K.F. (1992) *Global Methane Emissions from Livestock and Poultry Manure.* EPA 400–1–91–048. US Environment Protection Agency, Washington, DC.

Sass, R.L. and Fisher, F.M. (1995) Methane emissions from Texas rice fields: a five-year study. In: Peng, S., Ingram, K.T., Neüe, H.U. and Ziska, L.H. (eds) *Climate Change and Rice.* Springer, New York, pp. 46–59.

Sass, R.L. and Fisher, F.M. (1996) Methane from irrigated rice cultivation. *Current Topics in Wetland Biogeochemistry* 2, 24–39

Sass, R.L., Fisher, F.M., Harcombe, P.A. and Turner, F.T. (1991a) Mitigation of methane emissions from rice fields: possible adverse effects of incorporated rice straw. *Global Biogeochemical Cycles* 5, 275–287.

Sass, R.L., Fisher, F.M., Turner, F.T. and Jund, M.F. (1991b) Methane emission from rice fields as influenced by solar radiation, temperature, and straw incorporation. *Global Biogeochemical Cycles* 5, 335–350.

Sass, R.L., Fisher, F.M., Wang, Y.B., Turner, F.T. and Jund, M.F. (1992) Methane emission from rice fields: The effect of flood water management. *Global Biogeochemical Cycles* 6, 249–262.

Sass, R.L., Fisher, F.M., Lewis, S.T., Turner, F.T. and Jund, M.F. (1994) Methane emission from rice fields: effect of soil properties. *Global Biogeochemical Cycles* 8, 135–140.

Satpathy, S.N., Rath, A.K., Ramakrishnan, B., Rao, V.R., Adhya, T.K. and Sethunathan, N. (1997) Diurnal variation in methane efflux at different growth stages of tropical rice. *Plant and Soil* 195, 267–271.

Schlesinger, W.H. (1986) Changes in soil carbon storage and associated properties with disturbance and recovery. In: Trabalka, J.R. and Reichle, D.E. (eds) *The Changing Carbon Cycle: A Global Analysis.* Springer-Verlag, New York, pp. 194–220.

Schlesinger, W.H. (1990) Evidence for chronosequence studies for low carbon storage potential of soils. *Nature* 348, 232–234.

Shapouri, H., Duffield, J.A. and Graboski, M.S. (1995) Estimating the net energy balance of corn ethanol. ERS, USDA. *Agricultural Economic Report* 721, 1–16.

Shin, Y.K., Yun, S.H., Park, M.E. and Lee, B.L. (1996) Mitigation options for methane emission from rice fields in Korea. *Ambio* 25, 289–291.

Sigren, L.K., Byrd, G.T., Fisher, F.M. and Sass, R.L. (1997a) Comparison of soil acetate concentrations and methane production, transport, and emission in two rice cultivars. *Global Biogeochemical Cycles* 11, 1–14.

Sigren, L.K., Lewis, S.T., Fisher, F.M and Sass, R.L. (1997b) The effects of drainage on soil parameters related to methane production and emission from rice paddies. *Global Biogeochemical Cycles* 11, 151–162.

Wagner, G.H. (1989) Lessons in soil organic matter from Sanborn Field. In: Brown, J.R. (ed.) *Proceedings of the Sanborn Centennial: a Celebration of 100 years of agricultural research.* Publication No. SR-415. University of Missouri, Columbia, pp. 64–70.

Wang, B., Neüe, H.U. and Samonte, H.P. (1997a) The effect of controlled soil temperature on diel CH_4 emission variation. *Chemosphere* 35, 2083–2092.

Wang, B., Neüe, H.U. and Samonte, H.P. (1997b) Effect of cultivar difference (IR72, IR65598 and Dular) on methane emission. *Agriculture Ecosystems and Environment* 62, 31–40.

Wang, M.X., Dai, A., Shangguan, X., Ren. L., Shen. R., Schüts, H., Seiler. W., Rasmussen, R.A. and Khalil, M.A.K. (1994) Sources of methane in China. In: Minami, K., Mosier, A. and Sass R.L. (eds) *CH_4 and N_2O Global Emissions and Controls from Rice Fields and Other Agricultural and Industrial Sources.* Yokendo Publishers, Tokyo, pp. 9–26.

Wang, M.X., Dai, A., Shen, R.X., Wu, H.B., Schütz, H., Rennenberg, H. and Seiler, W. (1990) Methane emission from a Chinese paddy field. *Acta Meteorologica Sinica* 3, 265–275.

Wassmann, R., Shangguan, X.J., Cheng, D.X., Wang, M.X., Papen, H., Rennenberg, H. and Seiler, W. (1996) Spatial and seasonal distribution of organic amendments affecting methane emission from Chinese rice fields. *Biology and Fertility of Soils* 22, 191–195.

Yagi, K., Tsuruta, H., Kanda, K. and Minami, K. (1996) Automated monitoring of methane emission from a rice paddy field: the effect of water management. *Global Biogeochemical Cycles* 10, 255–267.

Yagi, K., Tsuruta, H. and Minami K. (1997) Possible options for mitigating methane emission from rice cultivation. *Nutrient Cycling in Agroecosystems* 49, 213–220

Yue, L., Lin, E. and Minjie, R. (1997) The effect of agricultural practices on methane and nitrous oxide emissions from rice field and pot experiments. *Nutrient Cycling in Agroecosystems* 49, 47–50

4 Crop Ecosystem Responses to Climatic Change: Wheat

DAVID W. LAWLOR AND ROWAN A.C. MITCHELL

Biochemistry and Physiology Department, IACR-Rothamsted, Harpenden, Herts AL5 2JQ, UK

4.1 Introduction

Wheat is the single most important crop on a global scale in terms of total harvested weight and amount used for human and animal nutrition (Anon., 1996; Evans, 1998). Global environmental change (GEC) is likely to be of great consequence for wheat production and thus human food supplies. Currently, about 600 Mt of wheat grain is produced annually, and wheat provides approximately 20% of the energy and 25% of the protein requirements of the world's 6.6 billion people. In addition, wheat makes a large contribution to the nutrition of animals that add milk and meat to the human diet. Any factors which affect the production and cost of wheat affect all societies, since wheat is the most traded agricultural product with a large proportion of world production entering international trade. Reviews of wheat production and utilization (Heyne, 1987; Evans, 1993; Gooding and Davies, 1997) provide more details of the wheat crop's relation to the environment and importance to humankind. This chapter considers the likely responses of wheat to the primary factors associated with GEC, increasing atmospheric CO_2 concentration ([CO_2]) and predicted global warming. It focuses particularly on more recent (post 1990) research, and readers are referred to reviews for details of earlier work.

4.1.1 Global wheat production

Wheat is grown from the Arctic Circle to the Equator (Fig. 5.1) and from sea level to 3000 m altitude. However, it is best suited to 30–50° N or 25–40° S latitudes (Briggle and Curtis, 1987; Gooding and Davies, 1997). Production of wheat has increased more than that of any other crop of consequence in the last century. In 1900 production was 90 Mt; by 1954 it was 200 Mt, due to

Fig. 4.1. Global distribution of wheat cultivation, indicated by black areas, based on data in Briggle and Curtis (1987) and data from Anon. (1996).

doubling the area under cultivation (90 to 190 Mha). By the early 1990s, total production was 600 Mt, achieved by increased productivity. Production is largely from Asia (China, 100 Mt; India 50 Mt), former USSR (100 Mt), North America (USA, 65 Mt; Canada, 23 Mt) and the European Union (100 Mt; including France, 65 Mt). There is little production in Africa, although it is a major crop in North Africa. Currently, 2.2×10^9 ha of land are used for wheat cultivation. Five exporters dominate world trade (USA, Canada, European Union, Australia and Argentina). Some relatively small producers are of considerable importance, because their exports balance poor harvests in other areas. Wheat has been the main provider of food for the increasing world population (1.5 billion in 1900; 5.5 billion in 1993; probably reaching 11 billion by the mid-21st century). Wheat is imported by many countries, including the former USSR, China, Japan, Egypt and Indonesia. Many developing economies with rapidly growing populations would be particularly susceptible to decreased wheat production. Increased world cereal production of 35–40% will be required, and wheat should make up a larger proportion than other cereals (Ansart, 1997).

Productivity of wheat was *c.* 1 t ha^{-1} as a global average during the first half of the 20th century, but has risen to 2.5 t ha^{-1} currently (Slafer *et al.*, 1996). In areas that are climatically favourable for wheat production, such as western Europe, with cool, moist, long growing seasons and days and fertile soils, productivity of modern, semi-dwarf varieties under intensive management (fertilizers, pest and disease control) has risen to over 10 t ha^{-1}, with peak yields to > 12 t ha^{-1} and country-wide yields of 7 t ha^{-1}. However, worldwide the crop is grown extensively with limited inputs and yields are small. In the last decade, the rate of increase in productivity may have slowed and the area under cultivation has decreased by 8%. This trend, if maintained, may result in shortages. Over the last two decades, investment in agriculture has decreased and prices to producers have dropped, both disincentives to production (Evans, 1998). Global storage of wheat was about 35% of production in the early 1980s, due to cumulative effects of worldwide increases in productivity resulting from governmental subsidies to research, development and to farmers. Stocks are now about 18% of consumption, just above the FAO recommended 15% (Ansart, 1997). There is urgent need to ensure greater investment in wheat production to meet the known demands of population increase and possible effects of GEC.

4.1.2 Global environmental change

Global [CO_2] has increased from *c.* 280 μmol mol^{-1} at the start of the industrial revolution to 360 μmol mol^{-1} now. It is expected to exceed 550 μmol mol^{-1} by around 2050 and may even reach 700 μmol mol^{-1} by 2100. Over the decade 1984–1993, [CO_2] increased by 1.5 μmol mol^{-1} per year on average (Schimel *et al.*, 1996; see Chapter 2, this volume). Major concerns of studies of the impacts of GEC on wheat include identifying the effects that changes in the primary conditions of [CO_2] and temperature will have on growth and yield. Less attention has been directed to the effects of secondary changes, although the magnitude of their impact on production could well be greater. Also, cultivation of wheat, because of the large areas devoted to it, may contribute to the altered global carbon balance (for example, by increasing oxidation of organic carbon or accumulation of carbon from roots). However, such impacts are not considered here.

4.1.3 Current assessments of effects of global environmental change on wheat production

Assessment of the impacts of GEC on global wheat production is based, of necessity, on modelling that uses limited information about crop responses. A number of forecasts of the impacts have been made. The 1995 Summary of the Intergovernmental Panel on Climate Change (IPCC) (Watson *et al.*, 1996) emphasized the considerable uncertainty in estimating impacts of doubling pre-industrial [CO_2] and increasing temperatures by 4°C. These authors wrote that variability in estimated yield impacts among countries, methods of analysis and crops, makes it difficult to generalize results across areas or for different climate scenarios. However, they tentatively concluded that wheat production will decrease substantially. Generally, modelling studies predict that wheat will be grown at higher latitudes (i.e. mainly further north) and that production in regions closer to the Equator will decrease (Leemans and Solomon, 1993; Rosenzweig and Parry, 1994). Adams *et al.* (1990) suggested a major decrease in wheat production in the USA. In Rosenzweig and Parry's study of the impacts of 555 μmol CO_2 mol^{-1} and approximately 4°C increase in temperature by 2060, simulated global agricultural production of major cereal crops declined, despite assuming farm-level adaptation and future technological improvements. With 2°C warming, yields might improve by 12% (due to longer crop duration) but a 4°C warming would decrease global production (this contrasts with Leemans and Solomon, 1993, who predict an increase). Thus, 'slight-to-moderate negative effects' were projected to result from GEC. As a consequence of GEC and world population increase, the proportion of the population at risk of hunger may increase, by 6–50%. GEC would increase the disparities between developed and underdeveloped agriculture. We have some doubts about the reliability of current simulation models for assessment of global production of wheat (see below) but, in any case, uncertainties about the nature of GEC probably outweigh the uncertainties in crop response. We share with Evans (1998) the view that feeding the greatly increased human population 'can be done, but to do so sustainably in the face of climatic change, equitably in the face of social and regional inequalities, and in time when few seem concerned, remains one of humanity's greatest challenges'.

4.2 Methods of Studying Global Environmental Change

Methods employed for studying wheat responses to [CO_2] and temperature have been discussed (Lawlor, 1996; Morison and Lawlor, 1999). The most relevant to agronomy for studying the effects of elevated [CO_2] is free-air carbon dioxide enrichment (FACE) (Hendry and Kimball, 1994), as other aspects of the field environment are unperturbed. Expense of the installation, and particularly of the large amounts of CO_2 required, has limited the number of wheat studies to one site in Arizona, USA. Results of FACE experiments may be difficult to extrapolate to other areas, where atmospheric and soil conditions differ. A major limitation is the current inability to test the interaction of [CO_2] and temperature. Small but significant temperature differences between control and elevated [CO_2] treatments arose due to omission of ventilation ('blowers') in the control [CO_2] plots in FACE experiments in 1992–1993 and 1994–1995 which hastened the rate of development and affected the interpretation of the impact of [CO_2]. Later experiments (1995–1996 and 1996–1997) were unaffected.

Open-top chambers (OTCs) are also widely used to maintain a given [CO_2] for a crop growing in the field. Such chambers generally decrease the radiation on the crop, increase humidity, slow air movement and, most importantly, are substantially warmer than the ambient air under a large radiation load (Leadley and Drake, 1993). By growing crops outside and inside OTCs, the temperature effect, but not its interaction with [CO_2], may be assessed. Kimball *et al.* (1997a) showed that the relative effects of [CO_2] and also the absolute wheat crop production were similar in OTC and FACE experiments in the cool winter period in Arizona. Wheat responses to [CO_2], temperature and their interactions have been widely investigated under controlled environment (CE) conditions; some simulate ambient field temperatures or allow a constant temperature differential (Lawlor *et al.*, 1993). Temperature gradient tunnels (Hadley *et al.*, 1995; Rawson, 1995) allow plants to be grown over a wide range of temperatures. Differential temperature may make interpretation complicated; however, the information is useful when combined with regulated [CO_2]. Usually, average and peak light intensities are reduced and the spectrum altered in all enclosed facilities and OTCs. Humidity (which is difficult and expensive to control) is often not regulated, despite its potential importance. True replication of facilities may not be possible. Pseudoreplication has to be applied, including frequent transfer of the basic treatments, and moving container-grown plants between rooms in controlled environments, minimizes bias. A major concern is the extent and depth of rooting of crops, because restricted growth may affect the supply of water and nutrients and limit the sink for assimilates (Chaudhuri, 1990a,b). Despite the wide range of methods, they provide a broad basis of agreement, with responses of wheat qualitatively similar in different systems, although quantitatively there is considerable difference.

In this chapter, we first assess the responses of basic processes to increases in [CO_2] and temperature, the most certain aspects of GEC. These are then related to the changes in biomass production and grain yield observed when wheat is grown under conditions of elevated [CO_2] and/or increased temperature. We then attempt to relate these processes and products to other aspects of the environment, particularly water status, which is also likely to change as part of GEC, but in a still uncertain manner for any given region. Grain quality and how wheat adaptation might be improved genetically for GEC are briefly considered. Finally, incorporation of the information into models simulating wheat production and their possible improvements are assessed.

4.3 Effect of [CO_2] and Temperature on Basic Processes

4.3.1 Effect of [CO_2] on photosynthesis

Elevated [CO_2] increases the rate of net photosynthesis (P_n) and decreases the rate of photorespiration (P_r) in C_3 plants such as wheat. The causes are well understood: the enzyme responsible for the reaction of CO_2 with ribulose bisphosphate (RuBP) is ribulose bisphosphate carboxylase-oxygenase (Rubisco). The reaction is not saturated at current [CO_2]; also oxygen reacts with RuBP, producing phosphoglycolate which is metabolized, ultimately releasing CO_2 in P_r. Doubling current [CO_2] saturates Rubisco and greatly decreases P_r, thus increasing P_n by 30–40% (in bright light and at 25–30°C) and the rate of carbon assimilates supply (sugars and starch) to the plant (Long, 1991). Although elevated [CO_2] also decreases the stomatal conductance (g_s), with potentially large reductions in water loss, the increased atmospheric [CO_2] is more than sufficient to compensate so that the internal CO_2 concentration, [CO_2]$_i$, is greatly increased, hence the increased P_n. The response of leaf P_n to

$[CO_2]_i$ for different light and temperature values is well predicted by the standard model of the process (Farquhar and von Caemmerer, 1982).

Wheat exhibits these typical responses and there is little varietal variation. Often the initial stimulation of P_n in C_3 plants at elevated $[CO_2]$ decreases in plants grown at elevated $[CO_2]$. This is associated with a general loss of photosynthetic capacity of plants grown in elevated $[CO_2]$, i.e. P_n measured at any given $[CO_2]$ is smaller than that of plants grown in ambient $[CO_2]$ (Sage, 1994). This is usually termed acclimation, but as increases also occur it is better to refer to it as negative or positive acclimation. The possible mechanisms of such acclimatory processes have been discussed by Lawlor and Keys (1993) and Drake *et al.* (1997). There are two explanations: (i) loss of photosynthetic components; or (ii) decreased activation of components ('down-regulation'). Both affect Rubisco. 'Up-regulation' is less frequently observed.

In studies on wheat grown under elevated $[CO_2]$, a range of different responses of photosynthetic capacity have been observed. Individual plants, grown under controlled environmental conditions, lost photosynthetic capacity with reduced amounts and activation states of Rubisco (McKee and Woodward, 1994a,b). In the later FACE studies, there was no acclimation due to growth under elevated $[CO_2]$ so that photosynthesis was stimulated when measured under elevated $[CO_2]$ (Garcia *et al.*, 1998). Earlier studies indicated substantial stimulation of P_n of leaves (31%) in well-watered and fertilized crops by 550 cf. 380 $\mu mol\ CO_2\ mol^{-1}$ (Nie *et al.*, 1995a). For leaves just emerged, elevated $[CO_2]$ during growth had no effect on photosynthetic capacity or on amounts of any of the major proteins in any leaf examined. For flag leaves during grain-fill, there was a substantial effect on photosynthetic capacity. Rubisco declined by 45% in control leaves but by 60% in elevated $[CO_2]$ (Nie *et al.*, 1995a,b). However, care must be exercised in interpreting these studies, due to the cooler control compared with elevated $[CO_2]$ treatments caused by omission of blowers (Garcia *et al.*, 1998). With correct controls, the effect of CO_2 was only seen during late grain-filling and particularly with N deficiency. Also, carboxylation efficiency was decreased, but by variable amounts in different leaves (Brooks *et al.*, 1996; Adam *et al.*, 1997; Wall *et al.*, 1997). Similar results were obtained in OTC experiments carried out in several European countries on the spring wheat cv. Minaret (Mitchell *et al.*, 1999). Here, there was also no effect of elevated $[CO_2]$ on photosynthetic capacity before anthesis, but there was a decrease in flag leaves photosynthesis during grain-fill, especially in crops with less applied N. Delgado *et al.* (1994) did not observe acclimation. However, in a subsequent, similar study on the same variety (Lawlor *et al.*, 1995 and unpublished) there was a progressive decrease in photosynthetic capacity in leaves of later insertion, and loss of Rubisco, demonstrating that subtle changes in environment can affect photosynthetic capacity responses to elevated $[CO_2]$.

The supply of N is critical in determining the response of photosynthetic capacity. Increased photosynthetic capacity and Rubisco content (positive acclimation) in response to elevated $[CO_2]$ have been observed in young plants grown with warm temperatures and unlimited N supply (e.g. Habash *et al.*, 1995). Possibly, such conditions stimulate extra root growth and N uptake.

Growth at elevated [CO_2] had no effect on photosynthetic capacity in young leaves of wheat with varying N supply but decreased it as leaves aged, and this effect was greater at lower N supplies (Theobald *et al.*, 1998). This work also showed that the critical factor is whether the N treatment results in a difference in leaf N content. This explains why elevated [CO_2] does not always result in decreased capacity even at low N supply (e.g. Delgado *et al.*, 1994).

There have been studies to separate the two explanations for the effects of [CO_2] on photosynthetic capacity. There is little evidence of down-regulation of photosynthetic capacity at high light (Delgado *et al.*, 1994; Lawlor *et al.*, 1995; Nie *et al.*, 1995a,b; Theobald *et al.*, 1998). McKee and Woodward (1994a) interpreted the down-regulation of Rubisco to occur when it is in excess compared with some other limitation on P_n. Changes in amounts of photosynthetic components, and Rubisco in particular, are therefore responsible for loss of photosynthetic capacity.

From all the above evidence, many of the observed effects of elevated [CO_2] on photosynthetic capacity in wheat can be explained in terms of changes in the sink–source balance, i.e. the result of more C assimilate (Morison and Lawlor, 1999), echoing the conclusions of Rogers *et al.* (1996). For example, if N supply is fixed, and canopy growth and leaf area are stimulated at elevated [CO_2], leaf N content and photosynthetic capacity are inevitably decreased. If elevated [CO_2] has stimulated ear and grain formation, they provide a greater sink for remobilization of N from photosynthetic tissue during grain-fill; thus, green area senescence may be accelerated. There is also the confounding effect of reduced stomatal conductance at elevated [CO_2] (see below), which may increase leaf temperature and accelerate senescence. When N is supplied, such that additional root growth at elevated [CO_2] can capture more N (and also use more assimilates), very different results are obtained (Habash *et al.*, 1995), or the effect of elevated [CO_2] on N content of leaves is not decreased (Rogers *et al.*, 1996). Therefore, the response of photosynthetic capacity to elevated [CO_2] supports the hypothesis that a general consequence of elevated [CO_2] is altered sink–source relations. The relationships between Rubisco content, thylakoid ATP-synthase and leaf N were unaltered in wheat leaves grown at elevated [CO_2]. A CO_2-specific response would be expected to decrease Rubisco preferentially (Theobald *et al.*, 1998).

The mechanism by which the changes in leaf N and photosynthetic capacity occur are still uncertain. Part of it possibly occurs as a consequence of the feedback regulation exerted by increased carbohydrates (hexoses or, specifically, glucose) on gene expression (van Oosten and Besford, 1996). However, in a FACE experiment, mRNA transcripts for a number of photosynthetic components, including the large and small subunits of Rubisco, were poorly correlated with the concentrations of carbohydrates (e.g. glucose-6-phosphate, sucrose and starch), which were different between the CO_2 treatments (Nie *et al.*, 1995b). Also, in wheat in a controlled-environment experiment, changes in amounts of Rubisco were not correlated with the concentrations of carbohydrates in leaves or stems (Lawlor *et al.*, 1995 and unpublished). Thus, the sugar repression model of photosynthetic acclimation

(Sheen, 1994) is not operating in a simple way in wheat grown in elevated [CO_2]. The complex effects observed are likely the result of several dynamic, interacting mechanisms operating to balance changing demand and supplies for C and N (Morison and Lawlor, 1999).

4.3.2 Effect of elevated [CO_2] on stomatal conductance

The other, well-established direct effect of elevated [CO_2] is on stomatal conductance (g_s). In contrast to the effect on P_n, the molecular mechanisms are not well understood. In studies on wheat, g_s is typically decreased by about 30–40% by doubling [CO_2] from current concentration (e.g. Samarakoon *et al.*, 1995). This is due to changes in aperture since stomatal density (stomata/unit leaf area) and stomatal index (ratio of stomata/leaf epidermal cells) are unaffected within a single season (Estiarte *et al.*, 1994). Effects resulting from continuous growth in elevated [CO_2] over generations have not been studied. There is no evidence of acclimation of g_s to elevated [CO_2], so that the [CO_2]$_i$/[CO_2] ratio was unaffected by [CO_2] during growth in simulated field conditions (Delgado *et al.*, 1994; Lawlor *et al.*, 1995) or in the FACE study (Wall *et al.*, 1997).

4.3.3 Effect of temperature on photosynthesis

Many studies show that P_n in wheat is small (25% of maximum) at 5°C, increases as temperature rises to a broad optimum before decreasing at high temperature (> 25°C) and ceasing at about 40°C. Sensitivity of P_n to [CO_2] is predicted to increase strongly with temperature due to the characteristics of Rubisco (Long, 1991; Lawlor and Keys, 1993). The temperature optimum increases by several degrees with elevated [CO_2] according to the Farquhar and von Caemmerer (1982) model (Long, 1991). This is due to the temperature dependence of the kinetic parameters of Rubisco and the relative solubility of CO_2 and O_2, which mean that photorespiration increases with temperature. An important effect of warmer temperature is accelerating leaf maturation and senescence, thus decreasing the duration of active photosynthesis.

4.3.4 Effects of [CO_2] and temperature on respiration

Dark respiration is a major component of plant productivity. Approximately 50% of total assimilates acquired by photosynthesis are respired (Amthor, 1997). An unexpected feature of respiration to GEC is an apparent decrease in its rate when expressed per unit of dry matter. Such a decrease is often observed in plants grown in elevated [CO_2] (Poorter *et al.*, 1992). Part, or all, of this decreased respiration is probably due to the decreased proportion of metabolic components per unit dry matter as carbohydrates accumulate (Drake *et al.*, 1997). However, this cannot explain why elevated [CO_2] also decreases the respiration rate of young wheat seedlings growing in darkness (Nátr *et al.*, 1996).

While respiration in wheat responds markedly to temperature in the short term (Mitchell *et al.*, 1991), the sensitivity is much less in the long term (Gifford, 1995). This suggests that division of respiration into growth and maintenance components, with the latter being highly temperature dependent, tends to exaggerate the temperature dependency. Assumption of a constant respiration : photosynthesis ratio regardless of temperature or [CO_2] could be more accurate (Gifford, 1995).

4.3.5 Effects of [CO_2] and temperature on crop development

It is difficult to disentangle with certainty direct (as CO_2 *per se*) and indirect (temperature increases due to smaller g_s) effects of [CO_2] on crop development. However, there is little evidence of any direct CO_2 effect on the rate of development in wheat at any stage. Slightly faster rates of leaf, ear, spike and tiller development are frequently observed in wheat grown at elevated [CO_2] and are probably indirect. The complete absence of an effect of elevated [CO_2] on wheat phenology in conditions of relatively low radiation also supports this conclusion (Mitchell *et al.*, 1995; Batts *et al.*, 1996, 1997; Wheeler *et al.*, 1996b). In the FACE experiments, N fertilizer altered crop architecture (Brooks *et al.*, 1996), but elevated [CO_2] had a minimal effect on rates of overall canopy development and senescence (Wall *et al.*, 1997). This was in contrast to accelerated development seen in earlier studies when temperatures differed (Garcia *et al.*, 1998). This emphasizes the great importance of small increases in temperature for wheat development and, ultimately, yield.

The rates of development of wheat are sigmoidal rather than linear functions of temperature (Shaykewich, 1995). Development, while dependent on variety, begins above a base temperature (1–5°C) and the rate rises slowly as temperature increases, then more rapidly to a maximum at about 30°C and rapidly slows thereafter. The effect of constant temperature increment on the duration of phenological phases is, therefore, dependent on absolute background temperature. However, the effects can often be modelled accurately with the thermal time approach, as shown for data from temperature gradient studies. These experiments provide a good range of temperature around the ambient. They indicate that a 1°C warming would reduce crop duration by about 21 days (8%) and the reproductive period by about 8 days (6%). The period from anthesis to maturity is shortened by about 3 days and the duration of grain-fill by about 2 days (5%) per °C. Similar results were obtained for an experiment with a single temperature treatment with a 4°C increment above ambient (Mitchell *et al.*, 1995). However, these responses show considerable differences among varieties (Batts *et al.*, 1998a,b). Rawson (1988) presented data for rapidly developing wheat, suggesting that, with increased radiation, ample water and nutrients, high temperatures may not be detrimental to production. Improved understanding of such responses may be of value for exploitation of hot environments.

4.3.6 Effects of temperature extremes

Quite separate from the effects of temperature on wheat phenology are the effects of temperature extremes. Extreme cold may kill wheat, and late frost induces sterility (Russell and Wilson, 1994). Chilling temperatures (below 5°C) and hot temperatures (above 30°C) at anthesis can damage pollen formation, which in turn reduces grain set and can decrease yield (Dawson and Wardlaw, 1989; Tashiro and Wardlaw, 1990). There is a wide range of susceptibility to these effects among varieties. GEC will probably cause lower yields due to temperature extremes, simply by changing the frequency of temperature extremes over the short periods of particularly sensitive stages of plant development. Tolerance to such conditions will require breeding and selection of better adapted varieties (Acevedo, 1991).

4.4 Effects of [CO_2] and Temperature on Biomass and Grain Yield

4.4.1 Effects of [CO_2] on biomass and grain yield

Biomass results from accumulation of carbon in plant products as the difference between photosynthesis and respiration, plus accumulation of minerals. Increasing biomass might, therefore, be expected to parallel stimulation of instantaneous P_n. However, in practice there are many complicating factors; for example, stimulation of root or canopy growth may allow additional resource capture. Conversely, increased C assimilation may result in nutrients being more limiting for growth, thus necessitating increased fertilizer applications, which may increase lodging and disease. At low temperatures, wheat growth is probably not limited by assimilation, but rather by sink capacity. These factors may explain the increases in wheat biomass that range from 0 to 40% in response to doubling [CO_2]. Early field studies (reviewed in Lawlor and Mitchell, 1991) indicated responses ranging from 30% increase for a doubling of [CO_2] to only a 20% increase for a quadrupling of [CO_2]. Biomass of winter wheat in several experiments simulating field temperatures in the UK gave a 15–27% response to doubling [CO_2] (Mitchell *et al.*, 1993, 1995; Batts *et al.*, 1997, 1998b), and a 6–34% increase when [CO_2] was doubled in temperature gradient tunnels. Rawson (1995), in a similar Australian system, obtained increases from 7 to 36%, depending on temperature. In 25 experiments on spring wheat in OTCs at nine European sites, stimulation of biomass with [CO_2] varied from 10% with a 320 μmol mol^{-1} enrichment to 75% with a 246 μmol mol^{-1} enrichment (Bender *et al.*, 1999). Regression analyses using all the data indicated a 13% increase in above-ground biomass resulting from a 100 μmol mol^{-1} enrichment. Under the FACE conditions, biomass increased by 10 and 15% in 2 years, with a 200 μmol mol^{-1} increase in [CO_2] (Kimball *et al.*, 1997a,b; Pinter *et al.*, 1996, 1997).

Response of wheat to extra C assimilate is flexible, and growth of all plant parts can be stimulated. Main stems are usually least affected. Tiller production and survival and root growth are stimulated most, but leaf size, grains per ear and grain size can all be increased, depending on environment and cultivar. Grain yield increase resulting from elevated [CO_2] is often similar to that of biomass, but not always (Mitchell *et al.*, 1995; Batts *et al.*, 1997). Grain yield is more sensitive to biomass production during the reproductive phase of growth, particularly around anthesis, than during the vegetative growth phase. It has been shown that effects of elevated [CO_2] on harvest index can be explained in terms of the relative increase of biomass in these different periods (Mitchell *et al.*, 1996). In multiple-site OTC experiments using spring wheat, cv. Minaret, the average increased grain yield was 11% per 100 μmol mol^{-1} enrichment (Bender *et al.*, 1999), which is similar to that from an experiment on winter wheat cv. Mercia (Mitchell *et al.*, 1995). The stimulation measured by Batts *et al.* (1997) varied greatly between seasons for a given temperature but was, on average, equivalent to 17% per 100 μmol mol^{-1} enrichment. All these increases are greater than the *c.* 7% from FACE experiments (Pinter *et al.*, 1996, 1997).

It would be expected (see arguments above on photosynthetic capacity) that yield increases resulting from elevated [CO_2] are less with smaller N supply. While this was the direction of the interaction in both FACE and the multiple-site OTC experiments, it was not significant. Since CO_2 stimulates root growth (Chaudhuri *et al.*, 1990b; Wall *et al.*, 1996), it may enable plants to capture more N in the field. When the total amount of N is strictly controlled, however, there is a strong interaction; for example, increased spring wheat grain yields resulting from elevated CO_2 in controlled environment experiments were 5, 10 and 18% with N supplies of 4, 9 and 24 g m^{-2}, respectively (Theobald *et al.*, unpublished). Theory also predicts that the degree of stimulation will be less at lower light intensities, but this has not been explicitly tested in wheat. Interactions with temperature and water status are more complex and are discussed below.

4.4.2 Effects of temperature on biomass and grain yield

Warmer temperature shortens the duration of all developmental stages. Thus, there is less time for capture of light, water and nutrients resources, and it is not surprising that biomass production decreases with increasing temperature. The magnitude of the effect depends on which growth stages are affected, since shortening the period of maximum growth rate will have more effect on final biomass than will shortening of early growth stages. In the UK, a constant increase of 3.5°C applied throughout the growing season reduced final winter wheat biomass by about 16% and grain yield by 35%, but early growth was greatly stimulated (Mitchell *et al.*, 1995). Generally, larger negative effects of increased temperature on biomass were found in polytunnel experiments with variable perturbation of ambient temperature (probably because the perturbations tended to be greater late in the season). However, this differed widely between seasons (Batts *et al.*, 1997) and cultivars (Batts *et al.*, 1998b).

The negative effect of increased temperature on grain yield tends to be larger than the effect on biomass; i.e. harvest index is reduced by warmer temperature (Mitchell *et al.*, 1995; Wheeler *et al.*, 1996a,b; Batts *et al.*, 1997). This is because of the particular sensitivity of grain yield to the duration of grain-fill. A 1°C increase in temperature during grain-fill will typically shorten it by 5% and reduce harvest index and grain yield proportionately. If high temperatures at anthesis induce partial sterility, this reduces yield (Mitchell *et al.*, 1993). A statistical analysis of the nine different sites (258 yield observations) growing spring wheat cv. Minaret in OTCs revealed a 6% decline in grain yield and biomass per degree Celsius increase during the growing season (emergence to maturity) (Bender *et al.*, 1999). This effect was quite well predicted by simulation models, including one where the only effect of temperature was on phenology (van Oijen and Ewert, 1999).

4.5 Atmospheric [CO_2] × Temperature Interactions

Since short-term stimulation of P_n is increased at higher temperatures (Long, 1991), a large positive interaction between effects of increased [CO_2] and temperature on biomass might be expected (Rawson, 1992). While a relatively consistent positive interaction was measured in wheat experiments conducted by Rawson (1995), no interaction was found by Mitchell *et al.* (1995) and Batts *et al.* (1997, 1998a,b). There was a non-significant positive interaction of [CO_2] with seasonal mean daily temperature in the multi-site OTC experiments (Bender *et al.*, 1999; van Oijen and Ewert, 1999). This is an area where simulation models have proved useful for analysis. While the Farquhar and von Caemmerer (1982) model predicts strong temperature × [CO_2] interaction for effects on P_n (see Long, 1991), wheat models which use this as a sub-model show that the interactive effect on biomass is often not large. This is because increasing temperature means the growing season ends earlier, so that less of the growth takes part in the hottest part of the year. Using 30-year averages of temperature and radiation for different European sites, van Oijen and Ewert (1999) showed that models predicted no interaction between temperature and [CO_2] in absolute yield terms. If the seasonal variation in temperature is less in the Australian climate, this would explain the [CO_2] × temperature interaction observed there.

4.6 Effects of [CO_2] Under Water-limiting Conditions

Much of the global wheat production is water-limited and so the effect of GEC on water availability is crucial. While the effect of elevated [CO_2] on g_s is fairly consistent (even for varieties which differ in g_s; Samarakoon *et al.*, 1995), the effects on transpiration are complex (Lawlor, 1998). A lower rate of transpiration and less latent cooling increases the water vapour pressure in the intercellular spaces of the leaves, and decreases the humidity of the air in the boundary layer. Consequently, the gradient of vapour pressure increases, and concomitantly transpiration. However, short-term exposure to elevated [CO_2] generally decreases transpiration. In a glasshouse experiment, doubling ambient [CO_2] decreased water use by about 20% under wet conditions and slightly increased it under dry conditions. Also, in a controlled environment, transpiration decreased by only 8% at double-ambient [CO_2] during vegetative growth (André and du Cloux, 1993). Thus, the magnitude of transpiration has a complex dependency on the environment. In the long term, increased leaf area and root growth result from elevated [CO_2], and higher temperature means that water use may decrease much less than is suggested by short-term controlled studies (Chaudhuri *et al.*, 1990a). Under water-limiting conditions, water loss may not decrease. In the FACE studies, a 200 μmol mol^{-1} increase in [CO_2] reduced evapotranspiration (ET) of wheat by about 5% in irrigated plots. With drought, ET was increased by about 3% (Hunsaker *et al.*, 1996, 1997). This may have resulted from the larger root system exploiting more water, greater leaf area, and/or slightly greater temperatures. Note that estimation of such small differences is difficult, because methods gave different values; a 5% reduction from water balance was obtained, 10% from latent energy exchange measurements and 7–23% from the sap flow measured with stem flow gauges (Senock *et al.,* 1996).

Despite the relatively small effect of [CO_2] on ET, elevated [CO_2] increased biomass substantially, and so increased water-use efficiency (WUE) by 145% in the dry treatment and 21% in the wet treatment with FACE. Water use efficiency was increased by about 64% by doubling [CO_2] under drought conditions in an Australian glasshouse experiment (Samarakoon *et al.*, 1995). The relative increase in biomass due to doubled [CO_2] was much greater under dry than wet conditions. In the FACE studies, a greater effect of CO_2 enrichment on biomass occurred under drought conditions vs. well-watered crop conditions, which was reflected in a 20% increase in grain yields in drought and a 10% increase in irrigated plots (drought decreased grain yield by 28%). The problem of temperature differences induced by the system (see earlier) may have affected the results, but the effect is similar to that observed in OTC studies (Grashoff *et al.,* 1995), if the smaller increment in [CO_2] is considered.

In summary, the relatively few wheat studies on the interaction of [CO_2] and drought under field-like conditions indicate that: (i) water use by wheat crops under elevated [CO_2] may decrease slightly in wet conditions, but may increase slightly under dry conditions; and (ii) the stimulation of biomass and grain yield by elevated [CO_2] tends to be greater under drought than in well-watered conditions.

4.7 Atmospheric Pollutants, [CO_2] and Temperature

Human activity is increasing the atmospheric concentrations of chemicals that are known or suspected to be toxic to plants (see Chapter 18, this volume) and De Kok and Stulen (1998) consider this topic in detail. Wheat is affected by gaseous pollutants, such as SO_3 and O_3, depending on their concentrations and nature, e.g. as dry or wet deposition (McKee *et al.*, 1995; Barnes and Wellburn, 1998; Bender *et al.*, 1998). The mechanisms of responses of wheat to pollution (concentration and duration of exposure) are complex. However, increased [CO_2] decreases the impacts of gaseous pollutants, partially due to smaller g_s, and also increased capacity for detoxification. UV-B radiation is increasing as a consequence of altered atmospheric chemistry (see Chapter 18, this volume). However, wheat has active protective mechanisms, such as accumulation of flavonoid pigments which reduce the UV-B transmissions into leaves. As wheat currently grows at high elevations, where fluxes of UV-B are large, it seems likely that there is sufficient genetic capacity to allow selection for future changes (Teramura, 1998).

4.8 Wheat Quality

Despite the great emphasis placed on quality of wheat for nutritional and industrial uses (Heyne, 1987), there is little understanding of the interaction between environmental and genetic factors determining composition of wheat grain under current conditions. Supply of nutrients, particularly N, determines (in conjunction with other factors such as water supply and temperature) protein amounts relative to starch and other non-N components. Nutrient supply thus alters N-concentration, but only affects amino acid and protein composition to a limited degree. However, sulphur supply does affect the latter.

Elevated [CO_2] tends to increase mass per grain and decrease percentage N because of the increased supply of carbohydrate from photosynthesis, either during grain-fill or from reserves (Mitchell *et al.*, 1993; Tester *et al.*, 1995; Batts *et al.*, 1997; Rogers *et al.*, 1998). Elevated [CO_2] does not greatly alter the composition of the carbohydrates, e.g. types of sugars or starches, sugar/starch ratio (Rogers *et al.*, 1998). It also does not affect protein composition (Shewry *et al.*, 1994). Increased [CO_2] interacted with nitrogen supply and temperature. Complex changes in lipid composition have been identified (Williams *et al.*, 1995); and the effects of 700 μmol CO_2 mol^{-1} are smaller than those of a 4°C temperature rise but the mechanisms are not understood. The non-polar neutral lipids di- and mono-galactosyldiacylglycerol increased where temperatures were 4°C above ambient and N was deficient at elevated [CO_2], but decreased in ambient temperature conditions. With elevated [CO_2] and temperature, only starch lipids were affected. Fatty acid composition was affected by warmer conditions, which decreased oleate but increased linoleate of non-polar lipids.

Analysis of grain from spring wheat, cv. Hereward (Tester *et al.*, 1995) showed that warmer temperatures substantially decreased mass per grain and starch content, due to fewer and smaller type A starch granules and fewer amyloplasts per endosperm. Starch gelatinization increased with temperature, but gelatinization enthalpy was unaffected. Also, total and lipid-free amylose increased, but amylolipid and lysophospholipids were not affected by warmth. The mechanisms of C and N accumulation related to grain filling are rather separate, and the biochemical pathways leading to the final stored products are complex and are likely to be affected by environmental conditions. In addition, variety/environment interactions are to be expected.

Given the dynamic nature of the processes and conditions, there is considerable scope for variation in final grain composition. Such alterations may affect the nutritional quality and industrial uses of wheat grain. It is surprising, therefore, that more attention has not been directed to assessing the impacts of GEC on wheat quality. Perhaps the perceived slowness of the environmental changes and the existing variability in wheat characteristics, coupled with the pragmatic approach of end-users, mitigate against scientific analysis of quality.

4.9 Designing Wheat Varieties for Global Environmental Change

Although much progress has been made in understanding determinants of wheat production, there are fundamental processes in metabolism that are not well understood but that are key to the response of wheat to environment (Feil, 1992). Wheat cultivars have been found to vary in their response to both increased temperature (Rawson and Richards, 1992; Batts *et al.*, 1998a,b) and, to a lesser extent, elevated [CO_2] (Rawson, 1995; Batts *et al.*, 1998b). The variation in effect of [CO_2] may be an indirect consequence of differences in phenology and sink size. Such varietal differences may be important for breeding for GEC conditions. Since the negative effects of warmer temperature are due to shortened duration, slower-developing varieties may be expected to overcome this effect when water is not limiting. Phenological models of wheat suggest that slower-developing varieties would be better adapted to a warmer climate than fast-developing ones, and would avoid late spring frosts and the onset of dry seasons (Miglietta and Porter, 1992). This would not hold if drought began earlier under GEC, since faster-developing varieties generally are selected in regions prone to late drought (Evans, 1993). In a [CO_2] × temperature experiment using wheat isolines differing in maturity date, the later line yielded more than the earlier line at the highest temperature regime in no-drought conditions (Rawson, 1995). This supports the notion that the negative effects of temperature could be overcome by breeding, but the expected genotype/temperature interaction was not clear in all experiments.

Increasing [CO_2] may present opportunities to breeders and possibly allow them to select for allocation of assimilates away from leaves toward roots. Since the increase in [CO_2] is gradual, this can probably be achieved simply by selecting empirically for yield. Potentially rapid changes in other aspects of GEC, such as temperature extremes or evaporative demand, would present more problems. If GEC leads to temperatures which exceed wheat crop 'design limits', then damage to production may be expected. High temperatures in winter would hasten development and excessive tillering, and make the crop subject to more frost damage. It may, therefore, be necessary to develop wheats with strong photoperiodic and vernalization control to slow growth and restrict development to more favourable conditions (Rawson and Richards, 1992). Also, there would be a need to ensure that crops have adequate protective systems to avoid photoinhibitory damage when growth is slow. Similarly, increased incidence of high temperatures around anthesis would pose problems which could be overcome by selection for tolerance.

Genetic manipulation of plant characteristics is currently in vogue, and wheat is a major target for transformation. Wheat has proved recalcitrant to transformation compared with other species, although considerable progress is being achieved (Lazzeri *et al.*, 1997). The methods are likely to result in over- or under-expression to individual endogenous genes, and introduction of novel genes (e.g. for disease resistance or grain composition). Changes to basic metabolism, such as in photosynthesis and growth as affected by temperature, are likely to require more time.

4.10 Modelling

Reliable robust models of wheat production that are applicable to a wide range of conditions are required to assess the possible impacts of GEC. Where wheat models have been tested in experiments using elevated [CO_2] and/or temperature treatments in controlled environments (Mitchell *et al.*, 1995, 1996), OTCs (van Oijen and Ewert, 1999) or FACE (Grossman *et al.*, 1995; Kartschall *et al.*, 1995), they could be modified to predict effects adequately. For [CO_2], this has been achieved by assuming only two sites of action: photosynthesis and stomatal conductance. While much experimental variation in the effect of [CO_2] on biomass and yield in field experiments cannot be explained by the models, average effects and dependence on temperature are adequately predicted (van Oijen and Ewert, 1999). Temperature acts mostly via effects on phenology (Butterfield and Morison, 1992), and possibly some increased maintenance respiratory costs, although this has been recently disputed (Gifford, 1995). The negative effects of increased temperature on grain yield (–6% $°C^{-1}$) in the multiple-site OTC experiments on spring wheat were well predicted by a model which assumed no effect of temperature on respiration, as well as by one which assumed a large effect on maintenance respiration. We are not aware of tests of wheat models' capacity to predict effects of temperature increases under water-limiting conditions, where the two dominant effects would be on phenology and evaporative demand. Such validation experiments would be valuable.

While it seems possible to incorporate the effects of [CO_2] into wheat models with reasonable security, the general suitability of current wheat models to predict the large-scale effects of GEC is doubtful. Crop models are largely empirical, and this will not be changed rapidly, given the complexity of the processes underlying such features as stomatal conductance and partitioning, which are poorly understood. The search, then, is for the most appropriate empiricisms for the particular application of predicting impact of GEC. Current wheat models, developed from field experiments and predicting many variables, perform adequately when tested with carefully controlled treatments at an experimental site (e.g. drought: Jamieson *et al.*, 1998). When tested over a wide range, however, they often do not give comparable results (Goudriaan, 1996), and exhibit different sensitivities to environmental conditions such as soil water-holding capacity (Landau *et al.*, 1998). Furthermore, models did not predict variation in wheat yields from UK trials, partly because the dominant weather effects were on management/disease and not physiology. Current crop simulation models are unlikely to be able to predict for environments for which they were not calibrated (Kabat *et al.*, 1995). In light of the above, the use of wheat models linked with socio-economic models to assess impact of GEC (Adams *et al.*, 1990; Rosenzweig and Parry, 1994) may be premature. We see the future of wheat modelling in the field of GEC as increasingly emphasizing the ability to predict broad-scale sensitivity of wheat yield to environment for diverse climates, soils, cultivars and agronomy. This will entail a move to more top-down models with fewer parameters (contrast with Kartschall *et al.*, 1995) that are tested with large yield sets and remote sensing data, rather than single-site experiments.

4.11 Summary and Conclusions

Wheat productivity (biomass and grain yield) will increase by some 7–11% per 100 μmol mol^{-1} increase in [CO_2] without other environmental changes under well-fertilized and watered conditions, but less where there are nutrient limitations. Under conditions of water limitation the benefit to production is potentially greater, increasing water-use efficiency, although water use will probably be relatively unaffected. All of the observed responses of wheat to elevated [CO_2] can be explained by two direct effects: (i) increased photosynthesis and decreased photorespiration; and (ii) decreased stomatal conductance. Effects of elevated [CO_2] on tillers, leaf area, ears, grain, photosynthetic capacity and development are all explicable as consequences of these and it is unnecessary to invoke other direct effects. A constant 1°C increase in temperature over the whole season would decrease yields by 6–10%, due to shorter duration of crop growth. This effect might be overcome by selection of slower-developing varieties where water is not limiting, but such varieties would be undesirable in today's environment. The overall effect of GEC on wheat productivity cannot be predicted with any certainty, partly because local changes in radiation and evaporative demand and interactions with changing technology are unknown. Wheat models have been used to suggest that GEC will reduce productivity, but the ability of such models to predict environmental effects on wheat yield over widely varying climate, varieties, soils and agricultural practices is unproven. The response to GEC will also require technological adjustments in wheat production by breeding, by incorporating new genetic resources and by agronomic techniques. Such changes will be required worldwide if the GEC problem and the demands arising from large increases in population are to be met.

Acknowledgements

Dr Bruce A. Kimball kindly provided information on the wheat FACE studies. Dr Jann Conroy is thanked for encouraging the review. IACR-Rothamsted is grant-aided by the UK Biotechnology and Biological Research Council; the authors also receive support from UK Ministry of Agriculture, Fisheries and Food and European Union research funds. The author's work on the effects of climate change on wheat formed part of the Global Change and Terrestrial Ecosystems Wheat Network.

References

Acevedo, E. (1991) Effects of heat stress on wheat and possible selection tools for use in breeding for tolerance. In: Saunders, D.A. (ed.) *Wheat for the Non-traditional, Warm Areas.* CIMMYT, México DF, pp. 401–421.

Adam, N.R., Wall, G.W., Brooks, T.J., Lee, T.D., Kimball, B.A., Pinter, P.J. Jr and LaMorte, R.L. (1997) Changes in photosynthetic apparatus of spring wheat in response to CO_2-enrichment and nitrogen stress. In: *Annual Research Report 1997.* US Water Conservation Laboratory, ASDA, ARS, Phoenix, Arizona, pp. 75–77.

Adams, R.M., Rosenweig, C., Peart, R.M., Ritchie, J.T., McCarl, B.A., Glyer, J.D., Curry, R.B., Jones, J.W., Boote, K.J. and Allen, L.H. Jr (1990) Global climate change and US agriculture. *Nature* 345, 219–224.

Amthor, J.S. (1997) Plant respiratory responses to elevated CO_2 partial pressure. In: Allen, L.H. Jr, Kirkham, M.H., Olszyck, D.M. and Whitman, C.E. (eds) *Advances in CO_2 Effects Research.* American Society of Agronomy, Madison, Wisconsin, pp. 35–77.

André, M. and du Cloux, H. (1993) Interaction of CO_2 enrichment and water limitations on photosynthesis and water use efficiency in wheat. *Plant Physiology and Biochemistry* 31, 103–112.

Anon. (1996) *FAO Yearbook. Production.* Vol. 49, FAO Statistics Series No. 130. Food and Agriculture Organization of the United Nations, Rome.

Ansart, C. (1997) Echanges internationaux des céréales. Peut-on miser une nouvelle croissance des marchés? *Perspectives Agricoles* 223, 4–14.

Barnes, J.D. and Wellburn, A.R. (1998) Air pollutant combinations. In: De Kok, L.J. and Stulen, I. (eds) *Responses of Plant Metabolism to Air Pollution and Global Change.* Backhuys Publishers, Leiden, The Netherlands, pp. 147–164.

Batts, G.R., Wheeler, T.R., Morison, J.I.L., Ellis, R.H. and Hadley, P. (1996) Developmental and tillering responses of winter wheat (*Triticum aestivum*) crops to elevated CO_2 concentration and temperature. *Journal of Agricultural Science, Cambridge* 127, 23–35.

Batts, G.R., Morison, J.I.L., Ellis, R.H., Hadley, P. and Wheeler, T.R. (1997) Effects of CO_2 and temperature on growth and yield of crops of winter wheat over several seasons. *European Journal of Agronomy* 7, 43–52.

Batts, G.R., Ellis, R.H., Morison, J.I.L. and Hadley, P. (1998a) Canopy development and tillering of field-grown crops of two contrasting cultivars of winter wheat (*Triticum aestivum*) in response to CO_2 and temperature. *Annals of Applied Biology* 133, 101–109.

Batts, G.R., Ellis, R.H., Morison, J.I.L., Nkemka, P.N., Gregory, P.J. and Hadley, P. (1998b) Yield and partitioning in crops of contrasting cultivars of winter wheat in response to CO_2 and temperature in field studies using temperature gradient tunnels. *Journal of Agricultural Science, Cambridge* 130, 17–27.

Bender, J., Hertstein, U. and Black, C. (1999) Growth and yield responses of spring wheat to increasing carbon dioxide, ozone and physiological stresses: a statistical analysis of 'ESPACE-wheat' results. *European Journal of Agronomy* 10, 185–195.

Briggle, L.W. and Curtis, B.C. (1987) Wheat worldwide. In: Heyne, E.J. (ed.) *Wheat and Wheat Improvement.* Agronomy Monograph No. 13, ASA/CSSA/SSSA, Madison, Wisconsin, pp. 1–32.

Brooks, T.J., Wall, G.W., Pinter, P.J. Jr, Kimball, B.A., Webber, A., Clark, D., Kartschall, T. and LaMorte, R.L. (1996) Effects of nitrogen and CO_2 on canopy architecture and gas exchange in wheat. In: *Annual Research Report 1996.* US Water Conservation Laboratory, ASDA, ARS, Phoenix, Arizona, pp. 83–86.

Butterfield, R.E. and Morison, J.I.L. (1992) Modelling the impact of climatic warming on winter cereal development. *Agricultural and Forest Meteorology* 62, 241–261.

Chaudhuri, U.N., Kirkham, M.B. and Kanemasu, E.T. (1990a) Carbon dioxide and water level effects on yield and water use of winter wheat. *Agronomy Journal, 82, 637–641.*

Chaudhuri, U.N., Kirkham, M.B. and Kanemasu, E.T. (1990b) Root growth of winter wheat under elevated carbon dioxide and drought. *Crop Science,* 30, 853–857.

Dawson, I.A. and Wardlaw, I.F. (1989) The tolerance of wheat to high temperatures during reproductive growth. III. Booting to anthesis. *Australian Journal of Agricultural Research* 40, 965–980.

De Kok, L.J. and Stulen, I. (eds) (1998) *Responses of Plant Metabolism to Air Pollution and Global Change.* Backhuys Publishers, Leiden, The Netherlands, 519 pp.

Delgado, E., Mitchell, R.A.C., Parry, M.A.J., Driscoll, S.P., Mitchell, V.J. and Lawlor, D.W. (1994) Interacting effects of CO_2 concentration temperature and nitrogen supply on the photosynthesis and composition of winter-wheat leaves. *Plant, Cell and Environment* 17, 1205–1213.

Drake, B.G., Gonzàlez-Meler, M.A. and Long, S.P. (1997) More efficient plants: a consequence of rising atmospheric CO_2? *Annual Review of Plant Physiology and Molecular Biology* 48, 609–639.

Estiarte, M., Peñuelas, J., Kimball, B.A., Idso, S.B., LaMorte, R.L., Pinter, P.J. Jr, Wall, G.W. and Garcia, R.L. (1994) Elevated CO_2 effects on stomatal density of wheat and sour orange trees. *Journal of Experimental Botany* 45, 1665–1668.

Evans, L.T. (1993) *Crop Evolution Adaptation and Yield.* Cambridge University Press, Cambridge, UK, 500 pp.

Evans, L.T. (1998) *Feeding the Ten Billion. Plants and Population Growth.* Cambridge University Press, Cambridge, UK, 247 pp.

Farquhar, G.D. and Caemmerer, S. von (1982) Modelling of photosynthetic response to environmental conditions. In: Lange, O.L., Nobel, P.S., Osmond, C.B. and Ziegler, H. (eds) *Physiological Plant Ecology,* Vol 12B. Springer Verlag, Berlin, pp. 549–587.

Feil, B. (1992) Breeding progress in small grain cereals – a comparison of old and modern cultivars. *Plant Breeding* 108, 1–11.

Garcia, R.L., Long, S.P., Wall, G.W., Osborne, C.P., Kimball, B.A., Nie, G.Y., Pinter, P.J. Jr, Lamorte R.L. and Wechsung, F. (1998) Photosynthesis and conductance of spring-wheat leaves: field response to continuous free-air atmospheric CO_2 enrichment. *Plant, Cell and Environment,* 21, 659–669.

Gifford, R.M. (1995) Whole plant respiration and photosynthesis of wheat under increased CO_2 concentration and temperature: long-term *vs* short-term distinctions for modelling. *Global Change Biology* 1, 385–396.

Gooding, M.J. and Davies, W.P. (1997) *Wheat Production and Utilization. Systems, Quality and the Environment*. CAB International, Wallingford, UK, 355 pp.

Goudriaan, J. (1996) Predicting crop yields under global change. In: Walker, B. and Steffen, W. (eds) *Global Change and Terrestrial Ecosystems*. Cambridge University Press, Cambridge, UK, pp. 260–274.

Grashoff, C., Dijkstra, P., Nonhebel, S., Schapendonk, A.H.C.M. and van de Geijn, S.C. (1995) Effects of climate-change on productivity of cereals and legumes, model evaluation of observed year-to-year variability of the CO_2 response. *Global Change Biology* 1, 417–428.

Grossman, S., Kartschall, T., Kimball, B.A., Hunsaker, D.J., LaMorte, R.L., Garcia, R.L., Wall, G.W. and Pinter, P.J. Jr (1995) Simulated responses of energy and water fluxes to ambient atmosphere and free-air carbon dioxide enrichment in wheat. *Journal of Biogeography* 22, 601–609.

Habash, D.Z., Paul, M.J., Parry, M.A.J., Keys, A.J. and Lawlor, D.W. (1995) Increased capacity for photosynthesis in wheat grown at elevated CO_2 – the relationship between electron-transport and carbon metabolism. *Planta* 197, 482–489.

Hadley, P., Batts, G.R., Ellis, R.H., Morison, J.I.L., Pearson, S. and Wheeler, T.R. (1995) Temperature gradient chambers for research on global environment change. II. A twin-wall tunnel system for low stature, field grown crops using a split heat pump: technical report. *Plant, Cell and Environment* 18, 1055–1063.

Hendry, G.R. and Kimball, B.A. (1994) The FACE Programme. *Agriculture and Forest Meteorology* 70, 3–14.

Heyne, E.J. (1987) *Wheat and Wheat Improvement,* 2nd edn. ASA/CSSA/SSSA, Madison, Wisconsin, 765 pp.

Hunsaker, D.J., Kimball, B.A., Pinter, P.J. Jr, Wall, G.W. and LaMorte, R.L. (1996) Wheat evapotranspiration as affected by elevated CO_2 and variable soil nitrogen. In: *Annual Research Report 1996*. US Water Conservation Laboratory, ASDA, ARS, Phoenix, Arizona, pp. 79–82.

Hunsaker, D.J., Kimball, B.A., Pinter, P.J. Jr, Wall, G.W. and LaMorte, R.L. (1997) Soil water balance and wheat evapotranspiration as affected by elevated CO_2 and variable soil nitrogen. In: *Annual Research Report 1997*. US Water Conservation Laboratory, ASDA, ARS, Phoenix, Arizona, pp. 67–70.

Jamieson, P.D., Porter, J.R., Goudriaan, J., Ritchie, J.T., Keulen, H. van and Stol, W. (1998) A comparison of the models AFRCWHEAT2, CERES-wheat, Sirius, SUCROS2 and SWHEAT with measurements from wheat grown under drought. *Field Crops Research* 55, 23–44.

Kabat, P., Marshall, B. and Broek, B.J. van den (1995) Comparison of simulation results and evaluation of parameterisation schemes. In: Kabat, P., Marshall, B., Broek, B.J. van den , Vos, J. and Keulen, H. van (eds) *Modelling and Parameterisation of the Soil–Plant–Atmosphere System*. Wageningen Pers, Wageningen, The Netherlands, pp. 439–502.

Kartschall, T., Grossman, S., Pinter, P.J. Jr, Garcia, R.L., Kimball, B.A., Wall, G.W. and LaMorte, R.L. (1995) A simulation of phenology, growth, carbon dioxide exchange and yields under ambient atmosphere and free-air carbon dioxide enrichment (FACE) Maricopa, Arizona, for wheat. *Journal of Biogeography* 22, 611–622.

Kattenberg, A., Giorgi, F., Grassl, H., Meehl, G.A., Mitchell, J.F.B., Stouffer, R.J., Tokioka, T., Weaver, A.J. and Wigley, T.M.L. (1996) Climate models – projections of future climate. In: Houghton, J.T., Meira Filho, L.G., Callendar, B.A., Harris, N., Kattenberg, A. and Maskell, K. (eds) *Climate Change 1995: The Science of Climate Change*. IPCC, Cambridge University Press, Cambridge, UK, 572 pp.

Kimball, B.A., Pinter, P.J. Jr, Wall, G.W., Garcia, R.L., LaMorte, R.L., Jak, P.M.C., Arnoud Fruman, K.F. and Vugts, H.F. (1997a) Comparisons of responses of vegetation to elevated carbon dioxide in free-air and open-top chamber facilities. In: *Advances in Carbon Dioxide Effects Research*. ASA Special Publication No. 61. ASA, CSSA and SSSA, Madison, Wisconsin, pp. 113–130.

Kimball, B.A., LaMorte, R.L., Seay, R., O'Brien, C., Pabian, D.J., Suich, R., Pinter, P.J. Jr, Wall, G.W., Brooks, T.J., Hunsaker, D.J., Adamsen, F.J., Clarke, T.R. and Rokey, R. (1997b) Effects of free-air CO_2 enrichment (FACE) and soil nitrogen on the energy balance and evapotranspiration of wheat. In: *Annual Research Report 1997*. US Water Conservation Laboratory, ASDA, ARS, Phoenix, Arizona, pp. 65–66.

Landau, S., Mitchell, R.A.C., Barnett, V., Colls, J.J., Craigon, J., Moore, K.L. and Payne, R.W. (1998) Testing winter wheat simulation models' predictions against observed UK grain yields. *Agricultural and Forest Meteorology* 89, 85–99.

Lawlor, D.W. (1996) Simulating plant-responses to the global greenhouse. *Trends In Plant Science* 1, 100–102.

Lawlor, D.W. (1997) Response of crops to environmental change conditions: consequences for world food production. *Journal of Agricultural Meteorology* 52, 769–778.

Lawlor, D.W. (1998) Plant responses to global change: temperature and drought stress. In: De Kok, L.J. and Stulen, I. (eds) *Responses of Plant Metabolism to Air Pollution and Global Change*. Backhuys Publishers, Leiden, The Netherlands, pp. 193–207.

Lawlor, D.W. and Mitchell, R.A.C. (1991) The effects of increasing CO_2 on crop photosynthesis and productivity: a review of field studies. *Plant, Cell and Environment* 14, 803–818.

Lawlor, D.W. and Keys, A.J. (1993) Understanding photosynthetic adaptation to changing climate. In: Fowden, L., Mansfield, T.A. and Stoddart, J. (eds) *Plant Adaption to Environmental Stress*. Chapman & Hall, London, pp. 85–106.

Lawlor, D.W., Mitchell, R.A.C., Franklin, J., Driscoll, S.P. and Delgado, E. (1993) Facility for studying the effects of elevated carbon dioxide concentration and increased temperature on crops. *Plant, Cell and Environment* 16, 603–608.

Lawlor, D.W., Delgado, E., Habash, D.Z., Driscoll, S.P., Mitchell, V.J., Mitchell, R.A.C. and Parry, M.A.J. (1995) Photosynthetic acclimation of winter wheat to elevated CO_2 and temperature. In: Mathis, P. (ed.) *Photosynthesis: From Light to Biosphere*, Vol. V. Kluwer Academic Publishers, Dordrecht, The Netherlands, pp. 989–992.

Lazzeri, P.A., Barcelo, P., Barro, F., Rooke, L., Cannell, M.E., Rasco-Gaunt, S., Tatham, A., Fido, R. and Shewry, P.R. (1997) Biotechnology of cereals: genetic manipulation techniques and their use for the improvement of quality, resistance and input use efficiency traits. *Aspects of Applied Biology* 50, 1–8.

Leadley, P.W. and Drake, B.G. (1993) Open top chambers for exposing plant canopies to elevated CO_2 concentration and for measuring net gas exchange. *Vegetatio* 104/105, 3–15.

Leemans, R. and Solomon, A.M. (1993) Modeling the potential change in yield and distribution of the earth's crops under a warmed climate. *Climate Research* 3, 79–96.

Long, S.P. (1991) Modification of the response of photosynthetic productivity to rising temperature by atmospheric CO_2 concentrations: has its importance been underestimated? *Plant, Cell and Environment* 14, 729–739.

McKee, I.F. and Woodward, F.I. (1994a) The effect of growth at elevated CO_2 concentrations on photosynthesis in wheat. *Plant, Cell and Environment* 17, 853–859.

McKee, I.F. and Woodward, F.I. (1994b) CO_2 enrichment responses of wheat: interactions with temperature, nitrate and phosphate. *New Phytologist* 127, 447–453.

McKee, I.F., Farage, P.K. and Long, S.P. (1995) The interactive effects of elevated CO_2 and O_3 concentration on photosynthesis in spring wheat. *Photosynthesis Research* 45, 111–119.

Miglietta, F. and Porter, J.R. (1992) The effects of climatic change on development in wheat: analysis and modelling. *Journal of Experimental Botany* 43, 1147–1158.

Mitchell, R.A.C., Lawlor, D.W. and Young, A.T. (1991) Dark respiration of winter wheat crops in relation to temperature and simulated photosynthesis. *Annals of Botany* 67, 7–16.

Mitchell, R.A.C., Mitchell, V.J., Driscoll, S.P., Franklin, J. and Lawlor, D.W. (1993) Effects of increased CO_2 concentration and temperature on growth and yield of winter-wheat at 2 levels of nitrogen application. *Plant Cell and Environment* 16, 521–529.

Mitchell, R.A.C., Lawlor, D.W., Mitchell, V.J., Gibbard, C.L., White, E.M. and Porter, J.R. (1995) Effects of elevated CO_2 concentration and increased temperature on winter-wheat – test of ARCWHEAT1 simulation-model. *Plant, Cell and Environment* 18, 736–748.

Mitchell, R.A.C., Gibbard, C.L., Mitchell, V.J. and Lawlor, D.W. (1996) Effects of shading in different developmental phases on biomass and grain-yield of winter-wheat at ambient and elevated CO_2. *Plant, Cell and Environment* 19, 615–621.

Mitchell, R.A.C., Black, C.R., Burkart, S., Burke, J.I., Donnelly, A., de Temmerman, L., Fangmeier, A., Mulholland, B.J., Theobald, J.C. and Oijen, M. van (1999) Photosynthetic responses in spring wheat grown under elevated CO_2 concentrations and stress conditions in the European, multiple-site experiment 'ESPACE-wheat'. *European Journal of Agronomy* 10, 205–214.

Morison, J.I.L. and Lawlor, D.W. (1999) Interactions between increasing CO_2 concentration and temperature on plant growth. *Plant, Cell and Environment* 22, 659–682.

Nátr, L., Driscoll, S. and Lawlor, D.W. (1996) The effect of increased CO_2 concentration on the dark respiration rate of etiolated wheat seedlings. *Cereal Research Communications* 24 (1), 53–59.

Nicholls, N., Gruza, G.V., Jouzel, J., Karl, T.R., Ogallo, L.A. and Parker, D.E. (1996) Observed climate variability and change. In: Houghton, J.T., Meira Filho, L.G., Callendar, B.A., Harris, N., Kattenberg, A. and Maskell, K. (eds) *Climate Change 1995: The Science of Climate Change*. IPCC, Cambridge University Press, Cambridge, UK, 572pp.

Nie, G.Y., Long, S.P., Garcia, R.L., Kimball, B.A., LaMorte, R.L., Pinter, P.J. Jr, Wall, G.W. and Webber, A.N. (1995a) Effects of free-air CO_2 enrichment on the development of the photosynthetic apparatus in wheat, as indicated by changes in leaf proteins. *Plant, Cell and Environment* 18, 855–864.

Nie, G.Y., Hendrix, D.L., Webber, A.N., Kimball, B.A. and Long, S.P. (1995b) Increased accumulation of carbohydrates and decreased photosynthetic gene transcript levels in wheat grown at an elevated CO_2 concentration in the field. *Plant Physiology* 108, 975–981.

van Oijen, M. and Ewert, F. (1999) The effects of climatic variation in Europe on the yield response of spring wheat cv. Minaret to elevated CO_2 and O_3: an analysis of open-top chamber experiments by means of two crop growth simulation models. *European Journal of Agronomy* 10, 249–264.

van Oosten, J.J. and Besford, R.T. (1996) Acclimation of photosynthesis to elevated CO_2 through feedback-regulation of gene-expression – climate of opinion. *Photosynthesis Research* 48, 353–365.

Pinter, P.J. Jr, Kimball, B.A., Wall, G.W., LaMorte, R.L., Adamsen, F. and Hunsaker, D.J. (1996) FACE 1995–1996: effects of elevated CO_2 and soil nitrogen on growth and yield parameters of spring wheat. In: *Annual Research Report 1997.* US Water Conservation Laboratory, ASDA, ARS, Phoenix, Arizona, pp. 75–78.

Pinter, P.J. Jr, Kimball, B.A., Wall, G.W., LaMorte, R.L., Adamsen, F. and Hunsaker, D.J. (1997) Effects of elevated CO_2 and soil nitrogen fertilizer on final grain yields of spring wheat. In: *Annual Research Report 1997.* US Water Conservation Laboratory, ASDA, ARS, Phoenix, Arizona, pp. 71–74.

Poorter, H., Gifford, R.M., Kriedemann, P.E. and Chin Wong, S. (1992) A quantitative analysis of dark respiration and carbon content as factors in the growth response of plants to elevated CO_2. *Australian Journal of Botany* 40, 501–513.

Rawson, H.M. (1988) Effects of high temperatures on the development and yield of wheat and practices to reduce deleterious effects. In: Klatt, A.R. (ed.) *Wheat Production Constraints in Tropical Environments.* CIMMYT, México DF, pp. 44–63.

Rawson, H.M. (1992) Plant responses to temperature under conditions of elevated CO_2. *Australian Journal of Botany* 40, 473–490.

Rawson, H.M. (1995) Yield responses of two wheat genotypes to carbon dioxide and temperature in field studies using temperature gradient tunnels. *Australian Journal of Plant Physiology* 22, 23–32.

Rawson, H.M. and Richards, R.A. (1992) Effects of high temperature and photoperiod on floral development in wheat isolines differing in vernalisation and photoperiod genes. *Field Crops Research* 32, 181–192.

Rogers, G.S., Milham, P.J., Gillings, M. and Conroy, J.P. (1996) Sink strength may be the key to growth and nitrogen responses in N-deficient wheat at elevated CO_2. *Australian Journal of Plant Physiology* 23, 253–264.

Rogers, G.S., Gras, P.W., Batey, I.L., Milham, P.J., Payne, L. and Conroy, J.P. (1998) The influence of atmospheric CO_2 concentration on the protein, starch and mixing properties of wheat flour. *Australian Journal of Plant Physiology* 25, 387–393.

Rosenzweig, C. and Parry, M.L. (1994) Potential impact of climate change on world food supply. *Nature* 367, 133–138.

Russell, G. and Wilson, G.W. (1994) *An Agro-pedo-climatological Knowledge-base of Wheat in Europe.* European Commission, Luxembourg, 160 pp.

Sage, R.F. (1994) Acclimation of photosynthesis to increasing atmospheric CO_2 – the gas-exchange perspective. *Photosynthesis Research* 39, 351–368.

Samarakoon, A.B., Muller, W.J. and Gifford, R.M. (1995) Transpiration and leaf-area under elevated CO_2 – effects of soil-water status and genotype in wheat. *Australian Journal of Plant Physiology* 22, 33–44.

Schimel, D., Ives, D., Enting, I., Heimann, M., Joos, F., Raynaud, D. and Wigley. T. (1996) CO_2 and the carbon cycle. In: Houghton, J.T., Meira Filho, L.G., Callendar, B.A., Harris, N., Kattenberg, A. and Maskell, K. (eds) *Climate Change 1995: The Science of Climate Change.* IPCC, Cambridge University Press, Cambridge, UK, 572 pp.

Senock, R.S., Ham, J.M., Loughin, T.M., Kimball, B.A., Hunsaker, D.J., Pinter, P.J. Jr, Wall, G.W., Garcia, R.L. and LaMorte, R.L. (1996) Sap flow in wheat under free-air CO_2 enrichment. *Plant, Cell and Environment* 19, 147–158.

Shaykewich, C.F. (1995) An appraisal of cereal crop phenology modelling. *Canadian Journal of Plant Science* 75, 329–341.

Sheen, J.F. (1994) Feedback-control of gene expression. *Photosynthesis Research* 39, 427–438.

Shewry, P.R., Broadhead, J., Harwood, J., Williams, M., Grimwade, B., Freedmore, R., Napier, J., Fido, R., Lawlor, D.W. and Tatum, A.S. (1994) Environmental and developmental regulation of gluten protein synthesis. In: *Wheat Kernel Proteins. Molecular and Environmental Aspects.* Proceedings of Meeting, University of Tuscany, Viterbo, Italy, pp. 273–275.

Slafer, G.A., Calderini, D.F. and Miralles, D.J. (1996) Yield components and compensation in wheat: opportunities for further increasing yield potential. In: Reynolds, M.P., Rajaram, S. and McNab, A. (eds) *Increasing Yield Potential in Wheat: Breaking the Barriers.* CIMMYT, México DF, pp. 101–133.

Tashiro, T. and Wardlaw, I.F. (1990) The response to high temperature shock and humidity changes prior to and during the early stages of grain development in wheat. *Australian Journal of Plant Physiology* 17, 551–561.

Teramura, A.H. (1998) Terrestrial plant responses to a changing solar UV-B radiation environment. In: De Kok, L.J. and Stulen, I. (eds) *Responses of Plant Metabolism to Air Pollution and Global Change.* Backhuys Publishers, Leiden, The Netherlands, pp. 209–214.

Tester, R.F., Morrison, W.R., Ellis, R.H., Piggott, J.R., Batts, G.R., Wheeler, T.R., Morison, J.I.L., Hadley, P. and Ledward, D.A. (1995) Effects of elevated growth temperature and carbon dioxide levels on some physicochemical properties of wheat starch. *Journal of Cereal Science* 22, 63–71.

Theobald, J.C., Mitchell, R.A.C., Parry, M.A.J. and Lawlor, D.W. (1998) Estimating the excess investment in Ribulose-1,5-bisphosphate carboxylase/oxygenase in leaves of spring wheat grown under elevated CO_2. *Plant Physiology* 118, 945–955.

Wall, G.W., Wechsung, F., Wechsung, G., Kimball, B.A., Kartschall, T., Pinter, P.J. Jr and LaMorte, R.L. (1996) Effects of free-air CO_2 enrichment (FACE) and two soil nitrogen regimes on vertical and horizontal distribution of root length density, surface area, and density of spring wheat. In: *Annual Research Report 1996.* US Water Conservation Laboratory, ASDA, ARS, Phoenix, Arizona, pp. 56–59.

Wall, G.W., Adam, N.R., Brooks, T.J., Kimball, B.A., Adamsen, F.J., Pinter, P.J. Jr and LaMorte, R.L. (1997) Acclimation of the photosynthetic apparatus of spring wheat grown under free-air CO_2 enrichment (FACE) and two soil nitrogen regimes. In: *Annual Research Report 1997.* US Water Conservation Laboratory, ASDA, ARS, Phoenix, Arizona, pp. 61–64.

Watson, R.T., Zinyowera, M.C. and Moss, R.H. (1996) *Climate Change 1995: Impacts, Adaptations and Mitigation. Summary for Policymakers.* Contribution of Working Group II to the Second Assessment Report. IPCC, Cambridge University Press, Cambridge, UK, 22 pp.

Wheeler, T.R., Hong, T.D., Ellis, R.H., Batts, G.R., Morison, J.I.L. and Hadley, P. (1996a) The duration and rate of grain growth, and harvest index, of wheat (*Triticum aestivum* L.) in response to temperature and CO_2. *Journal of Experimental Botany* 47, 623–630.

Wheeler, T.R., Batts, G.R., Ellis, R.H., Hadley, P. and Morison, J.I.L. (1996b) Growth and yield of winter wheat (*Triticum aestivum*) crops in response to CO_2 and temperature. *Journal of Agricultural Sciences, Cambridge* 127, 37–48.

Williams, M., Shewry, P.R., Lawlor, D.W. and Harwood, J.L. (1995) The effects of elevated temperature and atmospheric carbon dioxide concentration on the quality of grain lipids in wheat (*Triticum aestivum* L.) grown at two levels of nitrogen application. *Plant, Cell and Environment* 18, 999–1009.

5 Crop Ecosystem Responses to Climatic Change: Rice

TAKESHI HORIE[1], JEFFERY T. BAKER[2], HIROSHI NAKAGAWA[1], TSUTOMU MATSUI[1] AND HAN YONG KIM[3]

[1]Laboratory of Crop Science, Division of Agronomy, Graduate School of Agriculture, Kyoto University, Kitashirakawa Oiwake-Cho, Sakyo-ku, Kyoto 606-8502, Japan; [2]USDA-ARS/NRI/RSML, 10300 Baltimore Avenue, Bethesda, MD 20705-2350, USA; [3]Tohoku National Agriculture Experimental Station, Morioka, Iwate 020-0198, Japan

5.1 Introduction

Rice provides a substantial portion of the dietary requirements of nearly 1.6 billion people, with another 400 million relying on rice for one-quarter to one-half of their diet (Swaminathan, 1984). In order to keep pace with a rapidly increasing world population, global rice production levels seen in 1991 will need to increase by at least 46% by the year 2025 (IRRI, 1993; Fig. 5.1). This challenge has spurred researchers to seek innovative technologies to overcome the now common levelling off and even reduced rice yields seen in recent years in some areas of the world (Hossain, 1998). Potential future climate change resulting from increased atmospheric carbon dioxide concentration ([CO_2]) and other greenhouse gasses (Hansen *et al.*, 1984) further hampers our ability to predict global rice production and raises serious questions concerning food security for the near future. Accurate prediction of the impacts of climate change on rice productivity on a region-by-region basis is essential, not only for the estimation of world food security, but also for the continued sustainability of rice-farming societies, which in most countries consist of small hectarage subsistence farmers.

Several recent simulation studies have estimated the effects of global climate change on regional rice production using different climate change scenarios (Horie, 1993; Horie *et al.*, 1995b; Matthews *et al.*, 1995; Singh and Padilla, 1995). These studies predict that global climate change may have substantial positive or negative impacts on rice production, depending on the region of the world under consideration. However, these modelling studies point to the need for further research in order to test and improve our understanding of environmental effects on rice development, growth and yield. More specifically, research is needed to quantify further the interactive effects of [CO_2], air temperature and other environmental variables on rice growth and yield formation processes.

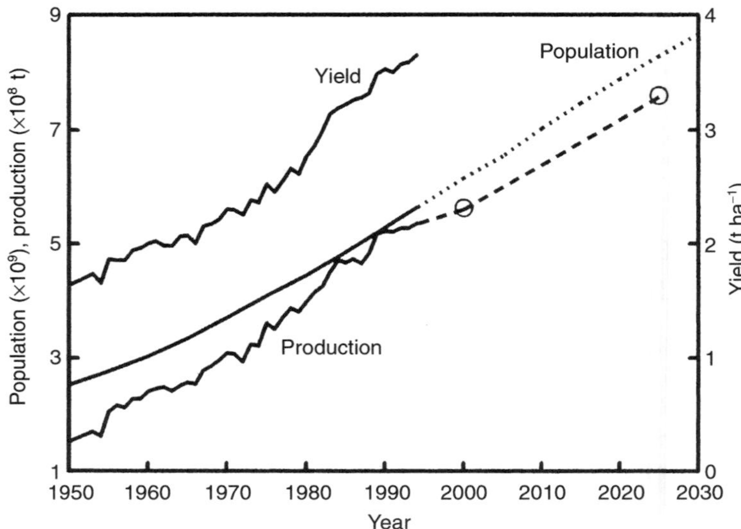

Fig. 5.1. Actual and projected trends in world population, rice production and yield during the period from 1950 to 2025. (Adapted from IRRI, 1993 and FAO, 1995.)

The objectives of this chapter are to summarize recent research findings on the effects of [CO_2], air temperature and other environmental variables on rice growth and yield. It then evaluates the current state of rice/climate simulation models with the intent of defining future research needs in both experimental and modelling studies.

5.1.1 Global rice production regions

Current regional rice hectarage and yields of the world are shown in Fig. 5.2. Over 90% of the world's rice is produced and consumed in South, Southeast and East Asia, where monsoonal climates dominate. In these areas, rice is not only the staple food but also provides the livelihood of most people. Unlike wheat and maize, rice is produced mainly by subsistence farmers in developing countries.

Cultivated rice consists of two distinct species of *Oryza*: *O. sativa* L. and *O. glaberrima* L. The dominant species in world rice production is *O. sativa; O. glaberrima* is limited mainly to west Africa. *O. sativa* consists of three subspecies: *indica*, *japonica* and *javanica*, which are genetically distant from each other. Crosses among these subspecies generally result in extensive or partial spikelet sterility. These three subspecies are also different physiologically and morphologically and occupy different ecological niches (cf. review by Takahashi, 1997). Compared with *indica* cultivars, many *japonica* cultivars exhibit higher tolerance to cool temperatures, while the reverse is true for drought resistance (Oka, 1953). These physiological differences define major production regions for these two subspecies. The Yangtze River in

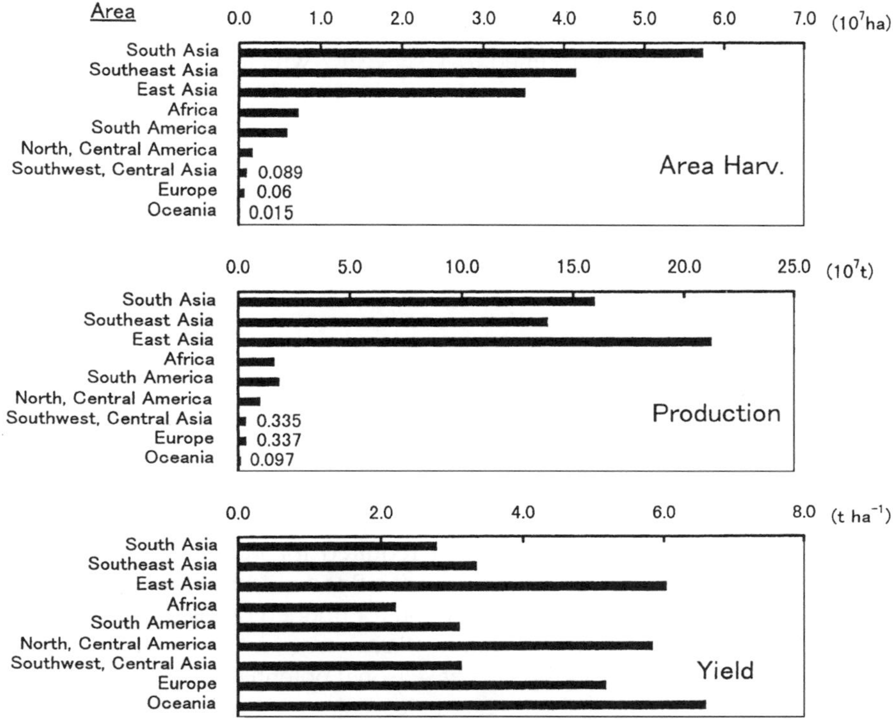

Fig. 5.2. Regional rice hectarage, production and yield in the world. (Adapted from FAO, 1995.)

China forms an approximate border between these two subspecies, with *japonica* cultivars predominating north of the river and *indica* cultivars south of the river. The *javanica* subspecies originated in island areas of Southeast Asia, including Indonesia. The three subspecies of *O. sativa* are further ecologically differentiated into upland, lowland or deep-water ecotypes, depending mainly on their adaptability to different water environments.

The large ecophysiological diversity of genotypes permits rice to be grown in widely different environments. Rice culture is generally classified into one of four main types, depending on water environment: irrigated rice, rainfed lowland rice, rainfed upland rice and deep-water rice cultures. Of the total 140–145 Mha of land area planted to rice annually, roughly half (about 53%) is grown as irrigated flooded-paddy rice, one-quarter (about 27%) as rainfed lowland rice, 12% as rainfed upland rice and 8% as deep-water rice (IRRI, 1993).

World rice production in the past three decades since the Green Revolution increased at a higher rate than did world population (Fig. 5.1) mainly due to yield increases. However, total rice production will need to increase by 46% by the year 2025 (IRRI, 1993) to keep pace with projected population growth, especially in Asia. Attainment of these production goals is likely to be very difficult since current increases in world rice production are slowing. For example, increases in world rice production during the 1985–1995 period declined to 1.7% year^{-1} which is lower than the population growth rate of 1.8% year^{-1} (Hossain, 1998). This stagnation in rice production is largely due to yield levelling for irrigated-rice cultures that use advanced technologies. Furthermore, rice produced under rainfed cultures has not benefited from the advances of the Green Revolution, due to environmental and infrastructural barriers to the adoption of new technologies. Clearly, technological innovations are needed to break through the current yield barriers in both irrigated and rainfed rice cultures.

5.1.2 Climatic requirements of rice

Rice is grown from 53° N in northeastern China to 35° S in New South Wales, Australia. Major environmental factors that limit rice cultivation are air temperature and water supply. A seasonal total of at least 2400 growing degree-days for a period with a daily mean temperature above 10°C, is the climatological temperature limit required for rice cultivation (Ozawa, 1964). Since irrigated rice is grown under flooded conditions during most of the growing season, evapotranspiration for flooded rice is nearly equal to potential evaporation at a given location (Sakuratani and Horie, 1985). Thus, a water supply equivalent to the sum of seasonal potential evaporation is the minimum water requirement for irrigated rice culture. However, the actual amount of water required for an irrigated rice crop is the sum of evapotranspiration, water lost during land preparation (100–200 mm) and water lost due to percolation, which ranges from 0.2 to 15.6 mm day^{-1} (Yoshida, 1981), depending on field infrastructure and soil properties.

The air temperature range for vegetative growth of rice is generally from 12 to 38°C, with an optimum between 25 and 30°C. However, during reproductive development, rice is particularly sensitive to low temperature at boot stage, during which pollen grains are in the early microspore phase (Satake and Hayase, 1970), and to both low and high temperatures at flowering (Abe, 1969). Temperatures below 20°C during the early microspore stage inhibit pollen development (Nishiyama, 1984). Temperatures below 20°C or above 35°C at flowering generally result in increases in spikelet sterility (Terao *et al.*, 1942; Abe, 1969; Satake and Hayase, 1970; Satake and Yoshida, 1976; Matsui *et al.*, 1997a) due to anther indehiscence (Satake and Yoshida, 1976; Matsui *et al.*, 1997a). Rice is more sensitive to cool temperatures and less sensitive to high temperatures during the early microspore stage than at flowering. For these reasons, low air temperature during the early microspore stage is the primary environmental factor that determines the northern limit of rice cultivation. Cool summers frequently result in severe yield reductions for rice grown in the northern parts of China, Korea and Japan.

High-temperature-induced spikelet sterility is typically found in tropical and subtropical areas, especially for rice crops grown during the dry season (Satake, 1995). Reductions in rice yields caused by high temperatures may become more severe under projected future global warming scenarios. In some regions of the world, potential future global warming could exacerbate this problem. However, relatively large genetic variability exists among rice cultivars with regard to tolerance of both high and low temperature damage.

The optimum temperature range for grain-fill in rice is between 20 and 25°C. This optimum is lower than that for vegetative growth (Yoshida, 1981). Temperatures above the optimum shorten the grain-filling period and reduce final yield.

5.2 Rice Responses to Elevated [CO_2] and Temperature

5.2.1 Experimental facilities

Three broad categories of experimental systems have been used to investigate rice responses to [CO_2], air temperature and other environmental factors. These include closed-chamber systems (Morrison and Gifford 1984a,b; Imai *et al.*, 1985; Baker *et al.*, 1990a), open-top chambers (Akita, 1980; Moya *et al.*, 1997) and temperature gradient chambers (TGCs) (Horie *et al.*, 1991, 1995c). Recently, a free-air CO_2 enrichment (FACE) system was constructed to study rice responses to [CO_2] (Kobayashi, 1999). In order to study effectively the response of rice to potential global climate change variables, it is essential to impose factorial combinations of treatments such as [CO_2], temperature and other environmental factors on the rice crop under field-like conditions. Additional desirable features include a high degree of robustness of the system, accuracy of environment control and reasonable construction and operating costs.

Rice responses to [CO_2] are quite different for plants grown under isolated conditions (e.g. in pots) compared with those measured under field conditions (Nakagawa *et al.*, 1994; see also section 5.2.5). Although rice experiments conducted using the FACE system utilize true field-grown conditions, obtaining factorial combinations of [CO_2] and air temperature treatments is difficult. The required factorial experiments are conducted in closed-chamber systems, open-top chambers and TGCs. Specifically, the TGCs (Fig. 5.3) have many of the desired experimental capabilities, including system robustness and the ability to maintain a wide range of [CO_2] and temperature regimes under field-like conditions with modest construction and operating costs (Horie *et al.*, 1995c).

5.2.2 Photosynthesis and respiration

Rice, being a C_3 plant, responds very well to increased [CO_2]. Several long- and short-term CO_2 enrichment studies show that doubling the current ambient [CO_2] increased leaf-level photosynthetic rate by 30–70%. The magnitude of this increase depended on the particular rice cultivar, growth stage and environment (Imai and Murata, 1978; Akita, 1980; Morrison and Gifford, 1983;

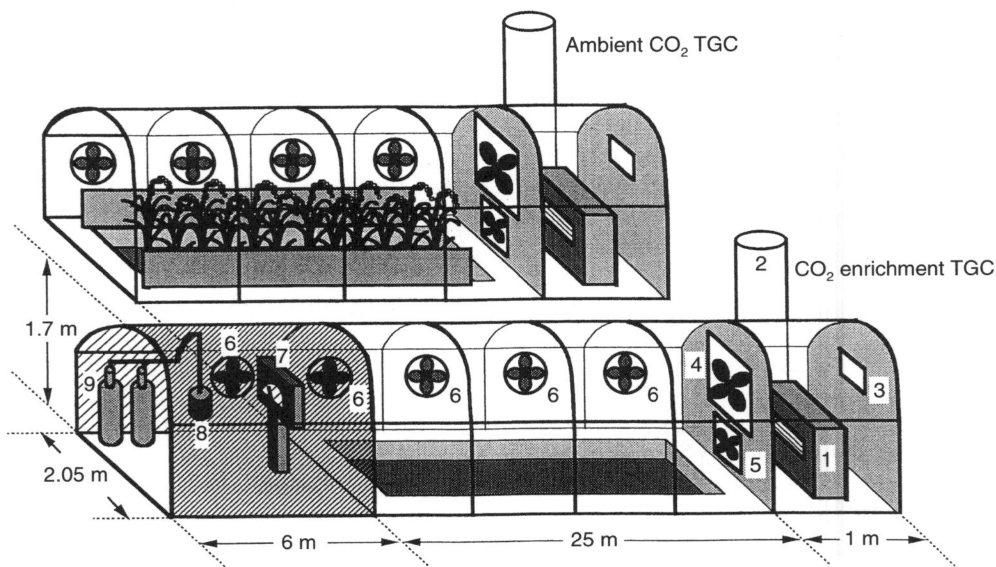

Fig. 5.3. Schematic drawing of the temperature gradient chamber (TGC) for CO_2 enrichment (bottom) and ambient [CO_2] (top) developed at Kyoto, Japan. (Adapted from Horie *et al.*, 1995c.) 1 = oil heater; 2 = stovepipe of oil heater; 3 = air exhaust window; 4 = variable speed exhaust fan; 5 = reversible exhaust fan; 6 = oscillating fan; 7 = CO_2 controller; 8 = CO_2 injection pipe; 9 = liquid CO_2 tanks.

Lin *et al.*, 1997). The effects of a wide range of [CO_2] on canopy photosynthesis of rice (cv. IR30) are shown in Fig. 5.4. The relative response of canopy net photosynthetic rate (P_n) to [CO_2], with 330 μmol mol^{-1} set to unity, was iteratively fitted to the following rectangular hyperbola (Baker and Allen, 1993):

$$P_n = (P_{max} [CO_2]) / ([CO_2] + K_m) + P_i \quad \text{(Eqn 5.1)}$$

where P_{max} is the asymptotic response limit of ($P_n - P_i$) at high [CO_2]; P_i is the intercept on the y-axis; and K_m is the value of [CO_2] at which ($P_n - P_i$) = 0.5P_{max}. Values of parameter estimates were 70.83 μmol mol^{-1}, 3.96 and −2.21 for P_{max}, K_m and P_i, respectively. Equation 5.1 indicates that the relative P_n reaches a ceiling value of 1.75 at infinite [CO_2] and that doubling [CO_2] from 330 to 660 μmol mol^{-1} increases rice canopy photosynthesis by 36%. Although this value of the relative response to doubled [CO_2] would be influenced by environmental conditions, cultivars and developmental stages of rice, it can be accepted as a reasonable value in comparison with the relative response in rice biomass production as described in section 5.2.5.

Since it has been shown that the effects of CO_2 enrichment on rice leaf-area development are small (Morrison and Gifford, 1984a; Imai *et al.*, 1985; Baker *et al.*, 1990c; Nakagawa *et al.*, 1993; Kim *et al.*, 1996a; Ziska *et al.*, 1997), the canopy photosynthetic response to CO_2 enrichment can be mainly

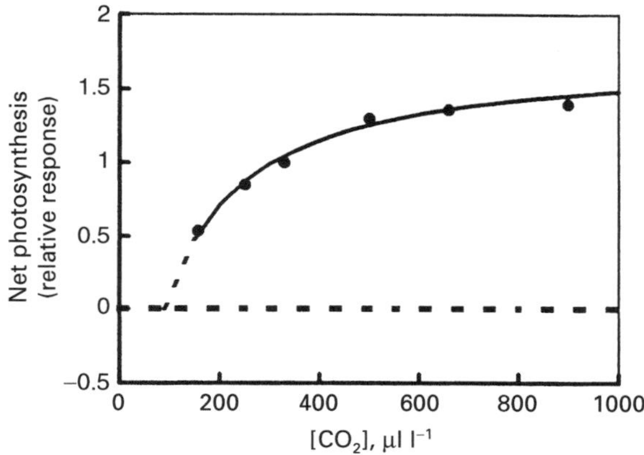

Fig. 5.4. Relative canopy net photosynthesis at a photon flux of 1500 μmol m^{-2} s^{-1} vs. season-long [CO_2] treatment. (Adapted from Baker and Allen, 1993.)

attributable to responses at the unit leaf-area level. This implies that responses to [CO_2] in terms of net assimilation rate and radiation use efficiency are similar to those previously described for canopy net photosynthesis.

A major question in the study of global climate change effects on rice is whether or not [CO_2] effects on rice photosynthesis are influenced by air temperature. Here, seemingly contradictory results have been reported. Lin *et al.* (1997) and Nakagawa *et al.* (1997) found that higher temperatures stimulated single-leaf photosynthesis of rice subjected to long-term [CO_2] treatments during the vegetative stages. In contrast, Baker and Allen (1993) reported that rice canopy photosynthesis was relatively unaffected by a wide range of air temperatures. Since one of the major effects of elevated [CO_2] on net photosynthesis is through the suppression of photorespiration, it could be expected that optimum temperature for photosynthesis shifts upward as [CO_2] increases. Indeed, this type of interaction of temperature and [CO_2] on leaf-level photosynthesis has been defined by Long (1991) for some C_3 species. This leaf-level response has been confirmed for rice by Lin *et al.* (1997) and Nakagawa *et al.* (1997).

Conversely, very small interactive effects of temperature and [CO_2] on canopy photosynthesis occur at the canopy level. This is consistent with the temperature effects on rice responses to [CO_2] in terms of biomass production obtained under field-like conditions (Baker *et al.*, 1992a; Horie, 1993; Kim *et al.*, 1996a; Nakagawa *et al.*, 2000; see also Fig. 5.7). Similarly, soybean (*Glycine max*, L.) canopy photosynthesis is also relatively insensitive to a rather wide range of air temperatures (Jones *et al.*, 1985). Carbon dioxide enrichment causes partial stomatal closure, increased stomatal resistance, reduced leaf and whole canopy transpiration and warmer leaf and whole canopy temperatures (Baker and Allen, 1993). However, the lowered canopy surface temperature caused by increased transpirational cooling at higher air temperature (Baker and Allen, 1993) may be one of the reasons for this differential photosynthetic response to temperature between single leaves and whole canopies. Further studies are needed to arrive at a definitive conclusion on this topic.

Potential photosynthetic acclimation (or down-regulation of photosynthesis) in response to elevated [CO_2] has been addressed in a few experiments conducted on rice. Baker *et al.* (1990a) grew rice (cv. IR-30) season-long at a wide range of [CO_2]: subambient (160 and 250 μmol mol^{-1}), ambient (330 μmol mol^{-1}) and superambient (500, 660 and 900 μmol mol^{-1}). They tested for canopy photosynthetic acclimation to long-term [CO_2] by comparing canopy photosynthesis at short-term [CO_2] of 160, 330 and 660 μmol mol^{-1}. When all long-term [CO_2] treatments were held at a common short-term [CO_2] of 160 μmol mol^{-1}, canopy photosynthesis was decreased by 44% across the long-term [CO_2] treatments from 160 to 900 μmol mol^{-1}. This decline in photosynthesis was accompanied by a 32% decrease in the amount of Rubisco protein relative to other soluble protein, and a 66% decrease in Rubisco activity (Rowland-Bamford *et al.*, 1991). However, the majority of this down-regulation of photosynthesis, as determined by canopy gas-exchange measurements, occurred from the subambient (160 μmol mol^{-1}) to ambient (330 μmol mol^{-1}) long-term [CO_2] treatments. The decline in canopy photosynthesis for the long-term ambient (330 μmol mol^{-1}) and twice ambient (660 μmol mol^{-1}) was only 3.6%.

In a subsequent experiment, Baker *et al.* (1997b) similarly tested for potential acclimation of rice (cv. IR-72) canopy photosynthesis to long-term [CO_2] growth treatments of 350 and 700 μmol mol^{-1}. They compared canopy photosynthesis across short-term [CO_2] ranging from 160 to 1000 μmol mol^{-1} and found no photosynthetic down-regulation. Here, photosynthetic rate was a function of current short-term [CO_2] rather than long-term [CO_2] growth treatment. However, in the same experiment, Vu *et al.* (1998) found reductions in leaf Rubisco content ranging from 6 to 22% for the CO_2-enriched treatments compared with ambient controls. Thus, while photosynthetic acclimation responses in terms of enzyme down-regulation may be detected at the single leaf biochemical level, these effects may or may not result in a detectable loss of canopy photosynthetic capacity when measured using gas-exchange techniques.

The response of leaf photosynthetic rate to intercellular [CO_2] was similar for rice plants grown at different [CO_2] and temperature regimes under field-like conditions during the vegetative phase of growth (Lin *et al.*, 1997). Thus, photosynthetic acclimation to elevated [CO_2] is not likely to occur for rice grown under field conditions within the [CO_2] range up to twice the current ambient level. Conversely, photosynthetic acclimation to enriched [CO_2] could result from plant exposures to much higher [CO_2] (above 1000 μmol mol^{-1}), as shown by Imai and Murata (1978), or from reduced sink size caused either by restricted root growth, in the case of potted plants (Arp, 1991), or by high-temperature-induced spikelet sterility (Lin *et al.*, 1997). From the experimental evidence and analysis, we conclude that, across the range from current ambient [CO_2] (near 360 μmol mol^{-1}) to the approximate doubling of [CO_2], projected for the mid to late 21st century, photosynthetic acclimation to elevated [CO_2] may not be a large or even significant factor governing rice photosynthetic responses to [CO_2].

Dark respiration rate of rice, expressed on a ground area basis, increased with increasing [CO_2], due to increased biomass, but specific respiration rate per unit biomass decreased (Baker *et al.*, 1992c). However, the differences in the specific respiration rate among rice crops subjected to different long-term [CO_2] were shown to be derived from differences in the N concentration of above-ground biomass. Since elevated [CO_2] can reduce plant tissue N concentration (Baker *et al.*, 1992c; Nakagawa *et al.*, 1993; Kim *et al.*, 1996a; Ziska *et al.*, 1996) and both the growth and maintenance components of respiration can be affected by tissue-N concentration (Penning de Vries *et al.*, 1974; Amthor, 1994), it is possible that elevated [CO_2] may reduce specific respiration.

5.2.3 Transpiration and water use

Leaf-level transpiration rate (E, mmol m^{-2} s^{-1}) can be expressed as:

$$E = K[e^*(T_L) - e_a] / (r_a + r_s) \qquad \text{(Eqn 5.2)}$$

where K is a slightly temperature-dependent, physical constant for the conversion of vapour pressure (kPa) to gas concentration (mmol mol^{-1}); $e^*(T_L)$ is the saturation vapour pressure at leaf temperature, T_L; e_a is the vapour pressure of air; and r_a and r_s are the boundary layer and stomatal diffusive resistances of water vapour (mol m^{-2} s^{-1}), respectively. A similar relationship to that shown in equation 5.2 also holds for canopy transpiration, with the substitution of bulk air (r_b) and canopy (r_c) resistances instead of r_a and r_s, respectively.

It is now well established that for many plants, including rice, an increase in [CO_2] increases the stomatal resistance (r_s) or, inversely, decreases the conductance ($g_s = 1/r_s$) through a reduction in stomatal aperture (Akita, 1980; Morrison and Gifford 1983, 1984a; Baker *et al.*, 1990a; Nakagawa *et al.*, 1997; Homma *et al.*, 1999). A 56% increase in r_s (Morrison and Gifford, 1984a) and 40–49% increase in r_c (Homma *et al.*, 1999) were reported for rice subjected to long-term doubled [CO_2] treatments. However, doubling [CO_2] does not reduce E to a similar extent as the r_s increase. This is because an increase in r_s causes a rise in T_L which leads to an increase in E caused by the increase in the vapour pressure gradient between leaf and air ($e^*(T_L) - e_a$). Thus, the effect of elevated [CO_2] on E depends not only on r_s but also on evaporative demand of the environment, which in turn is determined by factors such as temperature, solar radiation, humidity and wind speed.

Morrison and Gifford (1984a) found that daily E for rice grown in pots was similar for both ambient and doubled [CO_2]. However, Wada *et al.* (1993) measured season-long E for rice grown at different temperatures under field-like conditions in TGCs (Fig. 5.3). They found that CO_2 enrichment reduced seasonal total E by 15% at 26°C but increased E by 20% at 29.5°C. These results suggest that r_s in the ambient [CO_2] was not affected by air temperature, but r_s decreased with increasing temperature under CO_2 enrichment. Indeed, it has been shown from the energy budget analysis of remotely sensed canopy temperatures and microclimates for rice grown in TGCs that r_c declined at doubled [CO_2] with the rise in growing temperature, whereas r_c was unaffected by temperature at ambient [CO_2] (Homma *et al.*, 1999). Further, they also found a much stronger relationship between r_c and air temperature under CO_2 enrichment for the rice cultivar IR36, an *indica* type rice, compared with Akihikari, a *japonica* type rice. A reduction in r_s with increasing air temperature at elevated [CO_2] has also been reported by Imai and Okamoto-Sato (1991). These results indicate that although [CO_2] effects on E are much smaller than [CO_2] effects on r_s, at elevated [CO_2], r_s depends strongly on air temperature.

Rice, as with many other C_3 species, displays higher crop water-use efficiencies (WUE) under elevated [CO_2], due primarily to increased biomass production and partly to reduced transpiration. Morrison and Gifford (1984b) found that CO_2 enrichment for rice grown in pots increased WUE by 53–63%. For rice grown in a TGC, with [CO_2] enrichment at air temperatures between 24 and 26°C, WUE increased by 40–50% (Fig. 5.5). However, with further increases in air temperature, WUE decreased sharply to about a 20% enhancement at 30°C (Fig. 5.5). This declining CO_2 enrichment effect on WUE with increasing air temperature is similar to whole-canopy photosynthetic WUE reported by Baker and Allen (1993). Similar declines in the CO_2 enrichment effects on photosynthetic WUE at higher growth temperatures have also been reported for individual rice leaves: Nakagawa *et al.* (1997) found that leaf WUE at ambient [CO_2] was not affected by growing temperature; but at elevated [CO_2], WUE declined with increasing temperature.

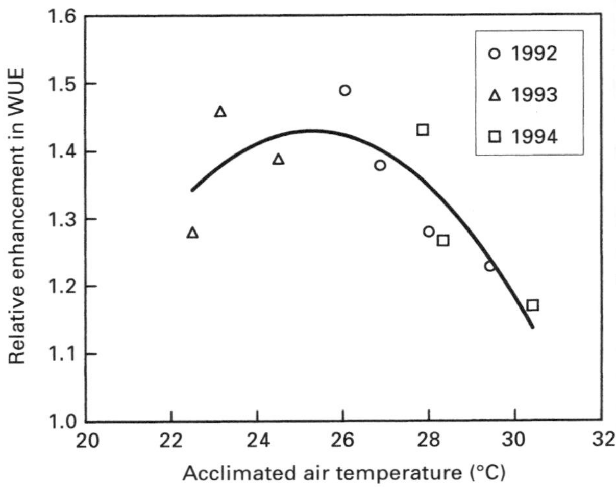

Fig. 5.5. Effect of daily mean temperature on relative enhancement of crop water use efficiency (WUE) for rice grown at ambient and enriched [CO_2] (350 and 700 μmol mol^{-1}, respectively) in TGCs. The WUEs at 700 μmol mol^{-1} are normalized to the values at ambient [CO_2]. (Adapted from Nakagawa *et al.*, 1997.)

We suggest that a simple modelling approach can help to explain these differential temperature responses in WUE for the elevated and ambient [CO_2]. Since net photosynthetic rate (P_n) per unit leaf area can also be expressed by a gas diffusion equation similar to equation 5.2, the transpirational WUE (P_n/E) as given by Sinclair *et al.* (1984) is:

$$P_n/E = KC_a\,(1 - C_i/C_a)\;/\;[e^*(T_L) - e_a] \qquad \text{(Eqn 5.3)}$$

where C_a and C_i are [CO_2] surrounding the leaf and in the leaf intercellular space, respectively. A constancy in the C_i/C_a ratio over a wide range of [CO_2] has been reported for many plant species (Goudriaan and van Laar, 1978; Wong *et al.*, 1979), including rice (Morrison and Gifford, 1983; Imai and Okamoto-Sato, 1991). This implies that stomatal aperture and hence r_s respond to [CO_2] in parallel with the response of leaf photosynthesis to [CO_2] (Wong *et al.*, 1979). At ambient [CO_2], the observed conservative response of P_n/E ratio to temperature indicates that the C_i/C_a ratio remains constant with temperature and, further, that both r_s and photosynthesis vary in tandem with temperature. However, at elevated [CO_2], the decline in both r_s and P_n/E ratio with increasing temperature indicates that the C_i/C_a ratio increased with temperature, and/or that $e^*(T_L) - e_a$ increased with temperature. Since the difference in T_L between elevated and ambient [CO_2] was very small at higher air temperatures (Homma *et al.*, 1999), it seems that the C_i/C_a ratio increased at higher temperatures despite the reduction in r_s. This suggests that at higher [CO_2] and temperatures, rice stomata open independently of photosynthetic activity. This effect could be a mechanism employed by rice to increase leaf transpirational cooling at high temperatures. In any event, it appears that doubling [CO_2] increases crop WUE in rice by about 50% over that at the optimal temperature range for growth. However, this increase in WUE declines sharply as temperature increases beyond the optimum.

5.2.4 Development

The temperature optimum for maximum phenological developmental rate of rice, in terms of the number of days to reach heading, is generally 27–30°C. Above or below this temperature optimum, developmental rate slows and a longer time period is required for the crop to reach heading (Horie, 1994). Elevated [CO_2] increases phenological developmental rate and hence reduces the number of days to heading (Baker *et al.*, 1990b; Nakagawa *et al.*, 1993; Kim *et al.*, 1996a). This [CO_2] effect on rice developmental rate has been shown to be temperature dependent. For the *japonica*-type rice cultivar Akihikari, CO_2 enrichment reduced the number of days to heading by 6% and 11% at air temperatures of 28°C and 30°C, respectively (Kim *et al.*, 1996a). There are two potential explanations for this accelerated crop developmental rate with [CO_2] enrichment. First, elevated [CO_2] could increase plant temperature by increasing r_s, thus reducing transpirational cooling. Second, it has been shown that rice grown under elevated [CO_2] generally has a higher carbon to nitrogen (C/N) ratio (Baker *et al.*, 1992c; Kim 1996;

Ziska *et al.*, 1996). A high C/N ratio has been shown to be associated with accelerated plant developmental rates (Zeevaart, 1976). However, CO_2 enrichment effects on leaf temperature as the primary mechanism seems unlikely since leaf temperature differences between ambient and doubled [CO_2] were very small at higher air temperatures (Homma *et al.*, 1999). Furthermore, developmental rates are slowed as air temperature increases beyond the optimum for development.

Baker *et al.* (1990b) found that [CO_2] did not significantly affect mainstem phyllochron intervals of rice. However, in that study, the final mainstem leaf number reduced with elevated [CO_2] due to an acceleration in the phenological development. Reports on the effects of [CO_2] on tiller production appear contradictory. Baker *et al.* (1990b, 1992b) found relatively minor effects of [CO_2] on tiller production across the [CO_2] range from 330 to 660 μmol mol^{-1}. In contrast, Kim *et al.* (1996a) and Ziska *et al.* (1997) found a marked increase in tiller number caused by a doubling of [CO_2]. This discrepancy is likely related to differences in plant densities among these studies. Kim *et al.* (1996a) and Ziska *et al.* (1997) utilized plant populations of 50 and 75 plants m^{-2}, respectively, whereas Baker *et al.* (1990b, 1992b) used a density of 235 plants m^{-2}, which is extraordinarily high. A high degree of mutual shading among plants at this high density would have suppressed the development of tiller primordia due to greater competition for light. Therefore, under planting densities usually practised in Asian rice cultures, elevated [CO_2] may substantially promote tiller production.

Elevated [CO_2] resulted in only minor effects on individual leaf size (Baker *et al.*, 1990c) and also on total leaf area of rice except during the initial growth stage (Morrison and Gifford, 1984a; Imai *et al.*, 1985; Baker *et al.*, 1990c; Nakagawa *et al.*, 1993; Kim *et al.*, 1996a; Ziska *et al.*, 1997). This appears to indicate that N uptake of rice was not influenced by elevated CO_2 (Baker *et al.*, 1992c; Ziska *et al.*, 1996) except at the initial stage (Kim, 1996), since rice leaf area increases proportionally with plant N uptake (Murata, 1961; Miyasaka *et al.*, 1975; Hasegawa and Horie, 1997).

Elevated [CO_2] increased specific leaf weight of rice during early growth, but less so at later growth stages (Baker *et al.*, 1990c). Also, the [CO_2] effects on the root to shoot (R/S) ratio were significant during the early growth stages, with a higher R/S ratio at elevated [CO_2] (Imai *et al.*, 1985; Baker *et al.*, 1990c; Ziska *et al.*, 1996). However, for later developmental stages, the [CO_2] effect on R/S ratio became negligibly small provided sufficient N fertilizer was applied (Ziska *et al.*, 1996; Kim *et al.*, 1996a).

Elevated [CO_2] markedly increased rice spikelet number per unit area over a wide range of air temperatures, through both increases in the number of productive tillers per unit area and spikelets per tiller (Imai *et al.*, 1985; Kim *et al.*, 1996b). In rice, the spikelet number per unit area is generally proportional to plant N content (Shiga and Sekiya, 1976; Horie *et al.*, 1997b) or to plant N concentration (Hasegawa *et al.*, 1994) at the spikelet initiation stage. The fact that elevated [CO_2] increased the spikelet number despite a reduction in plant N concentration suggests that elevated [CO_2] promoted spikelet production efficiency per unit of plant N. Elevated [CO_2] increases sink size by increasing spikelet number and this generally leads to increased yield, provided successful fertilization and grain-fill follows.

5.2.5 Biomass production

Crop growth rate (dw/dt; g m^{-2} day^{-1}) can be generally defined as the product of canopy light interception and radiation utilization efficiency (RUE; g MJ^{-1}) (Monteith, 1977; Horie and Sakuratani, 1985) and is represented by:

$$dw/dt = \text{RUE} \times S_0[1 - \exp(-k \times \text{LAI})] \qquad \text{(Eqn 5.4)}$$

where S_0 is incoming solar radiation (MJ m^{-2} day^{-1}), k is the radiation extinction coefficient of the canopy, and LAI is leaf area index. As previously noted, the effects of [CO_2] on LAI are small. Thus, differences in radiation interception among [CO_2] are also small for rice except during early growth. Therefore, CO_2 enrichment increases RUE and crop growth rate mainly by increasing photosynthesis and decreasing specific respiration rate. Rice biomass production is calculated by integrating equation 5.4 with respect to time for a given crop duration (d). Since elevated [CO_2] increases RUE but decreases d, the overall effect on crop biomass production is a result of the differential effects of elevated [CO_2] on these two parameters. Reported values of the relative enhancement of rice biomass caused by CO_2 enrichment vary widely. This is probably due to the difference in cultivars, temperatures, solar radiation levels, N or other fertilizer applications, and various experimental methodologies (e.g. field vs. potted plant experiments). Determination of the interactive effects of [CO_2] with these factors on rice biomass production is of primary importance.

With respect to temperature and [CO_2] effects on rice biomass production, Horie (1993) analysed published data for long-term CO_2 enrichment studies. Large differences in [CO_2] enrichment effects on biomass production occurred for potted plant experiments compared with those under near field-like environments. He concluded that, under field-like conditions, [CO_2] enrichment increases biomass production by 24% across a wide range of air temperature treatments. Nakagawa *et al.* (2000) conducted a similar analysis with more recently reported data (Fig. 5.6). This analysis revealed that [CO_2] effects on biomass production for rice grown in pots showed a strong temperature dependency while rice grown under field-like conditions was little affected by temperature. A temperature coefficient (e.g. the slope of the regression line) of only 1.8% °C^{-1} was obtained for the field-like conditions (Fig. 5.6). Since the value of this temperature coefficient was small and significantly different from zero only at $P \leq 0.05$, the average value of a 24% relative enhancement of biomass production for a doubling of [CO_2] for field-grown rice appears to be a reasonable estimate.

The reason pot-grown plants displayed a strong [CO_2] by temperature interaction for biomass production may be that pot-grown rice continues to produce tillers for a longer time than does field-grown rice, especially at higher temperatures. This is likely due to reduced mutual shading in potted plants

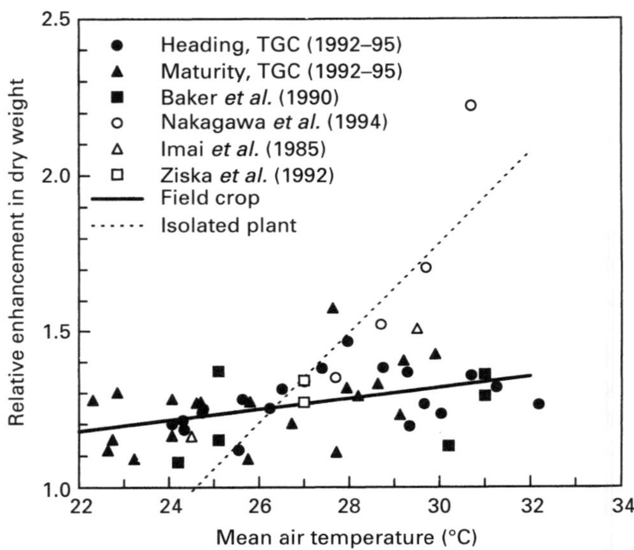

Fig. 5.6. Effects of mean air temperature on the relative enhancement of rice crop biomass. Plants were grown season long in ambient and nearly doubled [CO_2]. The open and closed symbols are for pot- and field-grown plants, respectively. The data are normalized to the values at ambient [CO_2] as unity. (Adapted from Nakagawa *et al.*, 2000.)

compared with field-grown plants and is similar to the light interception effects on dry matter production and tiller development described in section 5.2.4. In contrast with the nearly isolated pot-grown rice, plants grown under field-like conditions develop a closed canopy, where mutual shading suppresses positive feedback effects of increased photosynthesis on growth. Also, the shortening of growth duration by elevated [CO_2], especially at higher temperatures (Kim *et al.*, 1996a), contributes to a weakening of [CO_2] by temperature interactive effects on rice biomass production for field-grown plants, despite the appreciable interactive effects observed for single leaf photosynthesis (Nakagawa *et al.*, 1993; Lin *et al.*, 1997).

Since leaf photosynthesis depends strongly on leaf N concentration (Murata, 1961; Ishihara *et al.*, 1979), RUE also depends on leaf N (Sinclair and Horie, 1989). These facts suggest that the relative enhancement of rice biomass production with [CO_2] depends strongly on plant N content. In a fertilizer N by [CO_2] experiment conducted on a *japonica*-type rice in TGCs, Nakagawa *et al.* (1994) found enhanced biomass production due to CO_2 enrichment was 17, 26 and 30% at N applications rates of 40, 120 and 200 kg ha^{-1}, respectively. This finding is similar to that reported for an *indica*-type rice by Ziska *et al.* (1996) where the increased biomass production due to CO_2 enrichment was 0, 29 and 39% at N application rates of 0, 90 and 200 kg ha^{-1}, respectively. These results clearly illustrate that the relative enhancement in biomass production due to CO_2 enrichment depends strongly on N application rate. Furthermore, with adequate fertilizer N, the relative enhancement in rice biomass production due to CO_2 enrichment can reach or exceed 30%.

Since global climate change could involve shifts in precipitation patterns, potential interactive effects of [CO_2] and drought stress become another topic of interest. For pot-grown rice plants, CO_2 enrichment under drought-stress conditions resulted in increased relative biomass production compared with drought-stressed ambient controls (Morrison and Gifford, 1984b; Rogers *et al.*, 1984). This effect was attributed to higher leaf water potentials for the CO_2-enriched plants under drought-stress conditions. Due to the anti-transpirant effect of CO_2 enrichment on canopy evapotranspiration, Baker *et al.* (1997a) reported a modest reduction of about 10% in crop water use.

5.2.6 Yield

Increased biomass production of rice caused by elevated [CO_2] has the potential to increase yield, provided flowering and grain-fill are not disrupted by some environmental stress such as drought or high temperature. Baker and Allen (1993) showed that equation 5.1 can also be applied to relative yield responses of rice to [CO_2] for an *indica*-type rice (cv. IR30) subjected to long-term CO_2 at constant day/night temperature of 31°C under field-like conditions. The values of the parameters for the yield response (equation 5.1) were 284 μmol mol^{-1} for K_m, 2.24 for the asymptotic response parameter (P_{max}) and −0.13 for the *y*-intercept (P_i). Using these parameters, they calculated that doubling [CO_2] from 330 to 660 μmol mol^{-1} increased yield by 44%. Ziska *et al.* (1997) obtained a yield increase of 27% with CO_2 enrichment for the cultivar IR72 grown at ambient temperatures in the wet and dry seasons at Los Baños, the Philippines. The percentage yield increase due to a doubling of [CO_2] for a *japonica*-type rice (cv. Akihikari) grown at ambient temperatures in 1991 and 1992 in Kyoto, Japan, ranged from 20 to 40% (Kim *et al.*, 1996b). While there is considerable variation among these reports in the relative yield response to doubled [CO_2] it appears that a 30% enhancement in seed yield may be a reasonable estimate for rice exposed to long-term doubled [CO_2] under field conditions with moderate temperatures. This estimate for the relative yield increase also coincides with that for the relative biomass response under sufficient N application conditions.

In terms of predicting the effects of potential future global warming on rice yields, studies that examine the effects and interactions of both temperature and CO_2 enrichment are far more relevant than studies that examine only the effects of CO_2 enrichment. It has been well established that spikelet sterility in rice at mean daily temperatures below 20°C can be caused by a failure in pollen development at the microspore stage (Satake and Hayase, 1970) or injury sustained at flowering (Abe, 1969). Temperatures above 35°C at flowering can also cause spikelet sterility (Satake and Yoshida, 1976; Matsui *et al.*, 1997a). Whether or not elevated [CO_2] has any effect on ameliorating cool temperature damage on spikelets has not been studied. We consider spikelet sterility caused by high temperatures to be more important in the study of elevated [CO_2] and global warming effects on rice. The optimum daily mean temperature for grain-fill in rice is 20–25°C (Yoshida, 1981) which is lower than that for many other growth and developmental processes. In general, temperatures above 25°C result in poor grain-fill, caused in part by reduced grain-filling duration which in turn results in reduced grain yield.

Summarizing data from several experiments on the cultivar IR30, an *indica*-type rice, Baker *et al.* (1995) reported that grain yields declined with increasing temperatures above 26°C to zero yield near 36°C for both ambient and elevated [CO_2]. They estimated that rice yield declined by about 10% for each 1°C rise in daily mean temperature above 26°C. Ziska *et al.* (1997) reported significant yield reductions for the cultivar IR72 caused by a 4°C rise in air temperature above ambient temperatures in both the dry and wet seasons in the Philippines. They also reported a larger yield reduction caused by high temperature for elevated [CO_2] compared with ambient [CO_2] in the dry season.

Similar yield reductions at daily mean temperature above 26°C were reported for cultivar Akihikari a *japonica*-type rice under both doubled and ambient [CO_2] (Kim *et al.*, 1996b). In that experiment, rice grown at doubled [CO_2] also suffered more severe yield reductions with increasing temperature than plants in the ambient [CO_2]. They attributed the greater sensitivity of the [CO_2] enriched plants to high temperatures to having both a shorter grain-fill duration (cf. section 5.2.4) and a lower spikelet fertility than plants grown in ambient [CO_2] (Fig. 5.7). This reduction in spikelet fertility was mainly due to spikelet sterility induced by high temperatures during flowering. Since rice spikelets are most sensitive to high temperatures during flowering (Satake and Yoshida, 1976; Matsui *et al.*, 1997a) and because flowering in rice usually occurs at midday, the daily maximum temperature is usually more indicative of

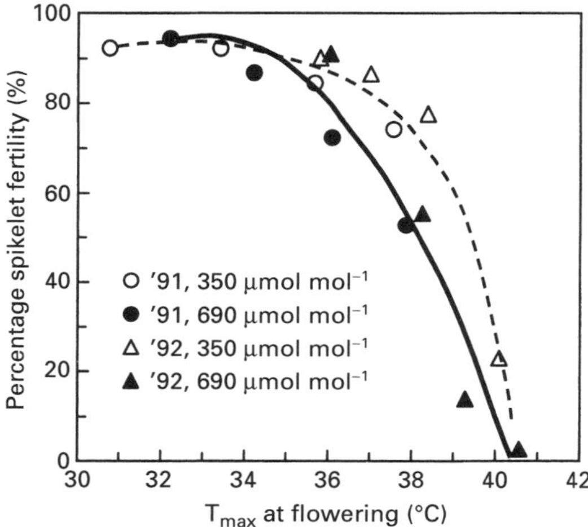

Fig. 5.7. Percentage fertility of rice spikelets vs. daily maximum air temperature (T_{max}) averaged over the flowering period (7 days) for rice grown at 690 and 350 μmol mol^{-1} [CO_2] in TGCs. (Adapted from Kim *et al.*, 1996b.)

high-temperature-induced spikelet sterility than is daily mean temperature. As a result, the relative enhancement in yield from a doubling of [CO_2] displays a sharp decline with increasing daily maximum air temperature averaged over the flowering period (usually about 7 days) and reaches negative values for air temperatures above 36.5°C (Fig. 5.8) (Kim *et al.*, 1996b). These results indicate that the effects of elevated [CO_2] on relative rice yield enhancement is strongly temperature dependent and may even become negative at extremely high temperatures during flowering.

This greater negative effect of elevated [CO_2] on high temperature damage to spikelets was also reported by Matsui *et al.* (1997b). They counted the number of pollen grains both shed and germinated on stigma of the spikelets that opened at high temperatures for cultivar IR72 (*indica* type) grown at both doubled and ambient [CO_2] at Los Baños, the Philippines. Shown in Fig. 5.9 are the percentages of spikelets having more than ten germinated pollen grains on the stigma as a function of temperature. This relationship was reduced sharply as temperature at flowering exceeded a specific threshold. This threshold temperature was 1–2°C lower for the CO_2-enriched plants compared with the ambient controls (Fig. 5.9). Since spikelets having more than ten germinated pollen grains on the stigma is a good criterion for predicting successful spikelet fertilization (Satake and Yoshida, 1976; Matsui *et al.*, 1997a), these results indicate that the elevated [CO_2] increased spikelet susceptibility to high-temperature-induced sterility. This reduction at high temperatures was caused by reductions in both the number of pollen grains shed and subsequent pollen germination.

Although the exact mechanism through which the elevated [CO_2] increased spikelet susceptibility to high-temperature-induced sterility is

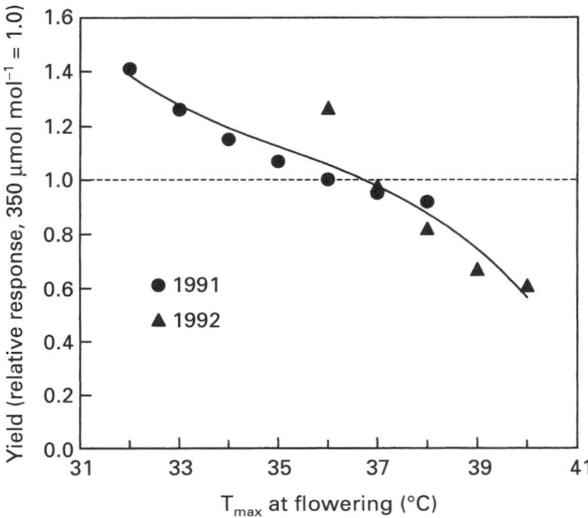

Fig. 5.8. Relative enhancement in yield from a doubling of [CO_2] vs. daily maximum air temperature (T_{max}) averaged over the flowering period (7 days) for rice grown in TGCs. (Adapted from Kim *et al.*, 1996b.)

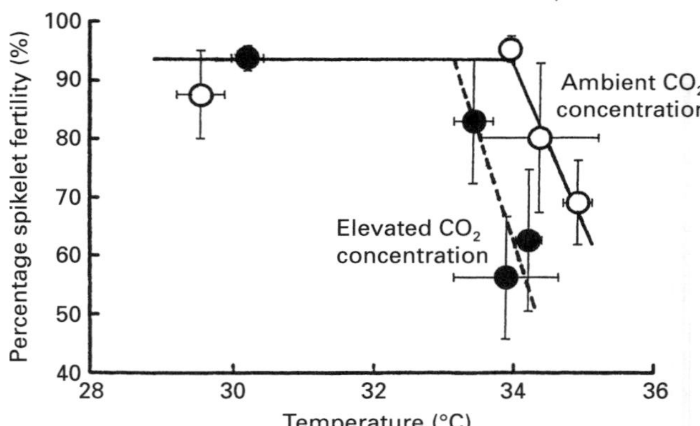

Fig. 5.9. Percentage of spikelets having ten or more germinated pollen grains vs. maximum temperature during flowering for rice grown at ambient and ambient + 300 μmol CO_2 mol^{-1} and at ambient and ambient + 4°C air temperatures. Vertical and horizontal bars indicate standard error of the means for spikelet sterility and temperature, respectively. (Adapted from Matsui *et al.*, 1997b.)

unknown, increased spikelet temperature under CO_2 enrichment due to suppression of transpiration is one possibility. However, temperatures during the few days prior to flowering (Satake and Yoshida, 1976; Matsui *et al.*, 1997a) as well as humidity and wind speed during flowering (Matsui *et al.*, 1997a) have been shown also to influence the magnitude of high-temperature-induced spikelet sterility. Clearly, further research is needed to delineate the specific mechanisms by which high temperatures induce spikelet sterility as well as the role of CO_2 enrichment in these processes.

A moderately large genetic variation in the tolerance to high-temperature-induced spikelet sterility has been reported both among and between *indica-* and *japonica*-type rice genotypes (Satake and Yoshida, 1976; Matsui *et al.*, 1997a). Some rice cultivars have the ability to flower early in the morning, thus potentially avoiding the damaging effects of higher temperatures later in the day (Imaki *et al.*, 1987). The continued search for rice genotypes having higher tolerance to, and/or the ability to avoid, high-temperature damage to spikelets is important in terms of finding options for mitigating potentially adverse effects of global climate change on rice production.

5.3 Model Simulation of Global Climatic Change Effects on Regional Rice Production

In order to describe the effects of [CO_2], temperature and other environmental factors on rice growth and yield formation processes, several models have been developed for simulating global climate change effects on rice to predict these effects at a given location and to develop strategies for production technologies adapted to climate change. These models include SIMRIW (Horie, 1993; Horie *et al.*, 1995a), CERES-RICE (Singh and Paddila, 1995) and ORYZA1 (Kropff *et al.*, 1995). Figure 5.10 summarizes the yield responses of a *japonica*-type rice to various air temperatures, solar radiation and [CO_2] as simulated by SIMRIW (Horie *et al.*, 1995a). This model adequately simulated observed year-to-year variations in rice yield for many locations in Japan (Horie, 1993) as well as the seasonal variation in rice yields at Los Baños, the Philippines (Horie *et al.*, 1995a). However, these simulations were based on current climates. Further improvements to existing rice models are needed to account for the interactive effects of [CO_2], temperature and N on rice growth and the yield formation processes described in the previous sections of this chapter.

Despite these limitations, work with both SIMRIW (Horie *et al.*, 1995b, 1997a) and ORYZA1 (Matthews *et al.*, 1995, 1997) has pointed out important implications of potential climate change on future rice yields in Asia. These simulations utilized climate scenarios from three global circulation models: GISS (Goddard Institute for Space Studies), GFDL (Geophysical Fluid Dynamics Laboratory) and UKMO (United Kingdom Meteorological Office). These simulations indicate that future climate change may result in substantially increased irrigated rice yields for the northern parts of China, Korea and Japan, and areas in tropical Indonesia and Malaysia. However, for southern Japan, as well as inland areas in subtropical Asia, reductions in rice yields are possible due to high-temperature-induced spikelet sterility and shortening of crop growth durations. These simulations also indicate that

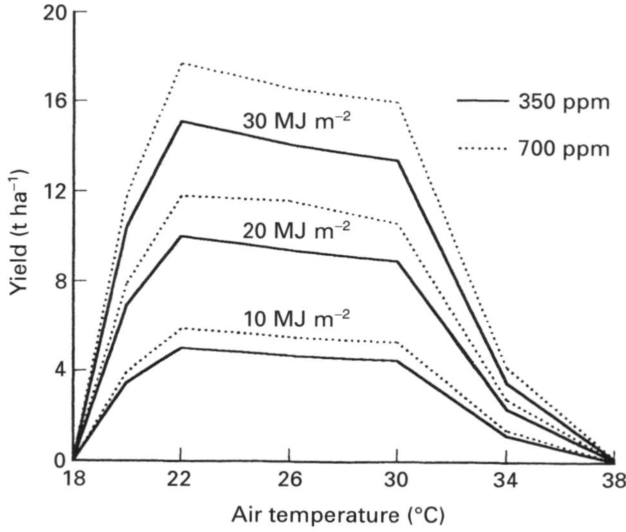

Fig. 5.10. Simulated yield of rice (cv. Nipponbare) at different daily mean temperatures, solar radiation and [CO_2] under constant environmental conditions. Daylength and diurnal temperature range were set at 12 h and 8°C, respectively. (Adapted from Horie *et al.*, 1995a.)

mitigation of the negative effects of climate change by altering planting dates in southern Japan, in order to avoid damagingly high temperatures, proved unsatisfactory because rice growth in the cooler season was limited by lower solar radiation due to the season of the year.

Rice models developed so far are relevant only to irrigated rice. Since rainfed rice cultures are considered more vulnerable to anticipated global climate change, synthesis of a reliable model for simulating growth and yield of rainfed rice is needed.

5.4 Conclusions and Future Research Needs

The ability to reliably predict effects of anticipated global climate change on rice production should provide an indispensable basis for forming a worldwide consensus to reduce greenhouse gas emission. If global climate change becomes unavoidable, reliable rice simulation models would greatly aid in the development of rice production technologies. Future food security, especially for rice-farming societies consisting of small-hectarage subsistence farmers, would be highly desirable. Over the past two decades, many research and modelling efforts have been conducted to determine the effects of [CO_2], temperature and other environmental factors on rice growth and yield formation processes. Major findings from these efforts include the following:

- Long-term doubling of [CO_2] promoted canopy net photosynthetic rate of rice by 30–40% and reduced specific respiration rate.
- Both increasing air temperature and CO_2 enrichment accelerated rice developmental rates and reduced the number of days to heading.
- Carbon dioxide enrichment resulted in minor effects on N uptake, which appeared to be associated with relatively minor effects of [CO_2] on leaf area growth of rice.
- Averaged over several different studies, a doubling of [CO_2] increased rice biomass production under field conditions by about 24% over a relatively wide range of air temperatures. However, this relative enhancement was strongly influenced by soil N availability. With sufficient N fertilizer application, relative enhancement of biomass production under CO_2 enrichment may reach 30% or more.
- Although a synergistic effect of [CO_2] and temperature on leaf-level photosynthesis has been reported for rice, such an effect on canopy photosynthesis and biomass production is likely to be slight under field conditions.
- With a doubling of [CO_2], rice WUE increased by 40–50% at optimum growth temperatures. This is mainly due to increased biomass production and partly to a reduction in transpiration. However, this effect of [CO_2] on WUE sharply declined with increases in air temperature above the optimum.
- A doubling of [CO_2] substantially increased the yield capacity of rice through enhanced tiller and spikelet production per unit ground area.

- The optimum daily mean temperature for rice yield was between 23 and 26°C. As air temperature deviated from this optimum, yield declined to near zero at temperatures below 15–18°C or above 36–40°C, depending on the cultivar. This effect is due mainly to spikelet sterility induced by either low or high temperatures. Rice yield increased by about 30% with CO_2 enrichment at optimum temperatures and declined with increases in temperature mainly because of high-temperature-induced spikelet sterility. Furthermore, it appears that CO_2 enrichment increased susceptibility to high-temperature-induced spikelet sterility.
- Rice genotypes differ in their ability to tolerate or avoid high-temperature-induced spikelet sterility at flowering and in stomatal responses to the interactive effects of [CO_2] and temperature.
- Rice responses to the effects and interactions of [CO_2] and temperature in terms of biomass production, WUE and yield differ considerably between pot- and field-grown plants, due mainly to difference in light interception. Therefore, special care is necessary for extrapolating experimental results from rice grown in pots to field conditions for global climate change studies.

Although a considerable amount of information has been accumulated concerning rice responses to [CO_2] and temperature, more experimental studies are needed to quantify the effects and interactions of [CO_2] and temperature on biomass production, stomatal conductance, WUE, spikelet sterility and final yield. Also, more detailed studies are needed to define the mechanism(s) governing the adverse effects of enriched [CO_2] on high-temperature-induced spikelet sterility. The genetic differences among rice cultivars in response to [CO_2] and temperature, especially in the areas of crop water use and high-temperature-induced spikelet sterility, need to be determined. This information is required to identify rice genotypes that are better adapted to potential future global climate changes.

The integration of information into rice models is indispensable for impact assessments of global climate change on regional rice production. Simulation studies based on existing rice models suggest that, while future climate change has the potential to increase irrigated rice yield in northern and tropical Asia, yield reductions are also likely in some inland areas of subtropical Asia, especially in rice produced in the dry season.

References

Abe, I. (1969) Agro-meteorological studies regarding the effect of easterly winds ('Yamase wind') on the growth of rice plants in Aomori prefecture. *Bulletin of Aomori Agricultural Experimental Station* 14, 40–138. (In Japanese.)

Akita, S. (1980) Studies on the difference in photosynthesis and photorespiration among crops. II. The differential response of photosynthesis, photorespiration and dry matter production to carbon dioxide concentration among species. *Bulletin of the National Institute of Agricultural Sciences* Series D, 31, 59–94. (In Japanese with English summary.)

Amthor, J.S. (1994) Respiration and carbon assimilate use. In: Boote, K.J. *et al.* (eds) *Physiology and Determination of Crop Yield.* American Society of Agronomy, Madison, Wisconsin, pp. 221–250.

Arp, W.J. (1991) Effects of source–sink relations on photosynthetic acclimation to elevated CO_2. *Plant, Cell and Environment* 14, 869–875.

Baker, J.T. and Allen, L.H. Jr (1993) Effects of CO_2 and temperature on rice: a summary for five growing seasons. *Journal of Agricultural Meteorology* 48, 575–582.

Baker, J.T., Allen, L.H. Jr, Boote, K.J., Jones, P. and Jones, J.W. (1990a) Rice photosynthesis and evapotranspiration in subambient, ambient and superambient carbon dioxide concentrations. *Agronomy Journal* 82, 834–840.

Baker, J.T., Allen, L.H. Jr, Boote, K.J., Jones, P. and Jones, J.W. (1990b) Developmental responses of rice to photoperiod and carbon dioxide concentration. *Agriculture and Forest Meteorology* 50, 201–210.

Baker, J.T., Allen, L.H. Jr and Boote, K.J. (1990c) Growth and yield responses of rice to carbon dioxide concentration. *Journal of Agricultural Science* 115, 313–320.

Baker, J.T., Allen, L.H. Jr and Boote, K.J. (1992a) Temperature effects on rice at elevated CO_2 concentration. *Journal of Experimental Botany* 43, 959–964.

Baker, J.T., Allen, L.H. Jr and Boote, K.J. (1992b) Response of rice to carbon dioxide and temperature. *Agricultural and Forest Meteorology* 60, 153–166.

Baker, J.T., Laugel, F., Boote, K.J. and Allen, L.H. Jr (1992c) Effects of daytime carbon dioxide concentration on dark respiration in rice. *Plant, Cell and Environment* 15, 231–239.

Baker, J.T., Boote, K.J., and Allen, L.H. Jr (1995) Potential climate change effects on rice: carbon dioxide and temperature. In: Rosenzweig, C. *et al.* (eds) *Climate Change and Agriculture: Analysis of Potential International Impacts.* Special Publication No. 59. American Society of Agronomy, Madison, Wisconsin, pp. 31–47.

Baker, J.T., Allen, L.H. Jr, Boote, K.J. and Pickering, N.B. (1997a) Rice responses to drought under carbon dioxide enrichment. I. Growth and yield. *Global Change Biology* 3, 119–128.

Baker J.T., Allen, L.H. Jr. Boote, K.J. and Pickering, N.B. (1997b) Rice responses to drought under carbon dioxide enrichment. II. Photosynthesis and evapotranspiration. *Global Change Biology* 3, 129–138.

FAO (1995) *Production Year Book* 49, 70–71.

Goudriaan, J. and Laar, H.H. van (1978) Relations between leaf resistance, CO_2-concentration, and CO_2-assimilation in maize, beans, lalang grass and sunflower. *Photosynthica* 12, 241–249.

Hansen, J., Lacis, A., Rind, D., Russell, G., Stone, P., Fung, I., Ruedy R. and Lerner, J. (1984) Climate sensitivity: analysis of feedback mechanisms. In: Hansen, J. *et al.* (eds) *Climate Processes and Climate Sensitivity.* Maurice Ewing Series, 5. American Geophysical Union, Washington, DC, pp. 130–164.

Hasegawa, T. and Horie, T. (1997) Modelling the effect of nitrogen on rice growth and development. In: Kropff, M.J. *et al.* (eds) *Applications of System Approaches at the Field Level,* Vol. 2. Kluwer Academic Publishers, Dordrecht, pp. 243–258.

Hasegawa, T., Koroda, Y., Seligman, N.G. and Horie, T. (1994) Response of spikelet number to plant nitrogen concentration and dry weight in paddy rice. *Agronomy Journal* 86, 673–676.

Homma, K., Nakagawa, H., Horie, T., Ohnishi, H., Kim H.Y. and Ohnishi, M. (1999) Energy budget and transpiration characteristics of rice grown under elevated CO_2 and high temperature conditions as determined by remotely sensed canopy temperatures. *Japanese Journal of Crop Science* 68, 137–145. (In Japanese with English abstract.)

Horie, T. (1993) Predicting the effects of climate variation and elevated CO_2 on rice yield in Japan. *Journal of Agricultural Meteorology* 48, 567–574.

Horie, T. (1994) Crop ontogeny and development. In: Boote, K.J. *et al.* (eds) *Physiology and Determination of Crop Yield.* CSSA-ASA-SSSA, Madison, Wisconsin, pp. 153–180.

Horie, T. and Sakuratani, T. (1985) Studies on crop–weather relationship model in rice. I. Relation between absorbed solar radiation by the crop and dry matter production. *Journal of Agricultural Meteorology* 40, 231–242. (In Japanese with English summary.)

Horie, T., Nakano, J., Nakagawa, H., Wada, K., Kim, H.Y. and Seo, T. (1991) Effect of elevated CO_2 and high temperature on growth and yield of rice. 1. Development of temperature gradient tunnels. *Japanese Journal of Crop Science*, 60 (extra issue 2), 127–128. (In Japanese.)

Horie, T., Nakagawa, H., Centeno, H.G.S. and Kropff, M. (1995a) The rice crop simulation model SIMRIW and its testing. In: Matthews, R.B. *et al.* (eds) *Modelling the Impact of Climate Change on Rice in Asia.* CAB International, Wallingford, UK, pp. 51–66.

Horie, T., Nakagawa, H., Ohnishi, M. and Nakano, J. (1995b) Rice production in Japan under current and future climates. In: Matthews, R.B. *et al.* (eds) *Modelling the Impact of Climate Change on Rice in Asia.* CAB International, Wallingford, UK, pp. 143–164.

Horie, T., Nakagawa, H., Nakano, J., Hamotani, K. and Kim, H.Y. (1995c) Temperature gradient chambers for research on global environment change. III. A system designed for rice in Kyoto, Japan. *Plant, Cell and Environment* 18, 1064–1069.

Horie, T., Centeno, H.G.S., Nakagawa, H. and Matsui, T. (1997a) Effects of elevated carbon dioxide and climate change on rice production in East and South East Asia. In: Oshima, Y. *et al.* (eds) *Proceedings of International Scientific Symposium on Asian Paddy Field.* College of Agriculture, University of Saskatchewan, Canada, pp. 49–58.

Horie, T., Ohnishi, M., Angus, J.F., Lewin, L.G., Tsukaguchi, T. and Matano, T. (1997b) Physiological characteristics of high-yielding rice inferred from cross-location experiments. *Field Crops Research* 52, 55–67.

Hossain, M. (1998) Sustaining food security in Asia: economic, social and political aspects. In: Dowling, N.G. and Fisher, K.S. (eds) *Sustainability of Rice in the Global Food System.* International Rice Research Institute, Los Baños, the Philippines, pp. 19–43.

Imai, K. and Murata, Y. (1978) Effect of carbon dioxide concentration on growth and dry matter production of crop plants. IV. After-effects of carbon dioxide-treatments on the apparent photosynthesis, dark respiration and dry matter production. *Japanese Journal of Crop Science* 47, 330–335.

Imai, K. and Okamoto-Sato, M. (1991) Effects of temperature on CO_2 dependence of gas exchange in C3 and C4 crop plants. *Japanese Journal of Crop Science* 60, 139–145.

Imai, K., Colman, D.F. and Yanagisawa, T. (1985) Increase of atmospheric partial pressure of carbon dioxide and growth and yield of rice (*Oryza sativa* L.). *Japanese Journal of Crop Science* 54, 413–418.

Imaki, T., Tokunaga, S. and Obara, S. (1987) High temperature-induced spikelet sterility of rice in relation to flowering time. *Japanese Journal of Crop Science* 56 (extra issue 1), 209–210. (In Japanese.)

IRRI (1993) *IRRI Almanac.* International Rice Research Institute, Manila, the Philippines, 142 pp.

Ishihara, K., Kuroda, E., Ishii, R. and Ogura, T. (1979) Relationships between nitrogen content in leaf blades and photosynthetic rate in rice plants measured with an infrared gas analyzer and an oxygen electrode. *Japanese Journal of Crop Science* 48, 551–556. (In Japanese with English summary.)

Jones, P., Allen, L.H. Jr. and Jones, J.W. (1985) Responses of soybean canopy photosynthesis and transpiration to whole-day temperature changes in different CO_2 environments. *Agronomy Journal* 77, 242–249.

Kim H.Y. (1996) Effects of elevated CO_2 concentration and high temperature on growth and yield of rice. PhD Thesis submitted to Graduate School of Agriculture, Kyoto University, Kyoto 606, Japan, 90 pp. (In Japanese with English summary.)

Kim, H.Y., Horie, T., Nakagawa, H. and Wada, K. (1996a) Effects of elevated CO_2 concentration and high temperature on growth and yield of rice. I. The effect on development, dry matter production and some growth characters. *Japanese Journal of Crop Science* 65, 634–643. (In Japanese with English abstract.)

Kim, H.Y., Horie, T., Nakagawa, H. and Wada, K. (1996b) Effects of elevated CO_2 concentration and high temperature on growth and yield of rice. II. The effect on yield and its component of Akihikari rice. *Japanese Journal of Crop Science* 65, 644–651. (In Japanese with English abstract.)

Kobayashi, K., Okada, M. And Kim, H.Y. (1999) The free-air CO_2 enrichment (FACE) with rice in Japan. In: Horie *et al.* (eds) *World Food Security and Crop Production Technologies for Tomorrow.* Graduate School of Agriculture, Kyoto University, Kyoto, Japan, pp. 213–215.

Kropff, M.J., Matthews, R.B., Laar, H.H. van and Berge, H.F.M. ten (1995) The rice model ORYZA1 and its testing. In: Matthews, R.B. *et al.* (eds) *Modelling the Impact of Climate Change on Rice in Asia.* CAB International, Wallingford, UK, pp. 27–50.

Lin, W., Ziska, L.H., Namuco, O.S. and Bai, K. (1997) The interaction of high temperature and elevated CO_2 on photosynthetic acclimation of single leaves of rice in situ. *Physiologia Plantarum* 99, 178–184.

Long, S.P. (1991) Modification of the response of photosynthetic productivity to rising temperature by atmospheric CO_2 concentration: has its importance been underestimated? *Plant, Cell and Environment* 14, 729–739.

Matsui, T., Omasa, T. and Horie, T. (1997a) High temperature-induced spikelet sterility of japonica rice at flowering in relation to air temperature, humidity and wind velocity. *Japanese Journal of Crop Science* 66, 449–455.

Matsui, T., Namuco, O.S., Ziska, L.H. and Horie, T. (1997b) Effects of high temperature and CO_2 concentration on spikelet sterility in Indica rice. *Field Crops Research* 51, 213–219.

Matthews, R.B., Horie, T., Kropff, M.J., Bachelet, D., Centenom, H.G., Shin, J.C., Mohandas, S., Singh, S., Zhu, D. and Lee, M.H. (1995) A regional evaluation of the effect of future climate change on rice production in Asia. In: Matthews, R.B. *et al.* (eds) *Modelling the Impact of Climate Change on Rice Production in Asia.* CAB International, Wallingford, U.K., pp. 95–139.

Matthews, R.B., Kropff, M.J., Horie, T. and Bachelet, D. (1997) Simulating the impact of climate change on rice production in Asia and evaluating options for adaptation. *Agricultural Systems* 54, 399–425.

Miyasaka, A., Murata, Y. and Iwata, T. (1975) Leaf area development and leaf senescence in relation to climatic and other factors. In: Murata, Y. (ed.) *Crop Productivity and Solar Energy Utilization in Various Climates in Japan.* University of Tokyo Press, Tokyo, pp. 72–85.

Monteith, J. (1977) Climate and the efficiency of crop production in Britain. *Philosophical Transactions Royal Society, London*, Series B, 281, 277–294.

Morrison, J.I.L. and Gifford, R.M. (1983) Stomatal sensitivity to carbon dioxide and humidity. A comparison of two C3 and two C4 grass species. *Plant Physiology* 71, 789–796.

Morrison, J.I.L. and Gifford, R.M. (1984a) Plant growth and water use with limited water supply in high CO_2 concentrations. I. Leaf area, water use and transpiration. *Australian Journal of Plant Physiology* 11, 361–374.

Morrison, J.I.L. and Gifford, R.M. (1984b) Plant growth and water use with limited water supply in high CO_2 concentrations. II. Plant dry weight, partitioning and water use efficiency. *Australian Journal of Plant Physiology* 11, 375–384.

Moya, T.B., Ziska, L.H., Weldon, C., Quilang, J.E.P. and Jones, P. (1997) Microclimate in open-top chambers: implication for predicting climate change effects on rice production. *Transaction of the American Society of Agricultural Engineers* 40, 739–747.

Murata, Y. (1961) Studies on the photosynthesis in rice plants and its cultural significance. *Bulletin of National Institute of Agricultural Sciences*, D. 9, 1–169. (In Japanese with English summary.)

Nakagawa, H., Horie, T., Nakano, J., Kim, H.Y., Wada, K. and Kobayashi, M. (1993) Effect of elevated CO_2 concentration and high temperature on growth and development of rice. *Journal of Agricultural Meteorology* 48, 799–802.

Nakagawa, H., Horie, T. and Kim, H.Y. (1994) Environmental factors affecting rice responses to elevated carbon dioxide concentrations. *International Rice Research Note* 19, 45–46.

Nakagawa, H., Horie, T., Kim, H.Y., Ohnishi, H. and Homma, K. (1997) Rice responses to elevated CO_2 concentrations and high temperatures. *Journal of Agricultural Meteorology* 52, 797–800.

Nakagawa, H., Horie, T., Goudriaan, J., Kim, H.Y., Ohnishi, H., Homma, K. And Wada, K. (2000) Effects of elevated CO_2 concentration and increased temperature on leaf area growth and dry matter production of rice as found in TGC study. *Plant Production Science* (in press).

Nishiyama, I. (1984) Climatic influence on pollen formation and fertilization. In: Tsunoda, S. and T. Takanashi (eds) *Biology of Rice.* Japan Science Society Press, Tokyo/Amsterdam, pp. 153–171.

Oka, H.I. (1953) Phylogenetic differentiation of the cultivated rice plant. I. Variations in respective characteristics and their combinations in rice cultivars. *Japanese Journal of Breeding* 3, 33–43. (In Japanese.)

Ozawa, Y. (1964) Agriculture and climate. In: Tsuboi *et al.* (eds) *Hand Book of Agricultural Meteorology.* Yokendo, Tokyo, pp. 266–292. (In Japanese.)

Penning de Vries, F.W.T., Brunsting, A.H.M. and Laar, H.H. van (1974) Products, requirements and efficiency of biosynthesis: a quantitative approach. *Journal of Theoretical Biology* 45, 339–377.

Rogers, H.H., Sionit, N., Cure, J.D., Smith, J.M. and Bingham, G.E. (1984) Influence of elevated carbon dioxide on water relations of soybean. *Plant Physiology* 74, 233–238.

Rowland-Bamford, A.J., Baker, J.T., Allen, L.H. Jr and Bowes, G. (1991) Acclimation of rice to changing atmospheric carbon dioxide concentration. *Plant, Cell and Environment* 14, 577–583.

Sakuratani, T. and Horie, T. (1985) Studies on evapotranspiration from crops. (1) On seasonal changes, varietal difference and the simplified methods of estimate in evapotranspiration of paddy rice. *Journal of Agricultural Meteorology* 41, 45–55. (In Japanese with English summary.)

Satake, T. (1995) High temperature injury. In: Matsuo, T. *et al.* (eds) *Science of the Rice Plant,* Vol. 2, *Physiology.* Food and Agriculture Policy Research Center, Tokyo, pp. 805–812.

Satake, T. and Hayase, H. (1970) Male sterility caused by cooling treatment at the young microspore stage in rice plants. 5. Estimations of pollen development stage and the most sensitive stage to coolness. *Proceedings of Crop Science Society of Japan* 39, 468–473.

Satake, T. and Yoshida, S. (1976) High temperature-induced sterility in Indica rice at flowering. *Japanese Journal of Crop Science* 47, 6–17.

Sinclair, T.R. and Horie, T. (1989) Leaf nitrogen, photosynthesis and crop radiation use efficiency: a review. *Crop Science* 29, 90–98.

Sinclair, T.R., Tanner, C.B. and Bennet, J.B. (1984) Water-use efficiency in crop production. *Bio Science* 34, 36–40.

Singh, U. and Padilla, J.L. (1995) Simulating rice response to climate change. In: Rosenzwseig, C. *et al.* (eds) *Climate Change and Agriculture: Analysis of Potential International Impacts.* American Society of Agronomy, Madison, Wisconsin, pp. 99–122.

Swaminathan, M.S. (1984) Rice. *Scientific American* 250, 81–93.

Takahashi, N. (1997) Differentiation of ecotypes in cultivated rice. 1. Adaptation to environments and ecotypic differentiation. In: Matsuo, T. *et al.* (eds) *Science of the Rice Plant,* Vol. 3, *Genetics.* Food and Agriculture Policy Research Center, Tokyo, pp. 112–118.

Terao, H., Otani, Y., Doi, Y. and Izumi, S. (1942) Physiological studies of the rice plant with special reference to the crop failure caused by the occurrence of unseasonable low temperatures. 8. The effect of various low temperatures on the panicle differentiation, heading and ripening in the different stages after transplanting to heading. *Proceedings of Crop Science Society of Japan* 13, 317–336. (In Japanese.)

Vu, J.C.V., Baker, J.T., Pennanen, A.H., Allen, L.H. Jr, Bowes, G. and Boote, K.J. (1998) Elevated CO_2 and water deficit effects on photosynthesis, ribulose bisphosphate carboxylase-oxygenase, and carbohydrate metabolism in rice. *Physiologia Plantarum* 103, 327–339.

Wada, K., Horie, T., Nakagawa, H., Kim, H.Y. and Nakano, Y. (1993) Effects of elevated CO_2 concentration and high temperature on water use efficiency of rice. *Report of the Society of Crop Science and Breeding in Kinki* 38, 49–50. (In Japanese.)

Wong, S.C., Cowan, R.R. and Farquhar, G.D. (1979) Stomatal conductance correlations with photosynthetic capacity. *Nature* 282, 424–426.

Yoshida, S. (1981) *Fundamentals of Rice Crop Science.* International Rice Research Institute, Los Baños, the Philippines, 269 pp.

Zeevaart, J.A.D. (1976) Physiology of flower formation. *Annual Review of Plant Physiology* 27, 321–348.

Ziska, L.H., Weerakoon, W., Namuco, O.S. and Pamplona, R. (1996) The influence of nitrogen on the elevated CO_2 response in field-grown rice. *Australian Journal of Plant Physiology* 23, 45–52.

Ziska, L.H., Namuco, O., Moya, T. and Quilang, J. (1997) Growth and yield response of field-grown tropical rice to increasing carbon dioxide and air temperature. *Agronomy Journal* 89, 45–53.

6 Crop Ecosystem Responses to Climatic Change: Maize and Sorghum

KEVIN J. YOUNG AND STEVE P. LONG

Departments of Crop Sciences and Plant Biology, University of Illinois at Urbana-Champaign, 190 Edward R. Madigan Laboratories, 1210 W. Gregory Drive, Urbana, IL 61801, USA

6.1 Introduction

Maize or corn (*Zea mays* L.) and sorghum (*Sorghum bicolor* (L.) Moench.) are, respectively, the third and fourth most important food crops globally in terms of sources of energy and protein in human nutrition; only wheat (*Triticum aestivum* L.) and rice (*Oryza sativa* L.) are more important. Among all uses of grains, maize and sorghum rank third and fifth in terms of global production, with wheat first, rice second, and barley (*Hordeum vulgare* L.) fourth. Maize and sorghum are the only major food crops that have C_4 photosynthesis (Table 6.1). Over the past 40 years the total area sown to sorghum has changed little, while that sown to maize has increased by *c.* 40% (Fig. 6.1a). In the same period, average grain yields have risen linearly from less than 2 t ha^{-1} to 4 t ha^{-1} for maize. By contrast, average sorghum yields showed a modest increase, from 1 t ha^{-1} in 1960 to about 1.4 t ha^{-1} in 1980, and no increase over the last 20 years (Fig. 6.1b). Globally, the lower yields for sorghum may reflect that the crop is commonly grown in semi-arid areas where maize cannot be grown, particularly in the regions bordering the Sahara (Purseglove, 1972; Maiti, 1996). The USA and Canada account for more of the global maize production than any other area, while sub-Saharan Africa accounts for the largest proportion of sorghum production (Fig. 6.2). Because yields vary tremendously among regions, total production is a poor indication of the area sown to the crop in a region. For example, 30 Mha were sown to maize in the USA and Canada in 1998 compared with a slightly smaller 24 Mha in sub-Saharan Africa (FAO, 1998). However, the total production in the USA and Canada was more than eight times that of sub-Saharan Africa (Fig. 6.2).

Table 6.1. Total global production of indicated C_3 and C_4 crops and the areas that they occupied in 1997 (FAO, 1998).

Crop	Area (10^6 km^2)	Total production (Mt)	Average yield (t ha^{-1})
Wheat (C_3)	2.292	613.6	2.68
Rice (C_3)	1.498	579.7	3.87
Maize (C_4)	1.423	588.0	4.13
Sorghum (C_4)	0.441	61.5	1.40
Millets (C_4)	0.281	37.9	0.74
Quinoa (C_4)	0.007	0.052	0.77

6.1.1 Maize climatic requirements

The oldest records of maize use are from the Tehuacan valley of Mexico, where the oldest harvested cobs are dated at 7200 years before present (BP). There is no convincing evidence of maize in the Old World before 1492. At this time cultivation of maize occurred throughout the Americas, except where it was too cold. Following introduction into the Old World, its cultivation spread widely between 50° N and 40° S. The bulk of the crop is grown in the warmer temperate regions and humid subtropics. It is not suited to hot semi-arid climates, or to tropical rainforest climates (Purseglove, 1972). The optimum temperature for germination is 18–21°C; germination is very slow below 13°C and fails at about 10°C. Selection of more low-temperature-tolerant lines has allowed viable cultivation of grain crops above 50° N in northern France and of silage crops to *c.* 55° N in Denmark, the Netherlands and England (Miedema *et al.*, 1987; Long, 1999). Flowering is temperature and day-length dependent, and so cultivars and landraces from low latitudes show delayed flowering when moved northwards (Purseglove, 1972).

6.1.2 Sorghum climatic requirements

Sorghum was probably domesticated in Ethiopia about 5000 BP. It spread to West Africa and later into Asia, but only reached the New World with the slave trade. Sorghum is perhaps of special importance to world food security. It produces useful yields of grain under conditions within tropical climates where most other cereals would fail. It tolerates hotter and drier conditions than the other major cereals and is generally now limited in Africa to areas too hot and dry for maize (Doggett, 1988). The optimum temperature for growth is about 30°C. It is killed by frost and is even less tolerant of chilling temperatures than maize (Long, 1999). Because both maize and sorghum are short-day species, tropical cultivars may fail to flower in temperate regions (Doggett, 1988).

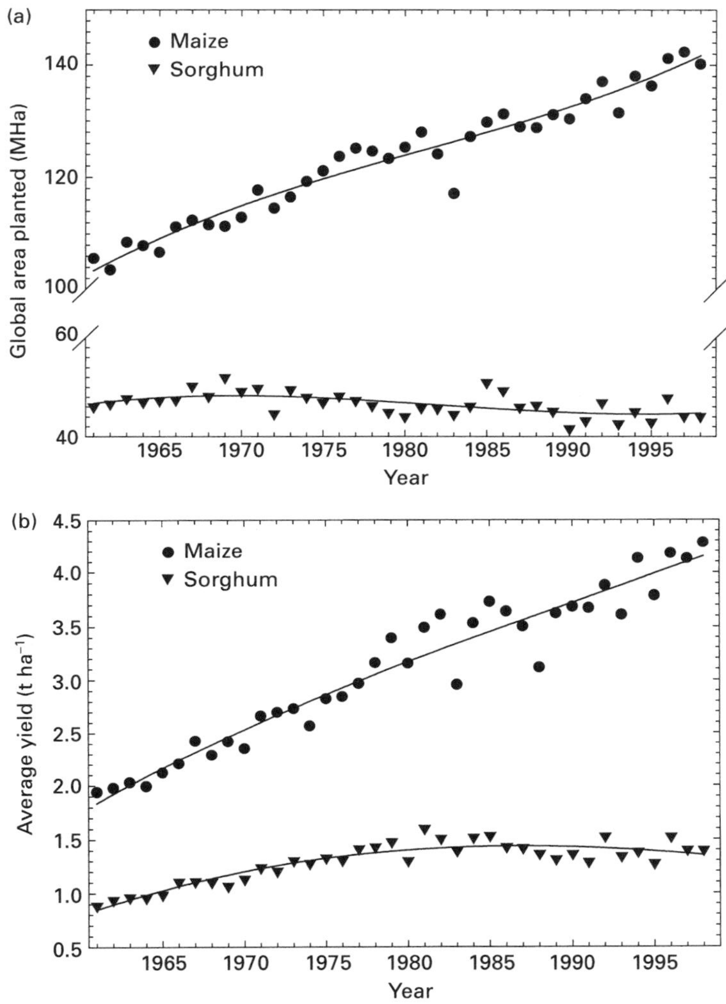

Fig. 6.1. Changes in (a) the global area planted and (b) average global yields for maize and sorghum over the 1961–1998 period. (FAO, 1998.)

6.2 Photosynthesis and Respiration

Maize and sorghum differ from the other arable crops reviewed in this book because they use the C_4 photosynthetic pathway. This pathway confers potentially more efficient use of CO_2, solar radiation, water and N in photosynthesis relative to C_3 crops. The possession of C_4 photosynthesis explains why these crops differ substantially from many C_3 crops in their photosynthetic

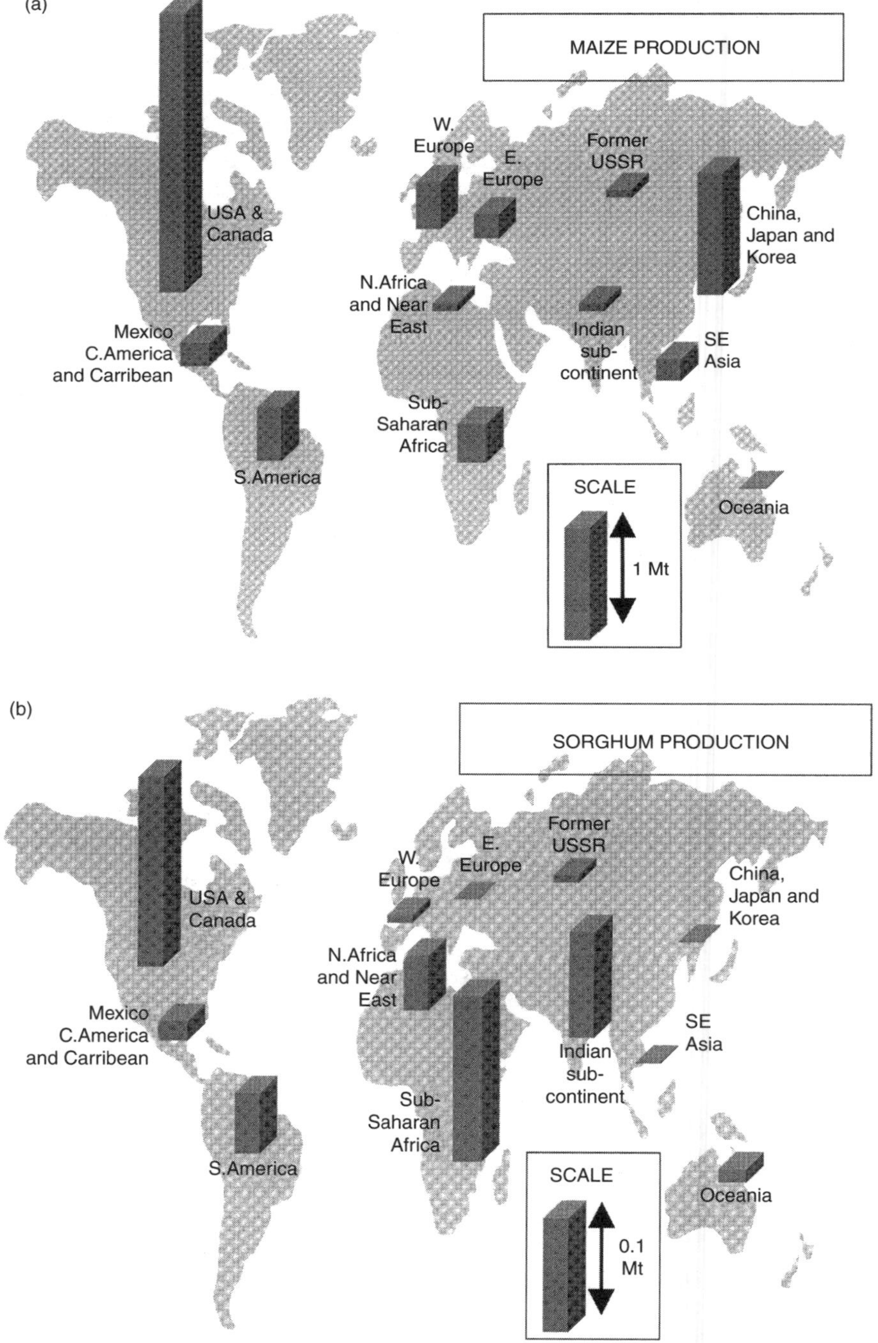

Fig. 6.2. Total production of (a) maize and (b) sorghum by global region in 1997. (FAO, 1998.)

and production responses to CO_2, solar radiation, water and N deficits (Brown, 1999; Long, 1999).

6.2.1 Carbon dioxide

The primary physiological effect of the unique combination of photosynthetic metabolism and leaf anatomy that characterizes C_4 plants (including maize and sorghum) is elevation of the CO_2 concentration at the site of Rubisco in the bundle sheath (Hatch, 1992). The C_4 dicarboxylate cycle in the these plants serves to concentrate CO_2 from an atmospheric partial pressure of CO_2 (pCO_2) of about 35 Pa at sea level to an equivalent pCO_2 of about 700 Pa (*c.* 20 ×) at the site of RubP carboxylase-oxygenase (Rubisco) in the bundle sheath. This elevated concentration has two effects. Firstly, competitive inhibition of the oxygenase reaction of Rubisco eliminates most of the C_2-photosynthetic oxidative or photorespiratory pathway (PCO) activity and its expenditure of energy associated with photorespiration. Secondly, it allows Rubisco to approach its maximum rate of catalysis despite its low affinity for CO_2. Phosphoenolpyruvate carboxylase (PEPc) from C_4 photosynthetic tissues has a high affinity for CO_2, such that photosynthesis is saturated at a low pCO_2 of typically 15–20 Pa (Long, 1999).

The pCO_2 of the atmosphere is rising by 0.4–0.5% per year, and should approximately double by the year 2100 (Houghton *et al.*, 1996). While this increase is expected to increase leaf photosynthesis of C_3 species by *c.* 58% in the absence of limitation on rooting volume, no direct effect of this increase in pCO_2 is expected in C_4 species (Drake *et al.*, 1997). Nitrogen deficiency and drought stress can increase leaking of CO_2 from the bundle sheath cells of C_4 leaves, thus possibly allowing a direct response of photosynthesis to elevated pCO_2 (Long, 1999). Some long-term field studies have measured increased photosynthesis of C_4 grasses, including sorghum and maize, under elevated pCO_2 (Samarakoon and Gifford, 1996). However, a complication is that elevated pCO_2 lowers stomatal conductance (g_s) in C_4 as in C_3 species. Thus, C_4 plants grown at elevated pCO_2 show improved water status and this in turn will allow increased rates of CO_2 assimilation whenever there is any shortage of water. Particularly significant is the observation that leaf and canopy photosynthesis have not been affected by doubled pCO_2 over 10 years in the C_4 grass *Spartina patens* growing on a marsh in Maryland, USA. The absence of any significant effect on photosynthesis in this environment is consistent with the hypothesis that the apparent response of photosynthesis of C_4 species to elevated pCO_2 in other environments is an indirect response to decreased transpiration (Long, 1999).

There are several reports that mitochondrial respiration is partially and significantly inhibited, on average by *c.* 20%, with a doubling of pCO_2 above the current ambient level (Drake *et al.*, 1997). The basis of this decrease is uncertain, with several potential sites of CO_2 effect suggested. Both C_3 and C_4 species show this response (Drake *et al.*, 1999).

6.2.2 Solar radiation

One of the first physiological features noted of C_4 plants, following the discovery of their unique biochemistry in 1965/66, was their high rate of photosynthesis at full sunlight under tropical conditions (Hatch, 1992). By avoiding photorespiration, C_4 species have potentially higher rates of net photosynthesis in full sunlight. Although additional energy is required to assimilate CO_2 via the C_4 pathway, this becomes irrelevant at light-saturation since light will by definition be in excess of requirements. In dim light when photosynthesis is linearly dependent on the photon flux, the rate of CO_2 assimilation depends entirely on the energy requirements of carbon assimilation (Long *et al.*, 1993). The additional two molecules of ATP required for assimilation of one molecule of CO_2 in photosynthesis in sorghum and maize, compared with C_3 plants, increases their photon requirement (Hatch, 1992). However, in C_3 species at 30°C, the amount of light energy diverted into photorespiration in photosynthesis will considerably exceed the additional energy required for CO_2 assimilation in C_4 photosynthesis.

The energy expended in photorespiration as a proportion of photosynthesis rises with temperature. At 25°C and below, the energy required for the net assimilation of one CO_2 molecule is higher in C_4 than in C_3 photosynthesis, but above 25°C the situation is reversed (Ehleringer and Monson, 1993). Thus, under warm conditions, the efficiency of photosynthetic CO_2 uptake should always be higher in maize and sorghum compared with a similar C_3 crop canopy. This difference is apparent in the efficiency of light use at the level of production (Monteith, 1978).

6.2.3 Water deficits

Because C_4 photosynthesis is saturated with CO_2 at partial pressures well below the current ambient level, some stomatal closure can occur without any effect on assimilation (A). In maize, following witholding of water, there was a progressive decrease in g_s, evaporation (E) and leaf intercellular pCO_2 (p_i) with little decrease in A until p_i was approximately half that of the controls. Further analysis suggested that decreased A resulted from reduced physical conductance of CO_2 during drought stress (Lal and Edwards, 1996).

In theory, because C_4 plants lack photorespiration, they could be more prone to photoinhibition of photosynthesis when drought-induced closure of the stomata prevents CO_2 uptake and the absence of photorespiration prevents internal cycling of CO_2. In practice, significant internal cycling of CO_2 in droughted leaves of maize is implied in the study by Lal and Edwards (1996). Even though sorghum subjected to drought did show increased photoinhibition of photosynthesis in sunlight compared with controls, there was no significant contribution of this photoinhibition to dry matter production (Ludlow and Powles, 1988).

6.2.4 Nutrient deficits

As a direct result of CO_2 concentration at the site of Rubisco in C_4 species, their theoretical requirement for N in photosynthesis is lower than that in C_3 species. Furbank and Hatch (1987) calculated a CO_2 concentration at the site of Rubisco in C_4 species of 10–100 times that found in C_3 species. At these CO_2 concentrations a C_4 leaf at 30°C would require 13.4–19.8% of the amount of Rubisco in a C_3 leaf to achieve the same A_{sat} (Long, 1999). The decreased requirement for N in Rubisco will be partially offset by the requirement of N for the enzymes of the photosynthetic C_4 dicarboxylate cycle, in particular PEPc. However, because of its tenfold higher maximum catalytic rates (k_{cat}) of carboxylation, much smaller concentrations of PEPc are required, relative to Rubisco. Together, Rubisco and PEPc in C_4 species constitute less than half the amount of N invested in Rubisco in C_3 species (Sage *et al.*, 1987).

Lower leaf N and higher leaf photosynthetic rates of C_4 species result in a photosynthetic N use efficiency (PNUE) which is about twice that of C_3 species (Long, 1999). Most comparisons of NUE have been undertaken in well-fertilized conditions. On N-deficient soils, NUE of C_4 crops can be twice that of C_3 (Brown, 1999). By decreasing the requirement for Rubisco in C_3 species, rising atmospheric pCO_2 may erode the PNUE advantage of maize and sorghum (Hocking and Meyer, 1991).

6.2.5 Temperature

In much of the developing world sorghum has been replaced by maize, except in areas that are too hot, dry or infertile for present maize cultivars. Consequently both crops are grown at temperatures closer to their upper limit. Progressive temperature increases associated with climatic change increase the risk of temperature stress for these crops. C_4 photosynthesis is more tolerant of high temperature than is C_3, due to the absence of photorespiration, which increases rapidly with temperature. However, C_4 photosynthetic efficiency declines with temperature above *c.* 35° C, with some inactivation at 40°C and above (Maiti, 1996). In much of the developed world, growth is at higher latitudes and these risks are much reduced. At the highest latitudes, elevated temperature may increase photosynthetic efficiency during cooler periods, extending the range and economic viability of these crops in temperate climates (Wittwer, 1995).

6.2.6 Ozone and UV-B radiation

Ozone and other atmospheric pollutants gain access to and damage the photosynthetic apparatus following their entry into the intercellular air space via the stomata. Because C_4 species have inherently lower stomatal conductances than equivalent C_3 species, they will show a lower rate of pollutant uptake. This explains why they are generally among the more pollutant-tolerant plants, such that the effect of the same level of ozone on C_4 maize is about half the effect on C_3 wheat (Rudorff *et al.*, 1996).

While C_4 crops may be more resistant to increases in tropospheric ozone, a decline in the stratospheric ozone layer and the concomitant rise in surface UV-B will affect their photosynthesis. Sorghum grown in the field under supplemental levels of UV-B, simulating a 20% reduction in the stratospheric ozone column, showed significant decreases in photosynthetic CO_2 uptake that were apparently linked to increased stomatal resistance. Enhanced UV-B also caused reductions in chlorophyll and carotenoid pigments and caused increases in UV-B absorbing pigments and peroxidase activity after 60 days of exposure (Ambasht and Agrawal, 1998). These changes suggest that damage was accompanied by partial acclimation. Similar reductions in photosynthetic CO_2 uptake in maize have been observed with similar UV-B doses (Mark and Tevini, 1997).

6.3 Water Use

In theory, C_4 species should require less water per gram of carbon assimilation than equivalent C_3 species within the same environment. In leaves, water vapour passes from the mesophyll cell wall surface through the internal air space, stomata, leaf and canopy boundary layers to the bulk atmosphere through the same pathway as the photosynthetic pCO_2 influx. Therefore, both gases are subject to the same diffusive limitations.

The primary carboxylase in maize and sorghum, PEPc, is pCO_2-saturated by a low p_i, of typically 10–15 Pa (Ehleringer and Monson, 1993). Reductions in stomatal or boundary layer conductance cannot affect A until they are sufficient to decrease p_i below 15 Pa. In contrast, because of the relatively low affinity of Rubisco for CO_2, A is not saturated in C_3 species until p_i approaches 100 Pa and shows very large decreases with decrease in p_i. For example, calculating from the A/p_i responses described for wheat by Evans (1989), a decrease in p_i from 36 Pa to 15 Pa would decrease A by 68% if Rubisco was limiting photosynthesis. In a C_4 species this reduction in p_i would have no effect on A.

Stomatal closure in drought will cause an initial decrease in p_i. Any decrease in p_i below the current atmospheric pCO_2 of 36 Pa will depress A in C_3 species but will not affect A in C_4 species until p_i is 0.4 of the external concentration. Species with C_4-type photosynthesis can therefore impose a considerable resistance on gas-phase diffusion and transpiratory water loss without loss of photosynthetic capacity, while any increase in gas-phase diffusion resistance in C_3 plants is likely to decrease A. Stomatal behaviour in C_4 species appears to have evolved in response to this property of C_4 photosynthesis since their conductance is typically about 50% of that in C_3 species within the same environment (Long, 1985).

Water use efficiency is used to describe water use at both the leaf (WUE_l) and canopy (WUE_c) levels. At the leaf level it is typically defined as the ratio of pCO_2 assimilated per unit of water transpired. At the crop level, it is commonly defined as the amount of dry matter produced per unit of water lost in evapotranspiration from the same area of ground. In a given environment, leaf water use efficiency is directly proportional to the gradient of external to intercellular pCO_2 ($p_a - p_i$) that a leaf can maintain for a given net rate of pCO_2 assimilation. Typically, given a similar leaf to air water vapour pressure deficit, $p_a - p_i$ in a C_4 leaf at light saturation will be *c.* 20 Pa, compared with about 10 Pa in a C_3 leaf (Long, 1985). This results in C_4 crops having twice the WUE_l of C_3 crops.

The following studies indicate that the theoretical and observed higher WUE_l of C_4 over C_3 species is translated into improved WUE_c. For grasses grown at 30°C in the same environment, WUE_c was *c.* 1.2×10^{-3} for C_3 species but *c.* 3.6×10^{-3} for C_4 species (Downes, 1969; Ehleringer and Monson, 1993; Brown, 1999).

6.3.1 Carbon dioxide

As in C_3 species, the stomata of maize and sorghum respond to rising pCO_2 with decreased stomatal aperture, thus allowing a significant improvement in WUE_l. In a comparison between soybean (C_3) and sorghum (C_4) at twice-ambient pCO_2, Dugas *et al.* (1997) found that evapotranspiration (ET) decreased by 43% in soybean and by 31% in sorghum under elevated pCO_2. Kimball and Idso (1983) calculated a 45% decrease in ET for maize grown at twice ambient pCO_2. Samarakoon and Gifford (1996) found that under continually wet soil conditions the ET of maize was on average 29% lower for plants grown at twice ambient as opposed to ambient pCO_2. Witholding water resulted in a 30% slower rate of soil drying, such that the soil was wet longer in the elevated pCO_2 treatment. This allowed assimilation and production to remain higher at elevated pCO_2.

6.3.2 Solar radiation

Solar radiation, directly or indirectly, provides the energy needed to drive ET. At a constant pCO_2, ET in maize and sorghum is linearly related to irradiance but decreases strongly with increasing pCO_2 (Louwerse, 1980). Future climate change scenarios suggest significant regional changes in cloudiness and therefore growing season solar radiation. Brown and Rosenberg (1997) assessed the impacts of this potential change in a simulation model. They predicted that for a 15% increase in solar radiation and constant pCO_2, transpiration rates will increase ET by 2% for rain-fed maize, 4% for irrigated maize and 1% for rain-fed sorghum; these changes are reversed when solar radiation is decreased by 15%. However, the strength or direction of these changes can be modified considerably when the interactive effects of elevated pCO_2 and rising temperature, together with corresponding changes in precipitation and water vapour pressure, are considered. For example, considering a possible future scenario for rain-fed maize when temperature is increased by 3°C, pCO_2

elevated to 55 Pa and precipitation decreased by 15%, the net result is a 7% decline in transpiration.

6.3.3 Temperature

For both C_3 and C_4 crops, ET increases with increasing temperature. The driving force for transpiration is the gradient in water vapour concentration between the air spaces within the leaf and the surrounding air (see Chapter 14, this volume). Because the amount of water needed to saturate a body of air rises exponentially with temperature, increased air temperature commonly results in an increase in water-vapour pressure deficit (VPD) and the potential for transpiration, and a decrease in WUE_c. In C_4 plants, the higher carboxylation efficiency of PEPc in comparison with Rubisco facilitates the maintenance of a much higher diffusive gradient for pCO_2 between the atmosphere and the site of carboxylation. This enables C_4 plants to maintain optimal p_i at a far lower g_s than in C_3 plants, so that the rate of increase of ET with increasing temperature is much reduced. Also, elevated temperature reduces the crop growth period and yield for both maize and sorghum, decreasing crop water use.

6.3.4 Water deficits

Sorghum is by far the more drought-resistant of the two crops, due to several morphological and physiological properties (Purseglove, 1972). They are:

1. The plant above ground grows slowly until the root system is established.
2. It produces twice as many secondary roots as maize.
3. Silica deposits in the endodermis of the root prevent tissue collapse during drought.
4. Leaf area is about half that of equivalent maize crops.
5. Leaves have a thicker cuticle and they in-roll completely in drought.
6. Evapotranspiration from sorghum is about half that of maize.
7. Sorghum requires about 20% less water per unit mass of dry matter gain.
8. The plant can remain dormant during drought and resume growth when favourable conditions return.

Sorghum leaves subjected to wilting for 1 week recover rapidly after watering, with normal diurnal stomatal rhythm being restored in 5 days. By contrast, in maize normal stomatal function is permanently impaired, with no restoration of normal diurnal patterns (Doggett, 1988).

Where the crop is dependent on stored soil water, deeper root systems result in greater yield. Here sorghum has a particular advantage, by penetrating the soil faster and to greater depths. When grown on the same field and irrigated prior to sowing in Yemen, maize roots grew to about 1 m depth whilst sorghum roots penetrated to more than 2 m, thus allowing extraction of substantially more water (Squire, 1990). In such water-limited environments, the higher WUE_c of C_4 crops would be critical to yield. The WUE of maize and sorghum was about double that of C_3 crops such as cotton and beans grown at the same sites.

Maize is also more susceptible than sorghum to salt stress, with leaf water potential declining more rapidly with increasing external osmolality (Nagy *et al.*, 1995).

Sorghum genotypes vary considerably in their response to drought stress. Sorghum genotypes exhibiting the glossy leaf trait show higher WUE and better growth under drought-stressed conditions than did non-glossy types (Maiti, 1996). Higher root growth under both heat and saline stress is suggested as a possible mechanism of resistance. In drought-resistant sorghum lines, leaf-rolling may reduce the effective leaf area of the uppermost leaves by about 75% (Matthews *et al.*, 1990).

6.3.5 Waterlogging

If climatic change involves increased incidence of high rainfall events, then increased tolerance of periodic waterlogging will be important for stable crop yields. Sorghum growth and development are more tolerant of waterlogging than in maize, though tolerance varies considerably among cultivars (Doggett, 1988). In trials throughout lowland eastern Africa, sorghum outyielded maize not only at sites where rainfall was less than 380 mm but also at sites where rainfall exceeded 750 mm. Greater tolerance of transient waterlogging probably explains the latter (Doggett, 1988).

6.3.6 Nutrient deficits

Where elevated temperature is accompanied by reduced precipitation then soil dehydration is likely, though this may to some extent be mitigated by increased WUE_c resulting from rising pCO_2. Decreasing soil water potential initiates stomatal closure that results in reduced plant transpiration, bulk flow of solutes to the roots and subsequently mineral nutrient uptake. However, as dry matter production also declines in these conditions, this decreased nutrient supply with decreased water use may have little effect on dry matter composition (Kramer and Boyer, 1995). Application of N during drought stress is likely to prove less effective or even ineffective unless accompanied by irrigation. For example, in trials of a hybrid grain sorghum, N-fertilizer addition had no significant effect on grain yield in unirrigated crops, but was beneficial under mild water stress conditions, i.e. on plants that had received a single irrigation 60 days after sowing (Khannachopra and Kumari, 1995).

Compared with maize, sorghum N-uptake efficiency is higher, particularly when soil N is low; therefore it is less affected by drought stress (Lemaire *et al.*, 1996), perhaps reflecting the larger and deeper root system of this crop. Thus, while the production potential of maize is greater than that of sorghum, realized yields of sorghum are likely to be higher under conditions of combined drought and nutrient stress.

6.4 Growth and Development

6.4.1 Carbon dioxide

As noted previously, there is uncertainty as to whether or not there is any direct effect of elevated pCO_2 on photosynthetic carbon gain in C_4 plants, but considerable evidence of a direct effect on respiration (Drake *et al.*, 1999). There is still debate as to whether or not elevated CO_2 directly enhances plant growth in C_4 species. Increased growth under elevated pCO_2 has been reported for a number of C_4 species, including maize and sorghum (Wong, 1979; Poorter, 1993; Amthor *et al.*, 1994). However, others found no effect (Hocking and Meyer, 1991; Ziska *et al.*, 1991; Ellis *et al.*, 1995), or an effect only when increased pCO_2 was accompanied by drought stress (Samarakoon and Gifford, 1996).

Samarakoon and Gifford (1996) found a significant positive growth response in maize to twice-ambient pCO_2 but only when the plants were drought-stressed. Plants grown in greenhouses in soil drying from field capacity, and at elevated as opposed to ambient CO_2, accumulated 35% more leaf area and 50% more dry matter. Enhanced growth responses became apparent early in the drying cycle when the soil was still close to field capacity and when drying was confined to the topsoil. In contrast, well-watered plants showed no significant growth response to elevated pCO_2 (Table 6.2). Soil water content of well-watered plants was just under field capacity throughout the experiment. These results indicate a pronounced sensitivity of maize growth to elevated pCO_2 in soil water contents well above wilting point. Samarakoon and Gifford (1996) concluded that reports of maize growth responses to elevated pCO_2 might have involved unrecognized minor water deficits. However, Wong (1979) reported a 20% increase in dry matter production over 30 days in maize grown at twice-ambient pCO_2 and to which water was added at 3 h intervals to balance transpiration losses. Thus, if drought stress is involved, the response is to very subtle changes in crop water status.

Elevated pCO_2 may be expected to increase the rate of development of C_4 crops indirectly by increasing plant temperature through decreased transpiration and latent heat transfer. However, where elevated pCO_2 has allowed a crop to conserve soil moisture for a longer time, ET and latent heat loss eventually may become greater by comparison with controls that have exhausted extractable soil moisture. In practice, growth under elevated CO_2 has been found to delay plant development and affect plant morphogenesis in sorghum. For two sorghum genotypes, a Sudanese landrace IS 22365 and US cv. RS 610, an increase in pCO_2 from 21 to 72 Pa increased the period from sowing to panicle initiation by about 20 days. This could be disastrous to sorghum production in dry years and delay provision of food at a time of year

Table 6.2. Average net assimilation rate (NAR), leaf area ratio (LAR) and relative growth rate (RGR) for maize grown at ambient and twice-ambient pCO_2 in wet soil (moisture content held close to field capacity) and in dry soil (allowed to dry from field capacity). Drying commenced after germination. Growth rates are for 26–37 days after germination. (Samarakoon and Gifford, 1996.)

	NAR (g m^{-2} day^{-1})	LAR (m^2 kg^{-1})	RGR (mg g^{-1} day^{-1})
Wet soil			
Ambient CO_2	13.4	10.8	135.8
Twice ambient CO_2	13.8	10.4	133.8
Change	+3%	−4%	−1%
Dry soil			
Ambient CO_2	5.1	10.2	50.6
Twice ambient CO_2	7.9	9.6	73.3
Change	+56%	−6%	+45%

when food reserves may be very low (Ellis *et al.*, 1995). Elevated pCO_2 did result in increased plant height, leaf number and apical extension in the same crops. This negative effect of elevated pCO_2 on development may be more than offset by the anticipated concomitant temperature increases (see below).

6.4.2 Solar radiation

Both species flower in response to short days, but some lines are daylength-insensitive. Landraces from low latitudes show delayed flowering when moved to high latitudes. However, there is an interaction with temperature, such that days from sowing to flowering (f) may be described by:

$$1/f = a + bT + cP,$$

where a, b and c are genotype-specific constants, T is mean temperature and P is photoperiod (h day^{-1}) (Ellis *et al.* 1990). In both species, adaptation of the crop to higher latitudes has therefore depended on the development of short-season cultivars through decreases in b and c. Rising temperature will allow the use of longer-season cultivars at higher latitudes than at present, not only because of a longer growing season but also because of decreased photoperiodic requirement at higher temperatures.

In C_3 crops, light may be saturated at values well under those experienced in full sunlight. In C_4 species, light levels are generally not saturated even in full sunlight, so they are likely to be more affected by changes in irradiance at moderate to high light levels. In the absence of moisture, temperature and nutrient limitation, dry matter production is closely linked to incident radiation once canopy closure has been achieved. Dry matter accumulation increases linearly with accumulated solar radiation. This may appear surprising since the response of photosynthesis of individual leaves to increasing light is hyperbolic. However, in sorghum and maize canopies, much of the leaf surface is inclined. This both decreases the light flux per unit surface area on the upper leaves and allows light to penetrate to lower leaves in the canopy. As a result, canopy photosynthetic CO_2 uptake responds linearly to radiation up to full sunlight (Long, 1985). Monteith (1978) reviewed maximum amounts of dry matter formed per unit of total solar radiation intercepted for a range of healthy crops growing under optimum conditions. Several C_3 species, including wheat, potato (*Solanum tuberosum* L.) and beet (*Beta vulgaris* L.), showed a maximum of 1.4 $\mu g\,J^{-1}$, compared with 2.0 $\mu g\,J^{-1}$ for C_4 species including maize, sorghum, elephant grass (*Pennisetum purpureum* Schum.) and sugar cane (*Saccharum officinarum* L.). Assuming that photosynthetically active radiation is 50% of the total and an average energy content for plant dry mass of 17 MJ kg^{-1}, this equates to efficiencies of conversion (ε_c) of photosynthetically active radiation into biomass of 0.070 (C_4) and 0.049 (C_3). The rate of dry matter production both within the tropics and in temperate zones during the summer can approach the assumed maximum for C_4 species of about 2.0 $\mu g\,J^{-1}$ (Monteith, 1978; Long *et al.*, 1990; Brown, 1999). As noted earlier, growth of these C_4 crops will therefore be particularly sensitive to changes in total radiation that result from regional changes in cloud cover.

6.4.3 Temperature

Temperature affects all stages of plant development, i.e. germination, emergence, vegetative growth, flowering and grain-fill. Three cardinal temperatures for these stages of development of each crop will be referred to: base (T_b), below which development ceases; optimum (T_o); and ceiling (T_c), above which development ceases. All three temperatures are higher in sorghum and maize than in C_3 cereals, with sorghum showing significantly higher values than maize. Considerable breeding effort has focused on extending the range of both crops into colder climates by selecting for lower T_b and T_o (Miedema *et al.*, 1987; Greaves, 1996). In temperate areas, where temperatures for maize and sorghum are often suboptimal, rising temperature with global change will result in increased rates of germination and emergence, accelerated development and increased viability for both crops (Ellis *et al.*, 1990). At lower temperate latitudes and in the tropics, global warming may result in temperatures that are supra-optimal for development, resulting in reduced germination and emergence, accelerated development and reduced viability, particularly for maize.

Minimum and optimum germination temperatures for maize are generally lower than for sorghum, though considerable variation exists among cultivars of both species. Germination rates for 13 sorghum hybrids increased with increasing temperature from a T_b of 15.5 to a T_o of 26.5–38°C depending on cultivar. As temperature increased above 15.5°C, average time to germination decreased significantly, with 80% emergence requiring 7 days at 15.5°C and just 1 day at 37.5°C. Above 38°C, germination declines, with a T_c in excess of 48°C (Kasalu *et al.*, 1993; Brar and Stewart, 1994; Maiti, 1996). However, some lines selected for Europe have a T_b as low as 10°C (Anda and Pinter, 1994).

The cardinal temperatures for maize seed germination, even in tropical lines, are about 3–10°C below those of sorghum. For Kenyan cultivars, Itabari *et al.* (1993) reported 6.1°C T_b, 33.6 T_o and 42.9°C T_c. In more moist conditions T_b is closer to 10°C, with 13°C required for a high percentage germination (Purseglove, 1972).

In suboptimal climates, phenological development of both species is closely linked to the product of time and temperature above a basal temperature for growth and development. In maize, the rate of development increases roughly linearly from a base temperature of 10°C to about 30°C (Stewart *et al.*, 1998). Therefore the rate of development may be assessed in thermal time or growing degree-days (θ) with units of °Cd, such that:

$$\theta = \sum_{i=1}^{n} (T_i - T_b)$$

where i is the ith day from sowing, T_i is the mean temperature for that day, and n is the number of days in the growing season (Ellis *et al.*, 1990).

For maize in North America, approximately 2100 °Cd, 2400 °Cd, and 2800 °Cd are required for the maturation of short-, medium- and long-season hybrids (Roth and Yocum, 1997). Essentially, just under 4 months are required at an average temperature of 20°C to mature a crop. Approximately 400 °C d of additional heat units are required to mature a sorghum crop (Maiti, 1996). Except at high altitudes, these requirements are easily met within the tropics. However, increased temperature in the tropics may have detrimental effects on the crop. As temperature rises, the rate of grain-fill in cereals increases and the duration of grain-fill decreases until a point is reached at which the rate of filling fails to compensate for the decrease in duration. In maize, mature kernel dry weight was maximal at 22°C and declined linearly with increase in temperature to 36°C, where mass was 45% lower. This decline was attributed to a premature decline in the activities of ADP glucose pyrophosphorylase and starch synthase, two of the key enzymes of starch synthesis in the developing kernel (Singletary *et al.*, 1994). Rising temperature within the tropics will increase the probability of these crops experiencing damaging high temperatures. Here the ability of sorghum to acclimate to high temperatures within a few days, via the production of heat shock proteins, may be critical (Howarth and Skot, 1994) and may explain the viability of sorghum at temperatures that would cause failure of crops of the other major cereals.

6.4.4 Water deficits

Developmental events particularly sensitive to water availability in the tropics are seedling emergence, the initiation and duration of the rapid descent of the root system, and the determination and filling of the grains. In relation to these critical stages of growth, the timing of rainfall is critical, and alteration of this timing with climate change could be as damaging to crop viability as change in the annual total.

There is considerable variation among sorghum genotypes in their ability to cope with drought in the tropics. For some, development is accelerated or unaffected, while for others it is delayed or suspended (Squire, 1990). The former strategy may allow the crop to avoid the worst of the drought, while the latter may allow the crop to resume growth after sufficient subsequent rainfall.

Drought stress reduces seedling germination and emergence (Anda and Pinter, 1994). After emergence, plants respond to rapid dehydration by reducing stomatal conductance, so restricting water loss. Where dehydration occurs over a longer period, developmental changes may be initiated. Reduced leaf turgor rapidly inhibits leaf expansion, while photosynthesis is less affected. This reduced leaf growth, with near-normal photoassimilate production, results in increased assimilate supply to the roots and increased root development. The rate at which the leading roots penetrate the soil is slow if the surface soil layer is repeatedly rewetted. If the surface layer dries, then the roots will descend rapidly, at about 30 to 40 mm day^{-1} (Squire, 1990). The reproductive stages of both maize and sorghum are particularly sensitive to water stress (Boyer, 1992; Craufurd and Peacock, 1993). Drought stress delays anthesis and maturation, thereby increasing crop duration (Donatelli *et al.*,1992; Khannachopra and Kumari, 1995). In maize, 100% reproductive failure may result even from brief periods of drought stress at a critical phase in development. This failure is not a direct result of dehydration, but possibly is due to failure in the supply of assimilate, since infusing the stem with a growth medium restored reproduction. Infusion of water alone did not restore reproduction (Boyer, 1992). Drought stress during reproductive growth in grain sorghum resulted in reduced grain yield, with yield reductions greatest (up to 87%) when stress was imposed during booting and late flowering. However, no reduction in yield resulted when stress was only applied in the vegetative stage (Craufurd and Peacock, 1993).

After flowering, yield depends on the number of grains set and the rate and duration of grain-fill (Maiti, 1996). In cereals, the potential number of grains is set early in development, when the reproductive apex forms, and reflects assimilate supply at that time. If conditions change (e.g. drought) the capacity of the photosynthetic tissue may be insufficient to provide assimilate needed to fill the grain. In these circumstances, yield can be very small and its resilience depends on the ability of the plant to retranslocate reserves. In many maize and some sorghum genotypes, no retranslocation occurs; however, in other sorghum cultivars, some 40% of dry matter can be relocated (Garrity *et al.*, 1983).

6.5 Yield

Current global yields are presented in Table 6.1. To predict how these will change with global atmospheric and climatic change, we are limited almost entirely to model estimates. These are predominantly empirical and depend on predicting beyond our experience, particularly with respect to interactions for which few experiments have been conducted. Because of the difficulty of containing pCO_2, experimentation has been limited to glasshouses and small open-top chambers. Whilst these provide an indication of direction of response and potential interactions, they are not suited to predicting absolute changes (McLeod and Long, 1999). Predicting crop yields in the future is rather like betting on a horse in a race, with all its uncertainties and potential surprises and with the added handicap of only having seen the horse move in its box and never on the race course. Morison and Lawlor (1999) have recently highlighted how many of the limited studies of the interactive effects of rising pCO_2 and temperature on crop production and yield have shown changes in production that contradict model expectations. This highlights the need for increased understanding of the mechanisms underlying these interactions in controlled environments and advancing systems for realistic field evaluation at multiple sites.

Technology exists to examine how crops will respond to elevated pCO_2 in the open air via free air carbon dioxide enrichment (FACE) experiments (McLeod and Long, 1999), or how elevated pCO_2 and temperature will interact via advanced climate-controlled glasshouses or gradient tunnels (Morison and Lawlor, 1999). The expense of such facilities seems to have deterred the development of any significant network to adapt and test the models. This will be essential for providing adequate guidance for managing future world food supplies and to start selecting or developing genotypes for these changed conditions. Regional climatic responses to a doubling of atmospheric pCO_2 will vary significantly (Watson *et al*., 1998). Crop yields are likely to be even more variable and difficult to predict, because they depend on complex interactions among plant physiological processes, weather and the soil environment, which can be subject to extensive agronomic modification (Brown and Rosenberg, 1997). These interactive processes determine plant photosynthesis and respiration, water use, growth and phenology, all of which impact plant yield. Reported enhanced growth under elevated pCO_2 in C_4 species (see above) has generally been attributed to increased WUE. However, there are reports of photosynthetic stimulation of C_4 under elevated pCO_2, although this enhancement may be limited to conditions where elevated pCO_2 alleviates mild drought stress (Samarakoon and Gifford, 1996). Overall predictions of maize and sorghum yields suggest a reduction in areas where temperatures are currently optimal (i.e. southern and central USA and Southern China, and much of sub-Saharan Africa) and an increase in areas where temperatures are suboptimal (i.e. northern USA, the European Union and northern China).

Models generally assume non-limiting water and nutrients and no change in agronomic practices (Houghton *et al*., 1996; Wang and Erda, 1996; Brown and Rosenberg, 1997; Buan *et al*., 1996). Elevated pCO_2 is generally assumed to have little or no effect on yield. This assumption seems increasingly flawed, particularly for rain-fed crops (Samarakoon and Gifford, 1996).

Brown and Rosenberg (1997) simulated crop yield with the Erosion Productivity Impact Calculator (EPIC) for five representative farms in the central USA. The individual and combined effects of a range of environmental and physiological factors were used to predict yields of rain-fed sorghum in Nebraska, wheat in Kansas, maize in Missouri and Iowa, and of irrigated maize in Nebraska. Single factor effects on crop yield (Fig. 6.3) varied between irrigated and rain-fed maize and rain-fed sorghum. In all cases, increase in temperature decreased yield by shortening the duration of growth, whilst increase in pCO_2 increased predicted yield. Reductions in solar radiation, humidity and (with the exception of irrigated maize) precipitation resulted in reduced crop yields, the magnitude of these reductions varying between crops.

Though modelling provides useful insight into the potential effects of a range of single factors on yield, actual yields will be determined by interaction among these and other variables. Brown and Rosenberg (1997) modelled interactive effects in 16 cases, which included elevation of pCO_2 to 45 Pa (C450) and 55 Pa (C550) (Fig. 6.3). In isolation, elevated pCO_2 increased yield of rain-fed crops, but when this was combined with the other expected changes in climate the net result was a decrease in yield of 2–35% for Missouri rain-fed maize and 3–15% for Nebraska irrigated maize. Increased yields of 6–16% for Iowa maize and of 23–37% for Nebraska sorghum were predicted for scenarios which included increased pCO_2 and precipitation together with reduced solar radiation. Generally, for all five crops examined, reductions in yield resulting from increased temperature were to some extent mitigated by elevated pCO_2 and increased precipitation. Mitigation was most pronounced in sorghum. Yield reductions in this region may be smaller if adaptation of cultivation is taken into account. Simulation studies predicted that changes in agronomic practices such as earlier planting of longer-season cultivars and moisture conservation measures could offset some of the yield losses induced by climatic change in maize, in the region of Missouri, Iowa, Nebraska and Kansas (Easterling *et al.*,1992).

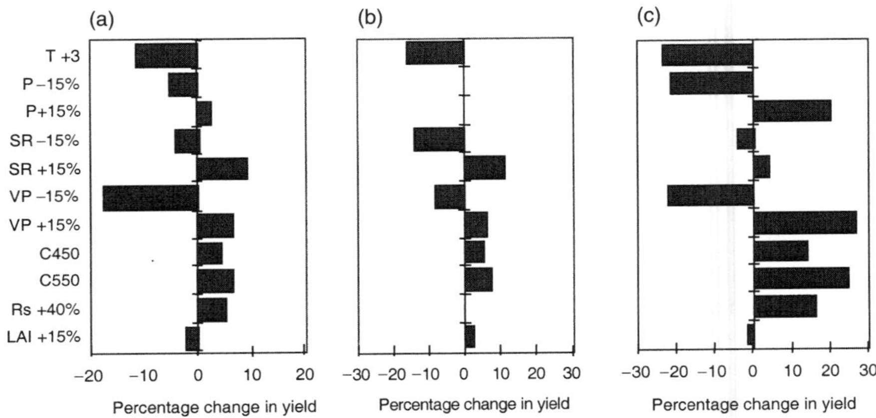

Fig. 6.3. Percentage alteration of yield from the baseline climate of 1951 to 1980 to changes in individual climatic variables: (a) Missouri rain-fed maize; (b) Nebraska irrigated maize; (c) Nebraska rain-fed sorghum. T = temperature + 3°C; P = precipitation; SR = solar radiation; VP = vapour pressure; C = CO_2 (Pa × 10); Rs = stomatal resistance; LAI = leaf area index. (Brown and Rosenberg, 1997.)

Wang and Erda (1996) considered the impact of climatic change and variability on simulated maize production in China. The CERES-Maize model was run for 35 sites that were representative of the main maize-growing regions. Simulated yields at most sites were reduced, primarily as a result of the reduced growth and grain-filling periods. In a few northern sites, simulated yields increased because maize growth at these higher latitudes is currently temperature-limited. Similar increases have been predicted for the northern edge of the North American corn belt. Singh *et al.* (1998) suggested that a twice-normal CO_2 climate would increase maize and sorghum yields in Quebec by 20% and decrease wheat and soybean yields by 20–30%. Wang and Erda (1996) noted the critical point that these models made assumptions that were likely to result in an overestimation of yield. This identifies once more the need for models that correctly simulate the effect of climate factors and soil on crop responses, including yield. They conclude that at present these model results should not be regarded as predictions, but rather as plausible assessments of the potential direction of climatic change-induced changes in maize production.

Although rice is the principal crop throughout most of tropical Asia, maize is an important secondary crop in many countries, particularly the Philippines, where the yield is about half that of rice (Watson *et al.*, 1998). The vulnerability of rice and corn to climatic change in the Philippines was assessed by applying the predicted climate change for a doubling of atmospheric pCO_2 from four general circulation models (GCMs) (Table 6.3) (Buan *et al.*, 1996). The effects

Table 6.3. Predicted change in yield of two corn hybrids for selected climatic change scenarios for three study sites (ISU, Isabela State University; CMU, Central Mindanao University; USM, University of Southern Mindanao) in the Philippines, using the CERES-Corn model. GCM models: CCCM = Canadian Climate Centre Model; GFDL = Geophysical Fluid Dynamics Laboratory model; GISS = Goddard Institute for Space Studies model; UKMO = United Kingdom Meteorological Office model. (Buan *et al.*, 1996.)

Study sites	Corn varieties	Cropping season	Yield base ($t\ ha^{-1}$)	Percentage changes in yield			
				CCCM	GFDL	GISS	UKMO
ISU	P3228	1st	6.7	−11.1	−8.6	−13.1	−5.8
		2nd	5.4	−3.7	0.4	−11.2	−3.0
	SWEET	1st	5.6	−10.5	−8.9	−11.4	−6.8
		2nd	4.9	−7.9	−4.7	−17.7	−9.2
CMU	P3228	1st	9.5	−11.4	−8.4	−13.1	−12.7
		2nd	8.0	−1.4	4.7	−7.2	2.8
	SWEET	1st	8.5	−11.7	−9.0	−12.3	−12.4
		2nd	7.4	−8.4	−1.8	−15.1	−2.0
USM	P3228	1st	7.1	−15.3	−17.9	−17.8	−14.6
		2nd	6.9	−16.1	−16.1	−17.3	−8.1
	SWEET	1st	6.0	−15.9	−22.9	−18.4	−15.9
		2nd	5.6	−18.2	−18.7	−18.2	−6.6
MEAN			6.8	−11.0	−9.3	−14.4	−7.8

of climatic change on crop productivity were simulated using the CERES-Corn model from Decision Support System for Agrotechnology Transfer version 3 (DSSAT 3). All GCMs indicated the same trends in temperature, but varied in their estimates for rainfall and solar radiation. The Canadian Climate Centre Model predicted a slight decrease in solar radiation, while other models predicted slight increases. Estimates for changes in rainfall were much more varied. For example, for the first cropping at Isabela State University, the Geophysical Fluid Dynamics Laboratory model predicted a 24.6% increase, whilst the Goddard Institute for Space Studies model predicted a 27.6% decrease. Despite these differences, a reduction in maize yield was predicted for all climate change scenarios (Table 6.3). This reduction results largely from a temperature-induced reduction in the growth period. As rainfall is generally high, a reduction in rainfall of 10% would not affect the water requirement significantly; however, an increase by the same magnitude could affect yield significantly through the effects of flooding. This could be exacerbated by decreased transpiration at elevated $p\mathrm{CO}_2$.

6.6 Conclusions and Future Research Directions

As global temperatures rise, both maize and sorghum will become increasingly viable at locations that currently lack adequate heat. However, as cultivation moves to higher latitudes and altitudes, these crops will still face the low temperature limitations of today at the current edges of their range. Poor germination, low-temperature-dependent photoinhibition of photosynthesis and impaired water and nutrient uptake by seedlings will remain limitations. The ability of *Miscanthus* × *giganteus*, a highly productive C_4 relative of sugar cane, to grow rapidly in the cool temperate climate of southern England without the low temperature damage observed in maize shows that there is no inherent limitation to C_4 plants in cold climates (Long, 1999).

Effective and rapid screens for low-temperature photoinhibition and germination at low temperature have been developed and are already being applied in the adaptation of both crops to colder climates. In addition, further adaptation or removal of the photoperiod requirement will be needed as temperature increases allow cultivation at higher latitudes. Rapid advances in methods for adapting the maize crop to cool temperate conditions have been achieved over the past two decades. Genetic transformation to enhance protection against active-oxygen radicals offers the potential for further improvements. Coupled with global climatic change, this could allow the crop to extend into yet higher latitudes.

Far more pressing, in terms of global food supply of grains for humans, will be adaptation of the crops to the warmer and drier conditions that may occur in areas of the tropics. Increased evaporative demand and increased incidence of supra-optimal or lethal temperatures may make larger areas of the semi-arid tropics unsuitable for maize cultivation, perhaps necessitating a return to lower yielding but more tolerant sorghum landraces. Sorghum is perhaps a neglected resource in considering the impacts of global climate change within the tropics. Its ability to maintain viable yields on low-fertility soils, survive transient drought and acclimate to very high temperatures may be critical to ensuring food security in the tropics under an uncertain future. Great variability exists within the sorghum germplasm. Critical to the future will be accelerating the description, preservation and utilization of this variation. Much recent effort has been placed on maximizing production; however, more important throughout the developing world may be ensuring future yield stability.

Doggett (1970) foresaw that sorghum grain yields of 20 t ha^{-1} 'should be attainable'. Thirty years on, average sorghum yields per unit of ground area worldwide have increased little and shown no improvement over the past 20 years (Fig. 6.2b).

References

Ambasht, N.K. and Agrawal, M. (1998) Physiological and biochemical responses of *Sorghum vulgare* plants to supplemental ultraviolet-B radiation. *Canadian Journal of Botany – Revue Canadienne De Botanique* 76, 1290–1294.

Amthor, J.S., Mitchell, R.J., Runion, G.B., Rogers, H.H., Prior, S.A. and Wood, C.W. (1994) Energy content, construction cost and phytomass accumulation of *Glycine max* (L) Merr and *Sorghum bicolor* (L) Moench grown in elevated CO_2 in the field. *New Phytologist* 128, 443–450.

Anda, A. and Pinter, L. (1994) Sorghum germination and development as influenced by soil temperature and water-content. *Agronomy Journal* 86, 621–624.

Boyer, J.S. (1992) Mechanisms for obtaining water use efficiency and drought resistance. In: Stalker, H.T. and Murphy, J.P. (eds) *Plant Breeding in the 1990s.* CAB International, Wallingford, UK, 539 pp.

Brar, G.S. and Stewart, B.A. (1994) Germination under controlled temperature and field emergence of 13 sorghum cultivars. *Crop Science* 34, 1336–1340.

Brown, R.H. (1999) Agronomic implications of C_4 photosynthesis. In: Sage, R.F. and Monson, R.K. (eds) *C_4 Plant Biology.* Academic Press, San Diego, pp. 473–507.

Brown, R.A. and Rosenberg, N.J. (1997) Sensitivity of crop yield and water use to change in a range of climatic factors and CO_2 concentrations: a simulation study applying EPIC to the central USA. *Agricultural and Forest Meteorology* 83, 171–203.

Buan, R.D., Maglinao, A.R., Evangelista, P.P. and Pajuelas, B.G. (1996) Vulnerability of rice and corn to climate-change in the Philippines. *Water, Air and Soil Pollution* 92, 41–51.

Craufurd, P.Q. and Peacock, J.M. (1993) Effect of heat and drought stress on sorghum (*Sorghum bicolor*). 2. Grain yield. *Experimental Agriculture* 29, 77–86.

Doggett, H. (1970) *Sorghum.* Longman, London, 403 pp.

Doggett, H. (1988) *Sorghum*, 2nd edn. Longman, Harlow, UK, 512 pp.

Donatelli, M., Hammer, G.L. and Vanderlip, R.L. (1992) Genotype and water limitation effects on phenology, growth, and transpiration efficiency in grain-sorghum. *Crop Science* 32, 781–786.

Downes, R.W. (1969) Differences in transpiration rate between tropical and temperate grasses under controlled conditions. *Planta* 88, 261–273.

Drake, B., Gonzalez-Meler, M. and Long, S.P. (1997) More efficient plants: a consequence of rising atmospheric CO_2. *Annual Review of Plant Physiology and Plant Molecular Biology* 48, 609–639.

Drake, B.G., Azcon-Bieto, J., Berry, J., Bunce, J., Dijkstra, P., Farrar, J., Gifford, R.M., Gonzalez-Meler, M.A., Koch, G., Lambers, H., Siedow, J. and Wullschleger, S. (1999) Does elevated CO_2 concentration inhibit mitochondrial respiration in green plants? *Plant, Cell and Environment* 22, 649–657.

Dugas, W.A., Prior, S.A. and Rogers, H.H. (1997) Transpiration from sorghum and soybean growing under ambient and elevated CO_2 concentrations. *Agriculture and Forest Meteorology* 83, 37–48

Easterling, W.E., Crosson, P.R., Rosenberg, N.J., McKenney, M.S., Katz, L.A. and Lemon, K.M. (1992) Agricultural impacts of and responses to climate-change in the Missouri–Iowa–Nebraska–Kansas (MINK) region. *Climatic Change* 24, 23–61.

Ehleringer, J.R. and Monson, R.K. (1993) Evolutionary and ecological aspects of photosynthetic pathway variation. *Annual Review of Ecology and Systematics* 24, 411–439.

Ellis, R.H., Hadley, P., Roberts, E.H. and Summerfield, R.J. (1990) Relations between temperature and crop development. In: Jackson, M., Ford-Lloyd, B.V. and Parry, M.L. (eds) *Climatic Change and Plant Genetic Resources.* Belhaven Press, London, pp. 85–115.

Ellis, R.H., Craufurd, P.Q., Summerfield, R.J. and Roberts, E.H. (1995) Linear relationship between carbon dioxide concentration and rate of development towards flowering in sorghum, cowpea and soyabean. *Annals of Botany* 75, 193–198.

Evans, J.R. (1989) Photosynthesis and the nitrogen relationship in leaves of C_3 plants. *Oecologia* 78, 9–19.

FAO (1998) *FAO Production Yearbook*, Vol. 52. Food and Agriculture Organization of the United Nations, Rome.

Furbank, R.T. and Hatch, M.D. (1987) Mechanism of C_4 photosynthesis: the size and composition of the inorganic carbon pool in bundle sheath cells. *Plant Physiology* 85, 958–964.

Garrity, D.P., Sullivan, C.Y. and Watts, D.G. (1983) Moisture deficit and grain sorghum performance: drought stress conditions. *Agronomy Journal* 75, 997–1004.

Greaves, J.A. (1996) Improving suboptimal temperature tolerance in maize – the search for variation. *Journal of Experimental Botany* 47, 307–323.

Hatch, M.D. (1992) C_4 photosynthesis – evolution, key features, function, advantages. *Photosynthesis Research* 34, 81–81.

Hocking, P.J. and Meyer, C.P. (1991) Effects of CO_2 enrichment and nitrogen stress on growth, and partitioning of dry-matter and nitrogen in wheat and maize. *Australian Journal of Plant Physiology* 18, 339–356.

Houghton, J.T., Meira Filho, L.G., Callander, B.A., Harris, N., Kattenberg, A. and Maskell, K. (eds) (1996) *Climate Change 1995: The Science of Climate Change.* Contribution of Working Group I to the Second Assessment Report of the Intergovernmental Panel of Climate Change. Cambridge University Press, Cambridge, UK, 572 pp.

Howarth, C.J. and Skot, K.P. (1994) Detailed characterization of heat-shock protein-synthesis and induced thermotolerance in seedlings of *Sorghum bicolor* L. *Journal of Experimental Botany* 45, 1353–1363.

Howell, T.A., Steiner, J.L., Schneider, A.D., Evett, S.R. and Tolk, J.A. (1997) Seasonal and maximum daily evapotranspiration of irrigated winter wheat, sorghum, and corn – Southern High Plains. *Transactions of the American Society of Agricultural Engineers* 40, 623–624.

Itabari, J.K., Gregory, P.J. and Jones, R.K. (1993) Effects of temperature, soil-water status and depth of planting on germination and emergence of maize (*Zea mays*) adapted to semiarid eastern Kenya. *Experimental Agriculture* 29, 351–364.

Kasalu, H., Mason, S.C. and Ejeta, G. (1993) Effect of temperature on germination and seedling emergence of grain-sorghum genotypes. *Tropical Agriculture* 70, 368–371.

Khannachopra, R. and Kumari, S. (1995) Influence of various amounts of irrigation water and nitrogen fertiliser on growth, yield and water-use of grain sorghum. *Journal of Agronomy and Crop Science – Zeitschrift für Acker und Pflanzenbau* 174, 151–161.

Kimball, B.A. and Idso, S.B. (1983) Increasing atmospheric CO_2-effects on crop yield, water-use and climate. *Agricultural Water Management* 7, 55–72.

Kramer, P.J. and Boyer, J.S. (1995) *Water Relations of Plants and Soils.* Academic Press, San Diego, 495 pp.

Lal, A. and Edwards, G.E. (1996) Analysis of inhibition of photosynthesis under water stress in the C_4 species *Amaranthus cruentus* and *Zea mays*: electron transport, CO_2 fixation and carboxylation capacity. *Australian Journal of Plant Physiology* 23, 403–412.

Lemaire, G., Charrier, X. and Hebert, Y. (1996) Nitrogen uptake capacities of maize and sorghum crops in different nitrogen and water-supply conditions. *Agronomie* 16, 231–246.

Long, S.P. (1983) C_4 photosynthesis at low temperatures. *Plant, Cell and Environment* 6, 345–363.

Long, S.P. (1985) Leaf gas exchange. In: Baker, J. and Baker, N.R. (eds) *Photosynthetic Mechanisms and the Environment.* Elsevier, Amsterdam, pp.453–499.

Long, S.P. (1999) Environmental responses. In: Sage, R.F. and Monson, R.K. (eds) *C_4 Plant Biology.* Academic Press, San Diego, pp. 215–250.

Long, S.P., Nugawela, A., Bongi, G. and Farage, P.K. (1987) Chilling dependent photoinhibition of photosynthetic CO_2 uptake. In: Biggins, J., (ed.), *Progress in Photosynthesis Research,* Vol. 4. Martinus Nijhof, Dordrecht, pp. 131–138.

Long, S.P., Farage, P.K., Aguilera, C. and Macharia, J.M.N. (1990) Damage to photosynthesis during chilling and freezing, and its significance to the photosynthetic productivity of field crops. In: Barber, J., Guerrero, M.G. and Medrano, H. (eds) *Trends in Photosynthesis Research.* Intercept, Andover, UK, pp. 344–356.

Long, S.P., Postl, W.F. and Bolharnordenkampf, H.R. (1993) Quantum yields for uptake of carbon-dioxide in C_3 vascular plants of contrasting habitats and taxonomic groupings. *Planta* 189, 226–234.

Long, S.P., Humphries, S. and Falkowski, P.G. (1994) Photoinhibition of photosynthesis in nature. *Annual Review of Plant Physiology and Plant Molecular Biology* 45, 633–662.

Louwerse, W. (1980) Effects of CO_2 concentration and irradiance on the stomatal behaviour of maize, barley and sunflower plants in the field. *Plant, Cell and Environment* 3, 391–398.

Ludlow, M.M. and Powles, S.B. (1988) Effects of photoinhibition induced by water-stress on growth and yield of grain-sorghum. *Australian Journal of Plant Physiology* 15, 179–194.

Maiti, R. (1996) *Sorghum Science.* Science Publishers, Lebanon, New Hampshire, 352 pp.

Makadho, J.M. (1996) Potential effects of climate-change on corn production in Zimbabwe. *Climate Research* 6, 147–151.

Mark, U. and Tevini, M. (1997) Effects of solar ultraviolet-B radiation, temperature and CO_2 on growth and physiology of sunflower and maize seedlings. *Plant Ecology* 128, 224–234.

Matthews, R.B., Azamali, S.N. and Peacock, J.M. (1990) Response of 4 sorghum lines to mid-season drought. 2. Leaf characteristics. *Field Crops Research* 25, 297–308.

McLeod, A.R. and Long, S.P. (1999) Free-air carbon dioxide enrichment (FACE) in global change research: a review. *Advances in Ecological Research* 28, 1–55.

Miedema, P., Post, J. and Groot, P. (1987) *The Effects of Low Temperature on Seedling Growth of Maize Genotypes.* Pudoc, Wageningen, The Netherlands.

Monteith, J.L. (1978) A reassessment of maximum growth rates for C_3 and C_4 crops. *Experimental Agriculture* 14, 1–5.

Morison, J.I.L. and Lawlor, D.W. (1999) Interactions between increasing CO_2 concentration and temperature on plant growth. *Plant, Cell and Environment* 22, 659–682.

Nagy, Z., Tuba, Z., Zsoldos, F. and Erdei, L. (1995) CO_2-exchange and water relation responses of sorghum and maize during water and salt stress. *Journal of Plant Physiology* 145, 539–544.

Poorter, H. (1993) Interspecific variation in the growth response of plants to an elevated ambient CO_2 concentration. *Vegetatio* 104/105, 77–97.

Purseglove, J.W. (1972) *Tropical Crops. Monocotyledons 1.* Longman, London, 334 pp.

Roth, G.W. and Yocum, J.O. (1997) Use of hybrid growing degree day ratings for corn in the northeastern USA. *Journal of Production Agriculture* 10, 283–288.

Rudorff, B.F.T., Mulchi, C.L., Lee, E.H., Rowland, R. and Pausch, R. (1996) Effects of enhanced O_3 and CO_2 enrichment on plant characteristics in wheat and corn. *Environmental Pollution* 94, 53–60.

Sage, R.F. and Pearcy, R.W. (1987) The nitrogen use efficiency of C_3 and C_4 plants. 2. Leaf nitrogen effects on the gas-exchange characteristics of *Chenopodium album* (L) and *Amaranthus retroflexus* (L). *Plant Physiology* 84, 959–963.

Sage, R.F., Pearcy, R.W. and Seemann, J.R. (1987) The nitrogen use efficiency of C3 and C4 plants. 3. Leaf nitrogen effects on the activity of carboxylating enzymes in *Chenopodium album* (L) and *Amaranthus retroflexus* (L). *Plant Physiology* 85, 355–359.

Samarakoon, A.B. and Gifford, R.M. (1996) Elevated CO_2 effects on water-use and growth of maize in wet and drying. *Australian Journal of Plant Physiology* 23, 53–62.

Singh, B., Elmaayar, M., Andre, P., Bryant, C.R. and Thouez, J.P. (1998) Impacts of a GHG-induced climate change on crop yields: effects of acceleration in maturation, moisture stress and optimal temperature. *Climatic Change* 38, 51–86.

Singletary, G.W., Banisadr, R. and Keeling, P.L. (1994) Heat-stress during grain filling in maize – effects on carbohydrate storage and metabolism. *Australian Journal of Plant Physiology* 21, 829–841.

Squire, G.R. (1990) Effects of changes in climate and physiology around the dry limits of agriculture in the tropics. In: Jackson, M., Ford-Lloyd, B.V. and Parry, M.L. (eds) *Climatic Change and Plant Genetic Resources.* Belhaven Press, London, pp. 116–147.

Stewart, D.W., Dwyer, L.M. and Carrigan, L.L. (1998) Phenological temperature response of maize. *Agronomy Journal* 90, 73–79.

Wang, J. and Erda, L. (1996) The impact of potential climate change and climate variability on simulated maize production in China. *Water, Air and Soil Pollution.* 92, 75–85.

Watson, R.T., Zinyowera, C.M., Moss, R.H. and Dokken, D.K. (eds) (1998) *The Regional Impact of Climate Change: an Assessment of Vulnerability.* Cambridge University Press, Cambridge, UK, 517 pp.

Wittwer, H.S. (1995) *Food, Climate and Carbon Dioxide: the Global Environment and World Food Production.* Lewis Publishers, New York, 236 pp.

Wong, S.C. (1979) Elevated atmospheric partial pressure of CO_2 and plant growth. *Oecologia* 44, 68–74.

Ziska, L.H., Hogan, K.P., Smith, A.P. and Drake, B.G. (1991) Growth and photosynthetic response of nine tropical species with long-term exposure to elevated carbon dioxide. *Oecologia* 86, 383–389.

7 Crop Ecosystem Responses to Climatic Change: Soybean

LEON HARTWELL ALLEN, JR[1] AND KENNETH J. BOOTE[2]

[1]*US Department of Agriculture, Agricultural Research Service, Crop Genetics and Environment Research Unit, IFAS Building No. 350, SW 23rd Street, University of Florida, Gainesville, FL 32611-0965, USA;* [2]*Agronomy Department, Institute of Food and Agricultural Sciences, University of Florida, Gainesville, FL 32611-0500, USA*

7.1 Introduction

Grain legumes are one of the primary sources of protein for humans and their animals. Grain legumes are basically the pulses, or the edible seeds of various leguminous crops, including beans, peas and lentils. Harlan (1992) provided a list of the most important cultivated plants grouped according to probable origin. Table 7.1 is a list of pulses (with scientific names) adapted from Harlan's inventory. Based on the United Nations Food and Agricultural Organization (FAO) Production Yearbooks, Harlan (1992) also provided a list of the world's 30 most important food crops ranked according to estimated dry matter of the edible product.

Production of soybean (*Glycine max* [L.] Merrill) was about 88 million metric tonnes (Mt) and ranked number five worldwide among the 30 leading food crops. Other pulses – bean, peanut and pea – ranked numbers 16, 17 and 18, respectively, with worldwide production of 14, 13 and 12 million Mt, respectively (Harlan, 1992). Although total dry matter production of the four major cereal grain crops – wheat (*Triticum aestivum* L.), maize (*Zea mays* L.), rice (*Oryza sativa* L.) and barley (*Hordeum vulgare* L.) – exceeds that of soybean (the leading pulse) the grain legumes are important because of their higher proteins and oils, which contribute to human and animal diets. Furthermore, the biochemical energy costs in producing storage protein and oils rather than carbohydrate results in greater cost per gram of seed produced and lower biomass of seed yield per hectare. However, the caloric energy per unit mass of oils is higher than that of stored carbohydrates, and the nutritional value of protein is greater.

The main focus of this chapter will be on soybean. Some of the information should be applicable to other grain legumes, such as groundnut (*Arachis hypogaea* L.), common bean (*Phaseolus vulgaris* L.), cowpea (*Vigna unguiculata* L.) and other pulses.

Table 7.1. Pulse crops (scientific and common names) grouped according to probable geographic origin. (Adapted from Harlan, 1992.)

Species names	Common name
The Near Eastern Complex (west of India)	
Cicer arietinum Linn.	Chickpea
Lathyrus sativus Linn.	Grasspea
Lens esculenta Moench.	Lentil
Lupinus albus Linn.	Lupin
Pisum sativum Linn.	Garden pea
Vicia ervilia Willd.	Bittervetch
Vicia faba Linn.	Broadbean
Africa	
Kirstingiella geocarpa Harms	Kirsting's groundnut
Lablab niger Medik.	Hyacinth bean
Vigna unguiculata [L.]	Cowpea
Voandzeia subterranea [L.] Thouars	Bambara groundnut
Chinese Region	
Glycine max [L.] Merrill	Soybean
Stizolobium hassjoo Piper et Tracy	Velvet bean
Vigna angularis [Willd.] Ohwi	Adzuki bean, Red bean
India and Southeast Asia	
Cajanus cajun [L.] Millsp.	Pigeonpea
Canavalia gladiata [Jacq.] DC	Jackbean
Cyamopsis tetragonolobus [L.] DC	Guar
Dolichos biflorus Linn.	Hyacinth bean
Psophocarpus tetragonolobus [L.] DC	Winged bean; New Guinea
Vigna aconitifolia	Mat bean
Vigna calcarata [Roxb.] Kurz	Rice bean
Vigna mungo [L.] Hepper	Urd, black gram
Vigna radiata [L.] Wilczek	Mung bean
The Americas	
Arachis hypogaea Linn.	Groundnut
Canavalia ensiformis [L.] DC	Sword bean
Canavalia plagiosperma Piper	Jackbean
Igna feuillei DC	Pacae
Lupinus mutabilis Sweet	Chocho
Phaseolus acutifolius A. Gray	Tepary bean
Phaseolus coccineus Linn.	Scarlet runner bean
Phaseolus lunatus Linn.	Lima bean and Seiva bean
Phaseolus vulgaris Linn.	Common bean

Soybean is produced on all of the populated continents. This grain legume is generally quite sensitive to photoperiod and it flowers in response to shortening of the dark period. The crop requires 100–150 days from sowing to maturity, although (as will be shown later) the duration of stages of development is temperature-dependent. As for most crops, except rice, soybean production requires aerobic soil conditions. Soybean can be produced over the mean daily air temperature range of 20–30°C, but low night-time temperatures (less than 12°C) and high daytime temperatures (greater than 36°C) can limit production seriously.

The scope of this review will be limited, for the most part, to studies that have been conducted over the whole life cycle of soybean plants grown in near-natural irradiance. These conditions have the most relevance for understanding crop yield responses to be expected from anticipated climatic changes.

7.2 Photosynthesis and Respiration

7.2.1 Carbon dioxide

(a) Photosynthesis

Probably more data have been collected worldwide on photosynthesis and respiration of soybean than of any other crop, certainly of any other grain legume. Cure (1985) provided a detailed review of responses of ten important crops worldwide – wheat, maize, rice, barley, soybean, sorghum (*Sorghum bicolor* [L.] Moench), lucerne (*Medicago sativa* L.), cotton (*Gossypium hirsutum* L.), potato (*Solanum tuberosum* L.) and sweet potato (*Ipomoea batatas* [L.] Lam.) – to a doubling of carbon dioxide concentration ([CO_2]). Many of the data sets had few observations from even fewer experiments. Only those responses calculated from three or more experiments will be considered in this chapter. Based on this criterion for elimination of responses, soybean ranked highest among all crops in percentage response to doubled [CO_2] for short-term CO_2 exchange rate (CER), (+70 ± 20%), acclimated CER (+42 ± 10%), initial net assimilation rate (NAR), (+35 ± 6%) and long-term NAR (+23 ± 5%). Cotton led in biomass accumulation and yield.

Based on earlier reports (Kramer, 1981), much attention was brought to bear on the acclimation of photosynthesis to elevated [CO_2] (i.e. photosynthetic inhibition, loss of photosynthetic capacity, down-regulation of photosynthesis) in certain plants (Oechel and Strain, 1985; Sasek *et al.*, 1985; Tissue and Oechel, 1987). However, much of the work on soybean photosynthesis suggests that soybean does not lose photosynthetic capacity with long-term exposure to elevated [CO_2].

Soybean photosynthetic rates increase with increasing [CO_2] at both the leaf (Valle *et al.*, 1985a) and canopy levels (Acock *et al.*, 1985; Jones *et al.*, 1985a). Valle *et al.* (1985a) exposed soybean leaflets grown at 330 μmol CO_2 mol^{-1} or 660 μmol CO_2 mol^{-1} to short-term concentrations ranging from 84 to 890 μmol CO_2 mol^{-1} under photosynthetic photon flux densities (PPFD) of 1100–1300 μmol (photon) m^{-2} s^{-1}. They fitted leaflet photosynthetic responses to [CO_2] using the following non-linear rectangular hyperbola:

$$CER = [(P_{maxc} \times CO_2)/(CO_2 + K_c)] + R_c, \qquad \text{(Eqn 7.1)}$$

where CER = photosynthetic CO_2 exchange rate (μmol CO_2 m^{-2} s^{-1}; CO_2 = carbon dioxide concentration (μmol CO_2 mol^{-1}); K_c = apparent Michaelis–Menten constant for CO_2; P_{maxc} = asymptotic maximum photosynthetic rate (μmol CO_2 m^{-2} s^{-1}); and R_c = the *y*-intercept parameter (μmol CO_2 m^{-2} s^{-1}). The parameter R_c can be regarded as the leaf respiration rate in high light at zero [CO_2].

The values of the parameters P_{maxc}, K_c, R_c, and the [CO_2] compensation point, Γ_c, which are derived from the CER response to [CO_2] of two leaves grown at 330 and two leaves grown at 660 μmol CO_2 mol^{-1}, are given in Table 7.2. Leakage from the leaf chambers was not a factor in these experiments because the leaf CERs were measured with the leaf chambers mounted inside the respective plant growth chambers. Nevertheless, the parameters might have been slightly different if exposures to [CO_2] had been extended to concentrations of at least 1200 μmol CO_2 mol^{-1}.

There was some evidence that elevated [CO_2] suppressed respiration rates (presumable photorespiration). The average Γ_c and R_c were 63.0 and −7.8 μmol CO_2 m^{-2} s^{-1}, respectively, for the leaves exposed to 330 μmol CO_2 mol^{-1}, but only 42.4 and −4.6 μmol CO_2 m^{-2} s^{-1}, respectively, for leaves exposed to 660 μmol CO_2 mol^{-1}.

On this individual leaf basis, there is no indication of a down-regulation of photosynthesis in soybean. In fact, Allen *et al.* (1990b) reported that soybean individual-leaf CER vs. [CO_2] was linear over the range of 300–800 μmol CO_2 mol^{-1} when plotted at PAR (photosynthetically active radiation) of 300, 600, 900 or 1500 μmol (photon) m^{-2} s^{-1}. When plotted on an intercellular [CO_2] basis, Campbell *et al.* (1988) reported that soybean leaflet CER grown at 660 μmol CO_2 mol^{-1} was about twofold that of CER of a leaflet grown at 330 μmol CO_2 mol^{-1}. Also, ribulose-1,5-bisphosphate carboxylase-oxygenase (Rubisco) protein of soybean was a rather constant fraction of leaf soluble protein (55.2 ± 1.3%) from leaves grown at [CO_2] ranging from 160 to 990 μmol mol^{-1}. Furthermore, Rubisco activity was nearly constant at 1.0 μmol CO_2 min^{-1} mg^{-1} protein across this range of [CO_2], and leaf-soluble protein (g m^{-2}) was almost constant (Campbell *et al.*, 1988). However, since specific leaf weight (SLW; leaf laminae weight/leaf area) increased from 20.3 to 30.5 g m^{-2} with [CO_2] increasing from 160 to 990 μmol mol^{-1}, the leaf soluble

Table 7.2. Mean asymptotic maximum photosynthetic rate (P_{maxc}) with respect to *y*-intercept parameter (R_c), apparent Michaelis–Menten constant for CO_2 (K_c), and CO_2 compensation point (Γ_c) for soybean leaves grown at two CO_2 concentrations and subjected to different short-term CO_2 exposures across the range of 84–890 μmol mol^{-1}. (Adapted from Valle *et al.*, 1985a.)

CO_2 concentration (μmol mol^{-1})	P_{maxc} (μmol m^{-2} s^{-1})	R_c (μmol m^{-2} s^{-1})	K_c (μmol mol^{-1})	Γ_c (μmol mol^{-1})
330[a]	52	−7.8	359	63
660	127	−4.6	1133	42

[a]Mean values within each CO_2 level of P_{maxc}, K_c and Γ_c were significantly different at $P = 0.05$ as tested by a *t*-test.

protein (expressed as mg g^{-1}, dry weight) decreased, and Rubisco activity (expressed as µmol CO_2 min^{-1} g^{-1}, dry weight), decreased concomitantly. Soybean leaf photosynthetic rates at elevated [CO_2] were probably maintained higher because of an increase of one layer of palisade cells in the leaves (Thomas and Harvey, 1983; Vu *et al.*, 1989). This modification of leaf structure provided more surface area of cells within the leaf with a similar amount of Rubisco per unit leaf area, as well as possibly providing better light distribution within the leaf.

An extensive set of data of soybean leaf photosynthesis versus [CO_2] was obtained by Harley *et al.* (1985), and included responses to temperature and PPFD. Leaf photosynthetic rates increased almost linearly with increasing intercellular [CO_2] up to 600–700 µmol mol^{-1}, and then increased slowly toward an asymptotic level above that concentration. Allen *et al.* (1990b) reported that soybean leaf photosynthetic rates increased linearly with external leaf [CO_2] to 800 µmol mol^{-1}, which is consistent with the data of Harley *et al.* (1985).

Increased photosynthetic rates per unit leaf area were also reflected in increased photosynthetic rates per unit ground area of soybean plant canopies (Jones *et al.*, 1984, 1985a,b,c). Jones *et al.* (1984) found that midday canopy CER was about 60 µmol m^{-2} s^{-1} for soybean grown at 330 µmol CO_2 mol^{-1}, but was about 50% higher (90 µmol m^{-2} s^{-1}) at 800 µmol CO_2 mol^{-1}. Similar results were found in other experiments. Likewise, the daytime total CO_2 exchange averaged 0.92 and 1.72 mol CO_2 m^{-2} for the 330 and 800 µmol CO_2 mol^{-1} treatments, respectively, for data collected during 8 clear days. Cumulative daytime CER of soybean for the season was 53.0 and 84.7 mol m^{-2} for canopies grown at 330 and 660 µmol CO_2 mol^{-1}, respectively (Jones *et al.*, 1985c).

Allen *et al.* (1987) used a non-linear model similar to equation 7.1 to express whole-canopy relative photosynthetic rates as a function of [CO_2]. This same formulation was also used to express relative biomass yield and relative seed yield. Since all photosynthetic rate data were expressed relative to ambient rates of soybean growing at 330 or 340 µmol CO_2 mol^{-1}, this equation was reformatted as:

$$R = [(R_{max} \times CO_2)/(CO_2 + K_m)] + R_{int}, \qquad \text{(Eqn 7.1a)}$$

where R = the relative photosynthetic response (or relative biomass or seed yield response, as discussed in the later section on crop yield) to CO_2; CO_2 = carbon dioxide concentration; R_{max} = the asymptotic relative response limit of $(R - R_{int})$ at high CO_2 concentration; R_{int} = the intercept on the y-axis; and K_m = the value of CO_2 concentration at which $(R - R_{int}) = 0.5\ R_{max}$. The model parameters for midday relative photosynthetic rate data shown in Allen *et al.* (1987) are R_{max} = 3.08, K_m = 279 µmol CO_2 mol^{-1}, and R_{int} = −0.68. From equation 7.1a, the value computed for the CO_2 compensation point, Γ, was 79 µmol CO_2 mol^{-1}.

(b) Respiration

It is well known that elevated [CO_2] decreases photorespiration or the photo-oxidative pathway of C_3 plants by increasing the reactivity of Rubisco for CO_2. In fact, this is the underlying mechanism for the response of C_3 plants, such as soybean, to increasing concentrations of CO_2, and the concepts of this mechanism have now been extended to modelling crop responses to elevated CO_2 (Boote *et al.*, 1997). Also, many studies have been conducted on the effects of elevated CO_2 on dark respiration of plants (e.g. Bunce and Ziska, 1996; Gonzàles-Meler *et al.*, 1996; Amthor, 1997). The magnitude of the short-term (direct) effect of elevated CO_2 on dark respiration is still being researched. Respiration rate per unit mass of plants appears to decrease with elevated CO_2, probably because of the greater relative accumulation of both structural and non-structural carbohydrates and the lesser relative accumulation of proteins. Protein content appears to be the controlling factor for the direct effect of elevated CO_2 on dark respiration. Baker *et al.* (1992) found that dark respiration of rice per unit land area was linearly related to the amount of plant N per unit land area.

7.2.2 Solar radiation

Valle *et al.* (1985a) reported leaf CER vs. PPFD for eight soybean leaves grown in 330 and 800 μmol CO_2 mol^{-1}. A Michaelis–Menten type of rectangular hyperbola was fitted to the CER data as follows:

$$\text{CER} = [(P_{\text{maxL}} \times \text{PPFD})/(\text{PPFD} + K_{\text{L}})] + RSP_{\text{L}}, \qquad \text{(Eqn 7.2)}$$

where P_{maxL} is the asymptotic maximum of CER with respect to the y-intercept parameter, RSP_L; and K_L is the apparent Michaelis–Menten constant. The parameter RSP_L can be viewed as the extrapolated respiration rate at zero light. These parameters are shown in Table 7.3. An apparent quantum efficiency, Q, was computed from the slope of the CER vs. PPFD curves. Values of Q evaluated at PPFD equal to zero are also shown in Table 7.3.

In summarizing another large data set, Allen *et al.* (1990b) found that soybean leaf CER vs. PPFD followed a similar trend to that reported by Valle *et al.* (1985a). However, Allen *et al.* (1990b) found that the leaf response to $[CO_2]$ was linear over the range of 330– 800 μmol mol^{-1}.

Table 7.3. Photosynthetic asymptotic ceiling (P_{maxL}) with respect to y-intercept parameter (RSP_L), apparent Michaelis–Menten constant (K_L), light compensation point (Γ_L) and apparent quantum yield (Q) for soybean leaves (cv. Bragg) grown and measured at two CO_2 concentrations. The parameters estimated by an overall analysis using data for four leaves at the same CO_2 are presented. Average temperature and standard deviations for leaves grown at 330 and 660 μmol mol^{-1} were 28.4°C ± 2.8 and 29.6°C ± 2.6, respectively. (Adapted from Valle *et al.*, 1985a.)

CO_2 concentration (μmol mol^{-1})	P_{max} (μmol CO_2 m^{-2} s^{-1})	RSP_L (μmol CO_2 m^{-2} s^{-1})	K_L (μmol photon m^{-2} s^{-1})	Γ_L (μmol photon m^{-2} s^{-1})	Q (mol CO_2 mol^{-1} photon)
330	37	−1.7	713	35	0.052
660	76	−2.4	844	27	0.091

Soybean leaf Rubisco activity (bicarbonate-magnesium activated) of 21- and 39-day-old plants decreased in proportion to elevated [CO_2] (450 and 800 μmol mol^{-1}) compared with plants grown at 330 μmol mol^{-1} (Vu *et al.*, 1983). However, leaves from 59-day-old plants showed no differences (Table 7.4). Light was necessary for full activation of Rubisco. When soybean plants were kept in darkness from pre-dawn to noon, both the non-activated and the bicarbonate-magnesium-activated Rubisco activities remained at the low pre-dawn values. As soon as the plants were uncovered, the Rubisco activity increased to about the same value as that of plants not kept in darkness. Bicarbonate-magnesium-activated Rubisco increased with PPFD up to 400–500 μmol (photon) m^{-2} s^{-1}. Campbell *et al.* (1988) found that leaf-soluble protein was relatively stable in soybean leaves, and that Rubisco protein was a constant 55% of leaf-soluble protein. These two reports showed that Rubisco activity per unit protein was not down-regulated in activity or amount in soybean grown in elevated CO_2. In studies of rice plants (Rowland-Bamford *et al.*, 1991), both Rubisco protein concentration and enzyme activity were down-regulated in leaves of plants grown in elevated CO_2.

7.2.3 Temperature

More studies have been conducted on the effects of low rather than of high temperature on photosynthesis of soybean. Harley *et al.* (1985) compiled extensive data on soybean leaf photosynthesis versus temperature (and CO_2 and PPFD). Their light-saturated and CO_2-saturated photosynthetic rates increased with temperature to about 40°C. From these data, Harley and Tenhunen (1991) developed equations for modelling leaf photosynthetic responses to both temperature and CO_2 concentration. Probably the most definitive studies on temperature effects on soybean canopy photosynthesis throughout the whole season were reported by Pan (1996). In general, whole-canopy photosynthesis rates (and vegetative growth) were maintained without failure up through at least 44/34°C sinusoidal maximum/minimum (day/night) temperatures.

Table 7.4. HCO_3^-/Mg^{2+}-activated Rubisco values from leaves of soybean grown under three CO_2 concentrations. Enzyme activities were determined on leaves sampled about 0830 EST at 21 and 39 days after planting (DAP). At 21 DAP, leaves were expanding; at 39 DAP, leaves were fully expanded. Data at 59 DAP were collected at four times during the day. (Adapted from Vu *et al.*,1983.)

CO_2 concentration (μmol mol^{-1})	Rubisco activity (μmol CO_2 mg^{-1} Chl h^{-1})		
	21 DAP	39 DAP	59 DAP
330	377 ± 18	559 ± 8	532 ± 20
450	359 ± 18	512 ± 11	541 ± 8
800	291 ± 13	448 ± 12	554 ± 19

Jones *et al.* (1985b) conducted experiments on photosynthetic rates of soybean canopies with short-term (1 day) changes in temperature. They found no differences in canopy CER from 28 to 35°C. Plots of CER vs. PPFD at different temperatures were indistinguishable within CO_2 levels of 330 or 800 μmol mol^{-1}.

7.2.4 Water deficits

Jones *et al.* (1985c) conducted a drought study on soybean and found that both an early and a late stress diminished the seed and biomass yield of beans grown at both 330 and 660 μmol CO_2 mol^{-1}. Plants grown under doubled [CO_2] and subjected to short but severe drought cycles yielded about 20% more than the non-stressed plants grown under ambient [CO_2]. Based on the same experiment, Allen *et al.* (1994) showed that leaf temperatures were about 2°C warmer when exposed to 660 compared with 330 μmol CO_2 mol^{-1}. As a 13-day drying cycle progressed, leaf temperatures of drought-stressed plants increased by about 7°C compared with non-stressed plants. The residual internal leaf conductance (so-called mesophyll conductance) for CO_2 of drought-stressed plants decreased throughout the drying cycle. However, the intercellular CO_2 concentration (C_i) remained nearly constant throughout the drought period. Furthermore, C_i/C_a ratios remained steady throughout the drought, where C_a is the ambient [CO_2].

During the seed-fill (R5–R6 stages of development; Fehr and Caviness, 1977) period, Allen *et al.* (1998) showed that soybean grown under elevated [CO_2] and exposed to a drying cycle maintained a higher leaf turgor pressure by maintaining a higher (less negative) water potential rather than by maintaining a lower osmotic potential. The small change in osmotic potential during seed-fill indicates that osmotic adjustment was suppressed, probably because leaf resources were being remobilized and transported to the seed.

Huber *et al.* (1984) reported that drought stress decreased CER more in soybean plants that were not enriched with CO_2 than in those that were enriched. As CER declined with increasing drought, the internal [CO_2] concentration remained relatively constant despite decreased stomatal conductance. This behaviour was similar to the results reported by Allen *et al.* (1994). Thus, internal factors, rather than stomatal closure itself, caused the decreases in CER under progressive drought stress.

7.2.5 Nutrient deficits

Nutrient deficiency for agricultural crops under potential climatic change conditions is probably of no consequence where fertilizers are heavily used. With regard to photosynthesis, leaf N concentration has long been known to govern leaf photosynthetic rates, presumably through its strong relationship to

Rubisco. In a 1983 experiment, soybean grown in pots without adequate N doubled leaf photosynthetic rates in 3 days after application of ammonium nitrate (R.R. Valle and L.H. Allen, Jr, unpublished).

Leaf total N and Rubisco per unit leaf area decreased under elevated [CO_2] in rice (Rowland-Bamford *et al.*, 1991) but not in soybean (Campbell *et al.*, 1988). Elevated [CO_2] seemed to have little effect on the C/N ratios of tissues at final harvest. DeWitt *et al.* (1983) measured N_2 fixation of whole soybean canopies throughout the growing season using the acetylene reduction method. The soybean canopies were grown in soil–plant–air research (SPAR) chambers at [CO_2] of 330, 450 and 800 μmol mol^{-1}. They found that long-term CERs, long-term acetylene reduction, root nodule numbers on the glass face of the rooting volume, nodule dry weights at final harvest and whole-plant N contents (g m^{-2}) at final harvest were proportional to CO_2 concentration. For these reasons and other experimental data, Allen *et al.* (1988) concluded that legume crops such as soybean would always supply a sufficient and balanced amount of N through symbiotic N_2 fixation to meet the requirements of vegetative growth and seed yield.

7.2.6 Pollutants such as ozone

Ozone is generally considered to be the primary gaseous pollutant of agricultural crops and forests (Heck *et al.*, 1982; Heagle *et al.*, 1983). In many regions of the USA, ozone concentrations are about double what they would be without anthropogenic influences (Heck *et al.*, 1984). In open-top chambers, Heagle *et al.* (1998a) found that increasing O_3 concentrations from 0.020 to 0.080 μl l^{-1} (12 h exposure day^{-1}) caused increased soybean foliar injury, decreased leaf chlorophyll and starch content, increased chlorophyll *a*/*b* ratio, and decreased SLW. In general, all these impacts on leaves became more severe with time from 43 to 113 days after planting (DAP).

Booker *et al.* (1997) found that the photosynthetic rates of soybean leaves near the mainstem terminal were the same (about 28 μmol CO_2 m^{-2} s^{-1}) at 54–62 DAP, when exposed to either the low ozone of charcoal-filtered air (0.020 μl O_3 l^{-1}) or air enriched to 1.5 times daytime ambient ozone (0.070 μl O_3 l^{-1}) in exposures of 12 h day^{-1} at ambient [CO_2] (364 μmol mol^{-1}). However, the photosynthetic rates in ambient [CO_2] were suppressed by 15% at 75–83 DAP and by 45% at 97–106 DAP by the 1.5 daytime ambient ozone level. Under 726 μmol CO_2 mol^{-1} treatments the adverse impacts of high ozone were much less on leaf photosynthesis, as was also shown by Reid and Fiscus (1998). Also, Rubisco activity decreased with level of ozone, especially with increasing leaf age (Reid *et al.*, 1998). Clearly, ozone has an increasingly suppressive effect on photosynthesis as the leaf ages.

Elevated [CO_2] generally decreased the impact of ozone on all leaf responses (Booker *et al.*, 1997; Heagle *et al.*, 1998a; Reid *et al.*, 1998). Allen (1990) speculated that partial stomatal closure or additional photoassimilate under elevated [CO_2] might decrease ozone stress on plants.

7.3 Water Use

7.3.1 Carbon dioxide

Carbon dioxide can influence water use of plants in two ways. Firstly, CO_2 directly controls stomatal conductance through its control of stomatal aperture. In most crops, including soybean, stomata are partially closed by elevated [CO_2]. Morison (1987) analysed 80 sets of data and found that doubled [CO_2] decreased stomatal conductance of leaves by about 40% ± 5%. On the other hand, [CO_2] can promote growth, and thus produce more leaf area that will increase the transpiring surface area per unit land area. This effect on water-use rates would be especially important before crops achieve complete ground cover, or a leaf area index (LAI) of about 2.0 is obtained.

Cure (1985) summarized available soybean transpiration rate data. The ratio of water use for plants grown in doubled [CO_2] compared with plants grown in ambient [CO_2] ranged from 0.60 in pots inside open-top chambers (Rogers *et al.*, 1994) to 0.94 for seasonal cumulative canopy transpiration of soybean canopies in SPAR chambers (Jones *et al.*, 1985c).

Most of the quantitative data available on the effect of [CO_2] on soybean canopy transpiration has been obtained in SPAR systems. Whole-day transpiration was calculated for several days of data at two times during the course of an experiment (Jones *et al.*, 1985b). The average daily transpiration rate was 344 and 352 mol H_2O m^{-2} for soybean grown at 800 and 330 µmol CO_2 mol^{-1}, respectively. These daily rates translate to a 2% decrease in transpiration caused by elevated [CO_2]. Data collected later in the season showed a daily average transpiration of 331 and 298 mol H_2O m^{-2} for [CO_2] of 800 and 330 µmol mol^{-1}, respectively. Over this later period, transpiration was 11% higher for the elevated [CO_2] treatment. The LAI in this experiment averaged 6.0 for the elevated [CO_2] treatments and 3.3 for the 330 µmol CO_2 mol^{-1} treatments. This example shows that the effect of greater leaf area can override the effect of decreased stomatal conductance resulting from elevated [CO_2] treatments. In a later study where LAI was 3.46 and 3.36 for treatments of 660 and 330 µmol CO_2 mol^{-1}, respectively, the seasonal cumulative water use decreased by 12% for doubled [CO_2] (Jones *et al.*, 1985c). This percentage decrease in transpiration is comparable to that simulated by whole-crop energy-balance models for closed canopies (Rosenberg *et al.*, 1990; Boote *et al.*, 1997).

Water-use efficiency (WUE), or the ratio of CO_2 uptake rate to transpiration rate (or evapotranspiration rate) of soybean, is always increased by elevated [CO_2] at both leaf (Valle *et al.*, 1985b) and canopy levels (Jones *et al.*, 1984, 1985a,b,c,d). In leaf-cuvette studies inside SPAR chambers, Valle *et al.* (1985b) found that individual leaf WUE was doubled with a doubling of [CO_2] from 330 to 660 µmol mol^{-1}, but leaf transpiration rate was decreased little because the effects of decreased stomatal conductance were offset by the effects of a concomitant increase in leaf temperature. Jones *et al.* (1985b) also showed that LAI over the range of 3.3 to 6.0 could have a noticeable effect on

WUE. Allen *et al.* (1985) developed a quantitative relationship among CER, transpiration rate and WUE which showed that CO_2 enrichment increased WUE mainly by an increase of CER and secondarily by a decrease of transpiration rate.

7.3.2 Solar radiation

Solar radiation received at the earth's surface is not likely to change much unless there are also extensive changes in cloudiness and rainfall. Differences in annual incident solar radiation were only about 3% among one set of general circulation model predictions for the southeastern USA (Peart *et al.*, 1989). Furthermore, changes in solar radiation *per se* may be overshadowed by other environmental changes such as temperature, air humidity, and water deficits mediated through changes in rainfall. Tanner and Lemon (1962) pointed out that water use, especially in humid climates, is very closely coupled to solar radiation. Their work preceded that of Pruitt (1964) and the Priestley–Taylor formulation (Priestley and Taylor, 1972).

7.3.3 Temperature

In theory, the temperature effect on plant water use is mediated primarily through its effect on saturation vapour pressure and vapour pressure deficit of the air. One formulation of saturation vapour pressure (e_{sat}) is given by:

$$e_{sat} = e_0 \exp[(T - 273)/(T - 35)] \qquad \text{(Eqn 7.3)}$$

where T is temperature in °K and e_0 is the saturation vapour pressure (0.611 kPa) at 273°K (0°C).

Jones *et al.* (1985b) grew soybean at 31°C daytime air temperature and 21°C dewpoint temperature. Daytime temperatures were changed to 28°C for several days and later to 35°C for several days. Evapotranspiration was 20% greater at 31°C (390 mol H_2O m^{-2} per day) than at 28°C (325 mol H_2O m^{-2} per day), and it was 30% greater at 35°C (384 mol H_2O m^{-2} per day) than at 28°C (295 mol H_2O m^{-2} per day) during the later comparison. Thus, over this range, transpiration increased by about 4% per °C increase in temperature. This change is less than the change in saturation vapour pressure and vapour pressure deficit over the range of 28–35°C, probably due to evaporative cooling of the leaves themselves (see Chapter 14, this volume, for further discussion on this topic).

7.3.4 Water deficits

Jones *et al.* (1985c) conducted an experiment with an early-season and a late-season drought-stress cycle in soybean grown at 330 and 660 μmol CO_2

mol^{-1}. Cumulative seasonal transpiration was decreased by the drought-stress cycles, but seasonal WUE was actually improved (Table 7.5). Aggregating data, the mass ratio of seed yield to transpired water was calculated to be 0.85 g g^{-1} for the low [CO_2] treatment and 1.36 g g^{-1} for the high [CO_2] treatment. The elevated [CO_2] caused a 60% increase in WUE (1.36/0.85). This was accomplished by a 40% increase in seed yield and a 12% reduction in water use.

7.3.5 Nutrient deficits

Few studies have been conducted on the effects of nutrient deficits on water use in soybean; however, one could hypothesize that nutrient deficits sufficient to decrease photosynthesis would decrease stomatal conductance and thereby decrease transpiration and water use. Knowledge of this type of limitation has little practical value or scientific concern.

7.3.6 Pollutants such as ozone

Few studies, if any, have been conducted on ozone effects on soybean water use. Based on foliar damage caused by ozone, one would expect that water use over the season would be decreased by ozone. However, the main issues are damage by ozone to the leaves and decreased final seed yields of soybean, rather than water use.

Table 7.5. Comparison of integrated season-long transpiration, daytime carbon dioxide exchange rate (CER), water-use efficiency (WUE) and final harvest seed yield of soybean under two drought treatments and no drought. Seasonal water-use efficiency was calculated by dividing cumulative daytime CER by cumulative transpiration. (Adapted from Jones *et al.*, 1985c.)

Total seasonal response (land area basis)	330 µmol CO_2 mol^{-1} treatment			660 µmol CO_2 mol^{-1} treatment		
	Late[a] drought	No drought	Early[a] drought	Late[a] drought	No drought	Early[a] drought
Transpiration (kmol H_2O m^{-2})	16.1	22.2	16.5	13.6	19.3	15.5
Daytime CER (mol CO_2 m^{-2})	47.1	53.0	47.0	62.8	84.7	63.3
WUE (mol CO_2 $kmol^{-1}$ H_2O)	2.93	2.39	2.85	4.62	4.39	4.08
Seed yield (g m^{-2})	271	316	254	335	457	388

[a]Late drought: irrigation was withheld from 70 to 83 DAP. Early drought: irrigation was withheld from 51 to 65 DAP.

7.4 Growth and Development – Phenology

7.4.1 Carbon dioxide

Elevated [CO_2] effects on soybean phenology appear to be small and not consistent. Allen *et al.* (1990a) summarized the effect of [CO_2] on soybean developmental stages for four experiments from 1981 to 1984. They fitted vegetative stage (V) vs. DAP to a linear regression equation as follows:

$$V = \beta_0 + \beta_1 \times (DAP), \qquad \text{(Eqn 7.4)}$$

where β_0 = intercept of the regression on the V-stage axis and β_1 = mainstem nodes per day. In general, β_1 increased slightly with increasing [CO_2], but the responses among the four experiments were variable.

The plastochron interval (days per mainstem node or mainstem trifoliate) tended to decrease with increasing [CO_2] so that there were more mainstem nodes and trifoliate leaflets at the end of vegetative growth. We fitted the following equation of plastochron interval ($1/\beta_1$) versus [CO_2] and obtained the following relationship for the data in Allen *et al.* (1990a):

$$1/\beta_1 = 4.25 - 0.00120 \times [CO_2] \qquad \text{(Eqn 7.5)}$$

Except for the slightly delayed initiation of the R1 and R2 phases of development under the severely limited [CO_2] of 160 μmol mol^{-1}, there appeared to be no effect of [CO_2] on reproductive stages of development. Although the overall effects of [CO_2] on phenological stages of development of soybean are minor, similar tendencies have been reported elsewhere (Hofstra and Hesketh, 1975; Rogers *et al.*, 1984, 1986; Baker *et al.*, 1989; also see Chapter 8, this volume).

7.4.2 Solar radiation

Phenology of soybean is not likely to be impacted directly by solar radiation. However, any shift in latitude zones or planting dates will affect the day length and night length, and will likely change the phenology of the soybean plant.

7.4.3 Temperature

Temperature exerts a major control on growth at the level of enzyme activity, protein synthesis and cell division. To a certain extent, these controls are similar at the whole organism level of the plant. Sionit *et al.* (1987a,b) found that the growth and development responses of soybean to CO_2 enrichment increased with increasing temperature within the rather cool range of 18/12°C to 26/20°C. Vegetative processes such as rates of leaf appearance, leaf expansion and branching are enhanced by high temperatures (at least up to some threshold level). Rates of soybean leaf appearance (Hesketh *et al.*, 1973) and leaf expansion (Hofstra and Hesketh, 1975) increase up to 30°C; relative growth rate increases up to 31°C (Hofstra and Hesketh, 1975); leaf photosynthesis rate increases up to 35°C (Harley *et al.*, 1985); and total biomass increases up to 28°C (Baker *et al.*, 1989) or 32°C (Pan, 1996).

Baker *et al.* (1989) investigated the effect of temperature and [CO_2] on rates of soybean development. Table 7.6 shows the plastochron interval and final mainstem node number for soybean as a function of temperature and [CO_2]. The results show that both increasing temperature and increasing [CO_2] decrease the plastochron interval. These results and those reported by Sionit *et al.* (1987a) and Hofstra and Hesketh (1975) indicate that temperature affects soybean developmental rate to a much greater degree than does [CO_2].

The rate of soybean node addition increases as mean temperature increases to about 28–30°C, with a computed base temperature of about 8°C (Hesketh *et al.*, 1973). Leaf photosynthesis of soybean also has a base temperature close to 8°C as computed from a linear projection of data (Harley *et al.*, 1985). However, there are few data near the base temperature and, as in most experiments, their study did not allow for temperature acclimation.

Although soybean reproductive development shows almost no response to [CO_2], it is strongly dependent on temperature (Hesketh *et al.*, 1973; Grimm *et al.*, 1994; Shibles *et al.*, 1975). The R1 stage is delayed by both low (less than 23°C) and high (greater than about 35°C) temperature. Likewise, the time to reach the R7 stage is delayed by mean daily temperature above 31°C, accompanied by decreased growth rates of seed and low yield (Pan, 1996). Similar results were found in cotton (see Chapter 8, this volume).

7.5 Growth and Development – Growth Rates of Organs

7.5.1 Carbon dioxide

Carbon dioxide affects the partitioning of dry matter to various plant organs. Acock and Allen (1985) summarized growth and development information for

Table 7.6. Plastochron interval and final mainstem node number for soybean in CO_2 and air temperature experiments in controlled-environment chambers. (Adapted from Baker *et al.*, 1989.)

CO_2 concentration (μmol mol^{-1})	Day/night temperature (°C)	Plastochron interval (d trifoliolate^{-1})	Final mainstem node number (no. plant^{-1})
300	26/19	4.2a*	10.3 ± 0.5
	31/24	3.3b	11.5 ± 0.9
	36/29	3.2b	12.0 ± 0.5
600	26/19	3.9a	11.2 ± 0.4
	31/24	2.7c	11.4 ± 0.3
	36/29	2.6c	12.1 ± 0.5

*Numbers followed by the same letter are not significantly different at the 0.05 probability level as determined by *t*-tests.

many species of vegetation, including soybean. In vegetative growth of soybean, Rogers *et al.* (1983, 1994, 1997) found that elevated [CO_2] increased dry matter partitioning in the order of roots > stems > leaves. Similar results were obtained by Allen *et al.* (1991). After beginning reproductive development, an increasing fraction of the total dry matter was found in the pods (Table 7.7).

The SLW increases with increasing [CO_2], probably due to increasing amounts of structural and non-structural carbohydrates (Rogers *et al.*, 1983; Allen *et al.*, 1988, 1991, 1998). Rogers *et al.* (1984) and Allen *et al.* (1991) found that SLW increased with elevated [CO_2] (Table 7.8).

The LAI generally increases with exposure to elevated [CO_2], due both to larger individual leaves and to more leaves produced per plant (Rogers *et al.*, 1983; Jones *et al.*, 1985a,c). Branching may increase with increasing [CO_2], so more sites exist for leaves to form (Rogers *et al.*, 1984).

Table 7.7. Soybean plant components as a percentage of total dry matter of plants grown at subambient to superambient concentrations of CO_2. (Adapted from Allen *et al.*, 1991.)

	(CO_2 concentration μmol mol^{-1})					
Component	160	220	280	330	660	990
13 DAP[a] (V2 Stage)[b]						
Root (%)	11	10	9	12	10	10
Cotyledon (%)	20	16	14	14	11	10
Stem (%)	23	25	26	26	26	27
Leaflet (%)	46	49	51	48	53	53
TDM[c] (g m^{-2})	14	13	15	16	20	24
34 DAP (V8 Stage)						
Root (%)	7	7	7	8	9	8
Stem (%)	23	25	27	26	28	29
Petiole (%)	12	13	15	14	16	16
Leaflet (%)	58	55	51	52	47	47
TDM (g m^{-2})	65	71	116	129	154	225
66 DAP (R5 Stage)						
Stem (%)	17	20	25	22	25	27
Petiole (%)	10	11	13	13	14	14
Leaflet (%)	35	35	35	32	32	30
Pod (%)	38	34	27	33	29	29
TDM (g m^{-2})	227	282	489	486	794	873
94 DAP (R7 Stage)						
Stem (%)	10	13	15	14	16	19
Petiole (%)	6	6	6	7	7	6
Leaflet (%)	17	16	15	14	12	111
Pod (%)	67	65	64	65	65	64
TDM (g m^{-2})	308	302	634	617	687	838

[a]Days after planting.
[b]Based on Fehr and Caviness (1977).
[c]TDM is total dry matter of the plant components listed at each DAP.

Table 7.8. Specific leaf weight[a] (g m^{-2}) for soybean grown at subambient to superambient concentrations of CO_2. (Adapted from Allen *et al.*,1991.)

Days after planting	Developmental stage[b]	CO_2 concentration (μmol mol^{-1}) 160	220	280	330	660	990
13	V2	23.9	23.5	22.8	22.3	25.9	26.0
24	V5	21.6	21.5	22.1	22.7	27.2	29.1
34	V8	21.5	22.7	23.0	23.8	28.4	30.1
46	R2	19.7	22.1	23.4	24.1	27.2	26.2
66	R5	23.8	26.2	22.7	25.8	28.8	29.0
81	R6	23.4	24.8	25.3	27.2	25.3	27.8
94	R7	22.8	21.8	24.5	24.8	23.6	27.4

[a]Leaf dry weight/leaf area.
[b]Based on Fehr and Caviness (1977).

Allen *et al.* (1991) developed a simple growth model and fitted it to dry matter data of soybean grown at [CO_2] of 160, 220, 280, 330, 660 and 990 μmol mol^{-1}. The plants were grown at day/night temperatures of 31/23°C under well-watered conditions and with optimum soil nutrients. The model was linear for DAP and a rectangular hyperbola for [CO_2]. Leaf area and dry weights of stems, petioles, leaves and the total plant were fitted to the model over the linear phase of growth from 24 to 66 DAP using the SAS non-linear least-squares iterative method (SAS, 1985). The model is:

$$Y = \beta_0 + \beta_1 \times [(\mathrm{DAP} - D) \times (\mathrm{CO_2} - G)]/(\mathrm{CO_2} + K_C), \qquad \text{(Eqn 7.6)}$$

where Y = growth variable (dry weight or leaf area); β_0 = intercept on y-axis when second term is zero; β_1 = maximum asymptotic value of growth rate when [CO_2] is no longer limiting; D = DAP offset parameter to remove time lag of seedling emergence and non-linear early plant growth; CO_2 = carbon dioxide concentration (μmol CO_2 mol^{-1}); G = apparent CO_2 compensation point; and K_C = hyperbolic function shape factor related to the apparent Michaelis constant. The partial derivative with respect to [CO_2] is:

$$\delta Y/\delta \mathrm{CO_2} = [\beta_1 \times (\mathrm{DAP} - D) \times (K_C + G)]/(\mathrm{CO_2} + K_C)^2, \qquad \text{(Eqn 7.7)}$$

and the partial derivative with respect to DAP is:

$$\delta Y/\delta \mathrm{DAP} = [\beta_1 \times (\mathrm{CO_2} - \mathrm{G})]/(\mathrm{CO_2} + K_C). \qquad \text{(Eqn 7.8)}$$

The fitted parameters of equation 7.6 based on 1984 soybean data are given in Table 7.9 (Allen *et al.*, 1991). These parameters were used to compute growth rates in Table 7.10 during the linear phase of vegetative growth (DAP 24 to 66). From this model, the relative enhancement resulting from [CO_2] at 660 vs. 330 μmol mol^{-1} was 1.62, 1.56, 1.36 and 1.21 for dry weight of stems, petioles, leaves, and for leaf area, respectively. Growth rate of individual soybean seeds (seed size at maturity) is much less affected by elevated [CO_2], except for smaller seeds when plants are grown at low concentrations of 160 μmol mol^{-1}

Table 7.9. Estimated parameters[a] of soybean growth model[b] fitted by non-linear regression procedures to CO_2 concentration and days after planting (DAP) for the 24–66 DAP linear growth phase. (Adapted from Allen *et al.*, 1991.)

Dependent variable	Parameter of growth model				
	β_0[c] ($g\ m^{-2}$)	β_1[d] ($g\ m^{-2}\ d^{-1}$)	D[e] (DAP)	G[f] ($\mu mol\ mol^{-1}$)	K_c[g] ($\mu mol\ mol^{-1}$)
Total dry weight	4.243	26.881	23.97	98.7	167.8
Stem dry weight	2.951	7.682	22.96	120.6	214.1
Petiole dry weight	0.987	3.930	22.14	116.3	183.8
Leaflet dry weight	4.556	7.206	19.53	101.6	84.2
Leaflet area	0.493[h]	0.214[i]	20.41	119.1	–34.6

[a]Computed originally on a per plant basis, and then adjusted to $g\ m^{-2}$ based on an average 28 plants m^{-2} over the 24–66 DAP interval.
[b]$y = \beta_0 + \beta_1 \times [(DAP - D) \times (CO_2 - G)]/(CO_2 + K_c)$.
[c]Intercept on *y*-axis when second term is zero.
[d]Maximum asymptotic growth rate as CO_2 increases.
[e]DAP offset parameter to remove time lag of seedling emergence and non-linear early plant growth.
[f]Apparent CO_2 compensation point.
[g]Apparent Michaelis constant.
[h]$m^2\ m^{-2}$.
[i]$m^2\ m^{-2}\ day^{-1}$.

Table 7.10. Growth rates[a] during the linear phase of soybean vegetative growth at subambient to superambient concentrations of CO_2. (Adapted from Allen *et al.*, 1991.)

	CO_2 concentration ($\mu mol\ mol^{-1}$)					
Plant component	160	200	280	330	660	990
Total shoot dry weight[b,c]	5.0	8.4	10.9	12.5	18.2	20.7
Stem dry weight[c]	0.8	1.7	2.4	2.9	4.7	5.5
Petiole dry weight[c]	0.5	1.0	1.4	1.6	2.5	2.9
Leaf dry weight[c]	1.7	2.7	3.4	3.9	5.3	5.8
Leaf area ($m^2\ m^{-2}\ day^{-1}$)	0.070	0.116	0.140	0.153	0.185	0.195

[a]Computed during 24 to 66 days after planting (DAP) from Eqn 7.8 and Table 7.9.
[b]Includes seed and podwalls, as well as stems, petioles, and leaves.
[c]Dry weight $g\ m^{-2}\ day^{-1}$.

(Allen *et al.*, 1991). Thus, the biggest factor for increasing seed yield appears to be the number of seeds harvested per plant.

7.5.2 Temperature

Vegetative growth is often stimulated by increasing temperature, as is leaf and canopy photosynthesis (Boote *et al.*, 1997). However, reproductive growth leading to seed yield is often depressed by the same increases of temperature that enhance vegetative growth. For soybean, individual seed growth rates and final mass per seed decline as temperature exceeds a daily average of about 23°C (Egli and Wardlaw, 1980; Baker *et al.*, 1989; Pan, 1996). (See Chapter 5, this volume, for temperature effects on rice.)

7.5.3 Water deficits and nutrient deficits

Available soil water is the primary determinant of plant growth, as was demonstrated by Briggs and Schantz (1914). Growth rates of all plants and all plant organs are decreased when soil water is limiting, as has been shown many times (Allen, 1999).

Boote *et al.* (1997) reviewed some of the literature regarding nutrient limitations on crop responses to elevated [CO_2]. In spite of early speculation that nutrient limitations might override the benefits of elevated [CO_2], this has not always been the case. Kimball *et al.* (1993) reported equal [CO_2]-stimulated growth responses of N-limited and adequately fertilized cotton grown in field soil. Elevated [CO_2] stimulated growth and yield of non-nodulated soybean plants cultured in either growth-limiting N supply (Cure *et al.*, 1988a) or growth-limiting P supply (Cure *et al.*, 1988b).

7.6 Crop Yields

7.6.1 Carbon dioxide

Cure (1985) summarized soybean seed yields from 12 elevated-[CO_2] experiments. Several of the studies were conducted with CO_2 enrichment above 1000 μmol mol^{-1}, and most yield responses were reported on a per plant rather than a land area basis. Also, several experiments were conducted in growth chambers rather than in sunlight. Nevertheless, the overall response for soybean was a yield increase of 29 ± 8% from a doubling of [CO_2].

Rogers *et al.* (1986) summarized yield response ratios and changes in harvest index (ratio of seed yield to total above-ground biomass yield) for six experiments in which the enrichment was more than 1000 μmol CO_2 mol^{-1} and for ten experiments in which the enrichment was less than 1000 μmol mol^{-1}. In the first cases, yield response ratios (1000/350 μmol CO_2 mol^{-1}] ranged from 0.97 to 1.78, with an average ratio of 1.28 ± 0.25. The changes in harvest index were not consistent, with more increases than decreases for CO_2 enrichment. In the second cases, yield response ratios (700/350 μmol CO_2 mol^{-1}) ranged from 0.93 to 2.34, with an average ratio of 1.31 ± 0.41. The harvest index was mainly decreased by elevated [CO_2]. The response ratios were much greater under drought-stress conditions.

Five experiments conducted in Gainesville, Florida, in SPAR chambers provided soybean seed yields (kg ha^{-1}) under various [CO_2] and these data were used to calculate yield response ratios at 660 vs. 330 μmol CO_2 mol^{-1}.

Results were: 3870/2770 = 1.40 for 1981 (Jones *et al.*, 1984); 3860/2730 = 1.41 for 1982 (Allen *et al.*, 1998); 4570/3160 = 1.45 for no drought stress, 3880/2540 = 1.53 for early-season drought stress and 3350/2710 = 1.24 for late-season drought stress in 1983 (Jones *et al.*, 1985c); 3570/2970 = 1.20 for 1984 (Allen *et al.*, 1991); and 3930/2700 = 1.46 for 26/19°C day/night temperatures, 3750/3030 = 1.24 for 31/24°C day/night temperatures and 3480/3030 = 1.15 for 36/29°C day/night temperatures (Baker *et al.*, 1989). The mean seed yields across the five experiments were 3807 ± 351 and 2849 ± 204 kg ha^{-1} for 660 and 330 μmol CO_2 mol^{-1}, respectively, while the mean yield response ratio was 1.34 ± 0.14 for the doubled [CO_2]. This yield response ratio is comparable to the 33 ± 6% increase calculated by Kimball (1983) from 490 observations of various crops.

Equation 7.1a was also used by Allen *et al.* (1987) to compute biomass yield and seed yield of soybean from 2 years of data at Gainesville, Florida, 2 years of data at Mississippi State, Mississippi, 1 year of data at Clemson, South Carolina, and 3 years of data at Raleigh, North Carolina. The fitted model parameters for the soybean relative biomass data were $R_{max} = 3.02$; $K_m = 182$ μmol CO_2 mol^{-1}; $R_i = -0.91$; and $\Gamma = 78$ μmol CO_2 mol^{-1}. Likewise, for soybean relative seed yield data, the fitted model parameters were $R_{max} = 2.55$; $K_m = 141$ μmol CO_2 mol^{-1}; $R_i = -0.76$; and $\Gamma = 60$ μmol CO_2 mol^{-1}. The parameters of the model provide a relative yield response ratio of 1.32 for a doubling of [CO_2], which is similar to the average yield response ratios discussed in the last paragraph. Of course, equation 7.1a is basically curve-fitting, and does not account for interactions with other environmental factors.

7.6.2 Solar radiation

In the absence of other stresses, crop yields are directly proportional to absorbed solar radiation. Various crops synthesize different compounds, and thus have different biomass conversion efficiencies.

7.6.3 Temperature

Pan (1996) reported that 'Bragg' soybean seed yields and seed quality decreased as day/night maximum/minimum air temperatures in SPAR chambers were increased from 32/22°C to 44/34°C, where seed yields were zero. However, photosynthesis and vegetative productivity were able to continue up to 44/34°C, probably because of evaporative cooling, but all of the plants except those adjacent to the end walls of the chamber failed to survive at 48/38°C (Table 7.11). Those few plants that did survive had distorted stems and leaves.

The ratios of biomass yield at 700 vs. 350 μmol CO_2 mol^{-1} were 1.52 and 1.54 at temperatures of 28/18°C and 40/30°C, respectively, and the 700/350 μmol CO_2 mol^{-1} ratio of seed yield was 1.32 and 1.30 at temperatures of 28/18°C and 40/30°C, respectively. There was no temperature/[CO_2]

Table 7.11. Response of soybean to diurnal 'sinusoidal' cycles of temperature with day/night, maximum/minimum of 28/18, 32/22, 36/26, 40/30, 44/34 and 48/38°C. Carbon dioxide concentrations were maintained at 700 μmol mol^{-1} in one chamber at each of the six temperatures, and at 350 μmol mol^{-1} in two chambers maintained at 28/18 and 40/30°C. (Adapted from Pan, 1996.)

CO_2 concentration (μmol mol^{-1})	Temperature, day/night maximum/minimum (°C)					
	28/18	32/22	36/26	40/30	44/34	48/38
Biomass yield[a] (g plant^{-1})						
700	23.0	24.1	27.0	25.5	26.1	1.7
350	15.1	–	–	16.6	–	–
(Ratio)	(1.52)			(1.54)		
Seed yield[a] (g plant^{-1})						
700	10.0	11.4	12.3	8.7	0.5	0.0
350	7.6	–	–	6.7	–	–
(Ratio)	(1.32)			(1.30)		
Harvest index						
700	0.43	0.47	0.45	0.34	0.02	–
350	0.50	–	–	0.40	–	–
(Ratio)	(0.86)			(0.85)		

[a]There were 21 plants m^{-2} at final harvest.
– Data not available.

interaction for either biomass or seed yield. The 30–32% increase in seed yield is similar to the percentage increases in yield for a doubling of [CO_2] that were predicted earlier for soybean (Allen *et al.*, 1987), and is close to the 33% increases reported in a wide-ranging review by Kimball (1983). Likewise, the 52–54% increase in biomass yield is similar to the predicted biomass yields and photosynthetic rates that were predicted earlier (Allen *et al.*, 1987).

The data in Table 7.11 show clearly the rather high limits of temperature for survival and successful vegetative growth of 'Bragg' soybean (with a growth failure threshold between the 44/34°C and 48/38°C treatments), and the peak of seed yield response between 32/22°C and 36/26°C, with a rapid decline at higher temperatures. These soybean responses to [CO_2] and temperature are similar to responses of rice (Baker and Allen, 1993), except that rice (cv. IR30) appeared to have both a temperature threshold for survival and a temperature maximum for yield that were about 2–4°C lower than for soybean.

7.6.4 Water deficits and nutrient deficits

Rogers *et al.* (1986) reported that the relative increase of soybean seed yield caused by elevated [CO_2] was greater under water deficit conditions than under well-watered conditions. Actually, a better interpretation of the data is that drought decreased the yields of plants grown under ambient [CO_2] more than it decreased the yields of plants under elevated [CO_2]. Nutrient deficits always decrease yields. However, as was discussed in the previous section, elevated [CO_2] still stimulated growth and yield relative to ambient [CO_2] in most cases (Boote *et al.*, 1997).

7.6.5 Pollutants such as ozone

Heagle *et al.* (1998b) reported that relative CO_2 stimulation of soybean seed yield was greater for plants stressed by ozone than for unstressed plants. Doubled [CO_2] increased the seed yield of 'Essex' by 16, 24 and 81% at ozone levels of 0.4 (charcoal-filtered air, CF), 0.9 and 1.5 times ambient, respectively. However, biomass yield was increased by 48, 55 and 75% by doubled [CO_2] at the same respective ozone levels (Miller *et al.*, 1998). Data extracted from Fiscus *et al.* (1997) showed that, under doubled [CO_2], soybean total biomass increased by 44% in CF air and 80% in 1.5 × ambient ozone air, whereas seed yield increased only 4% in CF air and 67% in 1.5 × ambient ozone air (Table 7.12). For unknown reasons, the partitioning to seed differed widely in the CF air, since the harvest index was high (0.53) in ambient [CO_2] but rather low (0.38) in doubled-ambient [CO_2]. Interactions of [CO_2] response of crops with ozone (or other factors of CF) on partitioning to seed need investigation.

7.7 Cultural and Breeding Strategies for Future Climate

Increasing [CO_2] will increase crop production potentials, but associated expected climatic changes may decrease potential production. Model predictions of soybean production in the southeastern USA were seriously impacted by both inadequate soil water (inadequate rainfall) and high temperatures (Curry *et al.*, 1990a,b, 1995). When simulations employed optimum irrigation, yields were always improved by elevated [CO_2]. Irrigation is a cultural practice that could alleviate both inadequate rainfall and higher

Table 7.12. Seed weight, total biomass and harvest index of soybean at two levels of daytime ozone exposure and two levels of CO_2 concentration. (Adapted from 1993 and 1994 data of Fiscus *et al.*, 1997.)

Ozone (nmol mol^{-1})	CO_2 concentration (μmol mol^{-1})	Seed yield (g plant^{-1})	Total biomass (g plant^{-1})	Harvest index
1.5 ×[a]	700	177	411	0.43
1.5 ×[a]	360	106	228	0.46
(Ratio)		(1.67)	(1.80)	(0.93)
CF[b]	700	172	449	0.38
CF[b]	360	166	312	0.53
(Ratio)		(1.04)	(1.44)	(0.72)

[a]Season-long 12 h average ozone concentration for the 1.5 × ambient air treatment was 92 and 70 nmol mol^{-1} in 1993 and 1994, respectively.
[b]Season-long 12 h average ozone concentration for the charcoal filtered air (CF) treatment was 22 and 25 nmol mol^{-1} in 1993 and 1994, respectively.

evaporation rates under higher temperatures, but competition for water and lack of available supplies might make irrigation difficult to implement. Changing crops to more favourable seasons or selecting more drought-tolerant crops, such as pigeon pea, rather than soybean might be an alternative.

New cultivars may need to be 'designed' for future climates (Hall and Allen, 1993). For the most part, the new cultivars would need to be more tolerant of both daytime and night-time high temperatures. To explore the opportunities for tolerance to higher temperatures, there is a need for more agronomists and plant breeders with a broad, worldwide base of knowledge to determine if heat-tolerant cultivars or wild types already exist, and if these traits can be incorporated into productive cultivars. Furthermore, if cultural patterns and planting dates are shifted, cultivars may have to be designed with different photoperiod sensitivities, or without photoperiod sensitivity.

Based on current knowledge, plants need to be selected for a higher harvest index. This may mean selection for greater utilization of carbohydrates that are produced in high quantities under CO_2 enrichment (Baker *et al.*, 1989; Allen *et al.*, 1998), more N fixation and storage for later translocation during seed-filling, and selection for greater pod loads.

7.8 Conclusions and Future Research Directions

Soybean does not exhibit a substantial photosynthetic acclimation to elevated $[CO_2]$ (down-regulation of Rubisco via loss of amount or activity of this enzyme), in contrast with other crops such as rice. Nevertheless, future research should focus on physiological mechanisms that are involved in down-regulation of Rubisco and the genetic controls on these processes (e.g. Vu *et al.*, 1997; Gesch *et al.*, 1998). The reasons for differences in down-regulation of Rubisco among crops need to be understood. Furthermore, genetic manipulations to decrease the intensity of down-regulation of Rubisco under elevated $[CO_2]$ should be explored as a potential mechanism for increasing crop seed yields as atmospheric $[CO_2]$ increases.

Since soybean is a legume, it might be possible that symbiotic N_2 fixation obviates the need for conservation of whole-plant N resources under conditions of elevated $[CO_2]$, as might be conveyed by down-regulation of Rubisco. Conservation of whole-plant N resources under elevated $[CO_2]$ by down-regulation of Rubisco has been proposed as an ecological adjustment in plants that are totally dependent on non-symbiotic soil N that allows them to fix and store more carbon per unit of plant N. Work should be expanded on the impact of elevated $[CO_2]$ on nitrogen fixation under field conditions and with other stresses (e.g. DeWitt *et al.*, 1983; Serraj *et al.*, 1998).

Although photosynthetic acclimation to elevated $[CO_2]$ does not appear to be prevalent in soybean, most studies show that large amounts of non-structural carbohydrates accumulate in leaf blade, petiole and stem tissues of soybean grown under elevated $[CO_2]$. Furthermore, most studies also show that the harvest index decreases when plants are grown under elevated $[CO_2]$. Future research should be undertaken to utilize available carbohydrates more readily, set more pods, and produce more seeds under elevated [CO_2] conditions.

Soybean, along with most warm-season seed crops, produces fewer viable seeds with decreasing quality as air temperatures increase above an optimum mean daily temperature (usually about 26°C). This progressive failure of seed yield with increasing temperature occurs despite the fact that photosynthetic rates and vegetative biomass growth can continue unabated to much higher temperatures (up to 44/34°C day/night maximum/minimum temperatures; Pan, 1996) before a threshold limit is reached. Probably the prime research problem to be addressed is the gamut of potential causes for the loss of both seed quantity and seed quality as a function of rising air temperature. These causes may exist in the area of floral initiation, floral development, pollination and fertilization, early embryo development, translocation and pod loading, and seed-filling processes. Source-to-sink ratio may play a role, although continued total above-ground biomass increase (vegetative growth) seems to indicate that source limitations may not be a factor at all (but the factors that control partitioning may be involved).

Eventually, genetic regulation and factors that promote or inhibit seed production physiology need to be investigated and understood. A global search for genetic materials that are more tolerant of high temperatures for seed production is needed for soybean as well as all other seed crops.

Stomatal regulation by elevated [CO_2] will provide small savings of water. However, changes in precipitation patterns or amounts, as well as rising temperatures, could completely dominate soil water availability for crops. Applied research on management strategies such as altered planting dates needs to be conducted, and assisted by the use of plant growth models.

Impacts of exposure to ozone and interactions with [CO_2] need further investigation (Fiscus *et al.*, 1997; Heagle *et al.*, 1998b). Especially perplexing are the differences in interaction of [CO_2] and ozone effects on soybean seed yield compared with total biomass production (Table 7.12). These differences are apparent in the relatively low seed yield and harvest index in charcoal-filtered, low-ozone air at high [CO_2] compared with the relatively high seed yield and harvest index in charcoal-filtered, low-ozone air at low [CO_2].

Acknowledgements

Contribution from US Department of Agriculture, Agricultural Research Service and the Institute of Food and Agricultural Sciences of the University of Florida (Florida Agricultural Experiment Station publication series, no. R-07142).

References

Acock, B. and Allen, L.H. Jr (1985) Crop responses to elevated carbon dioxide concentration. In: Strain, B.R. and Cure, J.D. (eds) *Direct Effects of Increasing Carbon Dioxide on Vegetation.* US Department of Energy, Carbon Dioxide Research Division, DOE/ER-0238, Washington, DC, pp. 53–97.

Acock, B., Reddy, V.R., Hodges, H.F., Baker, D.N. and McKinion, J.M. (1985) Photosynthetic responses of soybean canopies to full-season carbon dioxide enrichment. *Agronomy Journal* 77, 942–947.

Allen, L.H. Jr (1990) Plant responses to rising carbon dioxide and potential interactions with air pollutants. *Journal of Environmental Quality* 19, 15–34.

Allen, L.H. Jr (1999) Evapotranspiration responses of plants and crops to carbon dioxide and temperature. *Journal of Crop Production* (in press).

Allen, L.H. Jr, Jones, P. and Jones, J.W. (1985) Rising atmospheric CO_2 and evapotranspiration. In: *Advances in Evapotranspiration.* Proceedings of the National Conference, ASAF Publication no. 14–85, American Society of Agricultural Engineers, St Joseph, Michigan, pp. 13–27.

Allen, L.H. Jr, Boote, K.J., Jones, J.W., Jones, P.H., Valle, R.R., Acock, B., Rogers, H.H. and Dahlman, R.C. (1987) Response of vegetation to rising carbon dioxide: photosynthesis, biomass, and yield of soybean. *Global Biogeochemical Cycles* 1, 1–14.

Allen, L.H. Jr, Vu, C.V., Valle, R., Boote, K.J. and Jones, P.H. (1988) Nonstructural carbohydrates and nitrogen of soybean grown under carbon dioxide enrichment. *Crop Science* 28, 84–94.

Allen, L.H. Jr, Bisbal, E.C., Campbell, W.J. and Boote, K.J. (1990a) Carbon dioxide effects on soybean development stages and expansive growth. *Soil and Crop Science Society of Florida Proceedings* 49, 124–131.

Allen, L.H. Jr, Valle, R.R., Mishoe, J.W., Jones, J.W. and Jones, P.H. (1990b) Soybean leaf gas exchange responses to CO_2 enrichment. *Soil and Crop Science Society of Florida Proceedings* 49, 192–198.

Allen, L.H. Jr, Bisbal, E.C., Boote, K.J. and Jones, P.H. (1991) Soybean dry matter allocation under subambient and superambient levels of carbon dioxide. *Agronomy Journal* 83, 875–883.

Allen, L.H. Jr, Valle, R.R., Mishoe, J.W. and Jones, J.W. (1994) Soybean leaf gas-exchange responses to carbon dioxide and water stress. *Agronomy Journal* 86, 625–636.

Allen, L.H. Jr, Bisbal, E.C. and Boote, K.J. (1998) Nonstructural carbohydrates of soybean plants grown in subambient and superambient levels of carbon dioxide. *Photosynthetic Research* 56, 143–155.

Allen, L.H. Jr, Valle, R.R., Jones, J.W. and Jones, P.H. (1998) Soybean leaf water potential responses to carbon dioxide and drought. *Agronomy Journal* 90, 375–383.

Amthor, J.S. (1997) Plant respiratory responses to elevated carbon dioxide partial pressure. In: Allen, L.H. Jr, Kirkham, M.B., Olszyk, D.M. and Whitman, C.E. (eds) *Advances in Carbon Dioxide Effects Research.* ASA Special Publication No. 61, American Society of Agronomy, Madison, Wisconsin. pp. 35–77.

Baker, J.T. and Allen, L.H. Jr (1993) Contrasting crop species responses to CO_2 and temperature: rice, soybean, and citrus. *Vegetatio* 104/105, 239–260. Also in: Rozema, J., Lambers, H., van de Geijn, S.C. and Cambridge, M.L. (eds) *CO_2 and Biosphere.* Advances in Vegetation Science 14, Kluwer Academic Publishers, Dordrecht, The Netherlands, pp. 239–260.

Baker, J.T., Allen, L.H. Jr, Boote, K.J., Jones, P. and Jones, J.W. (1989) Response of soybean to air temperature and carbon dioxide concentration. *Crop Science* 29, 98–105.

Baker, J.T., Laugel, F., Boote, K.J. and Allen, L.H. Jr (1992) Effects of daytime carbon dioxide concentration on dark respiration of rice. *Plant, Cell and Environment* 15, 231–239.

Booker, F.L., Reid, C.D., Brunschön-Harti, S., Fiscus, E.L. and Miller, J.E. (1997) Photosynthesis and photorespiration in soybean [*Glycine max* (L.) Merr.] chronically exposed to elevated carbon dioxide and ozone. *Journal of Experimental Botany* 48, 1843–1852.

Boote, K.J., Pickering, N.B. and Allen, L.H. Jr (1997) Plant modeling: advances and gaps in our capability to predict future crop growth and yield. In: Allen, L.H., Jr, Kirkham, M.B., Olszyk, D.M. and Whitman, C.E. (eds) *Advances in Carbon Dioxide Effects Research*. ASA Special Publication No. 61, ASA-CSSA-SSSA, Madison, Wisconsin, pp. 179–228.

Briggs, L.J. and Schantz, H.L. (1914) Relative water requirements of plants. *Journal of Agricultural Research* 3, 1–63.

Bunce, J.A. and Ziska, L.H. (1996) Responses of respiration to increased carbon dioxide concentration and temperature in three soybean cultivars. *Annals of Botany* 77, 507–514.

Campbell, W.J., Allen, L.H. Jr and Bowes, G. (1988) Effects of CO_2 concentration on rubisco activity, amount, and photosynthesis in soybean leaves. *Plant Physiology* 88, 1310–1316.

Cure, J.D. (1985) Carbon dioxide doubling responses: a crop survey. In: Strain, B.R. and Cure, J.D. (eds) *Direct Effects of Increasing Carbon Dioxide on Vegetation.* US Department of Energy, Carbon Dioxide Research Division, DOE/ER-0238, Washington, DC, pp. 53–97.

Cure, J.D., Israel, D.W. and Rufty, T.W. Jr (1988a) Nitrogen stress effects on growth and seed yield of nonnodulated soybean exposed to elevated carbon dioxide. *Crop Science* 28, 671–677.

Cure, J.D., Rufty, T.W. Jr and Israel, D.W. (1988b) Phosphorus stress effects on growth and seed yield responses of nonnodulated soybean to elevated carbon dioxide. *Agronomy Journal* 80, 897–902.

Curry, R.B., Peart, R.M., Jones, J.W., Boote, K.J. and Allen, L.H. Jr (1990a) Simulation as a tool for analyzing crop response to climate change. *Transaction of the American Society of Agricultural Engineers* 33, 981–990.

Curry, R.B., Peart, R.M., Jones, J.W., Boote, K.J. and Allen, L.H. Jr (1990b) Response of crop yield to predicted changes in climate and atmospheric CO_2 using simulation. *Transaction of the American Society of Agricultural Engineers* 33, 1383–1390.

Curry, R.B., Jones, J.W., Boote, K.J., Peart, R.M., Allen, L.H. Jr and Pickering, N.B. (1995) Response of soybean to predicted climate change in the USA. In: Rosenzweig, C., Allen, L.H. Jr, Harper, L.A., Hollinger, S.E. and Jones. J.W. (eds) *Climate Change and Agriculture: Analysis of Potential International Impacts.* Special Publication No. 59, American Society of Agronomy, Madison, Wisconsin. pp. 163–182.

DeWitt, C.A., Waldron, R.E. and Lambert, J.E. (1983) *Response of Vegetation to Carbon Dioxide. Number 010: Effects of carbon dioxide enrichment on nitrogen fixation in soybeans (1982 Progress Report), Clemson University, Clemson, SC.* Joint Program of the US Department of Energy, Carbon Dioxide Research Division and the US Department of Agriculture, Agricultural Research Service, Washington, DC.

Egli, D.B. and Wardlaw, I.F. (1980) Temperature response of seed growth characteristics of soybean. *Agronomy Journal* 72, 560–564.

Fehr, W. and Caviness, E.C. (1977) *Stages of Soybean Development.* Iowa State University Cooperative Extension Service Special Report 80, Ames, Iowa.

Fiscus, E.L., Reid, C.D., Miller, J.E. and Heagle, A.S. (1997) Elevated CO_2 reduces O_3 flux and O_3-induced yield losses in soybeans: possible implications for elevated CO_2 studies. *Journal of Experimental Botany* 48, 307–313.

Gesch, R.W., Boote, K.J., Vu, J.C.V., Allen, L.H. Jr and Bowes, G. (1998) Changes in growth CO_2 result in rapid adjustment of ribulose-1,5-bisphosphate carboxylase/

oxygenase small subunit gene expression in expanding and mature leaves of rice. *Plant Physiology* 118, 521–529.

Gonzàles-Meler, M.A., Ribas-Carbó, M., Siedrow, J.N. and Drake, B.G. (1996) Direct inhibition of mitochondrial respiration by elevated CO_2. *Plant Physiology* 112, 1349–1355.

Grimm, S.S., Jones, J.W., Boote, K.J. and Herzog, D.C. (1994) Modeling the occurrence of reproductive stages after flowering for four soybean cultivars. *Agronomy Journal* 86, 31–38.

Hall, A.E. and Allen, L.H. Jr (1993) Designing cultivars for the climatic conditions of the next century. In: Buxton, D.R., Shibles, R., Forsberg, R.A., Blad, B.L., Asay, K.H., Paulsen, G.M. and Wilson, R.F. (eds) *International Crop Science I.* Crop Science Society of America, Madison, Wisconsin. pp. 291–297.

Harlan, J.R. (1992) *Crops and Man*, 2nd edn. American Society of Agronomy and Crop Science Society of America, Madison, Wisconsin.

Harley, P.C., Weber, J.A. and Gates, D.M. (1985) Interactive effects of light, leaf temperature, $[CO_2]$ and $[O_2]$ on photosynthesis in soybean. *Planta* 165, 249–263.

Harley, P.C. and Tenhunen, J.D. (1991) Modeling the photosynthetic response of C_3 leaves to environmental factors. In: Boote, K.J. and Loomis, R.S. (eds) *Modeling Crop Photosynthesis from Biochemistry to Canopy.* CSSA Special Publication 19, American Society of Agronomy, Madison, Wisconsin. pp. 17–39.

Heagle, A.S., Heck, W.E., Rawlings, J.O. and Philbrick, R.B. (1983) Effects of chronic doses of ozone and sulfur dioxide on injury and yield of soybeans in open-top field chambers. *Crop Science* 23, 1184–1191.

Heagle, A.S., Miller, J.E. and Booker, F.L. (1998a) Influence of ozone stress on soybean response to carbon dioxide enrichment. I. Foliar properties. *Crop Science* 38,113–121.

Heagle, A.S., Miller, J.E. and Pursley, W.A. (1998b) Influence of ozone stress on soybean response to carbon dioxide enrichment. III. Yield and seed quality. *Crop Science* 38, 128–134.

Heck, W.W., Taylor, O.C., Adams, R.M., Bingham, G., Miller, J.E., Preston, E.M. and Weinstein, L.H. (1982) Assessment of crop loss from ozone. *Journal of Air Pollution Control Association* 32, 353–361.

Heck, W.W., Cure, W.W., Rawlings, J.O., Zaragoza, L.J., Heagle, A.D., Heggestad, H.E., Kohut, R.J., Kress, L.W. and Temple, P.J. (1984) Assessing impacts of ozone on agricultural crops. I. Overview. *Journal of Air Pollution Control Association* 34, 729–735.

Hesketh, J.D., Myhre, D.L. and Willey, C.R. (1973) Temperature control of time interval between vegetative and reproductive events in soybeans. *Crop Science* 13, 250–254.

Hofstra, G. and Hesketh, J.D. (1975) The effects of temperature and CO_2 enrichment on photosynthesis in soybean. In: Marcelle, R. (ed.) *Environmental and Biological Control of Photosynthesis.* Dr W. Junk, The Hague, The Netherlands, pp. 71–80.

Huber, S.C., Rogers, H.H. and Mowry, F.L. (1984) Effects of water stress on photosynthesis and carbon partitioning in soybean (*Glycine max* [L.] Merr.) plants grown in the field at different CO_2 levels. *Plant Physiology* 76, 244–249.

Jones, P., Allen, L.H. Jr, Jones, J.W., Campbell, W.J. and Boote, K.J. (1984) Soybean canopy growth, photosynthesis, and transpiration responses to whole-season carbon dioxide enrichment. *Agronomy Journal* 76, 633–637.

Jones, P., Allen, L.H. Jr, Jones, J.W. and Valle, R. (1985a) Photosynthesis and transpiration responses of soybean canopies to short- and long-term CO_2 treatments. *Agronomy Journal* 77, 119–126.

Jones, P., Allen, L.H. Jr and Jones, J.W. (1985b) Responses of soybean canopy photosynthesis and transpiration to whole-day temperature changes in different CO_2 environments. *Agronomy Journal* 77, 242–249.

Jones, P., Jones, J.W. and Allen, L.H. Jr (1985c) Seasonal carbon and water balances of soybeans grown under CO_2 and water stress treatments in sunlit chambers. *Transaction of the American Society of Agricultural Engineers* 28, 2021–2028.

Jones, P., Jones, J.W. and Allen, L.H. Jr (1985d) Carbon dioxide effects on photosynthesis and transpiration during vegetative growth in soybeans. *Soil and Crop Science Society of Florida Proceedings* 44, 129–134.

Kimball, B.A. (1983) Carbon dioxide and agricultural yield: an assemblage and analysis of 430 prior observations. *Agronomy Journal* 75, 779–788.

Kimball, B.A., Mauney, J.R., Nakayama, F.S. and Idso, S.B. (1993) Effects of increasing CO_2 on vegetation. *Vegetatio* 104/105, 65–75.

Kramer, P.J. (1981) Carbon dioxide concentration, photosynthesis, and dry matter production. *BioScience* 31, 29–33.

Miller, J.E., Heagle, A.S. and Pursley, W.A. (1998) Influence of ozone stress on soybean response to carbon dioxide enrichment. II. Biomass and development. *Crop Science* 38, 122–128.

Morison, J.I.L. (1987) Intercellular CO_2 concentration and stomatal response to CO_2. In: Zeiger, E., Cowan, I. and Farquhar, G.D. (eds) *Stomatal Function*. Stanford University Press, Stanford, California. pp. 229–251.

Oechel, W.C. and Strain, B.R. (1985) Native species responses to increased atmospheric carbon dioxide concentration. In: Strain, B.R. and Cure, J.D. (eds) *Direct Effects of Increasing Carbon Dioxide on Vegetation*. US Department of Energy, Carbon Dioxide Research Division, DOE/ER-0238, Washington, DC, pp. 53–97.

Pan, D. (1996) Soybean responses to elevated temperature and CO_2. PhD dissertation, University of Florida, Gainesville, 224 pp.

Peart, R.M., Jones, J.W., Curry, R.B., Boote, K.J. and Allen, L.H. Jr (1989) Impact of climate change on crop yield in the Southeastern USA: a simulation study. In: Smith, J.B. and Tirpak, D.A. (eds) *The Potential Effects of Global Climate Change on the United States*. Appendix C, Agriculture, Vol. 1. EPA-230–05–89–053. US. EPA, Office of Policy, Planning and Evaluation (PM-221), Washington, DC, pp. 2–54.

Priestley, C.H.B. and Taylor, R.J. (1972) On the assessment of surface heat flux and evaporation using large-scale parameters. *Monthly Weather Review* 100, 81–92.

Pruitt, W.O. (1964) Cyclic relations between evapotranspiration and radiation. *Transaction of the American Society of Agricultural Engineers* 7, 271–275.

Reid, C.D. and Fiscus, E.L. (1998) Effects of elevated [CO_2] and/or ozone on limitations to CO_2 assimilation in soybean (*Glycine max*). *Journal of Experimental Botany* 49, 885–895.

Reid, C.D., Fiscus, E.L. and Burkey, K.O. (1998) Combined effects of chronic ozone and elevated CO_2 on Rubisco activity and leaf components in soybean (*Glycine max*). *Journal of Experimental Botany* 49, 1999–2011.

Rogers, H.H., Bingham, G.E., Cure, J.D., Smith, J.M. and Surano, K.A. (1983) Responses of selected plant species to elevated carbon dioxide in the field. *Journal of Environmental Quality* 12, 569–574.

Rogers, H.H., Cure, J.D., Thomas, J.F. and Smith, J.M. (1984) Influence of elevated CO_2 on growth of soybean plants. *Crop Science* 24, 361–366.

Rogers, H.H., Cure, J.D. and Smith, J.M. (1986) Soybean growth and yield response to elevated carbon dioxide. *Agricultural Ecosystems and Environment* 16, 113–128.

Rogers, H.H., Runion, G.B. and Krupa, S.V. (1994) Plant responses to atmospheric CO_2 enrichment with emphasis on roots and the rhizosphere. *Environmental Pollution* 83, 155–189.

Rogers, H.H., Runion, G.B., Krupa, S.V. and Prior, S.A. (1997) Plant responses to atmospheric carbon dioxide enrichment: implications in root–soil–microbe interactions. In: Allen, L.H. Jr, Kirkham, M.B., Olszyk, D.M. and Whitman, C.E. (eds) *Advances in Carbon Dioxide Effects Research*. Special Publication No. 61, ASA-CSSA-SSSA, Madison, Wisconsin, pp. 1–34.

Rosenberg, N.J., Kimball, B.A., Martin, F. and Cooper, C.F. (1990) From climate and CO_2 enrichment to evapotranspiration. In: Waggoner, P.E. (ed.) *Climate Change and US Water Resources*. John Wiley & Sons, New York, pp. 151–175.

Rowland-Bamford, A.J., Allen, L.H. Jr, Baker, J.T. and Bowes, G. (1991) Acclimation of rice to changing atmospheric carbon dioxide concentration. *Plant, Cell and Environment* 14, 577–583.

SAS (1985) *User's Guide: Statistics,* 5th edn. SAS Institute, Cary, North Carolina.

Sasek, T.W., DeLucia, E.H. and Strain, B.R. (1985) Reversibility of photosynthetic inhibition in cotton after long-term exposure to elevated CO_2 concentrations. *Plant Physiology* 78, 619–622.

Serraj, R., Sinclair, T.R. and Allen, L.H. Jr (1998) Soybean nodulation and N_2 response to drought under carbon dioxide enrichment. *Plant, Cell and Environment* 21, 491–500.

Shibles, R., Anderson, I.C. and Gibson, A.H. (1975) Soybean. In: Evans, L.T. (ed.) *Crop Physiology, Some Case Histories.* Cambridge University Press, Cambridge, pp. 151–189.

Sionit, N., Strain, B.R. and Flint, E.P. (1987a) Interaction of temperature and CO_2 enrichment on soybean: growth and dry matter partitioning. *Canadian Journal of Plant Science* 67, 59–67.

Sionit, N., Strain, B.R. and Flint, E.P. (1987b) Interaction of temperature and CO_2 enrichment on soybean: photosynthesis and seed yield. *Canadian Journal of Plant Science* 67, 629–636.

Tanner, C.B. and Lemon, E.R. (1962) Radiant energy utilized in evapotranspiration. *Agronomy Journal* 54, 207–212.

Thomas, J.F. and Harvey, C.N. (1983) Leaf anatomy of four species grown under continuous CO_2 enrichment. *Botanical Gazette* 144, 303–309.

Tissue, D.L. and Oechel, W.C. (1987) Physiological response of *Eriophorum vaginatum* to field elevated CO_2 and temperature in the Alaskan tussock tundra. *Ecology* 68, 401–410.

Valle, R.R., Mishoe, J.W., Campbell, W.J., Jones, J.W. and Allen, L.H. Jr (1985a) Photosynthetic responses of 'Bragg' soybean leaves adapted to different CO_2 environments. *Crop Science* 25, 333–339.

Valle, R.R., Mishoe, J.W., Jones, J.W. and Allen, L.H. Jr (1985b) Transpiration rate and water-use efficiency of soybean leaves adapted to different CO_2 environments. *Crop Science* 25, 477–482.

Vu, C.V., Allen, L.H. Jr and Bowes, G. (1983) Effects of light and elevated CO_2 on ribulose-1,5-bisphosphate carboxylase activity and ribulose-1,5-bisphosphate level of soybean leaves. *Plant Physiology* 73, 729–734.

Vu, J.C.V., Allen, L.H. Jr and Bowes, G. (1989) Leaf ultrastructure, carbohydrates, and protein of soybeans grown under CO_2 enrichment. *Environmental Experimental Botany* 29, 141–147.

Vu, J.C.V., Allen, L.H. Jr, Boote, K.J. and Bowes, G. (1997) Effects of elevated CO_2 and temperature on photosynthesis and Rubisco in rice and soybean. *Plant, Cell and Environment* 20, 68–76.

8 Crop Ecosystem Responses to Climatic Change: Cotton

K. RAJA REDDY[1], HARRY F. HODGES[1] AND BRUCE A. KIMBALL[2]

[1]Department of Plant and Soil Sciences, Mississippi State University, 117 Dorman Hall, Box 9555, Mississippi State, MS 39762, USA; [2]USDA-ARS, Environmental and Plant Dynamics Research Unit, US Water Conservation Laboratory, 4331 East Broadway, Phoenix, AZ 85040, USA

8.1 Introduction

Human activities have resulted in increased atmospheric levels of carbon dioxide and many other greenhouse gases. This increase and mankind's continued activity have the potential to warm the earth's climate over the next century (Houghton *et al.*, 1996). As human activities result in continued potential perturbations in climate, the increasing world population also puts pressure on agriculturists and sets new challenges for crop scientists to help to meet the larger population food and fibre needs. The world's population is forecast to increase from 5.3 billion in 1990 to 8.1 billion by the year 2025, with about 84% of this growth expected in the developing countries (Bos *et al.*, 1995). By the year 2050, experts believe that 12.4 billion people will inhabit the earth. Since there is essentially no new arable land that can be cultivated to meet the demand, the increased food supply must come primarily from more intensive cultivation of existing arable land.

As agriculture becomes more intensive, soil degradation will become a major concern. The world's water resource is also finite, and changing climate and increasing population demands will result in the availability of less water for agriculture. Metropolitan and urban communities nearly always have higher priorities than agricultural production for a scarce resource such as water. In many highly populated countries, food and fibre needs are being met by irrigating up to 75% of the arable land (Hoffman *et al.*, 1990). If a major climate change occurs as is forecast, agriculturalists will face a daunting challenge to produce, in an environmentally sustainable manner, enough food and fibre to satisfy the world's increasing population.

Cotton belongs to the genus *Gossypium* of the *Malvaceae* family. Of the 39 species of that genus, which are diverse in habitat, only four produce commercial lint and are grown commercially throughout the world. The

Upland and Acala varieties belong to *G. hirsutum* L. and the extra-long staple Pima and Sea island or Egyptian varieties belong to *G. barbadense* L. Upland cotton is grown on more than 5 Mha in the USA and more than 34 Mha worldwide (USDA, 1989). Most of the world's production is in arid and semi-arid climates and must be irrigated for commercial production. Major cotton-producing countries during the 1997/98 market year (million bales) were: USA, 18.8; China, 18.5; India, 12.5; Pakistan, 7.5; Uzbekistan, 5.4; Turkey, 3.3; and Australia, 2.9 (USDA, 1998).

Cotton is grown worldwide, but in a relatively narrow temperature range compared with many other species. The minimum temperature for growth and development of cotton is 12–15°C, optimum temperature is 26–28°C, and maximum temperature depends on the duration of exposure (K.R. Reddy *et al.*, 1997b). Even short periods of above-optimum canopy temperatures may cause injury to young fruit. Canopy temperatures do not necessarily follow air temperature closely, however, and in well-watered, low-humidity environments, cotton crop canopy temperatures may be several degrees cooler than air temperature (e.g. Idso *et al.*, 1987; Kimball *et al.*, 1992a).

Climatic conditions in the middle or latter part of the 21st century are expected to be different from those of today. Currently, atmospheric carbon dioxide concentration [CO_2] is about 360 μmol mol^{-1} (Keeling and Whorf, 1994) and there is general agreement among climatic and atmospheric scientists that [CO_2] could be in the range of 510–760 μmol mol^{-1} some time in the middle or latter part of the 21st century (Rotty and Marland, 1986; Trabalka *et al.*, 1986). Other greenhouse trace gases are also increasing rapidly and will contribute to climatic change. These changes are predicted to warm the earth by 2–5°C. Since plant growth and crop production are controlled by weather, it is important to understand the implications of such weather changes on agriculture.

Temperatures that routinely occur in many cotton-producing areas strongly limit many growth and developmental processes (K.R. Reddy *et al.*, 1992a, 1997a,b). Elevated [CO_2] generally enhances leaf and canopy CO_2 assimilation rates in plants because CO_2 is the substrate for photosynthesis, and also is a competitive inhibitor for photorespiration. Both of these factors result in increased growth and productivity (Kimball, 1983a,b, 1986; Bowes, 1993; K.R. Reddy *et al.*, 1995c, 1996a). Elevated [CO_2] often reduces stomatal aperture (e.g. Morison, 1987) and increases the ratio of CO_2 assimilated relative to water transpired. Some studies have also found that elevated [CO_2] causes altered partitioning of photoassimilate among plant organs (Rogers *et al.*, 1994).

This chapter will help to quantify these processes and their effects on crops and develop ways to manage crops effectively with the available climatic resources. In particular, we need to summarize and integrate knowledge of crop responses to weather factors and link those responses appropriately with management factors. In this chapter, we review cotton responses to global climate change.

8.2 Approach and Methodology

There are several approaches one might use to determine the impact of global change on crop production. This chapter will discuss the use of naturally sunlit chambers known as soil–plant–atmosphere research (SPAR) units, which are capable of controlling and monitoring temperature, water, nutrients and [CO_2] systematically. The SPAR facility has been described in detail elsewhere (K.R. Reddy *et al.*, 1992b; V.R. Reddy *et al.*, 1995b; V.R. Reddy and K.R. Reddy, 1998). It has the advantage of having near-natural levels of radiation incident on the plants, and excellent control of other physical factors. Crop data obtained in this manner are less ambiguous and allow understanding of the responses to environmental variables and nutrient status. SPAR units have the disadvantage of containing a limited number of plants, so experiments that require much destructive sampling during the growing season must be avoided. Experiments using SPAR units have shown unambiguously that cotton responds to several environmental factors of concern in a changing climate (Table 8.1). Original references should be checked to determine how individual experiments were conducted.

This chapter will also discuss some results from the experiments conducted in open-top CO_2-enrichment chambers (OTCs) and free-air carbon dioxide enrichment (FACE) settings in Arizona (Table 8.2). The OTCs enable plants to be grown in normal field soils and in aerial environments that approach open-field conditions but with elevated levels of CO_2 (Kimball *et al.*, 1992b, 1997; Kimball and Mauney, 1993). However, like other chamber methods, the walls shade the plants somewhat, and the wind-flow patterns and energy exchange processes are different from those outside. Generally, air temperatures are somewhat warmer and humidities are somewhat higher inside than outside (Kimball *et al.*, 1997).

The FACE approach allows experimental plants to be grown under conditions as representative of open fields in the future high-[CO_2] world as is possible to create them today (Hendrey, 1993; Dugas and Pinter, 1994). Although precise techniques have detected a slight disturbance of the microclimate if the crop is enriched at night (Pinter *et al.*, 2000), the FACE plants generally experience full sunlight and normal wind, air temperature and humidity conditions. The relatively large plot size enables frequent destructive sampling and also enables researchers from many different disciplines to make measurements on nearly identically grown plant material (Kimball *et al.*, 1997). A major disadvantage of the FACE technique is the high cost of the prodigious amounts of CO_2 required. However, because the relatively large plot size provides an economy of scale, the FACE approach is least expensive per unit of high-[CO_2]-grown plant material (Kimball, 1993).

8.3 Photosynthesis

Cotton canopies grown continuously in elevated [CO_2] responded to increasing solar radiation with increasing rates of photosynthesis (Fig. 8.1).

Table 8.1. Treatment structures for experiments conducted on cotton in naturally sunlit environment chambers (SPAR units) for the last 10 years.

Year	Cultivar	Temperatures (°C) day/night	$[CO_2]$ (μmol mol^{-1})	Comments and references
Expt 1-1989	DPL 50	15/7, 20/10, 25/15, 30/20, 35/25	350, 700	70 days from emergence, well-watered and fertilized (V.R. Reddy *et al.*, 1994a,b, 1995a,b)
Expt 2-1989	DPL 50	30/22	Several $[CO_2]$s	Short-term, few weeks during flowering, well-watered and fertilized (K.R. Reddy *et al.*, 1995b)
Expt 3-1989	DPL-50	Several high temperatures	600	4 weeks during the fruiting period, flower abscission study (V.R. Reddy *et al.*, 1994a,b, 1995a)
Expt 4-1990	Pima-S-6	20/12, 25/17, 30/22, 35/27, 40/32	350, 700	64 days from emergence, well-watered and fertilized (K.R. Reddy *et al.*, 1995a,b)
Expt 5-1990	Pima-S-6	25/17, 30/22, 35/27	350, 700	Flowering to maturity, well-watered and fertilized (K.R. Reddy *et al.*, 1995a,b,c)
Expt 6-1990	Pima-S-6	Four high temperatures	700	Flowering to end of season, fruit retention study (A.R. Reddy *et al.*, 1997; K.R. Reddy *et al.*, 1993)
Expt 7-1991	Pima-S-6	30/22	350, 450,700	95 days from emergence, three drought stress levels (K.R. Reddy and Hodges, 1998)
Expt 8-1991	DES 119	30/22	350	Four nitrogen levels (K.R. Reddy *et al.*, 1997b; V.R. Reddy *et al.*, 1997)
Expt 9-1992	DPL 5415	26/18, 31/23, 36/18	350, 450, 700	60 days from emergence, well-watered and fertilized (A.R. Reddy *et al.*, 1998)

Expt 10-1992	DPL 5415	26/18, 31/23, 36/18	350, 450, 700	Flowering to maturity, well-watered and fertilized.
Expt 11-1993	DES 119	30/22	350, 700	49 days from emergence, five N levels (K.R. Reddy *et al.*, 1997b)
Expt 12-1993	DES 119	30/22	350, 450, 700	80 days from emergence, three drought stress levels (K.R. Reddy *et al.*, 1997c)
Expt 13-1994	Acala Maxxa HS-26, DPL 51	20/12, 25/17, 30/22, 35/27, 40/32	360, 720	46 days from emergence, well-watered and fertilized (K.R. Reddy *et al.*, 1997d)
Expt 14-1994	DPL 51	Temperatures: long-term MS July mean, July mean −2, and July mean plus 2, 5, and 7	360, 700	4 weeks, flowering period, well-watered and fertilized (K.R. Reddy *et al.*,1997a)
Expt 15-1995	DPL 51	Temperatures: 1995 ambient, 1995 ambient −2, and 1995 ambient plus 2, 5 and 7	360, 720	Full-season, well-watered and fertilized (K.R. Reddy *et al.*, 1997a; 1998).
Expt 16-1996	NuCot33	30/22	360, 720	84 days, five K levels
Expt 17-1996	NuCott33	26/26	360	Manual de-leafing and de-fruiting study, well-watered and fertilized
Expt 18-1997	Nucot33	30/22	360, 720	Several water deficient studies
Expt 19-1997	NuCot33	Several short-term temperature treatments	Several [CO_2]s	Short-term, few days to weeks, well-watered and fertilized

Table 8.2. Treatment structures for experiments conducted on cotton in open-top chambers (OTCs) and free-air CO_2 enrichment (FACE) studies.

Year	Cultivar	Water supply	Nutrient supply	$[CO_2]$, (μmol mol^{-1})	Comments and references
Open-top chamber experiments					
1983	DPL 70	Well-watered	Well-fertilized	Ambient, 500, 650	Ambient-no-chamber treatment also included (Kimball *et al.*, 1992b; Kimball and Mauney, 1993)
1984	DPL 61	Well-watered Drought-stressed	Well-fertilized	Ambient, 500, 650	Ambient-no-chamber treatment also included (Kimball *et al.*, 1992b; Kimball and Mauney, 1993)
1985	DPL 61	Well-watered Drought-stressed	Well-fertilized	Ambient, 500, 650	Ambient-no-chamber treatment also included (Kimball *et al.*, 1992b; Kimball and Mauney, 1993)
1986	DPL 61	Well-watered Drought-stressed	High-N Low-N	Ambient, 650	Kimball *et al.* (1992b); Kimball and Mauney (1993)
1987	DPL 61	Well-watered Drought-stressed	High-N Low-N	Ambient, 650	Barley N-removal crop grown during winter of 1986–1987 (Kimball *et al.*, 1992b; Kimball and Mauney, 1993)
Free-air CO_2 enrichment (FACE) experiments					
1989	DPL 77	Well-watered	Well-fertilized	Ambient, 550	Hendrey (1993); Dugas and Pinter (1994)
1990	DPL 77	Well-watered Drought-stressed	Well-fertilized	Ambient, 550	Hendrey (1993); Dugas and Pinter (1994)
1991	DPL 77	Well-watered Drought-stressed	Well-fertilized	Ambient, 550	Hendrey (1993); Dugas and Pinter (1994)

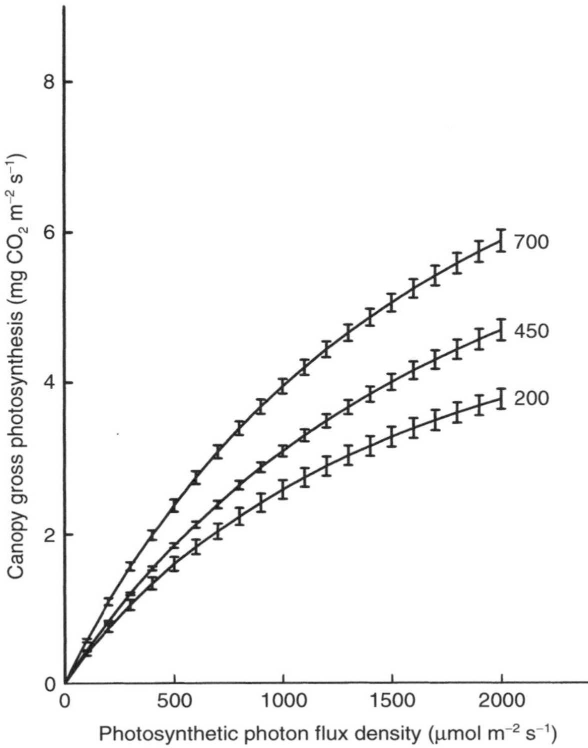

Fig. 8.1. Influence of carbon dioxide and solar radiation (photosynthetic photon flux density, PPFD) on cotton canopy gross photosynthesis (P_g) at 60 DAE on SPAR experiments. P_g = net photosynthesis plus respiration. (K.R. Reddy *et al.*, 1995c.)

Both the initial slopes of the light response curves and the asymptotes were greater in elevated [CO_2]. This indicates that in nearly optimum growth conditions, additional [CO_2] resulted in higher rates of photosynthesis due to the efficiency of the carboxylase fixation rates (higher rates where light is limiting) and diffusion of the gas to the fixation sites (higher rates where CO_2 is the primary limiting factor). This point was further illustrated by the photosynthetic (P_{net}) responses of cotton canopies to CO_2 and radiation at different air temperatures (Fig. 8.2). This experiment was conducted by growing plants in ambient and twice-ambient [CO_2] at five temperatures. The 1995 temperature in Mississippi, USA, was used as a reference, with the other temperatures being 1995 minus 2°C, and 1995 plus 2, 5 and 7°C. Daily and seasonal variation and amplitudes were maintained. Net photosynthesis was less at both higher and lower temperatures than at optimum. The trends of the responses to air temperature and [CO_2] were similar to those reported by K.R. Reddy *et al.* (1995c). At the three lower temperatures and ambient CO_2, P_{net} response to increasing radiation decreased as temperature increased. This apparently reflected the higher respiration caused by higher temperatures of the plant canopies. The opposite effect was found with high CO_2, but this was

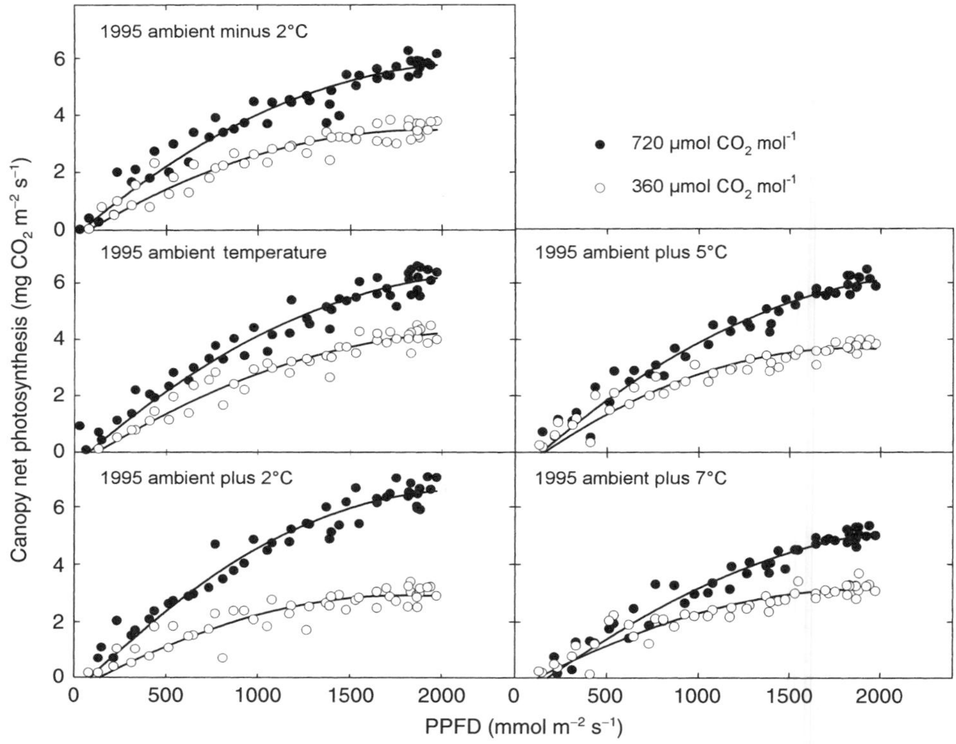

Fig. 8.2. Effect of solar radiation (PPFD), [CO_2] and temperature on cotton canopy net photosynthesis (P_n) in 1995 SPAR experiments. Plants were grown from emergence in their respective [CO_2] and temperature environments. Data represent 80 days after emergence when the canopies were intercepting about 95–98% of the incoming solar radiation.

the result of greater growth in the elevated [CO_2]-grown plants and more young leaves that contributed to higher photosynthetic rates.

In the 1995 plus 5 and plus 7°C temperature treatments, the photosynthetic rates of plants in the 760 μmol CO_2 mol^{-1} environments were also high. These plants abscised their bolls soon after anthesis. Cotton is a perennial plant that is managed for commercial cotton production as an annual. Cotton plants typically set their first fruit on a fruiting branch near the bottom of the plant (nodes 5–8) (K.R. Reddy *et al.*, 1997b). In optimum conditions, a new fruiting site is produced at the next higher mainstem node 3 days later and at the next node of the same fruiting branch 6 days later. Cotton plants progressively add bolls until the available photosynthetic supply will no longer support additional fruit. As the nutrient requirements for supporting a high population of bolls on the plant increases, the nutrients for vegetative growth become less available. This causes slowing and eventual cessation of stem and leaf growth. Reduced stem growth is reflected in both slower expansion of internodes and fewer nodes produced. As fewer nodes are produced, fewer fruiting branches and therefore fewer sites for additional fruit are available. Even under good conditions, nutrient requirements decline as bolls mature and vegetative growth resumes. In high-temperature environments (above 32°C), fruits were abscising 3–5 days after anthesis, thus voiding typical nutrient sinks and therefore promoting luxuriant vegetative growth. The high photosynthetic rates in these high-temperature environments (1995 plus 5 or plus 7°C) resulted from the relatively young population of leaves continuously being added to the top of the canopy.

The impact of [CO_2] on photosynthetic efficiency is shown in Fig. 8.3. By interpolating from that figure, one can determine that cotton plants were fixing CO_2 at 4.3 g CO_2 m^{-2} MJ^{-1} in ambient (360 μmol mol^{-1}) and 6.3 g CO_2 m^{-2} MJ^{-1} in twice-ambient [CO_2]. This represents a 25% increase in photosynthetic efficiency caused by doubling the [CO_2]. Both of these values represent highly efficient fixation rates in non-stressed environments. The photosynthetic efficiency was limited at the low [CO_2] by the available CO_2. The efficiency of CO_2 fixation increased as [CO_2] increased, but at a diminishing rate. At twice today's ambient [CO_2], essentially all the feasible gain in photosynthesis was accomplished.

The information provided in Fig. 8.4 shows changes in canopy photosynthesis throughout the season with normalized radiation of 1200 μmol m^{-2} s^{-1}, and naturally varying temperature. These data illustrate photosynthetic rates over time with solar radiation normalized so that seasonal variation in radiation was not a confounding factor. Photosynthesis increased until about 80 days after emergence, and then decreased throughout the rest of the growing season. Crops growing in ambient [CO_2] decreased their photosynthetic rates more than crops growing in twice-ambient [CO_2]. This was caused by the increasing age of the light-intercepting canopies. In the high CO_2 environment, photosynthesis was still proceeding at a very healthy

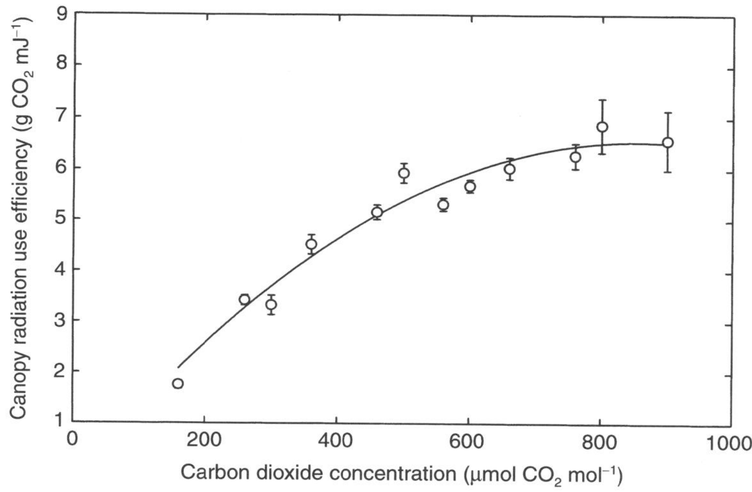

Fig. 8.3. Effect of atmospheric [CO_2] on radiation use efficiency of cotton canopies. (K.R. Reddy *et al.*, 1997c.)

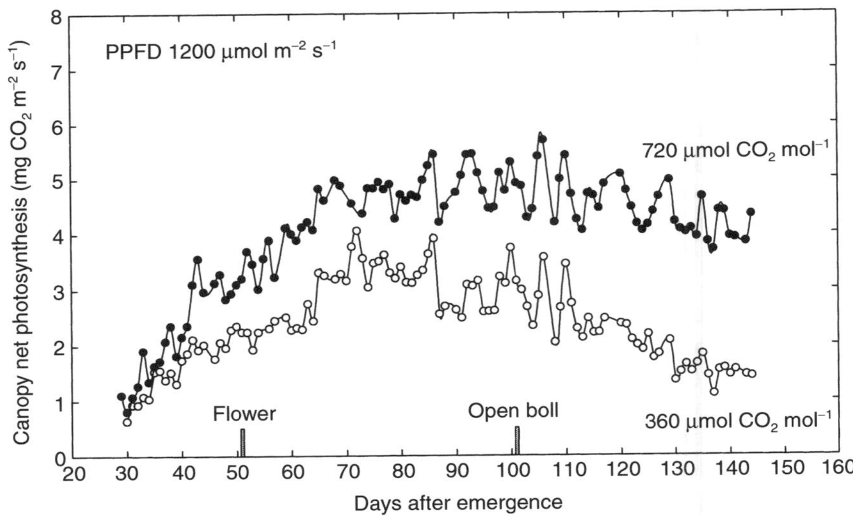

Fig. 8.4. Temporal trends in net photosynthesis (P_n) of cotton canopies grown in 1995 ambient temperatures in 360 and 720 µmol CO_2 mol^{-1} at Mississippi State, Mississippi. Photosynthesis was measured continuously, summarized at intervals of 900 s, and normalized at 1200 µmol m^{-2} s^{-1} each day throughout the season. Time of first flower and first mature boll is indicated. (K.R. Reddy *et al.*, 1998b.)

rate even at the end of the season and provided enough reduced carbon to support some new leaf growth during the fruiting period.

Average photosynthetic rates of the canopies at both ambient and twice-ambient [CO_2] were summarized as rates at 720/360 µmol CO_2 mol^{-1} and plotted against average seasonal temperatures (Fig. 8.5). The seasonal average photosynthetic rate at 720 µmol CO_2 mol^{-1} was about 140% of that at 360 µmol mol^{-1} at 20°C and 32°C, but the response increased to more than 180% at nearly optimum temperatures (26–28°C). Figure 8.6 provides the relative photosynthetic response for cotton canopies grown at elevated [CO_2] compared with crops grown at ambient [CO_2], but in a range of temperatures and water-deficient conditions. The data in this figure represent photosynthetic rates of both Pima and Upland cotton canopies, and although there was considerable variability among individual data points, the regression line indicated that photosynthesis at 720 µmol CO_2 mol^{-1} was 156% of that at 360 µmol CO_2 mol^{-1}. The photosynthetic response to twice-ambient [CO_2] appeared to be linear even at high rates.

These results from controlled-environment chambers are generally consistent with observations of the effects of elevated [CO_2] on photosynthesis of field-grown plants. Radin *et al.* (1987) reported that CO_2 enrichment to 650 µmol CO_2 mol^{-1} in the 1985 open-top chamber experiment (Table 8.2) increased leaf photosynthesis of well-watered cotton more than 70%. Hileman *et al.* (1994) found that both leaf and canopy photosynthetic rates in well-watered conditions were increased by about 27% when [CO_2] was

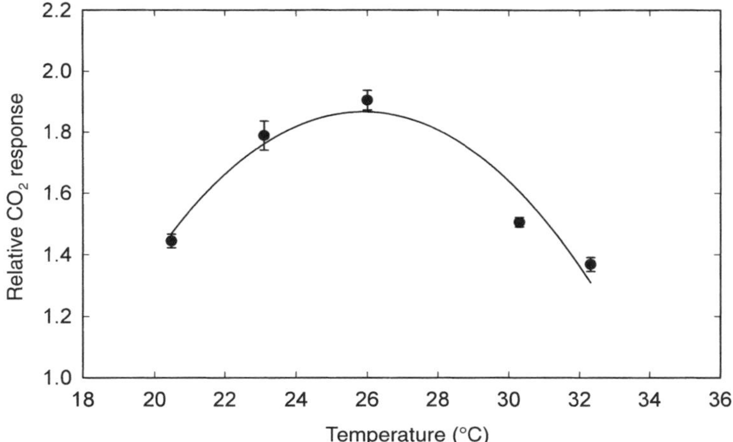

Fig. 8.5. Effect of temperature on the relative response of cotton canopy photosynthesis to doubling ambient [CO_2]. Temperature and photosynthetic rates were averaged over the season from the time 95% of the radiation was intercepted by the canopy until first open boll.

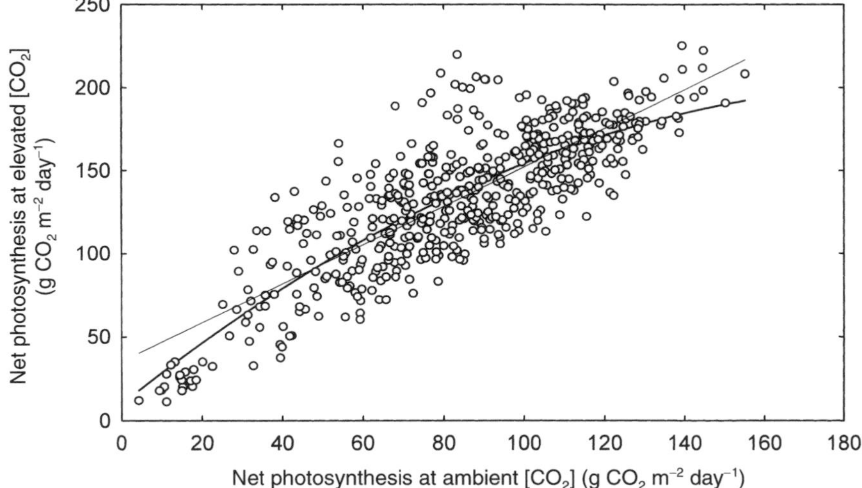

Fig. 8.6. Relation between daily total net canopy photosynthesis of cotton plants grown at ambient (360 μmol mol^{-1}) and elevated [CO_2]. The data are from five temperature treatments of Upland cotton and three water-deficit treatments of Pima cotton collected over several days during fruiting.

enriched to 550 μmol mol^{-1} in the 1989 FACE experiment (Table 8.2), and canopy rates were increased by about 31% in 1990.

Cotton plants grown in water-deficient conditions in controlled-environment chambers decreased their photosynthetic rates as midday leaf water decreased regardless of the [CO_2] (Table 8.3; K.R. Reddy *et al.*, 1997c). Plants grown in 720 μmol CO_2 mol^{-1} fixed an average of 48% more CO_2 than

Table 8.3. Parameters for equations regressing several developmental events/rates and growth of cotton grown in elevated (700 or 720 μmol CO_2 mol^{-1}) atmospheric CO_2 (*y*) as a function of plants grown in ambient (350 or 360 μmol CO_2 mol^{-1}) CO_2 (*x*) ($y = b_0 + b_1x - b_2x^2$). The data were obtained from plants grown at a range of temperatures and nutrient and water deficient conditions.

	Regression parameters			
Parameter	b_0	b_1	b_2	r^2
Developmental events				
Emergence to square (days)	−0.595	1.016	–	0.99
Square to flower (days)	−1.184	0.944	–	0.92
Flower to open boll (days)	−1.615	0.996	–	0.99
Mainstem nodes (no. $plant^{-1}$)	0.313	0.998	–	0.98
Fruiting sites (no. $plant^{-1}$)	−0.999	1.384	–	0.95
Roots (no. m^{-2} day^{-1})	−2.460	1.160	–	0.94
Stomatal density (no. mm^{-2})	−9.365	0.954	–	0.99
Growth and other processes				
Plant height (cm $plant^{-1}$)	2.419	1.026	–	0.93
Branch length (m $plant^{-1}$)	−103.88	1.168	–	0.96
Leaf area (m^2 $plant^{-1}$)	−0.022	1.248	–	0.98
Total plant weight (g $plant^{-1}$)	2.352	1.324	–	0.88
Stem weight (g $plant^{-1}$)	−0.0427	1.434	–	0.97
Leaf weight (g $plant^{-1}$)	1.936	1.240	–	0.97
Root weight (g $plant^{-1}$)	0.156	1.298	–	0.89
Fruit weight (g $plant^{-1}$)	0.216	1.256	–	0.98
Net photosynthesis (g m^{-2} day^{-1})	−107.129	9.012	−0.167	0.94
Transpiration (kg H_2O m^{-2} ground area day^{-1})	2.7799	0.923	–	0.83
Transpiration (mg H_2O m^{-2} leaf area s^{-1})	36.741	0.510	–	0.99
Leaf N (%)	−0.46	0.954	–	0.95

plants grown in 360 μmol mol^{-1} in several water-deficient conditions. The midday leaf water potential varied from approximately −1.2 MPa to −3.0 MPa. There was no significant interaction between the photosynthetic response of plants grown in elevated [CO_2] and water-deficient conditions. Plants grown in both ambient and twice-ambient [CO_2] decreased photosynthesis by about 43% as their midday leaf water potential decreased from −1.2 MPa to −3.0 MPa. Field results have been similar. Radin *et al.* (1987) measured a 52% increase in net leaf photosynthesis when [CO_2] was at 650 μmol CO_2 mol^{-1} under drought-stressed conditions in the 1995 open-top chamber experiment (Table 8.2). Hileman *et al.* (1994) reported a 21% increase in canopy photosynthesis with enrichment to 550 μmol CO_2 mol^{-1} under drought stress in the 1990 FACE experiment.

Cotton plant responses to N deficiency in controlled-environment chambers were somewhat similar to plant responses to water deficits (Table

8.3; K.R. Reddy *et al.*, 1997b,c). Photosynthesis decreased as the leaf nitrogen decreased. Plants grown in twice-ambient [CO_2] decreased their photosynthesis by about the same amount in N-deficient conditions as plants grown in ambient [CO_2]. Plants grown with 2% leaf N in both 360 and 720 μmol CO_2 mol^{-1} had 38% lower photosynthetic rates than plants with 4% leaf N, and plants grown at twice-ambient [CO_2] only fixed 41% more CO_2 than plants grown in ambient [CO_2]. Canopy-level photosynthesis was greater in plants grown in higher N with both ambient and elevated [CO_2]. Photosynthesis declined rapidly during the fruiting period when canopies became N deficient, but if N was sufficient (leaf N about 3%), the canopy photosynthetic rates maintained about 50% higher rates in twice-ambient [CO_2] than in ambient [CO_2] (K.R. Reddy and H.F. Hodges, unpublished data). Cotton canopies maintained 90% maximum photosynthetic rates with lower leaf N in a high-[CO_2] environment than in ambient [CO_2]. Similar gains in efficiency of N utilization have been found in rice (P.J. Conroy, NSW, Australia, personal communication, 1999). These results suggest that higher yields may be obtainable in high-[CO_2] environments with increasing N fertilization. Field-grown cotton exposed to 650 μmol CO_2 mol^{-1} in OTCs increased net leaf photosynthetic rates by 45–70%, depending on the year, and there were no significant effects of N stress on the response (Table 8.2; Kimball *et al.*, 1986, 1987).

Potassium deficiency caused 9.1% and 6.8% declines in photosynthetic rates as leaf K decreased from 4% to 2% in ambient and twice-ambient [CO_2] environments, respectively (K.R. Reddy *et al.*, 1997c). However, photosynthesis decreased more rapidly as leaf K decreased from 2% to 1%. Photosynthesis was 10.6% and 15.7% lower in the 1% leaf K plants compared with the 2% leaf K plants in ambient and twice-ambient [CO_2], respectively. Potassium nutrition and [CO_2] did not interact in their effect on photosynthesis when the leaf K concentration was above 2%, but photosynthesis declined to nearly zero in plants grown with the extremely low K. Thus, it appears that whether stresses are caused by water deficits or N or K deficiencies, there is little or no interaction with the effect of [CO_2] on photosynthesis. Barrett and Gifford (1995) found similar results with P nutrition and [CO_2] enrichment.

8.4 Transpiration and Water Use

Water is the key variable that affects cotton production, since the crop is grown mostly in arid and semi-arid regions of the world under rain-fed conditions. Transpiration from individual leaves growing in high [CO_2] is usually lower than that from leaves growing in ambient [CO_2] (e.g. Kimball and Idso, 1983; K.R. Reddy *et al.*, 1996b) because the elevated [CO_2] causes partial stomatal closure. Changes in stomatal density in herbarium samples collected over several centuries indicate that increasing [CO_2] may also be affecting stomatal density. However, there is insufficient information to have confidence that changes in stomatal density are occurring. Growing cotton plants for an entire season in elevated [CO_2] produced leaves throughout the season that had similar stomatal densities to leaves produced in ambient [CO_2] (K.R. Reddy *et al.*, 1998a). Therefore, if stomatal densities have changed because of elevated [CO_2], alterations must have occurred over many generations rather than within a single generation.

Several factors affect the degree to which a reduction in leaf transpiration will change water use per unit of land area. As the elevated [CO_2] causes partial stomatal closure, the resultant decreases in transpirational cooling increases the foliage temperature of cotton (Idso *et al.*, 1987; Kimball *et al.*, 1992a). The increased foliage temperature increases the partial pressure of water vapour inside the leaves and increases leaf transpiration, thereby partially counteracting the CO_2-induced stomatal closure. At the same time, the CO_2 stimulation of growth results in larger plants (e.g. Kimball, 1983a,b) with larger leaf areas, which would also tend to increase whole-plant transpiration. Samarakoon and Gifford (1995) recently conducted an interspecific comparison among cotton, wheat and maize, using glasshouses, that nicely illustrated the dependence of water use on the relative effects of CO_2 on changes in leaf area and stomatal conductance. Cotton had a large increase in leaf area and a small change in conductance so that water use per pot actually increased. Maize had very little photosynthetic or leaf area response to elevated [CO_2] and so the reduction in conductance resulted in significant water conservation. Wheat was intermediate between the other two species.

Cotton plants grown in ambient and twice-ambient [CO_2] in the SPAR units were gradually allowed to develop increasingly more severe water deficits while other conditions remained uniform (K.R. Reddy *et al.*, 1997c). Measurements taken at midday showed that leaf water potentials of plants grown in ambient and twice-ambient [CO_2] were not different. Also, when placed in environments of varying evaporative-demand, with adequate soil moisture, transpiration rates (per unit of land area) of canopies grown under the two CO_2 environments were similar (Fig. 8.7; V.R. Reddy *et al.*, 1995b). Kimball *et al.* (1993) found a slight decrease (4%) in seasonal water use of cotton as measured with lysimeters in OTCs at 650 μmol CO_2 mol^{-1} in 1983 (Table 8.2). Determinations of water use as a residual in the soil water balance indicated that water use changed from −5 to +28% in 1983 and 1984. Being devoid of walls, the FACE approach has the least disturbance of wind flow and other micrometeorological factors. For this reason, the most definitive data on the effects of elevated [CO_2] on cotton water use are probably those from the FACE experiment in 1991 (Table 8.2). In that experiment, cotton water use was determined from sap flow measurements (Dugas *et al.*, 1994), as a residual in the soil water balance (Hunsaker *et al.*, 1994) and as a residual in the energy balance (Kimball *et al.*, 1994). The conclusion from all three independent measurements was that, under ample water supply, elevated [CO_2] at 550 μmol mol^{-1} did not significantly affect water use of cotton per unit of land area under open-field conditions. These results are similar to those found for plants grown in the SPAR units.

Air temperature also affects transpiration rates. In the SPAR experiments, transpiration rates of individual leaves strongly interacted with [CO_2] and temperature. Transpiration rates increased linearly on a leaf area basis as

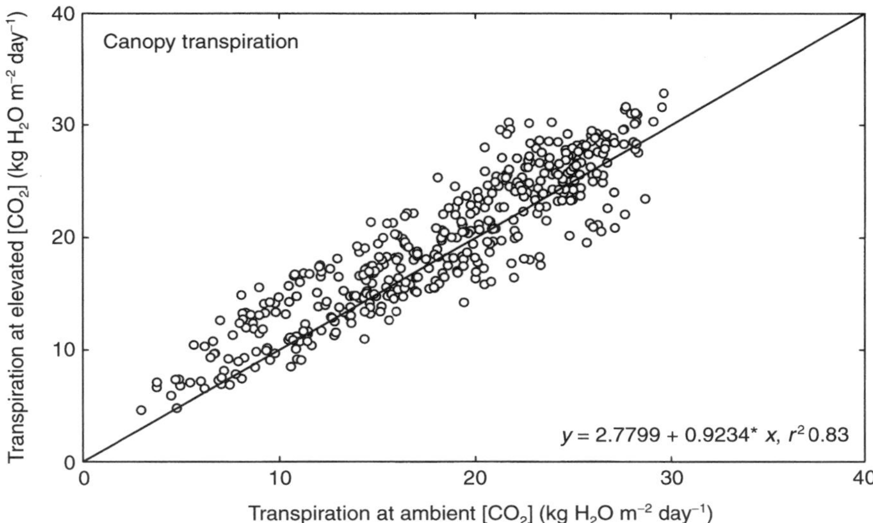

Fig. 8.7. Relation between daily total transpiration of cotton canopies grown at ambient (350 or 360 μmol mol^{-1}) and elevated [CO_2]. The data are from five temperature treatments of Upland cotton and three water-deficit treatments of Pima cotton collected over several days during fruiting.

temperature increased from 26 to 36°C (A.R. Reddy *et al.*, 1998). At 26°C, the transpiration rate of leaves in 700 μmol CO_2 mol^{-1} was only 70% as great as the transpiration rate of leaves in 350 μmol mol^{-1}. Leaves at 36°C and 700 μmol CO_2 mol^{-1} transpired only 62% as much as leaves at 350 μmol mol^{-1}. The average transpiration rates of individual leaves at 36°C were over twice those at 26°C. Therefore, one should avoid extrapolating [CO_2] effects on canopies from data on individual leaves.

8.5 Phenology

Developmental processes are defined in this context as the time between like and dissimilar events, or the duration of a process. Like events include the time intervals between mainstem leaves and branch leaves on a plant. Unlike events include the intervals between plant emergence and formation of a flower bud, flower or mature fruit. Duration of a process might include the period between unfolding of leaves and the time required for a leaf or internode to reach its maximum size.

Temperature and photoperiod are the two main environmental factors that determine flowering in young and established plants. Commercially grown cotton cultivars are not very sensitive to photoperiod, but are very sensitive to temperature (K.R. Reddy *et al.*, 1997b). The effects of temperature and [CO_2] were nicely illustrated (K.R. Reddy *et al.*, 1997a) by conducting an experiment at ambient and twice-ambient atmospheric [CO_2] at five temperatures. The

1995 temperature in Mississippi was used as a reference, with the other temperatures being 1995 minus 2°C, and 1995 plus 2, 5 and 7°C. Daily and seasonal variation and amplitudes were maintained. Developmental events occurred much more rapidly as temperatures increased. Number of days to the appearance of first square (flower bud), first flower and mature open boll decreased as the average temperature increased during development of the respective events (Fig. 8.8). However, the plants grown in chambers with an average daily temperature of 32.3°C did not produce any mature bolls unless there were a few cooler days immediately after flowering. The time required to produce squares, flowers and mature fruit was reduced by an average of 1.6, 3.1 and 6.9 days per degree of increased temperature, respectively. Thus,

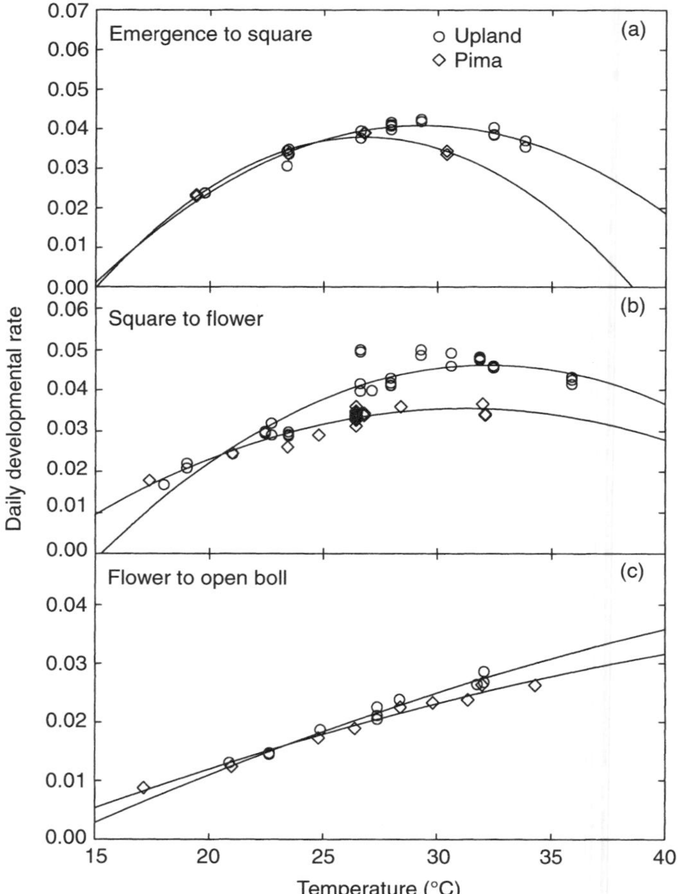

Fig. 8.8. Role of temperature on daily progress of various phenological events: (a) from cotton plant emergence to first flower bud formation (square); (b) square to flower; (c) flower to open boll. Some bolls survive average temperatures of 32°C if conditions are cooler for a few days after flowering. There were no differences between CO_2 levels. (K.R. Reddy *et al.*, 1997b.)

assuming the temperature increase will be equally distributed throughout the growing season, a 5°C increase in average global temperature should speed development from emergence to maturity by 35 days. Unfortunately, the data suggest that most of the shortening of developmental time occurs during the boll growth period and this results in smaller bolls, lower yields and poor quality lint (Hodges *et al.*, 1993). Doubling atmospheric [CO_2] did not affect the developmental rates (Figs 8.8 and 8.9).

8.6 Organ Growth Rates and Mass Partitioning

Cotton growth rates are very sensitive to temperature and somewhat sensitive to elevated [CO_2]. Growth rates of stems, leaves, bolls and roots are all responsive to both of these environmental factors. Typical growth responses (for example, stem extension rate) to temperature and [CO_2] are illustrated in Fig. 8.10. Stem growth elongation rates increased by 20% due to doubling [CO_2]. Leaf area was 25% greater at the end of the season on plants grown in elevated [CO_2], while roots and fruits were 30% and 26% greater, respectively (K.R. Reddy *et al.*, 1998a). The harvest index did not change because of elevated [CO_2].

Seedlings grown at 28 vs. 21°C under optimum water and nutrient conditions accumulated 4 to 6 times more biomass during the first 3 weeks after emergence (K.R. Reddy *et al.*, 1997a). Cotton plants grown in the field in elevated [CO_2] environments grew faster and intercepted 15–40% more solar radiation than plants grown in similar but ambient conditions during the first

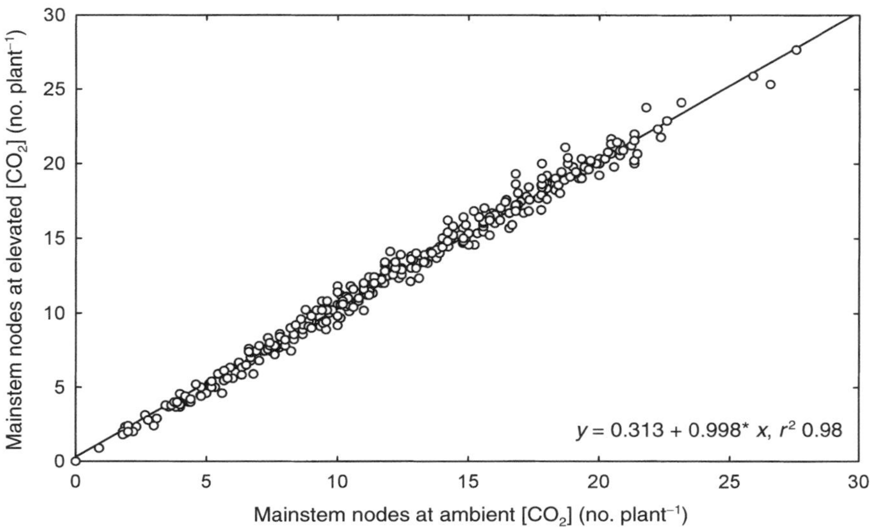

Fig. 8.9. Relation between rates of mainstem node formation of cotton plants grown at ambient and elevated [CO_2]. The data are from several temperature treatments of Upland and Pima cotton grown under optimum water and nutrient conditions.

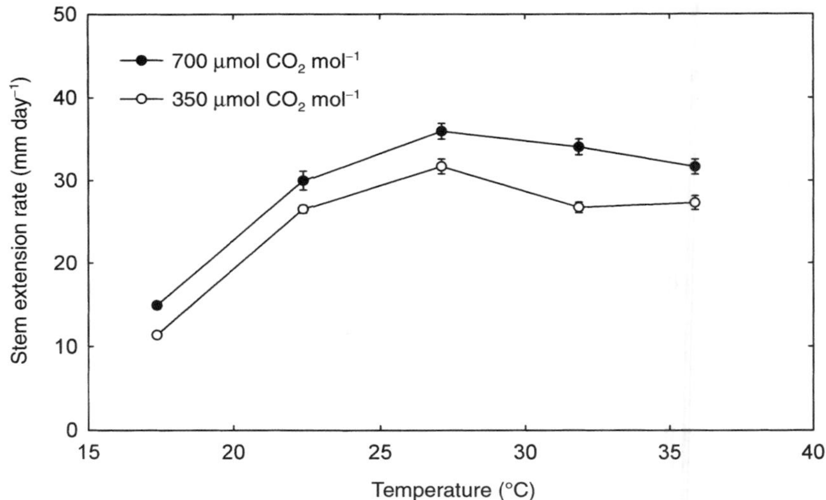

Fig. 8.10. Effect of temperature and [CO_2] on cotton mainstem elongation rate.

half of the growing season (Pinter *et al.*, 1994). This result is typical of plants grown in elevated [CO_2] environments. Such plants produce more and larger leaves, resulting in accumulated growth that intercepts more radiation. Consequently, they have a production advantage when other factors are equal. However, boll retention declines sharply at canopy temperatures above 28°C (K.R. Reddy *et al.*, 1997a). Young bolls are injured by exposure to only a few hours of such high temperatures.

Although height growth was much faster in twice-ambient [CO_2] in optimum conditions during the first 3 weeks, by the end of the season plants grown in the elevated [CO_2] were only 5% taller than those grown in ambient [CO_2] (K.R. Reddy *et al.*, 1997a). Obviously, during a large portion of the season, other factors determining height are more important than [CO_2]. For example, leaf N may limit stem growth rates. When leaf N was about 2.5 g m^{-2}, stem growth was 32–37 mm per day; but when the leaf N was only 1.5 g m^{-2}, stem growth was less than 50% of the higher N treatment (K.R. Reddy *et al.*, 1997b). Stem growth rates were about 17% faster in elevated [CO_2] than in ambient [CO_2] in the N-limited environment. Similarly, stem extension is limited by carbon during much of the season. Relative leaf expansion was also much slower when N was limited. Leaf expansion decreased from over 0.1 cm^2 cm^{-2} day^{-1} to zero as leaf N decreased from 2.25 to 1.5 g m^{-2}. Relative leaf expansion rates were not influenced by elevated [CO_2]. Boll growth rates are sensitive to temperature but not to elevated [CO_2]. At 20 and 32°C, individual boll growth was 82% and 66%, respectively, of boll growth rates at 25°C, but there were no differences in growth rate due to increased [CO_2].

In the SPAR experiments (Table 8.1), there were no significant changes in the partitioning coefficients due to elevated [CO_2]. Greater amounts of photosynthate were produced in elevated [CO_2], but this appeared to be used in forming additional structural mass of all kinds, and the ratios of one plant organ to another remained unchanged. Kimball and Mauney (1993) reported similar results from the open-top chamber experiments (Table 8.2). Elevated [CO_2] did not affect root/shoot ratio, harvest index or lint percentage at ample or limited supplies of water and N. On the other hand, Pinter *et al.* (1996) showed that there were some changes in partitioning due to the 550 μmol CO_2 mol^{-1} FACE treatment in 1991, including an increase in harvest index under well-watered conditions.

8.7 Yield and Yield Components

Atmospheric [CO_2] is an important factor influencing cotton yields. In a controlled-environment experiment (K.R. Reddy *et al.*, 1997a), boll and square production in five temperature conditions averaged 44% more in 720 than in the 360 μmol CO_2 mol^{-1} atmosphere. Excellent water and nutrient conditions were maintained throughout the season. In a series of OTC experiments under ample and limiting levels of water and N (Table 8.2), Kimball and Mauney (1993) found that [CO_2] of 650 μmol mol^{-1} increased the yield of seed cotton (lint plus seed) by 60% and above-ground biomass by 63% compared with yields of plants grown at ambient [CO_2] (about 350 μmol mol^{-1}). Results from the FACE experiments (Table 8.2; Mauney *et al.*, 1994) showed that enrichment to 550 μmol mol^{-1} increased biomass production by about 35%. Whole boll yields increased by about 40% (Mauney *et al.*, 1994) and lint yield increased by about 60% (Pinter *et al.*, 1996).

Temperature is another important factor affecting cotton yields. The highest numbers of bolls and squares were produced in the highest temperature environments, 1995 plus 5°C and 1995 plus 7°C (K.R. Reddy *et al.*, 1997a). However, it must be emphasized that plants grown at these high temperatures retained very few of the bolls produced – an average of only one boll per plant in the 1995 plus 5°C condition and less than that in the 1995 plus 7°C environment. The plants grown in high [CO_2] retained 20% more bolls per plant than plants grown in ambient [CO_2]. Average boll weights also were 20% greater in the plants grown in high [CO_2].

Fruit production efficiency, defined as dry weight of boll per total dry weight, increased as temperatures increased to 29°C, then declined rapidly at temperatures above 29°C (Fig. 8.11). Since the rate of fruit retention dropped dramatically at temperatures above 29°C (K.R. Reddy *et al.*, 1997a), the fruit production efficiency was also expected to decline at higher temperatures. The number of bolls retained was strongly influenced by temperatures higher than July average plus 2°C (29.1°C). At the two higher temperatures, the number of bolls retained or fruit production efficiency was drastically reduced.

The upper limit for cotton fruit survival is approximately 32°C, or July average plus 5°C for the Midsouth US cottonbelt. Such a statement may be misleading, because the survival and growth of bolls are not equally sensitive to high temperature throughout their development. Bolls are usually abscised within 2–4 days after flowering when exposed to high temperatures. However,

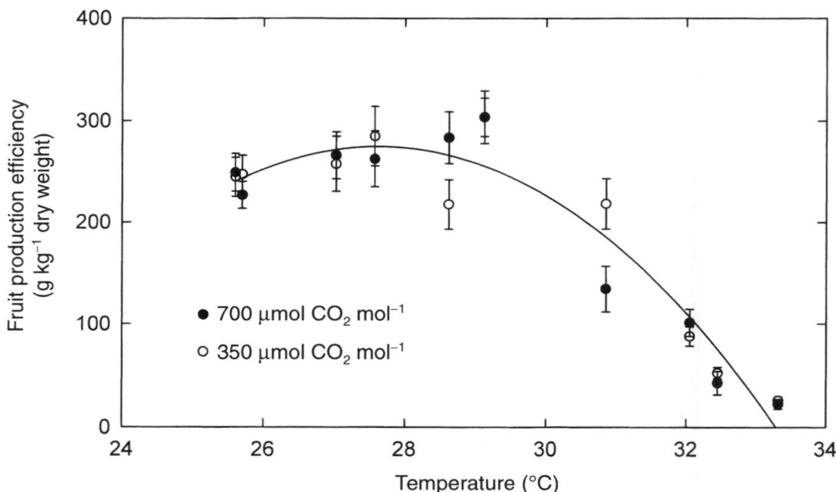

Fig. 8.11. Fruit production efficiency for cotton (boll mass/total dry weight produced) at several temperatures and two [CO_2]. (K.R. Reddy *et al.,* 1997a.)

flowers which developed at lower temperatures did not abscise even when temperatures exceeded 32°C a few days after anthesis. There appears to be a short time period prior to, during and after the flowers formed when they are most vulnerable to high temperature. If the developing bolls escape high temperature during that time, many can survive unfavourable, high-temperature conditions during the rest of their growth period. However, we have yet to delineate the exact definition of the developmental stages and timing of the temperature-sensitive periods.

8.8 Historical Trends: Cotton Yields, [CO_2], and other Technological Advances

Historical US cotton yields (USDA, 1998) are presented in Fig. 8.12 along with changes in [CO_2] (Friedli *et al.*, 1986; Keeling and Whorf, 1998). There has not been any attempt to relate long-term yield responses to temperatures because temperature data are much more complex. Cotton yields prior to 1940 were approximately 200 kg ha^{-1} and remarkably stable from year to year. Yields began to increase by about 1940 and reached 800 kg ha^{-1} in the mid-1990s. Numerous changes in technology occurred during that period. These include development of improved varieties, improved weed and insect control, increased use of fertilizers, and irrigation of sizable acreages. The number of acres grown decreased dramatically from the mid 1930s to the early 1950s. As acreage declined, yields began to improve, probably because the lowest-producing land was removed from production. Low-cost fertilizer became more widely available in the 1950s and crop production research increased

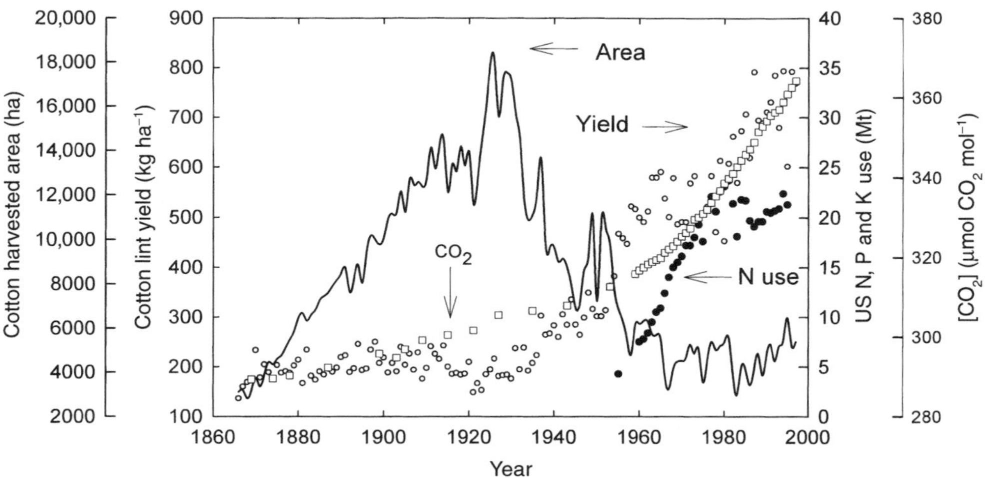

Fig. 8.12. Historical US cotton lint yields, acreage (× 1000), N, P and K use (USDA, 1998) and $[CO_2]$ measured at Mauna Loa, Hawaii (Keeling and Whorf, 1998.)

dramatically after World War II, resulting in many changes in technology that improved cotton yields.

During this period, atmospheric $[CO_2]$ also increased dramatically due to the burning of fossil fuels and other human activity; from 306 μmol mol^{-1} in 1930 to 364 μmol mol^{-1} in 1997 (Fig. 8.12). From the responses of cotton growth and other physiological processes to increasing atmospheric $[CO_2]$, it appears reasonable to estimate that yields probably increased by 19% because of $[CO_2]$ increases during that period. Results from the previous section suggest that cotton yields will increase about 60% when $[CO_2]$ increases by 300 μmol mol^{-1} above today's ambient $[CO_2]$, provided average daily canopy temperatures do not exceed about 30°C. If it is reasonable to assume that the yield response to CO_2 has been linear below today's ambient concentration, and that an approximately 60 μmol mol^{-1} increase in $[CO_2]$ has occurred since 1940, this may have caused cotton yields to increase by about 12%. However, the relative changes in canopy photosynthesis are larger below than above 350 μmol CO_2 mol^{-1} (Fig. 8.12) and so it is likely that the growth response has been greater below than it will be above 350 μmol mol^{-1}. Therefore, the historical cotton yield increase due to increasing $[CO_2]$ may have been larger than 12%. Experiments at subambient $[CO_2]$ have been conducted by Allen *et al.* (1987) on soybean and by Mayeux *et al.* (1997) on wheat. These authors estimated that the yields of soybean and wheat have increased by about 13% and 54%, respectively, due to the increase in atmospheric $[CO_2]$ since pre-industrial times (about 1850).

A number of studies have been conducted to determine the influence of genetic improvement on cotton yields. The most common approach is to use modern cultural practices to relate cultivar yields from a particular year backward to the year of the cultivar's release. Results vary by region, the years the comparisons were made, and the cultivars selected for comparison. The yield improvements range from 5.6 to 11.5 kg ha^{-1} $year^{-1}$ (Meredith *et al.*, 1997, and references cited therein). Assuming that all cotton acreage currently being cultivated is planted in the most modern varieties, it may be calculated that about 50% of the yield increase may be attributable to improved varieties during the period from 1938 to 1993. This leaves considerable yield increases that may be attributed to other causes, some of which are fertilizer use, improved tillage practices, pest control and land selection. Meredith *et al.* (1997) also compared cultivars grown using the N fertility practices recommended in 1940 and 1993. They found a 27% improvement because of N use efficiency among modern cultivars compared with the obsolete cultivars. The change in N use resulted in a 10% increase in yields.

8.9 Summary and Conclusions

Cotton crops grown in future environments will be subjected to projected climatic changes for which they were not bred. Our series of studies, using SPAR, OTC and FACE experimental technologies and facilities, provided detailed insight into how cotton will respond to a changing environment. We believe that several important conclusions regarding the effects of elevated [CO_2], temperature, water and nutrients on plants can be drawn from our cotton experiments. The direct physiological effects of higher [CO_2] on cotton photosynthesis and transpiration will have a myriad of secondary effects. More carbon was fixed in plants grown in high [CO_2] at all levels of water and nutrient deficient conditions and across a wide range of temperatures. Plants grown in high [CO_2] have greater stomatal resistance. However, increased green leaf area offsets that effect, resulting in virtually no difference in canopy water use. The developmental events of cotton plants such as floral initiation, flowering, boll opening and leaf initiation are relatively insensitive to high [CO_2]. Since cotton is very plastic in its growth, additional carbon available in a high [CO_2] environment will favour more vegetative and reproductive growth across a wide range of conditions. This will result in more fruit and, if temperature conditions are favourable for fruit retention, higher yields. Developmental processes are very temperature dependent, because increases in temperature due to greenhouse gases or small changes in canopy temperatures due to effects induced by high [CO_2] will be reflected in overall crop development. Average temperatures above 30°C caused young bolls to abscise. Doubling [CO_2] did not ameliorate the adverse effects of high temperatures on cotton fruit retention. If temperature increases, farmers will likely modify their cultural practices by planting earlier. In that way, plants may escape some of the adverse effects of temperature extremes by completing flowering prior to the onset of injurious high temperatures. Also, one would expect the geographical distribution of individual crops to shift. One would expect the areas of the world that are marginally too cool to become more productive, and areas that are marginally too warm to become less productive. No difference was found in tolerance to high temperatures among the widely grown upland cotton cultivars. However, in the less tolerant Pima cotton species (K.R. Reddy *et al.*, 1992a; Lu *et al.*, 1998), some cultivars were found to escape the effects of high temperature by transpirational cooling. Because young fruits of all cotton cultivars are particularly vulnerable to heat, increasing crop tolerance to high temperatures and short-term heat-shock would be useful to sustain crop productivity in a warmer world. Water and nutrients are scarce resources that will be required in greater quantity to utilize high [CO_2] environments fully for cotton production. Resource-rich economies will benefit more from the greater yield potential resulting from increased [CO_2] because they can more adequately provide these yield-limiting variables.

Acknowledgements

Appreciation is expressed for the excellent technical assistance provided by Gary Burrell, Kim Gourley, Wendell Ladner and Sam Turner, and to Dr James McKinion for providing the SPAR facility. Part of the research was funded by the USDOE National Institute for Global Environment Change through the South Central Regional Center at Tulane University (DOE cooperative agreement no. DE-FC03–90ER 61010).

References

Allen, L.H. Jr, Boote, K.J., Jones, J.W., Jones, P.H., Valle, R.R., Acock, B., Rogers, H.H. and Dahlman, R.C. (1987) Response of vegetation to rising carbon dioxide: photosynthesis, biomass, and seed yield of soybean. *Global Biogeochemical Cycles* 1, 1–14.

Barrett, D. and Gifford, R.M. (1995) Photosynthetic acclimation to elevated CO_2 in relation to biomass allocation in cotton. *Journal of Biogeography* 22, 331–339.

Bos, E., Vu, M.Y., Massiah, E. and Bulatao, R.A. (1995) *World Population Projections 1994–95 Edition: Estimates and Projections with Related Demographic Statistics.* Johns Hopkins University Press, Baltimore, Maryland.

Bowes, G. (1993) Facing the inevitable: plants and increasing atmospheric CO_2. *Annual Review of Plant Physiology and Molecular Biology* 44, 309–332.

Dugas, W.A., and Pinter, P.J. Jr (eds) (1994) Agricultural and forest meteorology: a new field approach to assess the biological consequences of global change. *Agricultural and Forest Meteorology* 70, 1–342.

Dugas, W.A., Heuer, M.L., Hunsaker, D.J., Kimball, B.A., Lewin, K.F., Nagy, J. and Johnson, M. (1994) Sap flow measurements of transpiration in open-field-grown cotton under ambient and enriched CO_2 concentrations. *Agricultural and Forest Meteorology* 70, 231–245.

Friedli, H., Lotscer, H., Oeschger, H., Siegenthaler, U. and Stuffer, B. (1986) Ice core record of $^{13}C/^{12}C$ ratio of atmospheric CO_2 in the past two centuries. *Nature* 324, 220–223.

Hendrey, G.R. (ed.) (1993) *FACE: Free-Air CO_2 Enrichment for Plant Research in the Field.* C.K. Smoley, Boca Raton, Florida. 308 pp.

Hileman, D.R., Huluka, G., Kenjige, P.K., Sinha, N., Bhattacharya, N.C., Biswas, P.K., Lewin, K.F., Nagy, J. and Hendrey, G.R. (1994) Canopy photosynthesis and transpiration of field-grown cotton exposed to free-air CO_2 enrichment (FACE) and differential irrigation. *Agricultural and Forest Meteorology* 70, 189–207.

Hodges, H.F., Reddy, K.R., McKinion, J.M. and Reddy, V.R. (1993) *Temperature Effects on Cotton.* Mississippi Agricultural and Experimental Station Bulletin no. 990, Mississippi State University, 15 pp.

Hoffman, G.J., Howell, T.A. and Soloman, K.H. (1990) *Management of Farm Irrigation Systems.* American Society Agricultural Engineers, St Joseph, Michigan, 1040 pp.

Houghton, J.T., Meira Filho, L.G., Callander, B.A., Harris, N., Kattenburg, A. and Maskell, K. (eds) (1996) *Climate Change 1995: Summary for Policy Makers.* Cambridge University Press, Cambridge, UK, pp. 1–7.

Hunsaker, D.J., Hendrey, G.R., Kimball, B.A., Lewin, K.F., Mauney, J.R. and Nagy, J. (1994) Cotton evapotranspiration under field conditions with CO_2 enrichment and variable soil moisture regimes. *Agricultural and Forest Meteorology* 70, 247–258.

Idso, S.B., Kimball, B.A. and Mauney, J.R. (1987) Atmospheric carbon dioxide enrichment effects on cotton midday foliage temperature: implications for plant water use and crop yield. *Agronomy Journal* 79, 667–672.

Keeling, C.D., and Whorf, T.P. (1994) Atmospheric CO_2 records from sites in the SIO air sampling network. In: Boden, T.A., Kaiser, D.P., Sapanski, R.J. and Stoss, F.W. (eds) *Trends '93: A Compendium of Data on Global Change. ORNL/CDIAC-65.* Carbon Dioxide Information Analysis Center, Oak Ridge National Laboratory, Oak Ridge, Tennessee, pp. 16–19.

Keeling, C.D. and Whorf, T.P. (1998) Http//cdiac.esd.ornl.gov/trends/co2/sio-mlo.htm.

Kimball, B.A. (1983a) Carbon dioxide and agricultural yield: an assemblage and analysis of 430 prior observations. *Agronomy Journal* 75, 779–788.

Kimball, B.A. (1983b) *Carbon Dioxide and Agricultural Yield: an assemblage and analysis of 770 prior observations.* Water Conservation Laboratory Report 14, US Water Conservation Laboratory, USDA-ARS, Phoenix, Arizona, 71 pp.

Kimball, B.A. (1986) Influence of elevated CO_2 on crop yield. In: Enoch, H.Z. and Kimball, B.A. (eds) *Carbon Dioxide Enrichment of Greenhouse Crops.*, Vol. 2, *Physiology, Yield, and Economics.* CRC Press, Boca Raton, Florida, pp. 105–115.

Kimball, B.A. (1993) Cost comparisons among free-air CO_2 enrichment, open-top chamber, and sunlit controlled-environment chamber methods of CO_2 exposure. *Critical Reviews in Plant Science* 11, 265–270.

Kimball, B.A., and Idso, S.B. (1983) Increasing atmospheric CO_2: effects on crop yield, water use and climate. *Agricultural Water Management* 7, 55–72.

Kimball, B.A. and Mauney, J.R. (1993) Response of cotton to varying CO_2, irrigation, and nitrogen: yield and growth. *Agronomy Journal* 85, 706–712.

Kimball, B.A., Mauney, J.R., Radin, J.W., Nakayama, F.S., Idso, S.B., Hendrix, D.L., Akey, D.H., Allen, S.G., Anderson, M.G. and Hartung, W. (1986) *Effects of Iincreasing Atmospheric CO_2 on the Growth, Water Relations, and Physiology of Plants Grown under Optimal and Limiting Levels of Water and Nitrogen.* Number 039, *Responses of Vegetation to Carbon Dioxide.* US Department of Energy, Carbon Dioxide Research Division and the US Department. of Agriculture, Agricultural Research Service, Washington, DC, 125 pp.

Kimball, B.A., Mauney, J.R., Akey, H.H., Hendrix, D.L., Allen, S.G., Idso, S.B., Radin, J.W. and Lakatos, E.A. (1987) *Effects of Increasing Atmospheric CO_2 on the Growth, Water Relations, and Physiology of Plants Grown under Optimal and Limiting Levels of Water and Nitrogen.* No. 049, *Response of Vegetation to Carbon Dioxide,*

US Department of Energy, Carbon Dioxide Research Division, and the US Department of Agriculture, Agricultural Research Service, Washington, DC, 124 pp.

Kimball, B.A., Pinter, P.J. Jr and Mauney, J.R. (1992a) Cotton leaf and boll temperatures in the 1989 FACE experiment. *Critical Reviews in Plant Sciences* 11, 233–240.

Kimball, B.A., Mauney, J.R., LaMorte, R.L., Guinn, G., Nakayama, F.S., Radin, J.W., Lakatos, E.A., Mitchell, S.T., Parker, L.L., Peresta, G., Nixon III, P.E., Savoy, B., Harris, S.M., MacDonald, R., Pros, H. and Martinez, J. (1992b) *Carbon Dioxide Enrichment: data on the response of cotton to varying CO_2, irrigation, and nitrogen.* ORNL/CDIAC-44, NDP-037, Oak Ridge National Laboratory, Oak Ridge, Tennessee, 592 pp.

Kimball, B.A., Mauney, J.R., Nakayama, F.S. and Idso, S.B. (1993) Effects of increasing atmospheric CO_2 on vegetation. *Vegetatio* 104/105, 65–75.

Kimball, B.A., LaMorte, R.L., Seay, R.S., Pinter, P.J. Jr, Rokey, R.R., Hunsaker, D.J., Dugas, W.A., Heuer, M.L., Mauney, J.R., Hendrey, G.R., Lewin, K.F. and Nagy, J. (1994) Effects of free-air CO_2 enrichment on energy balance and evapotranspiration of cotton. *Agricultural and Forest Meteorology* 70, 259–278.

Kimball, B.A., Pinter, P.J. Jr, Wall, G.W., Garcia, R.L., LaMorte, R.L., Jak, P., Frumau, K.F.A. and Vugts, H.F. (1997) Comparisons of responses of vegetation to elevated CO_2 in free-air and open-top chamber facilities. In: Allen, L.H. Jr, Kirkham, M.B., Olszyk, D.M. and Whitman, D.M. (eds) *Advances on CO_2 Effects Research.* Special Publication No. 61, American Society of Agronomy, Madison, Wisconsin, pp. 113–130.

Lu, Z.M., Percy, R.G., Qualset, C.O. and Zeiger, E. (1998) Stomatal conductance predicts yields in irrigated Pima cotton and bread wheat grown at high temperatures. *Journal of Experimental Botany* 49, 453–460.

Mauney, J.R., Kimball, B.A., Pinter, P.J. Jr, LaMorte, R.L., Lewin, K.F., Nagy, J. and Hendrey, G.R. (1994) Growth and yield of cotton in response to a free-air carbon dioxide enrichment (FACE) environment. *Agricultural and Forest Meteorology* 70, 49–68.

Mayeux, H.S., Johnson, H.B., Polley, H.W. and Malone, S.R. (1997) Yield of wheat across a subambient carbon dioxide gradient. *Global Change Biology* 3, 269–278.

Meredith, W.R. Jr, Heitholt, J.J., Pettigrew, W.T. and Rayburn, S.T. Jr (1997) Comparison of obsolete and modern cotton cultivars at two nitrogen levels. *Crop Science* 37, 1453–1477.

Morison, J.I.L. (1987) Intercellular CO_2 concentration and stomatal response to CO_2. In: Zeiger, Z. and Farquhar, G.D. (eds) *Stomatal Function.* Stanford University Press, Stanford, California, pp. 229–251.

Pinter, P.J. Jr, Kimball, B.A., Mauney, J.R., Hendry, G.R., Lewin, K.F. and Nagy, J. (1994) Effects of free-air carbon dioxide enrichment on PAR absorption and conversion efficiency by cotton. *Agricultural and Forest Meteorology* 70, 209–230.

Pinter, P.J. Jr, Kimball, B.A., Garcia, R.L., Wall, G.W., Hunsaker, D.J. and LaMorte, R.L. (1996) Free-air CO_2 enrichment: responses of cotton and wheat crops. In: Koch, G.W. and Mooney, H.A. (eds) *Carbon Dioxide and Terrestrial Ecosystems.* Academic Press, San Diego, California, pp. 215–249.

Pinter, P.J. Jr, Kimball, B.A., Wall, G.W., Hunsaker, D.J., Adamsen, F.J., Frumau, K.F.A., Vugts, H.F., Hendrey, G.R., Lewin, K.F. Nagy, J., Johnson, H.B., Wechsung, F., Leavitt, S.B., Thompson, T.L., Matthias, A.D. and Brooks, T.J. (2000) Free-air CO_2 enrichment (FACE): blower effects on wheat canopy microclimate and plant development. *Agriculture and Forest Meteorology* (in press).

Radin, J.W., Kimball, B.A., Hendrix, D.L. and Mauney, J.R. (1987) Photosynthesis of cotton plants exposed to elevated levels of carbon dioxide in the field. *Photosynthesis Research* 12, 191–203.

Reddy, A.R., Reddy, K.R. and Hodges, H.F. (1997) Dynamics of canopy photosynthesis in Pima cotton as influenced by growth temperatures. *Indian Journal of Experimental Biology* 35, 1002–1006.

Reddy, A.R., Reddy, K.R. and Hodges, H.F. (1998). Interactive effects of elevated carbon dioxide and growth temperature on photosynthesis in cotton leaves. *Plant Growth Regulation* 22, 1–8.

Reddy, K.R., Hodges. H.F. and Reddy, V.R. (1992a) Temperature effects on cotton fruit retention. *Agronomy Journal* 84, 26–30.

Reddy, K.R., Hodges, H. F. and Reddy, V.R. (1992b) Temperature effects on early season cotton growth and development. *Agronomy Journal* 84, 229–237.

Reddy, K.R., Hodges, H.F. and McKinion, J.M. (1993) A temperature model for cotton phenology. *Biotronics* 2, 47–59.

Reddy, K.R., Hodges, H.F. and McKinion, J.M. (1995a) Carbon dioxide and temperature effects on Pima cotton growth. *Agriculture, Ecosystems and Environment* 54, 17–29.

Reddy, K.R., Hodges, H.F. and McKinion, J.M. (1995b) Carbon dioxide and temperature effects on Pima cotton development. *Agronomy Journal* 87, 820–826.

Reddy, K.R., Hodges, H.F. and McKinion, J.M. (1995c) Cotton crop responses to a changing environment. In: Rosenzweig, C., Allen, L.H. Jr, Harper, L.A., Hollinger, S.E. and Jones, J.W. (eds) *Climate Change and Agriculture: Analysis of Potential International Impacts.* American Society of Agronomy Special Publication no. 59, Madison, Wisconsin, pp. 3–30.

Reddy, K.R., Hodges, H.F. and McKinion, J.M. (1996a) Food and agriculture in the 21st century: a cotton example. *World Resource Review* 8, 80–97.

Reddy, K.R., Hodges, H.F., McCarty, W.H. and McKinion, J.M. (1996b) *Weather and Cotton Growth: present and future.* Bulletin no. 1061, Mississippi Agricultural and Forestry Experiment Station, Mississippi State University, Mississippi State, 23 pp.

Reddy, K.R., Hodges, H.F. and McKinion, J.M. (1997a) A comparison of scenarios for the effect of global climate change on cotton growth and yield. *Australian Journal of Plant Physiology* 24, 707–713.

Reddy, K.R., Hodges, H.F. and McKinion, J.M. (1997b) Crop modeling and applications: a cotton example. In: Sparks, D.L. (ed.) *Advances in Agronomy,* Vol. 59. Academic Press, San Diego, pp. 225–290.

Reddy, K.R., Hodges, H.F. and McKinion, J.M. (1997c) Water and nutrient deficits, crop yields and climate change. *World Resource Review* 10, 23–43.

Reddy, K.R., Hodges, H.F. and McKinion, J.M. (1997d) Modeling temperature effects on cotton internode and leaf growth. *Crop Science* 37, 503–509.

Reddy, K.R., Robana, R.R., Hodges, H.F., Liu, X.J. and McKinion, J.M. (1998a) Influence of atmospheric CO_2 and temperature on cotton growth and leaf characteristics. *Environmental and Experimental Botany* 39, 117–129.

Reddy, K.R., Hodges, H.F. and McKinion, J.M. (1998b) Photosynthesis and environmental factors. *Proceedings of the Beltwide Cotton Conference,* Vol. 2. National Cotton Council, Memphis, Tennessee, pp. 1443–1450.

Reddy, V.R. and Reddy, K.R. (1998) Soil–Plant–Atmosphere Research (SPAR) facility: a unique source of data for process-level crop modeling. *Proceedings of International Agricultural and Engineering Conference, Bangkok, Thailand,* pp. 639–649

Reddy, V.R., Reddy, K.R., Acock, M.C. and Trent, A. (1994a) Carbon dioxide enrichment and temperature effects on root growth in cotton. *Biotronics* 23, 47–57.

Reddy, V.R., Reddy, K.R. and Acock, B. (1994b) Carbon dioxide and temperature effects on cotton leaf initiation and development. *Biotronics* 23, 59–74.

Reddy, V.R., Reddy, K.R. and Acock, B. (1995a) Carbon dioxide and temperature interactions on stem extension, node initiation and fruiting in cotton. *Agriculture, Ecosystems and Environment* 55, 17–28.

Reddy, V.R., Reddy, K.R. and Hodges, H.F. (1995b) Carbon dioxide enrichment and temperature effects on canopy cotton photosynthesis, transpiration, and water use efficiency. *Field Crops Research* 41, 13–23.

Reddy, V.R., Reddy, K.R. and Wang, Z. (1997). Cotton responses to nitrogen, carbon dioxide, and temperature interactions. *Soil Science and Plant Nutrition* 43, 1125 -1130.

Rogers, H.H., Runion, G.B. and Krupa, S.V. (1994) Plant responses to atmospheric CO_2 enrichment with emphasis on roots and the rhizosphere. *Environmental Pollution* 83, 155–189.

Rotty, R.M., and Marland, G. (1986) Fossil fuel combustion: recent amounts, patterns, and trends of CO_2, In: Trabalka, J.R. and Reichle, D.E. (eds) *The Changing Carbon Cycle: a Global Analysis.* Springer-Verlag, New York, pp. 474–490.

Samarakoon, A.B. and Gifford, R.M. (1995) Soil water content under plants at high CO_2 concentration and interactions with direct CO_2 effects: a species comparison. *Journal of Biogeography* 22, 193–202.

Trabalka, J.R., Edmonds, J.A., Reilly, J.M., Gardner, R.H., and Reichle, D.E. (1986) Atmospheric CO_2 projections with globally averaged carbon cycle models, In: Trabalka, J.R. and Reichle, E. (eds) *The Changing Carbon Cycle: a Global Analysis.* Springer-Verlag, New York, pp. 534–560.

USDA (1989) *Agricultural Statistics 1989.* US Government Printing Office, Washington, DC.

USDA (1998) *Agricultural Statistics 1997.* US Government Printing Office, Washington, DC.

9 Crop Ecosystem Responses to Climatic Change: Root and Tuberous Crops

FRANCO MIGLIETTA[1], MARCO BINDI[2], FRANCESCO P. VACCARI[1], A.H.C.M. SCHAPENDONK[3], JUSTUS WOLF[4] AND RUTH E. BUTTERFIELD[5]

[1]Instituto Nazionale Analisi e Protezione Agro-ecosistemi, IATA-CNR, P. le delle Cascine 18, 50144 Firenze, Italy; [2]Department of Agronomy and Land Management, University of Firenze, P. le delle Cascine 18, 50144 Firenze, Italy; [3]Research Institute for Agrobiology and Soil Fertility (AB-DLO), PO Box 14, 6700 AA Wageningen, The Netherlands; [4]Department of Theoretical Production Ecology, Wageningen Agricultural University, PO Box 430, 6700 AK Wageningen, The Netherlands; [5]Environmental Change Unit, University of Oxford, 1a Mansfield Road, Oxford OX1 3TB, UK

9.1 Introduction

Root and tuber crops are highly important food resources. They comprise several genera and supply the main part of the daily carbohydrate intake of large populations. These carbohydrates are mostly starches found in storage organs, which may be enlarged roots, corms, rhizomes or tubers. Many root and tuber crops are grown in traditional agricultural systems or are adapted to unique ecosystems and do not enter world trade; however, some are grown worldwide. With some important exceptions, root and tuber crops are more important for food production in tropical than in temperate climates.

There are many crop species that produce edible roots and tubers. Food and Agriculture Organization (FAO) statistics indicate that in 1997 root and tuber crops were cultivated over more than 49 M ha and the total production was greater than 650 Mt (FAOSTAT, 1998). Potato (*Solanum tuberosum* L.), cassava (*Manihot esculenta*), sweet potato (*Ipomea batatas*), yam (*Dioscorea* subsp.), radish (*Raphanus sativus)*, carrot (*Daucus carota* subsp.) and kohlrabi (*Brassica oleracea*) are the most widespread root and tuber crops used for food production. Among these, potato and cassava are by far the most widely cultivated species in the world as their current production accounts for 72% of the total harvest of root and tuber crops. Potato makes the largest contribution to the total production of tuber and root crops, but its global production share decreased from about 60% in the 1960s to 45% in the 1990s. Cassava became more important and its global production share grew from 15 to 25% over the same 37 years (Fig. 9.1). This reflects the faster growth of populations and their food demand in developing vs. developed countries. Cassava is mainly cultivated in tropical regions, whereas potato is an important crop in the temperate zones. Production of these crops steadily increased by about 4 Mt per year from 1961 to 1997 (Fig. 9.2). This increase was due mainly to increased yields per unit of land, as the global area cultivated did not change (Fig. 9.3). Several factors may have caused these yield increases, including improved crop varieties, crop management, increased use of fertilizers, reduced losses from pest and disease infestations, improved harvesting and conservation methods and extended irrigation. The continuous increase in atmospheric carbon dioxide concentration ([CO_2]) may also have contributed.

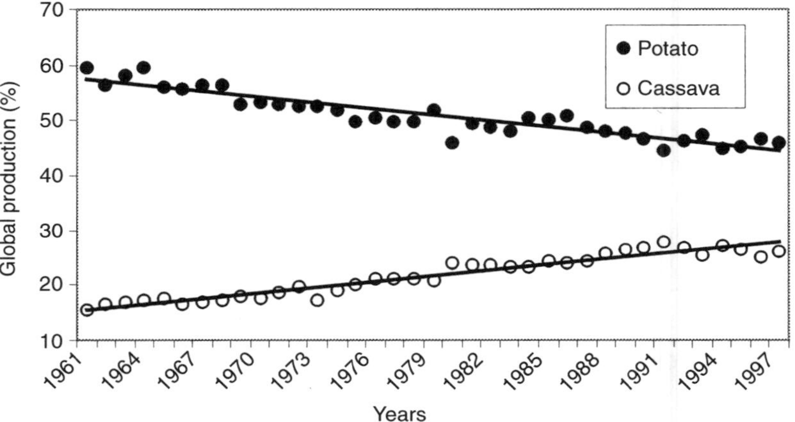

Fig. 9.1. Changes in relative contribution of potato and cassava as percentage of total tuber and root crop global production from 1961 to 1997.

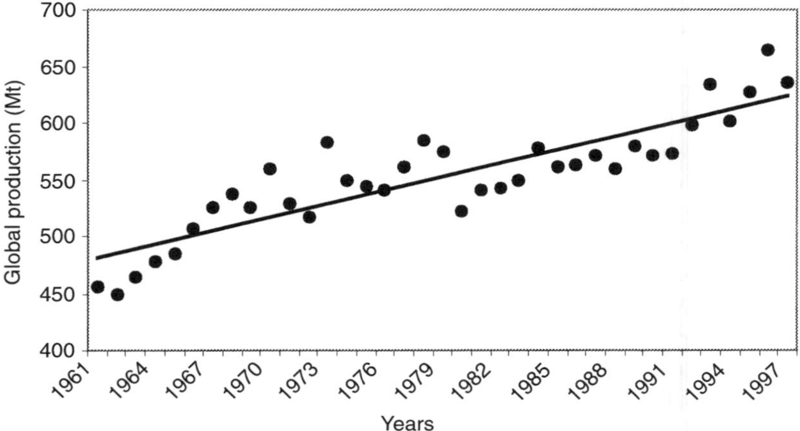

Fig. 9.2. Global production (Mt) of tuber and root crops from 1961 to 1997.

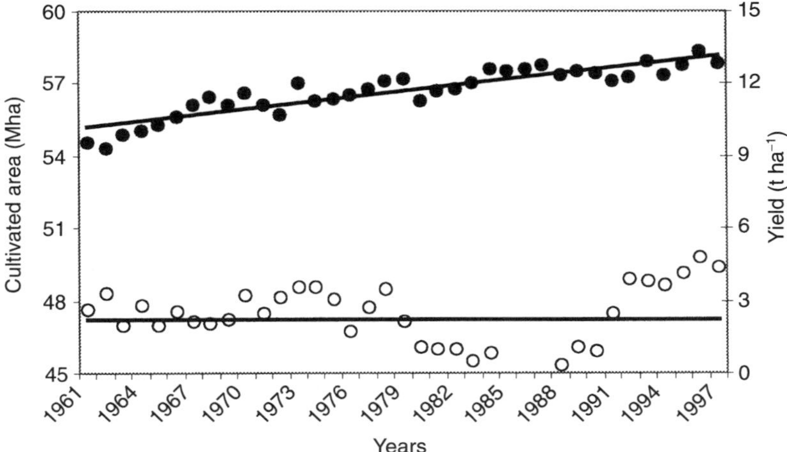

Fig. 9.3. Average yield per unit land (closed circles, t ha^{-1}) and the total cultivated area (open circles, Mha) of tuber and root crops in the world, from 1961 to 1997.

9.2 Expected Effects of Climatic Change on Root and Tuber Crops

Despite their economic and global importance, little research has been done on the potential effects of climatic change on tuber and root crops with potato as a single exception. For the other crops, most of the limited studies have addressed specific questions and failed to provide broader significant information. In general, research focused on key questions that had to do with the response of different species/varieties to increasing air temperature and to rising [CO_2] in the atmosphere. In particular, many studies have addressed the responses of various crops to [CO_2], because changes in [CO_2] have a significant impact on crop yields. Another reason is that the increase in [CO_2] is, as well as ozone and SO_2, a clearly detectable, measurable and unavoidable effect of anthropogenic activity on earth's atmosphere. There is clear evidence since the 1950s (Keeling *et al.*, 1995) that atmospheric [CO_2] is increasing, and plant physiologists have repeatedly demonstrated that such increases likely have already caused substantial increases in leaf photosynthesis of C_3 species (Sage, 1994). Finally, projections based on reliable scenarios of industrial development, trading and transport are indicating that [CO_2] as high as 550 μmol mol^{-1} will be unavoidably reached in the middle of the 21st century, (Houghton *et al.*, 1992). This does not imply that those changes in temperature, precipitation, tropospheric ozone concentrations and nitrogen depositions have lower impacts, but explains why the 'CO_2 effect' has been considered a major research topic. Scientists have repeatedly recognized that species with large below-ground sinks for carbon (Chu *et al.*, 1992; Körner *et al.*, 1995; Farrar, 1996) and with apoplastic mechanisms of phloem loading (Komor *et al.*, 1996) are likely to be the best candidates for a large response to rising atmospheric [CO_2]. There is also an increasing consensus among scientists that growth or storage sink limitations are possibly major factors constraining responses of plants to elevated [CO_2]. This explains why the large CO_2 response often observed in crop species with a large sink capacity (Kimball, 1983) could not be reproduced with natural vegetation or in nutrient-poor plant associations (Poorter, 1993; Körner and Miglietta, 1994). The potential effects of rising atmospheric [CO_2] and temperature on physiological processes and the ecology of root and tuber crops will be analysed. The main part of the experimental results for potato was derived from a FACE (free air CO_2 enrichment) experiment that was made in Italy in 1995 and from a series of experiments with open-top chambers (OTC) in the Netherlands. In the FACE experiment, a potato crop was grown in the field under natural conditions except for being exposed to elevated [CO_2] by means of a sophisticated fumigation device (Miglietta *et al.*, 1998), and plants were exposed to a gradient of [CO_2] ranging from ambient to 660 μmol mol^{-1}. The OTC experiments allowed potato cultivars of different earliness to be compared in 700 μmol mol^{-1} CO_2 (A.H.C.M. Schapendonk, personal communication, 1995). Further analysis of the potential effects of climatic change on potato production will use the results from simulation studies made at site and regional scales within the European Union-funded CLIVARA Project (Downing *et al.*, 1999). Obviously, it is not always possible to extrapolate all the considerations and conclusions made for potato to the entire root and tuber crop category, but the information will guide the reader to a critical analysis of the issue and will help to identify the most important topics and where and why more research is needed.

9.2.1 Photosynthesis

Increased atmospheric [CO_2] reduces photorespiration of C_3 plants by promoting carboxylation and diminishing oxygenation of the photosynthetic enzyme Rubisco (Sage, 1994). This change, together with the increased CO_2 diffusion gradient into the leaves, leads to enhanced photosynthetic rates. However, in some cases this effect appeared to be not maintained in the long term. Leaves of C_3 plants that were exposed over prolonged periods of time to enhanced [CO_2] have sometimes shown decreased photosynthetic capacity. This phenomenon, called acclimation or down-regulation of photosynthetic capacity, has been the subject of several investigations (Long and Drake, 1992; Sage, 1994), but many of the results remain contradictory. This issue is of interest as it affects the prediction of the likely consequences of rising [CO_2] on carbon uptake and storage by crops. In the OTC experiments conducted in the Netherlands (A.H.C.M. Schapendonk, personal communication, 1995), signs of acclimation or down-regulation of photosynthesis were observed as the initially large CO_2 stimulation of photosynthesis decreased during the period of active tuber filling.

To understand and analyse the potential consequences of such acclimation for growth and yield, a mechanistic simulation model of potato growth (Spitters *et al.*, 1986; Schapendonk *et al.*, 1995) was used. Leaf gross CO_2 assimilation was simulated according to the biochemical model described by

Farquhar *et al.* (1980) and von Caemmerer and Farquhar (1981), which is based on Michaelis–Menten kinetics and simulates the kinetics of electron flow and carboxylation rates. From these relationships, the effects of temperature, photosynthetic photon flux density (PPFD) and [CO_2] on gross leaf photosynthesis were calculated. Gross leaf photosynthesis is known to be determined by two rate-limiting processes: the production of reducing equivalents in the electron-transport chain and the rate of carbon fixation in the Calvin cycle. At low PPFD, the electron-transport rate, equivalent to energy delivery, is limiting. Elevated [CO_2] has a positive effect due to more efficient energy utilization by suppression of photorespiration. At high PPFD the positive effect is enhanced because the carboxylation rate, linked with the availability of CO_2, is rate-limiting. In fact, in the OTC experiment, doubling [CO_2] resulted in 24–40% increases in tuber yields. Simulations showed that the yield effect of increased [CO_2] would have been almost two times as large, if acclimation had not down-regulated the photosynthetic rates. This shows that acclimation may be responsible for a major reduction of the CO_2 effect on biomass and tuber production (Fig. 9.4).

Acclimation was also observed in the FACE experiment, but this was caused by earlier senescence of the plants exposed to elevated [CO_2].

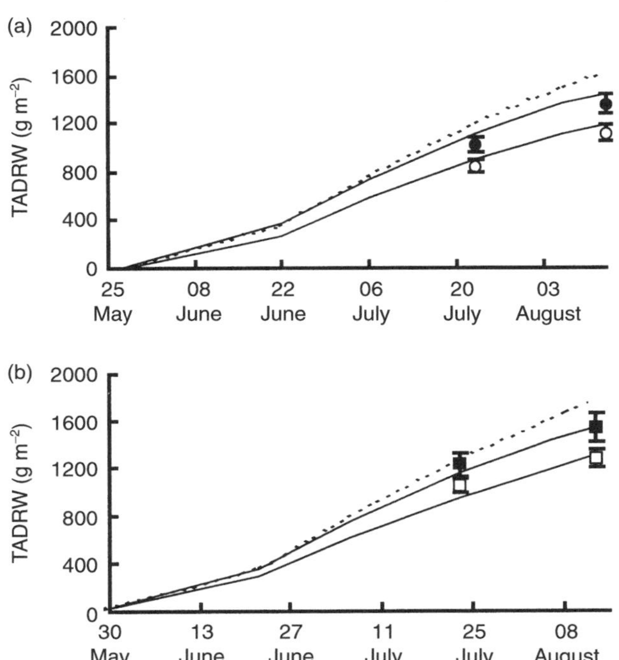

Fig. 9.4. Simulated (dotted line) and observed total biomass (minus roots) dry matter production for ambient (open symbols) and doubled (closed symbols) [CO_2] treatments measured in the open-top chambers (OTC) experiment in Wageningen in 1995 in two different maturity class of potatoes: (a) early variety and (b) late variety.

Photosynthetic acclimation in the FACE experiment was investigated by gas exchange. The leaves of potato plants grown in the field at ambient, 460, 560 and 660 μmol mol^{-1} were periodically collected to determine rates of assimilation (*A*) versus intercellular [CO_2] (C_i). Ribulose 1,5-bisphosphate (RuBP) saturated (V_{cmax}, maximum carboxylation rate) and RuBP regeneration limited (J_{max}, electron transport-mediated regeneration capacity) carboxylation rates were calculated for each A/C_i curve by fitting a deterministic photosynthesis model (Farquhar and von Cammerer, 1982) to the data. The measurements were made during the period between flowering and maximum leaf area expansion and later when the tubers were growing.

Leaf senescence was also investigated at the time of the second round of photosynthesis measurements. It was assumed that as broad leaves senesce, their reflectance increases in the green wavelength region, peaking at 550 nm because of chlorophyll degradation (Knipling, 1970). Hence, leaf reflectance was assessed on upper canopy leaves by means of a laboratory spectroradiometer (Benincasa *et al.,* 1988; Malthus and Madeira, 1993).

Measurements in the first part of the growing season showed that the response of intercellular [CO_2] of potato leaves was not affected by long-term elevated [CO_2] exposure in the FACE (Fig. 9.5). Subsequent measurements made later in the season showed that leaf photosynthesis decreased progressively in the higher [CO_2] environments (data not shown). This decrease in photosynthetic capacity was due to senescence, which was shown by leaf reflectance in leaves grown in high [CO_2] (Fig. 9.6). The reflectance measured at 550 nm was well correlated (the regression equation is $y = 0.0248 + 7.788x$, with $r = 0.99$) with the loss of photosynthetic capacity.

Leaf N concentration decreased faster in the leaves grown under high [CO_2] than in those grown under ambient [CO_2], thus further supporting the conclusion that leaf senescence was accelerated in plants grown in high [CO_2]. Such loss of leaf N led to a reduction in leaf photosynthetic capacity. Data from this experiment clearly indicated that there was no down-regulation of photosynthetic capacity in potato leaves exposed to increasing [CO_2], at least during the main part of the growing season. When tubers are actively growing, they provide a large sink for carbon fixed by the leaves. This likely prevents negative feedback effects of carbohydrate accumulation in leaves that otherwise might lead to a permanent loss of photosynthetic capacity.

Plants grown under elevated [CO_2] may continuously fix a larger amount of carbon than those grown at ambient concentration and, not surprisingly, increasing [CO_2] has large and progressive positive effects on tuber yields (see section 9.2.3). However, accelerated senescence of the leaves in the three elevated [CO_2] FACE plots was a likely consequence of the higher canopy temperatures during daytime hours with elevated [CO_2], or a result at the end of the growing season of the carbohydrate storage capacity of the tubers being more rapidly exhausted in the plants grown in high [CO_2] than in those grown in ambient [CO_2], or both. Whatever the case, tuber growth under elevated [CO_2] had indirect consequences on the photosynthetic properties of the leaves, mediated by this acceleration of leaf senescence. Overall, the results obtained in these experiments with potato suggest that photosynthetic

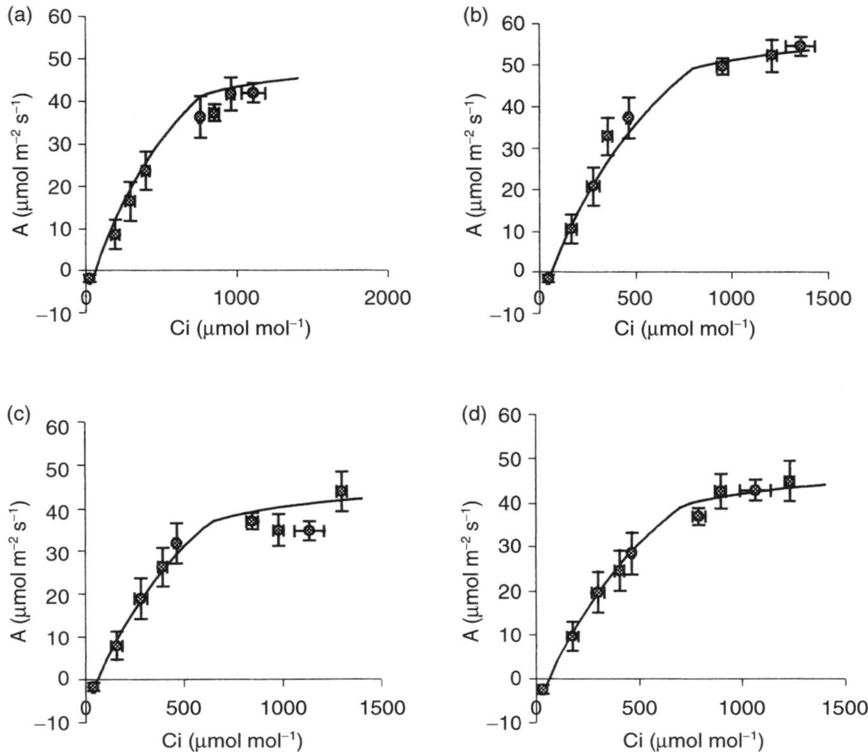

Fig. 9.5. Relationship between carbon assimilation (A) and internal leaf carbon dioxide concentration (Ci) measured in July on potato leaves growing in different CO_2 concentrations. Error bars indicate standard deviation of assimilation (A) and CO_2 intercellular concentrations (Ci). (a) Leaves grown at ambient [CO_2]; (b) Leaves grown at 460 µmol mol^{-1}; (c) Leaves grown at 560 µmol mol^{-1}; (d) Leaves grown at 660 µmol mol^{-1}.

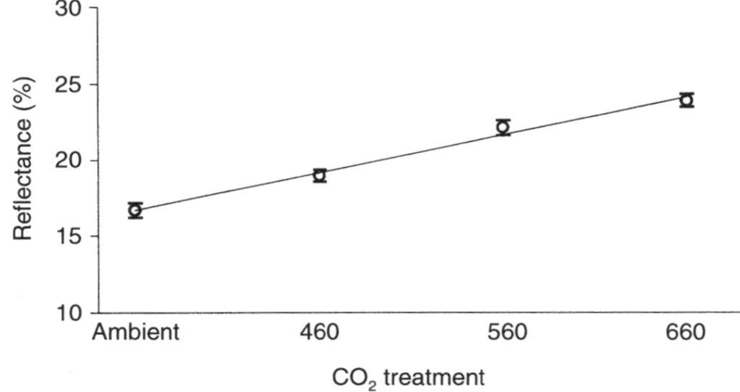

Fig. 9.6. Leaf reflectance (%) of potato leaves grown at different [CO_2] treatments measured at 550 nm in late August (significant at $P < 0.01$). Ambient [CO_2] was approximately 360 µmol mol^{-1}.

acclimation is a complex process involving interactive effects of elevated [CO_2] on source/sink relations in the plants (Ziska *et al.*, 1995).

Photosynthetic carbon supply may interact with the availability of inorganic nutrients and with the canopy energy balance as reflected by canopy temperature. It remains doubtful if the decreased leaf N concentrations that have been observed in leaves grown at elevated [CO_2], are *per se* a sign of decreased photosynthetic capacity. Tuber and root crops do not make an exception to this observation, as decreased leaf N was also observed in plants grown under elevated [CO_2] on kohlrabi (Sritharan *et al.*, 1992), radish (McKeehen *et al.*, 1995) and potato (Miglietta *et al.*, 1998).

9.2.2 Development and phenology

Climatic change may influence crop phenology by changes in [CO_2], air temperature and drought. The expected increase in [CO_2] may potentially affect crop phenology in two different ways. It may cause an increase in the surface temperature of the crop, and higher photosynthetic rates that can be realized under elevated [CO_2]. Higher photosynthetic rates will likely occur because of a higher CO_2 gradient from the source to the chloroplast and a higher reduced-carbon gradient from the leaves to the sink organs. This will cause faster filling of the sinks and perhaps the occurrence of earlier leaf senescence.

Changes in surface temperature of a crop grown under elevated [CO_2] are caused basically by the existence of a so-called physiological feedback effect of stomatal conductance on the surface energy balance of the crop. It is known that an increase in the external [CO_2] causes an increase in the intercellular or substomatal [CO_2] (C_i) (Farquhar *et al.*, 1980) and that increased C_i reduces the aperture of the stomatal pores (Morison, 1987). It has been theoretically demonstrated that, for most weather conditions, this physiological feedback effect favours increased surface temperature of the crop (Raupach, 1998).

Phenological development of a crop is strongly determined by changes in canopy temperature. It has been widely shown that, for most plant species, changes in air temperature affect their development rate, with generally an acceleration of development at higher temperatures. In most crops, development is associated with the production of leaves. Leaf production or, better, the initiation of leaf primordia in the apices is a temperature-driven process, but it may also be sensitive to severe nutrient or water deficiencies. Potato is no exception and its leaf production appeared to be accelerated when the temperature increased from 9°C to 25°C (Kirk and Marshall, 1992). However, the same study showed that the leaf production rate did not increase further when the temperature was above 25°C. Varietal differences were not investigated specifically, but it may be assumed that temperature increases, at least in warmer climates, may have less influence on potato phenology than on the development rates of other crops such as winter cereals (Miglietta and Porter, 1992; Miglietta *et al.*, 1995). This suggests that temperature increases in cooler climates are more likely to hasten crop phenology than in warmer climates. Confirmation of such an effect comes from the results of the potato FACE experiment that was made in a warm climate during the summer. In that experiment, surface temperature of the crops exposed to elevated [CO_2] increased (Miglietta *et al.*, 1998), but phenological development and leaf production rates were not affected, at least until flowering. Tuber initiation in potato is determined largely by photoperiod, being retarded by long days (Kooman, 1995). Temperature increases that result from climatic change may allow earlier planting dates and this may result in initial growth during shorter days, and thus accelerated crop development and earlier crop senescence.

Finally, the length of the growing season in cool areas may limit the growth period of a crop such as potato and thus its production potential. With climatic change in such areas, increased temperature may result in a longer growing season and higher production for potato and other crops (Carter *et al.*, 1996; Harrison and Butterfield, 1996). Of course, generalization of the data shown and considerations made from potato to other root and tuber crops are not possible. Each species deserves specific investigations, because the functional similarity in the mechanisms of phenological development among these crops is probably not that strong.

9.2.3 Biomass and yield

The presence of large sinks for assimilates in tuber and root crops makes these crops good candidates for large growth and yield responses to rising [CO_2]. In the early 1980s, literature on the effects of rising [CO_2] on growth and yields of several crop species were reviewed and summarized (Kimball, 1983). These data confirmed that root and tuber crops are very responsive to elevated [CO_2]. Yield increases were higher than those of the other crop groups except for those reported for fibre crops (Fig. 9.7). More recent studies have also reported substantial increases in yield for root and tuber crops grown with CO_2 enrichment. For example, both CO_2 enrichment experiments with potato in the Netherlands and in Italy showed considerable yield increases. In the OTC experiment (A.H.C.M. Schapendonk, personal communication, 1995), tuber growth was stimulated by 24–49%, depending on maturity class and weather conditions. Leaf area effects, however, were only small or even negative. In that case, in contrast to the predictions of the basic model, the benefit of elevated [CO_2] on tuber yield was higher for the late cultivar than for the early maturing varieties and higher in a temperate year than in a warm year. Thus the extent of the CO_2 effect is not determined by earliness or lateness of the cultivar but mainly by interaction effects of [CO_2] and temperature/drought on: early leaf area development, senescence of leaves later in the season, the remaining buffer capacity of photosynthetically active leaves and partitioning of carbon to different plant organs. In the FACE experiment, tuber growth and final yield were also stimulated by rising [CO_2] levels (Fig. 9.8). Yield stimulation by CO_2 was as large as 10% for each 100 $\mu mol\ mol^{-1}$ increase, which was equal to 40% yield increase with doubling of ambient [CO_2]. This

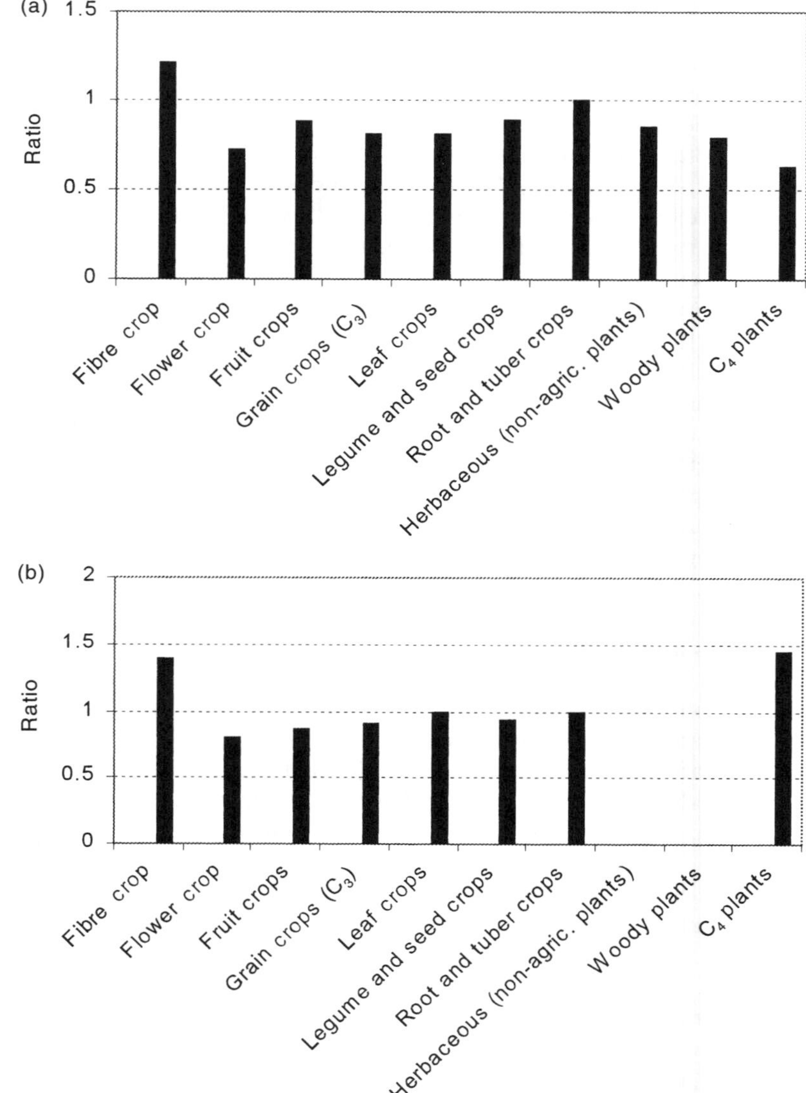

Fig. 9.7. Mean relative yield increase (ratios) of CO_2-enriched to control crops for (a) immature and (b) mature plants. Immature plants are agricultural crops for which yield was taken as total plant height or weight. (For more detail, see Kimball, 1983.)

was due to more tubers per plant (1.5 tubers for each 100 μmol mol^{-1} increase in $[CO_2]$) rather than to a greater mean tuber mass or size. Studies in the USA to assess the suitability of this crop for growth under very high $[CO_2]$ in the CELSS (Controlled Ecological Life Support System; Wheeler *et al.*, 1994) showed that

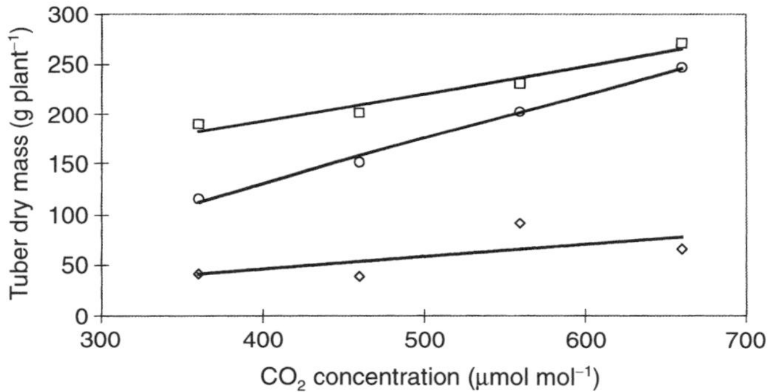

Fig. 9.8. Relationship between tuber dry mass (g plant[-1]) and atmospheric [CO_2] during growing season, measured at three different time intervals. □ = day 248, $y = 83.5 + 0.275x$, $r = 0.97$; ○ = day 228, $y = 49.1 + 0.447x$, $r = 0.99$; ◇ = day 200, $y = 2.83 + 0.123x$, $r = 0.65$. (Slope coefficients all significant at $P < 0.01$.)

potato yield response to [CO_2] was continuously increased up to 1000 μmol mol^{-1}. The positive effects of CO_2 enrichment on tuber yields may decrease if tuber sink strength is limited because of a down-regulation of the assimilation rate (Wheeler *et al.*, 1991). In this experiment, CO_2 enrichment increased tuber yield and total biomass dry weight by 39% and 34%, respectively, under short-day and low-light conditions, by 27% and 19% under short-day and high-light conditions, by 9% and 9% under long-day (24 h) and low-light conditions, and decreased dry weight by 9% and 9% under long-day and high-light conditions. This shows that where tuber growth was limited, whether by longer days in which the tuber initiation and growth were retarded or by higher irradiation (from high light intensity and/or long days), the response to elevated [CO_2] was also limited. In other studies, yields of carrot and kohlrabi (Sritharan *et al.*, 1992; Mortensen, 1994; Wheeler *et al.*, 1994) increased with [CO_2] doubling, despite significant interactions between the availability of phosphorus and the [CO_2].

The optimal temperature range for tuber growth (between 16 and 22°C) is small (Kooman, 1995). Daily temperatures that are outside this optimal range result in reduced tuber growth, a lower allocation of assimilates to the tubers, and thus a lower harvest index and tuber yield. With climatic change, the prevailing temperature during tuber growth will likely be different, but whether higher or lower and how much is uncertain.

9.3 Herbivory

There is evidence that increased [CO_2] may modify growth and composition of plants and may change the C/N ratio of leaves (Wong, 1979; Norby *et al.*, 1986; William *et al.*, 1986; Curtis *et al.*, 1989; Kuehny *et al.*, 1991). Plants grown under elevated [CO_2] have a lower leaf N concentration, due to the increase in carbohydrate production that 'dilutes' the protein content of the leaf (Lambers, 1993). This increased carbohydrate supply in plants exposed to elevated [CO_2] tends to increase the concentration of secondary compounds in leaves. Such compounds play an important role in distinct ecological functions, including allelopathy and the deterrence of herbivores (Baas, 1989; Dicke and Sabelis, 1989; Lambers, 1993). These changes in plant composition alter interactions between plant and herbivory (see Chapter 16, this volume). Prolonged development, increased food consumption, decreased food processing efficiency and general growth reduction are some of the typical responses of insects to reduced leaf N concentration (Roth and Lindroth, 1995). Hence, changes in plant composition in response to rising [CO_2] may influence the feeding habits and spread of insect populations. This aspect was investigated in the potato FACE experiment for Colorado potato beetle (*Leptinotarsa decemlineata* Say), which is considered one of the most important worldwide pests of potato. The experiment addressed questions concerning the intensity of beetle attacks on potato under elevated [CO_2] conditions, and the effect of changes in leaf composition on the growth rates of beetle larval populations and on their winter survival (Hare, 1990).

The Colorado potato beetle was originally confined to the semi-desert areas in Colorado, USA, where it fed on wild species of *Solanum*, especially *S. rostratum*. It spread rapidly when the country was opened – helped by human transport and by cultivation of potato, which proved to be an excellent host. Colorado potato beetle adults emerge from the soil in late spring and lay eggs on potato leaves. The eggs hatch; the larvae develop, by consuming a considerable amount of green tissues, and then pupate in the soil. Adult emergence may occur in the same season or, more often in temperate climates, in the following season. For instance, in northern Germany and Holland there is usually only one generation per year, but in southern Europe two or three generations per year have been observed.

The Colorado potato beetle's feeding behaviour was investigated in the potato FACE experiment by collecting a large number of larvae younger than 2 days in areas of the field that were at ambient [CO_2]. Larvae were subdivided in groups and fed with leaves collected from plants exposed to a range of [CO_2]. Larval growth rates (i.e. biomass gained per day) and consumption rates (i.e. food ingested per day) were determined. Results clearly indicated that growth of *L. decemlineata* larvae was sensitive to changes in leaf composition (N concentration and C/N ratio). Larvae grew faster when feeding on leaves grown in ambient [CO_2] than on leaves grown in high [CO_2] (Fig. 9.9), but differences between mean daily growth rates of larvae fed on leaves of plants exposed to 460, 560 and 660 μmol CO_2 mol^{-1} were not appreciable. Larval size at the end of the experiment was affected by the quality of the foliage ingested, with larvae from the ambient treatments having 23.8% larger dry mass than those fed from leaves grown in high [CO_2] (ambient = 29.54 ± 3.2 mg larva^{-1}, high [CO_2] = 22.51 ± 3.7 mg larva^{-1}). However, the total amount of food ingested by the larvae and the leaf consumption rates were the same for all the treatments. Lower protein intake

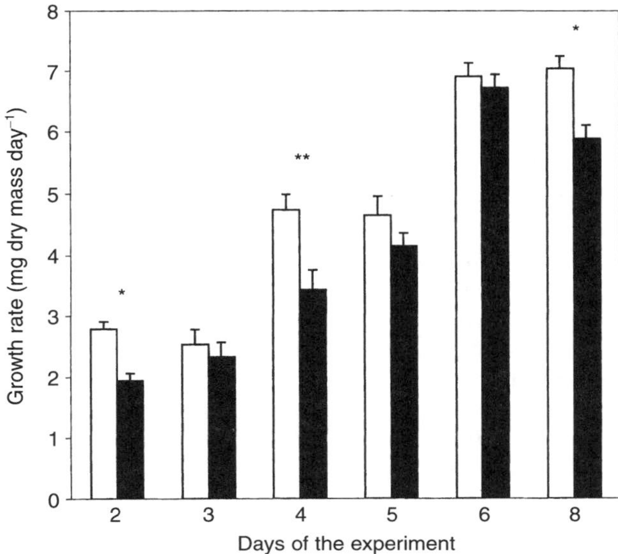

Fig. 9.9. Growth rate (mg dry weight day^{-1}) of Colorado potato beetle (*L. decemlineata*) larvae fed with leaves grown at ambient [CO_2] (open bars) and elevated [CO_2] (solid bars). (Difference in growth rate significant at $P < 0.05$ indicated by * or at $P < 0.01$ by **.)

as a result of changed leaf composition decreased the growth rates of Colorado beetle larvae feeding on potato leaves and, possibly, on other wild relatives of this species. Reduced growth of the larvae may result in lower larvae reserves at the time of pupation, with possible consequences for the ability of the insect to survive winter conditions while diapausing into the soil.

9.4 Carbon Dioxide × Climatic Change Interactions in Potato

The positive effect of [CO_2] on crop growth may be counteracted by the effect of a concomitant temperature rise. Such effects cannot be traced back to one or a few critical physiological and morphological characteristics because the linkages between plant productivity and physiological characteristics are concealed by feedback control mechanisms by interactions among varying environmental conditions, and changes in crop characteristics during its phenological development. Simulation models are probably the only tools for interpreting and predicting the impact of an environmental change on plant production or evaluating the best strategy for further experimental research. Several crop growth models have been formulated for potato that differ in approach and detail (Fishman *et al.*, 1984; Ng and Loomis, 1984; MacKerron and Waister, 1985).

A simulation model of potato growth was used to highlight the importance of CO_2/temperature interactions on potato productivity and their implications for varietal choice (Schapendonk *et al.,* 1995). The model was previously described (Spitters and Schapendonk, 1990). Briefly, that model calculates the light profile within the canopy on the basis of a leaf area index and the extinction coefficients for both the flux of direct solar radiation and diffuse sky light. The rates of photosynthesis at various heights within the canopy are calculated using the von Caemmerer and Farquhar (1981) model. Discrimination was made between shaded leaf area receiving diffuse radiation only, and sunlit leaf area receiving both diffuse and direct radiation (Spitters *et al.*, 1986). Daily gross canopy assimilation rates are obtained from instantaneous photosynthetic rates of individual leaves and the light profile in the canopy, integrated over the canopy and day (Goudriaan, 1986). A death rate of leaves due to senescence depends on the maturity class (Spitters and Schapendonk, 1990).

Simulations were made with this model to calculate total dry matter production and tuber yield of different potato cultivars, growing on a sandy loam under Dutch weather conditions. Averaged over the years 1988–1991, a simulated increase of the [CO_2] from 350 μmol mol^{-1} to 700 μmol mol^{-1} increased the tuber dry matter production by 22% for late cultivars and 29% for early cultivars. The effects were smaller for late cultivars, irrespective of the occurrence of a drought period. Elevated temperature reduced the positive effects of elevated [CO_2], because the stimulation of leaf area expansion in the juvenile stage was offset later in the season by earlier foliage senescence. Temperature increases had only a net positive effect on production in situations with optimal water supply and high irradiance during the juvenile stage, combined with a severe late drought. In those situations, the larger foliage prevented a rapid decrease in light interception during senescence. Recently, a simulation study was performed within an EU-funded project (CLIVARA) to determine the effect of climatic change on potato yields at several sites and on regional scales. A detailed potato model (NPOTATO) was used to analyse the effects of climatic change, climatic variability and increased [CO_2] on potato production throughout Europe (Wolf, 1999a). In addition, a simplified potato model (POTATOS) was used to analyse the effects of climatic change on potato at a regional scale (Great Britain: Downing *et al.*, 1999; Wolf, 1999b). The model simulations for both studies were made for baseline climatic and future climatic scenarios (corresponding with the period 2035–2064) obtained using the HADCM2 (Hadley Centre unified model climate change experiment) greenhouse gas experiment (HCGG, greenhouse gas only integration) and the HADCM2 greenhouse gas with sulphate aerosol experiment (HCGS, sulphate and greenhouse gas integration), both with an increase in climatic variability (HCGGv, greenhouse gas only integration, including changes in climatic variability; and HCGSv, sulphate and greenhouse gas integration including changes in climatic variability) and without. For the baseline climate, the [CO_2] was set at 353 μmol mol^{-1} and for the four scenarios at 515 μmol mol^{-1}.

9.4.1 Site scale

Simulated climatic change in northern Europe resulted in small to considerable increases in a mean tuber yield for both irrigated and non-irrigated potato crops (Fig. 9.10a,c). In central and southern Europe, climatic change resulted in both small decreases and increases in tuber yields for both irrigated and

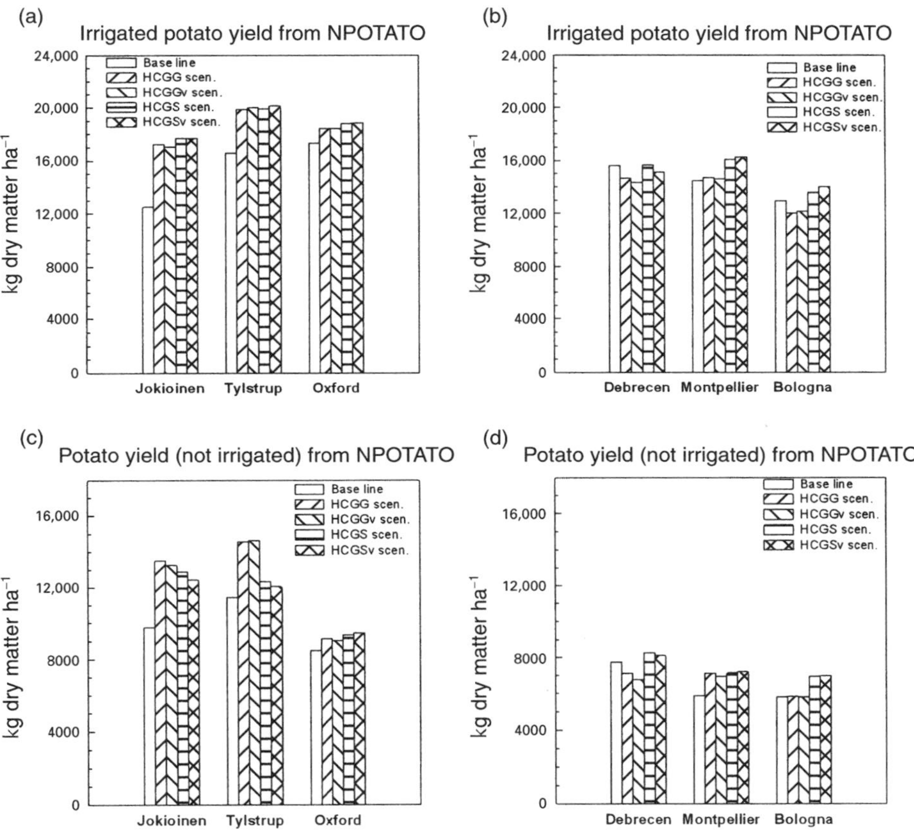

Fig. 9.10. Simulated potato tuber yields (mid variety) using NPOTATO. Present and future climate conditions at sites in northern (Jokioinen, Tylstrup and Oxford) and southern Europe (Debrecen, Montpellier and Bologna), (a, b) with and (c, d) without irrigation. Results refer to 30 years of generated weather data for baseline climate and for four climatic change scenarios. HCGG = Hadley Centre unified model climate change experiment with the greenhouse gas only integration; HCGSv = Hadley Centre unified model climate change experiment with the sulphate and greenhouse gas integration including changes in climatic variability; HCGS = Hadley Centre unified model climate change experiment with sulphate and greenhouse gas integration; HCGGv = Hadley Centre unified model climate change experiment with greenhouse gas only integration, including changes in climatic variability. For the baseline climate, the atmospheric [CO_2] was set at 353 μmol mol^{-1} and for the four scenarios at 515 μmol mol^{-1}.

non-irrigated crops, depending on the selected future climatic change scenario (Fig. 9.10b,d). Variability of irrigated tuber yields slightly increased with climatic change in northern Europe, and it increased slightly to moderately in southern Europe (Fig. 9.11a,c). Under a baseline climate, the variability of water-limited tuber yields was much higher than that of irrigated yields, particularly in southern Europe and the UK. With climatic change, yield variability was essentially zero to moderately decrease in northern Europe and slightly to moderately lower in southern Europe (Fig. 9.11b,d).

An evaluation of the effectiveness of changes in crop management (i.e. variety, planting dates, irrigation) in response to climatic change was also performed. The results showed that at a site in northern Europe (i.e. Oxford)

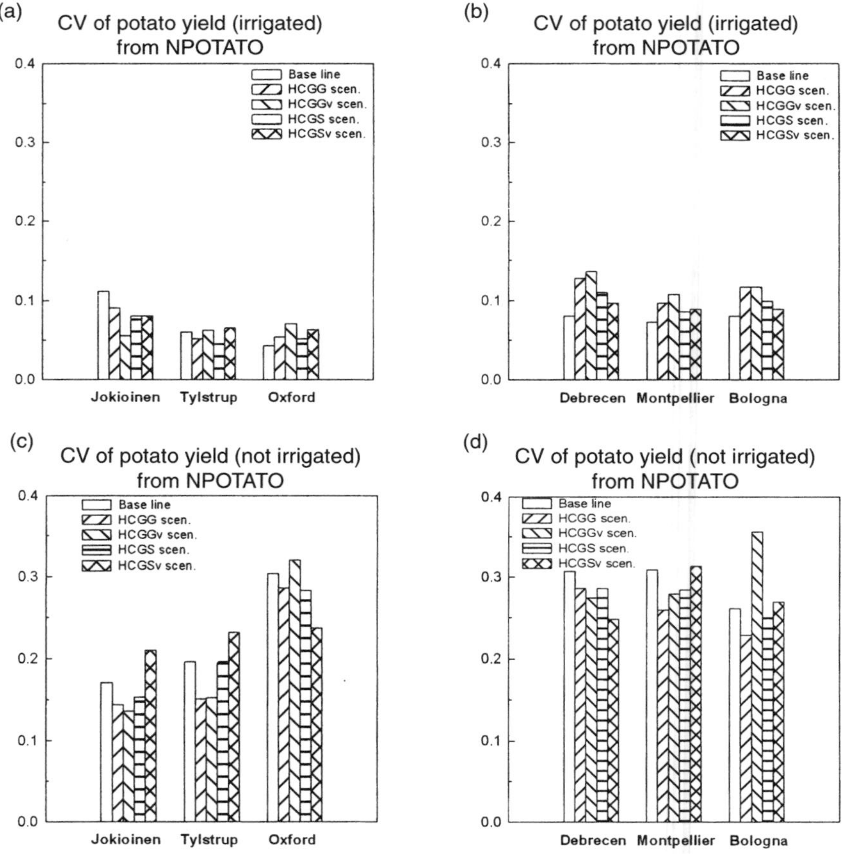

Fig. 9.11. Coefficient of variation (CV) of simulated tuber yields (mid variety) calculated with NPOTATO for present and future climate conditions at sites in northern and southern Europe, (a, b) with and (c, d) without irrigation. Results refer to 30 years of generated weather data for baseline climate and for four climatic change scenarios. (See legend to Fig. 9.10.)

the impact of crop variety on the calculated change in tuber yields of irrigated crops under the climatic change scenarios was nil, although the absolute yield differed among crop varieties (Fig. 9.12a). The increases in crops with water-limited yields under the climatic change scenarios were larger for the earlier varieties (Fig. 9.13a). Cultivation of earlier varieties resulted in more positive or fewer negative simulated tuber yield changes under the climatic change scenarios in southern Europe (e.g. Bologna), because the hot summer period was avoided, both with and without irrigation (Figs 9.12b and 9.13b). An advanced planting date resulted in a higher yield. This yield increase became large with climatic change in northern Europe (Fig. 9.14a,b); it was already large with the present climate in southern Europe and it increased further under the climatic change scenarios (Fig. 9.14c,d). Irrigation requirements may increase or decrease with climatic change, even without changes in rain. The extent that irrigation will be required in the future varies with the site, climatic change scenario, planting date and crop variety (data not shown). However, both an early crop variety and earlier planting dates considerably reduced irrigation requirements.

9.4.2 Regional scale

The results of the simple potato model applied for the unirrigated potato crop across Great Britain showed that under the HCGG scenarios (Fig. 9.15a), which are wetter and hotter than the HCGS scenarios, potato tuber yields were

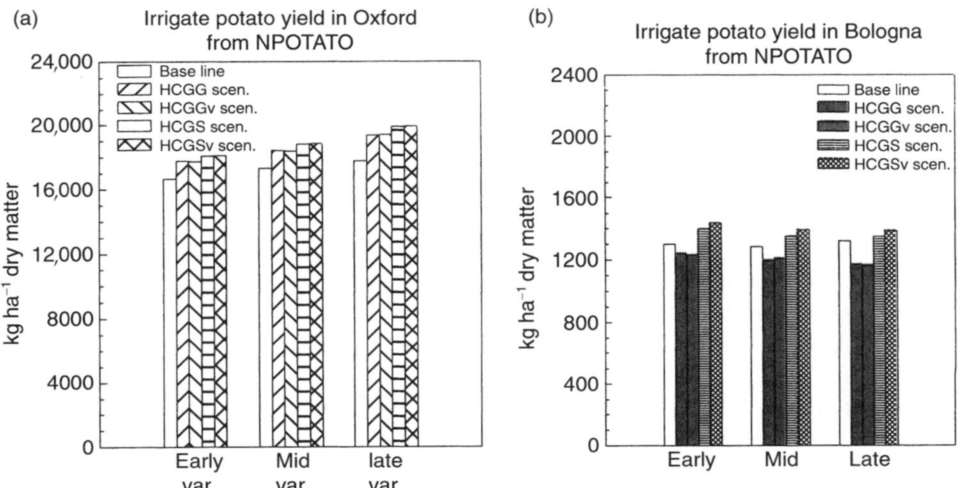

Fig. 9.12. Simulated potato tuber yields of different irrigated potato varieties with NPOTATO for present and future climate conditions at (a) Oxford and (b) Bologna. Results refer to 30 years of generated weather data for baseline climate and for four climatic change scenarios. (See legend to Fig. 9.10.)

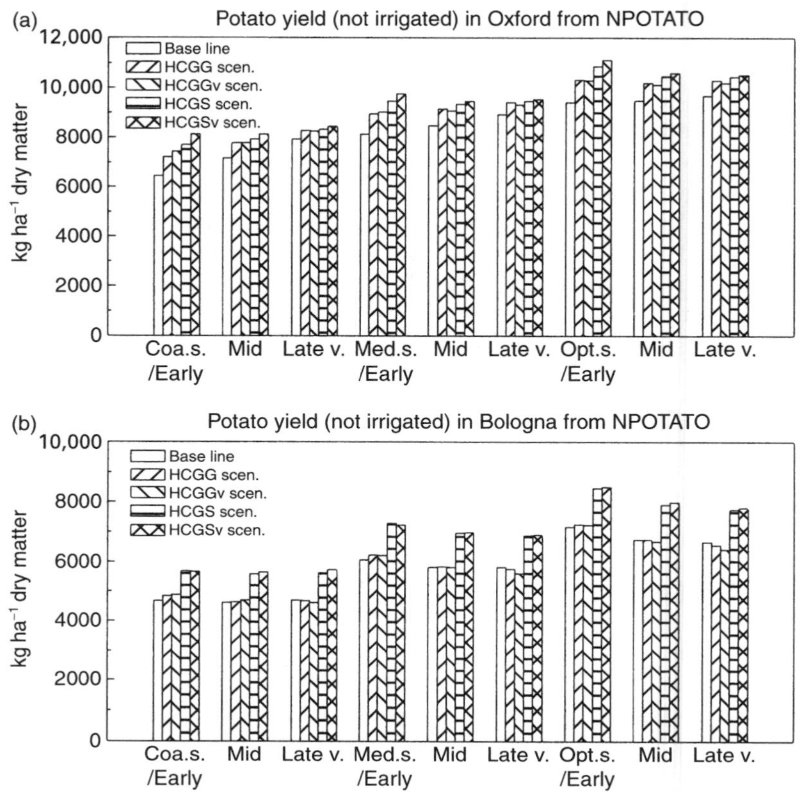

Fig. 9.13. Simulated potato tuber yields of three potato varieties (early/mid/late; non-irrigated) on three soil types (coarse/medium/optimal), calculated with NPOTATO for present and future climate conditions at (a) Oxford and (b) Bologna. Results refer to 30 years of generated weather data for baseline climate and for four climatic change scenarios. (See legend to Fig. 9.10.)

lower in practically all regions of the UK. These lower yields were caused by the temperature rise, which speeded the phenological development of the crop and reduced the time for growth and biomass production. Under the HCGG scenarios, central England mainly had simulated yield increases (Fig. 9.15b). The temperature increase under this scenario is smaller than under the HCGG scenario, which accounts for the reduction in the area showing lower yields.

These results reinforce the belief that climatic change will have negative effects on yields because of increased temperature and lower rainfall. These negative effects may be more or less compensated for by higher photosynthetic rates caused by increasing atmospheric [CO_2] and by appropriate crop management strategies (i.e. variety, planting dates, irrigation).

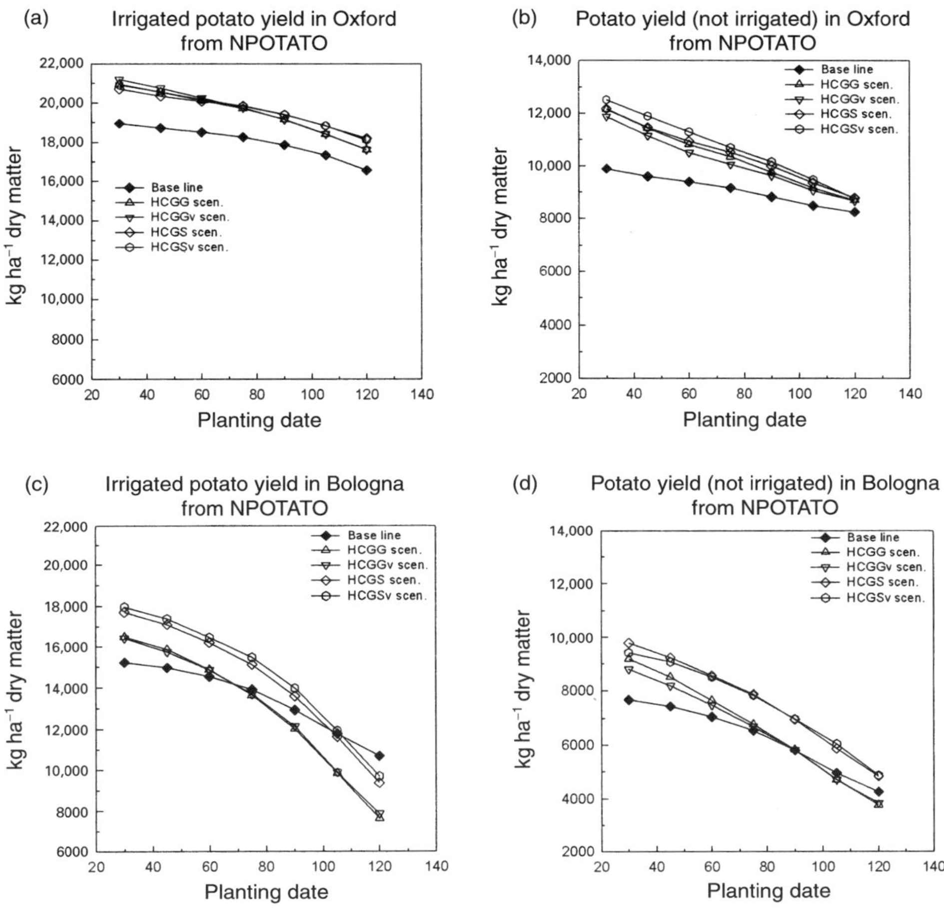

Fig. 9.14. Sensitivity to changes in planting date (Julian day) of the potato tuber yields (mid variety), calculated with NPOTATO for present and future climate conditions at Oxford, (a) with and (b) without irrigation; and at Bologna, (c) with and (d) without irrigation. Results refer to 30 years of generated weather data for baseline climate and for four climatic change scenarios. (See legend to Fig. 9.10.)

9.5 Conclusions

Weather is a most significant factor determining plant growth and productivity. The IPCC estimates that under the business-as-usual scenario global mean temperatures will increase by 0.2°C per decade (Houghton *et al.*, 1996). Such a temperature rise, and concurrent changes in other weather variables such as rainfall and radiation, might alter the crop yield potentials by changing the degree of high-temperature stress or frost-free period, the degree of growth

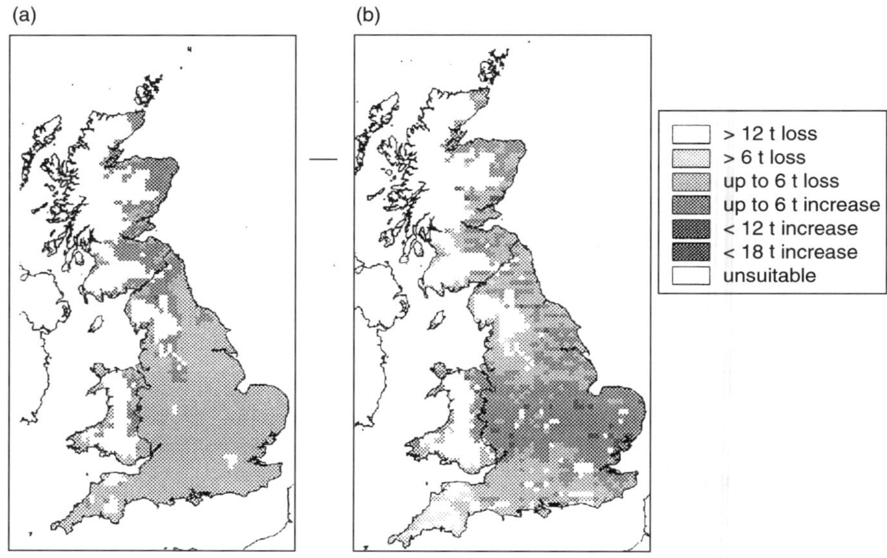

Fig. 9.15. Changes in simulated potato tuber yields in UK from baseline under (a) the HCGG and (b) the HCGS climatic change scenarios. (See legend to Fig. 9.10.)

reduction by water shortage or nutrient deficiency, and the risk for yield losses by hail, diseases and pests. Changes in the climate may also result in an increase in climatic variability and extremes. Such changes in the climate will affect growth and production of tuber and root crops, but the possible impact of climatic change has not yet been studied for most species of this crop group. Increased [CO_2] may also have significant effects on crop phenology, productivity and physiology. Unfortunately, potential interactions among effects of increased [CO_2] on growth and yield and those of changes in climatic variables such as temperature and solar radiation are complex. Because of this complexity, there is not enough information to draw exhaustive conclusions about the potential impact of climatic change on the productivity of root and tuber crops. The effects of climatic change on world agriculture remain uncertain because of ignorance of responses of plant and soil processes to several weather variables.

Acknowledgements

The climatic change impact assessments for potato have been conducted within the EU-funded CLIVARA project (European Commission's Environment Programme, contract number ENV4-CT95–0154).

References

Baas, W.J. (1989) Secondary plant compounds, their ecological significance and consequences for the carbon budget. Introduction of the carbon/nutrient cycle theory. In: Lambers, H., Cambridge, M.L., Konings, H. and Ponds, T.L. (eds) *Causes and Consequences of Variation in Growth Rate and Productivity of Higher Plants.* SPB Academy Publishing, The Hague, pp. 331–340.

Benincasa, F., Fasano, G. and Materassi, A. (1988) Analisi della radiazione nella gamma fotosinteticamente attiva e infrarossa vicina tramite banco ottico. *Quaderno Metodologico,* 10. IPRA, Rome.

Caemmerer, S. von and Farquhar, G.D. (1981) Some relations between the biochemistry of photosynthesis and the gas exchange of leaves. *Planta* 153, 376–387.

Carter, T.R., Saarikko, R.A. and Niemi, K.J. (1996) Assessing the risks and uncertainties of regional crop potential under a changing climate in Finland. *Agricultural and Food Sciences in Finland* 5, 329–350.

Chu, C.C., Coleman, J.S. and Mooney, H.A. (1992) Controls of biomass partitioning between roots and shoots – atmospheric CO_2 enrichment and the acquisition and allocation of carbon and nitrogen in wild radish. *Oecologia* 89, 580–587.

Curtis, P.S., Drake, B.G. and Whigham, D.F. (1989) Nitrogen and carbon dynamics in C_3 and C_4 estuarine marsh plants grown under elevated CO_2 in situ. *Oecologia* 78, 297–301.

Dicke, M. and Sabelis, M.W. (1989) Does it pay to advertise for body guards? In: Lambers, H., Cambridge, M.L., Konings, H. and Ponds, T.L. (eds) *Causes and Consequences of Variation in Growth Rate and Productivity of Higher Plants.* SPB Academy Publishing, The Hague, The Netherlands, pp. 341–358.

Downing, T.E., Harrison, P.A., Butterfield, R.E. and Lonsdale K.G. (eds) (1999) *Climate Change, Climate Variability and Agriculture in Europe: an integrated assessment.* Research Report No. 21, Environmental Change Unit, University of Oxford, Oxford, UK, 330 pp.

FAOSTAT (1998) On-line and multilingual database. FAO, Rome.

Farquhar, G.D. and Caemmerer, S. von (1982) Modelling of photosynthetic response to environmental conditions. In: Lange, O.L., Nobel, P.S., Osmond, C.B. and Ziegler, H. (eds) *Encyclopedia of Plant Physiology, New Series,* Vol. 12B, *Physiological Plant Ecology II.* Springer-Verlag, Berlin, pp. 549–587.

Farquhar, G.D., Caemmerer, S. von and Berry, J.A. (1980) A biochemical model of photosynthetic CO_2 assimilation in leaves of C_3 species. *Planta* 149,78–90.

Farrar, J.F. (1996) Sinks, integral parts of a whole plant. *Journal of Experimental Botany* 47, 1273–1280.

Ferris, R., Wheeler, T.R., Ellis, R.H., Hadley, P., Wollenweber, B., Porter, J.R., Karacostas, T.S., Papadopoulos, M.N. and Schellberg, J. (1999) Effects of high temperature extremes on wheat. In: Downing, T.E., Harrison, P.A., Butterfield, R.E. and Lonsdale K.G. (eds). *Climate Change, Climate Variability and Agriculture in Europe: an integrated assessment.* Research Report No. 21, Environmental Change Unit, University of Oxford, Oxford, UK, pp. 31–55.

Fishman, S., Talpaz, H., Dinar, M., Levy, M., Arazi, Y., Rozman, Y. and Varshavsky, S. (1984) A phenomenological model of dry matter partitioning among plant organs for simulation of potato growth. *Agricultural Systems* 14, 159–169.

Goudriaan, J. (1986) A simple and fast numerical method for the computation of daily totals of crop photosynthesis. *Agricultural and Forest Meteorology* 38, 249–254.

Hare, J.D. (1990) Ecology and management of the Colorado potato beetle. *Annual Review of Entomology* 35, 81–100.

Harrison, P.A. and Butterfield, R.E. (1996) Effects of climate change on Europe-wide winter wheat and sunflower productivity. *Climate Research* 7, 225–241.

Houghton, J.T., Callander, B.A. and Varney, S.K. (eds) (1992) *Climate Change 1992: The Supplementary Report to the IPCC Scientific Assessment.* Cambridge University Press, Cambridge, UK, 200 pp.

Houghton, J.T., Meira Filho, L.G., Callander, B.A., Kattenberg, A. and Maskell, K. (eds) (1996) *Climate Change 1995: The Science of Climate Change.* Cambridge University Press, Cambridge, UK, 572 pp.

Keeling, C.D., Whorf, T.P., Whalen, M. and van der Plicht, J. (1995) Interannual extremes in the rate of rise of atmospheric carbon dioxide since 1980. *Nature* 375, 666–670.

Kimball, B.A. (1983) Carbon dioxide and agricultural yield: an assemblage and analysis of 430 prior observations. *Agronomy Journal* 75, 779–788.

Kirk, W.W. and Marshall, B. (1992) The influence of temperature on leaf development and growth in potatoes in controlled environments. *Annals of Applied Biology* 120(3), 511–525.

Knipling, E.B. (1970) Physical and physiological basis for the reflectance of visible and near-infrared radiation from vegetation. *Remote Sensing of Environment* 1,155–159

Komor, E., Orlich, G., Weig, A. and Kockenberger, W. (1996) Phloem loading – not metaphysical, only complex: towards a unified model of phloem loading. *Journal of Experimental Botany* 47, 1155–1164.

Kooman, P.L. (1995) Yielding ability of potato crops as influenced by temperature and daylength. PhD thesis, Wageningen Agricultural University, Wageningen, The Netherlands.

Körner, C. and Miglietta, F. (1994) Long-term effects of naturally elevated CO_2 on Mediterranean grassland and forest trees. *Oecologia* 99, 343–351.

Körner, C., Pelaez-Riedl, S. and Vanbel, A.J.E. (1995) CO_2 responsiveness of plants. A possible link to phloem loading. *Plant, Cell and Environment* 18, 595–600.

Kuehny, J.S., Peet, M.M., Nelson, P.V. and Willits, D.H. (1991) Nutrient dilution by starch in CO_2-enriched *Chrysanthemum*. *Journal of Experimental Botany* 42, 711–716.

Lambers, H. (1993) Rising CO_2, secondary plant metabolism, plant–herbivore interactions and litter decomposition. *Vegetatio* 104/105, 263–271.

Long, S.P. and Drake, B.G. (1992) Photosynthetic CO_2 assimilation and rising atmospheric CO_2 concentrations. In: Baker, N.R. and Thomas, H. (eds) *Crop Photosynthesis: Spatial and Temporal Determinants.* Elsevier, Amsterdam, pp. 69–101.

MacKerron, D.K.L. and Waister, P.D. (1985) A simple model of potato growth and yield. Part I: Model evaluation and sensitivity analysis. *Agricultural and Forest Meteorology* 34, 241–252.

Malthus, T.J. and Madeira, A.C. (1993) High resolution spectroradiometry: spectral reflectance of field bean leaves infected by *Botrytis fabae*. *Remote Sensing Environment* 45, 107–116.

McKeehen, J.D., Smart, D.J., Mackowiak, C.L., Wheeler, R.M. and Nielsen, S.S. (1995) Effect of CO_2 levels on nutrient content of lettuce and radish. *Advances in Space Research* 18, 85–92.

Miglietta, F. and Porter, J.R. (1992) The effects of climatic change on development in wheat – analysis and modelling. *Journal of Experimental Botany* 43, 1147–1158.

Miglietta, F., Tanasescu, M. and Marica, A. (1995) The expected effects of climate change on wheat development. *Global Change Biology* 1, 407–415.

Miglietta, F., Giuntoli, A. and Bindi, M. (1996) The effect of free air carbon dioxide enrichment (FACE) and soil nitrogen availability on the photosynthetic capacity of wheat. *Photosynthesis Research* 47, 281–290.

Miglietta, F., Lanini, M., Bindi, M. and Magliulo, V. (1997) Free air CO_2 enrichment of potato (*Solanum tuberosum,* L.): design and performance of the CO_2-fumigation system. *Global Change Biology* 3, 417–427.

Miglietta, F., Magliulo, V., Bindi, M., Cerio, L., Vaccari, F., LoDuca, V. and Peressotti, A. (1998) Free air CO_2 enrichment of potato (*Solanum tuberosum,* L.): development, growth and yield. *Global Change Biology* 4, 163–172.

Morison, J.I.L. (1987) Plant growth and CO_2 history. *Nature* 327, 560.

Mortensen, L.M. (1994) Effects of elevated CO_2 concentrations on growth and yield of 8 vegetable species in a cool climate. *Scientia Horticulturae* 58, 177–185.

Ng, N. and Loomis, R.S. (1984) Simulation of growth and yield of the potato crop. *Simulation Monographs*. Pudoc, Wageningen, The Netherlands.

Norby, R.J., O'Neill, E.G. and Luxmoore, R.J. (1986) Effects of atmospheric CO_2 enrichment on the growth and mineral nutrition of *Quercus alba* seedlings in nutrient-poor soil. *Plant Physiology* 82, 83–89.

Poorter, H. (1993) Interspecific variation in the growth response of plants to an elevated CO_2 concentration. *Vegetatio* 104, 77–97.

Raupach, M.R. (1998) Influences of local feedbacks on land–air exchanges of energy and carbon. *Global Change Biology* 4, 477–494.

Roth, S.K. and Lindroth, R.L. (1995) Elevated atmospheric CO_2: effects on phytochemistry, insect performance and insect–parasitoid interactions. *Global Change Biology* 1, 173–182.

Sage, R.F. (1994) Acclimation of photosynthesis to increasing atmospheric CO_2. The gas exchange perspective. *Photosynthesis Research* 39, 351–368.

Schapendonk, A.H.C.M., Pot, C.S. and Goudriaan, J. (1995) Simulated effects of elevated carbon dioxide concentration and temperature on productivity of potato. In: Haverkort, A.J. and MacKerron, D.K.L. (eds) *Potato Ecology and Modelling of Crops under Conditions Limiting Growth*. Kluwer, Amsterdam, pp. 101–113.

Semenov, M.A., Wolf, J., Evans, L.G., Eckersten, H. and Iglesias A. (1996) Comparison of wheat simulation models under climate change. II. Application of climate change scenarios. *Climate Research* 7, 271–281.

Spitters, C.J.T. and Schapendonk, A.H.C.M. (1990). Evaluation of breeding stategies for drought tolerance in potato by means of crop growth simulation. *Plant and Soil* 123, 193–203.

Spitters, C.J.T., Toussaint, H.A.J.M. and Goudriaan, J. (1986) Separating the diffuse and direct component of global radiation and its implications for modelling canopy photosynthesis. I. Components of incoming radiation. *Agricultural and Forest Meteorology* 38, 231–242.

Sritharan, R., Caspari, H. and Lenz, F. (1992) Influence of CO_2 enrichment and phosphorus supply on growth, carbohydrates and nitrate utilization of kohlrabi plants. *Gartenbauwissenschaft* 57, 246–251.

Wheeler, R.M., Tibbitts, T.W. and Fitzpatrick, A.H. (1991) Carbon dioxide effects on potato growth under different photoperiods and irradiance. *Crop Science* 31, 1209–1213.

Wheeler, R.M., Mackowiak, C.L., Sager, J.C. and Knott, W.M. (1994) Growth of soybean and potato at high CO_2 partial pressure. *Advances in Space Research* 14, 251–255.

William, W.E., Garbutt, K., Bazzaz, F.A. and Vitousek, P.M. (1986) The response of plants to elevated CO_2. IV. Two deciduous-forest communities. *Oecologia* 69, 454–459.

Wolf, J. (1999a) Modelling climate change impacts at the site scale on potato. In: Downing, T.E., Harrison, P.A., Butterfield, R.E. and Lonsdale K.G. (eds) *Climate Change, Climate Variability and Agriculture in Europe: an integrated assessment.* Research Report No. 21, Environmental Change Unit, University of Oxford, Oxford, UK, pp. 135–154.

Wolf, J. (1999b) Modelling climate change impacts on potato in central England. In: Downing, T.E., Harrison, P.A., Butterfield, R.E. and Lonsdale K.G. (eds) *Climate Change, Climate Variability and Agriculture in Europe: an integrated assessment.* Research Report No. 21, Environmental Change Unit, University of Oxford, Oxford, UK, pp. 239–261.

Wong, S.C. (1979) Elevated atmospheric partial pressure of CO_2 and plant growth. I. Interactions of nitrogen nutrition and photosynthetic capacity in C_3 and C_4 plants. *Oecologia* 44, 68–74.

Ziska, L.H., Sicher, R.C. and Kremer, D.F. (1995) Reversibility of photosynthetic acclimation of swiss-chard and sugar-beet grown at elevated concentrations of CO_2. *Physiologia Plantarum* 3, 355–364.

10 Crop Ecosystem Responses to Climatic Change: Vegetable Crops

MARY M. PEET[1] AND DAVID W. WOLFE[2]

[1]*Department of Horticultural Science, North Carolina State University, Raleigh, NC 27695-7609, USA;* [2]*134-A Plant Science Building, Cornell University, Tower Road, Ithaca, NY 14853-5908, USA*

10.1 Introduction

10.1.1 Importance of vegetables

Worldwide, over 37 Mha of vegetables were harvested in 1997 for a total production of over 596 Mt (FAOSTAT, 1997). This represents a significant contribution to the dietary needs of the world's human population, especially since many vegetables are good sources of proteins, vitamins and minerals. In addition, many have been associated with protective effects against cancer and heart disease.

10.1.2 State of knowledge on effects of global climate change on vegetables

Data that should hold predictive value for climatic change responses of agronomic crops and some natural ecosystems are available from open-air or naturally lit compartments. Even for vegetable crops, such as potatoes (*Solanum tuberosum* L.) and cabbage (*Brassica oleracea* L. var. *capitata* L.) grown on 18 and 2 Mha worldwide, respectively, little is known about reactions to global climatic change. The available data are mostly based on greenhouse and growth chamber studies. Greenhouse studies of carbon dioxide concentration ([CO_2]) response are particularly problematic, since typical CO_2 enrichment levels (1000 to 1200 μmol mol^{-1}) are much higher than [CO_2] predicted for the next century. Another problem with these studies is that greenhouses are not CO_2-enriched during sunny weather because they must be ventilated for cooling, which prevents control of [CO_2]. This problem also eliminates much of the value of CO_2 enrichment for greenhouse production.

Recent review books on the projected effects of global climatic change on crop yields include Kimball (1990), Rosenzweig, (1995), Allen *et al.* (1997) and Rosenzweig and Hillel (1998). Recent review chapters on temperature and/or [CO_2] effects on plants from climatic change include: Wolfe and Erickson (1993); Paulsen (1994), Wolfe (1994) and Crawford and Wolfe (1999). Nederhoff (1994) reviewed the literature on use of [CO_2] to increase the growth of greenhouse vegetable crops and also conducted extensive experimentation in this area. However, these reviews have included little information on field vegetable crop responses to elevated [CO_2] and temperature. A recently published volume, *The Physiology of Vegetable Crops* (Wien, 1997a), describes crop responses to many environmental factors, though not specifically in the context of global climatic change. Much of the specific information on crop responses to temperatures in this review is taken from these chapters.

10.1.3 Chapter goals

The objective of this chapter is to review some of the known responses of vegetable crops to increases in [CO_2], temperature and pollutants and to extrapolate vegetable crop responses where data from only agronomic crops is available. Case studies on potatoes, carrots and lettuce are also presented as examples of potential responses.

Because quality is often more important than total yield, in vegetable crops some assessment will be made in this chapter of how global climatic changes may affect vegetable quality. For example, sweet-corn (*Zea mays* L. subsp. *mays*) and field maize (*Zea mays* L.) are classified as the same genus and species, but higher temperatures during sweet-corn ripening may cause premature decline in sugar levels and poor tip-fill. Either problem can make fresh-market sweet-corn unmarketable but only reduces yield in field maize, if there is an effect at all.

10.2 Effects of Global Climatic Change on Crop Plants

10.2.1 Direct effects of carbon dioxide

Introduction

Carbon dioxide concentration is increasing by at least 1.5 μmol mol^{-1} $year^{-1}$ and will continue to rise even more rapidly for the next 50–100 years unless current patterns of fossil fuel consumption change. Wolfe (1994) and Drake *et al.* (1997) reviewed effects on plant physiology and growth under elevated [CO_2].

Transpiration

In most crops, increased [CO_2] improves water-use efficiency (WUE) because of declines in stomatal conductance (e.g. Rogers and Dahlman, 1993), potentially decreasing drought susceptibility and reducing irrigation requirements.

However, the effect of decreased transpiration on vegetable crop yields is unlikely to be large since vegetables are irrigated in most production areas. Physiological disorders such as tipburn in lettuce (*Lactuca sativa* L. var. *capitata* L.) and cole crops and blossom-end rot in tomato (*Lycopersicon esculentum* Mill syn. *Lycopersicon lycopersicum* (L.) Karsten.), pepper (*Capsicum annuum* L. Grossum group) and watermelon (*Citrullus lanatus* (Thunb.) Matsum & Nakai) are sometimes associated with excessive transpiration, so the incidence of these disorders may be reduced (Table 10.1).

Respiration

Effects of elevated [CO_2] on respiration have been reviewed recently (Amthor, 1997; Drake *et al.*, 1997). Respiration of leaves and roots in the dark slows within minutes of an increase in ambient [CO_2], so night-time respiration would be lower at high [CO_2]. This direct, short-term and readily reversible effect of [CO_2] on respiration has been noted in tomatoes, lettuce, peppers, peas and maize. It is apparently caused by inhibited respiration *per se* rather

Table 10.1. Physiological disorders of vegetables caused or exacerbated by high or low temperatures. Data summarized from the respective crop chapters in Wien (1997a).

Crop	Disorder	Aggravating factor
Asparagus	High fibre in stalks	High temperatures
Asparagus	Feathering and lateral branch growth	Temperatures > 32°C, especially if picking frequency is not increased
Bean	High fibre in pods	High temperatures
Carrot	Low carotene content	Temperatures < 10°C or > 20°C
Cauliflower	Blindness, buttoning, ricy curds	Low temperatures
Cauliflower, broccoli	Hollowstem, leafy heads, no heads, bracting	High temperatures
Cole crops and lettuce	Tipburn	Drought, especially combined with high temperatures; high transpiration
Lettuce	Tipburn, bolting, loose, puffy heads	Temperatures > 17–28°C day and 3–12°C night
Maize	Poor kernel development, poor husk cover, tasselate ear	High temperatures, especially combined with drought
Onion	Bulb splitting	High temperatures
Pepper	Low seed production and off-shaped fruit	Low temperatures
Pepper	Sunscald	High temperatures
Potato	Secondary growth and heat sprouting	High temperatures
Tomato	Fruit cracking, sunscald	High temperatures
Tomato, pepper, watermelon	Blossom-end rot	High temperatures, especially combined with drought; high transpiration

than stimulated carboxylase activity, but the specific mechanism(s) is not known.

In theory, lower night-time respiration rates, especially when coupled with higher daytime photosynthetic rates at high [CO_2], should increase growth and yield. Relatively little data is available on long-term effects of elevated night-time [CO_2] on growth (Amthor, 1997), however. There can be no growth without respiration, so it is also possible that night-time inhibition of respiration could slow growth, as was seen in soybean given 2 days with elevated night-time [CO_2] (Bunce, 1995). Inhibition of maintenance processes by elevated [CO_2] might exacerbate the effects of some stresses by slowing acclimation and repair processes. For example, both high temperature in Cocklebur (*Xanthium strumarium*) and ozone stress in beans (*Phaseolus vulgaris* L.) may require enhanced leaf respiration for repair, and this repair might be impaired by elevated night-time [CO_2] (Gale, 1982; Amthor, 1988, 1997). Amthor (1997) concluded that respiratory decreases in elevated [CO_2] will benefit plants only if the respiratory components inhibited are unnecessary, rather than essential.

Root growth

In a recent review of 167 studies on root response to elevated [CO_2], Rogers *et al.* (1994) found that root dry weight increased in about 87% of the studies and plant roots were longer or more numerous in 77% of the studies. In cassava (*Manihot esculenta* Crantz), a tropical root crop, there was not only a large increase in growth (150%), but partitioning to the root was also increased. Overall, however, effects of CO_2 enrichment on root/shoot ratio and partitioning have been highly variable and may differ between C_3 and C_4 crops (Rogers *et al.*, 1997; Wolfe *et al.*, 1998). This topic is discussed further in the potato case study.

Whether or not root/shoot ratios in mature plants are increased by high [CO_2], rooting and establishment could be enhanced in both vegetatively propagated material and in transplants. Enhancement should result both from increased carbohydrate availability and from decreased stomatal conductance. For example, elevated [CO_2] during propagation increased root number and length in sweet potato slips (Bhattacharya *et al.*, 1985). Since propagation and establishment are critical phases for many vegetable crops, especially in terms of increasing uniformity and early yield, increases in these processes from elevated [CO_2] would be particularly valuable.

Nutrition

Rogers *et al.* (1997) and Boote *et al.* (1997) have recently reviewed effects of elevated [CO_2] on plant nutrition. By increasing plant size, elevated [CO_2] increases total nutrient uptake. Since nutrients are distributed over a larger plant, however, the concentration per unit weight is reduced. Nutrient use efficiency (unit of biomass produced per unit of nutrient) generally increases under elevated [CO_2], while nutrient uptake efficiency (unit of nutrient per unit weight of root) generally declines (e.g. Sritharan *et al.* (1992), in kohlrabi (*Brassica oleracea* L. var. *gongylodes* L.). Conroy (1992) and Wolfe *et al.* (1998) concluded that the greatest absolute increases in productivity as a result of elevated [CO_2] exposure will occur when soil N and P availability are high. Nitrogen shortage does not preclude an elevated [CO_2] effect on growth, but in some C_3 plants low P can eliminate a high CO_2 response. For maximum productivity of C_3 crops, higher P levels may be required but the N requirement may be lower. Sritharan *et al.* (1992) concluded that [CO_2] enrichment would negatively affect kohlrabi growth at a low P supply. Conversely, Boote *et al.* (1997) concluded that [CO_2] enrichment response will be similar over a range of fertility levels. Radoglou and Jarvis (1992) found that [CO_2] enrichment increased growth of bean seedlings even when plants were grown in a nutrient-poor medium. In any case, vegetables, which are considered high value crops, usually receive ample to excess fertilization, and nutrients are not likely to limit [CO_2] response in most production systems.

Salinity

Although no vegetables are true halophytes, celery and tomatoes can tolerate higher salinity levels compared with other vegetables. Salt tolerance appears to rise as [CO_2] is increased (see review by Rogers *et al.*, 1997). Ball and Munns (1992) summarized the literature as indicating: (i) that the [CO_2] growth response is higher under moderate salt stress than under optimal salinity; (ii) that water-use efficiency usually increases at higher [CO_2]; and (iii) that leaf salt concentrations are similar in plants grown in both enriched and ambient [CO_2]. Thus, it may be possible to grow vegetables with some salt tolerance in higher salinity.

Yields

Vegetable yields should show at least moderate increases with elevated [CO_2]. Responses of lettuce, carrot and potato are described in more detail in section 10.3. In addition, responses of four glasshouse crops – cucumber (*Cucumis sativus* L.), pepper, tomato and eggplant (*Solanum melongena* L.) – to CO_2 enrichment have been studied by Nederhoff (1994) in the Netherlands. In cucumber given [CO_2] of 620 μmol mol^{-1}, fresh-weight fruit harvest increased by 34% compared with a crop grown at 364 μmol mol^{-1}. Two-thirds of this increase was caused by a greater number of harvested fruits and one-third by an increased average fruit weight. In pepper, increasing [CO_2] from 300 to 450 μmol mol^{-1} increased fruit production by 46%. One-third of this increase was caused by a shift in allocation, and the remaining two-thirds was caused by increased CO_2 assimilation rate.

Disorders of leaf and shoot growth occurred in CO_2-enriched tomato and eggplant. Depending on the incidence of the short-leaf syndrome in tomatoes, yield increases varied from 0 to 31%. In eggplant, yield increases were 24% even though active leaf area was reduced. Fruit quality was not affected by [CO_2] enrichment in any of the greenhouse crops in the study. In a greenhouse tomato study in Raleigh, North Carolina, USA, cooling was used to increase the length of time the tomatoes could be [CO_2] enriched. While yield increases of up to 35% were seen, they occurred only if the duration of enrichment was at least 90% of the daylight hours (Willits and Peet, 1989). In two onion (*Allium cepa* L. Cepa group) cultivars (Daymond *et al.*, 1997), an increase in [CO_2]

shortened time to bulbing, but the time from bulbing to bulb maturity was delayed. For the two cultivars at elevated [CO_2], increases in bulb yield of 28.9–51% were due to an increase in the rate of leaf area expansion and the rate of photosynthesis during the pre-bulbing period as well as (or in addition to) the longer duration of bulbing.

10.2.2 Direct effects of temperature

Introduction

As summarized by Rosenzweig and Hillel (1998), based on a doubling of effective greenhouse gases (which include [CO_2], methane, nitrous oxide and chlorofluorocarbons), most global circulation models predict an increase in global temperature of 2–4°C. The projected temperature increase may already be well underway. Preliminary data (NOAA, 1999) indicate that the average temperature in the USA in 1998 was 12.6°C. Thus, based on records dating back to 1895, 1998 tied with 1934 as the warmest year in a 103-year period. Ten out of the last 13 years have averaged from nearly as warm as to much warmer than the long-term mean. High-temperature records were also set worldwide in 1998 (Fig. 10.1), which was the warmest year on record. The second warmest year was 1997, and 7 of the 10 warmest years have occurred in the 1990s. While anomalously warm temperatures are found throughout the tropics, the warmest anomalies occurred over North America and northern Asia (Fig. 10.2).

The predicted temperature rise may not be evenly distributed between day and night and between summer and winter. Most theories, models and observations suggest that night-time minima will increase more than daytime maxima (Karl *et al.*,1991) and winter temperatures will increase more than summer temperatures. Although the absolute amount of the temperature increase may be small, Mearns *et al.* (1984) suggested that relatively small changes in mean temperature can result in disproportionately large changes in

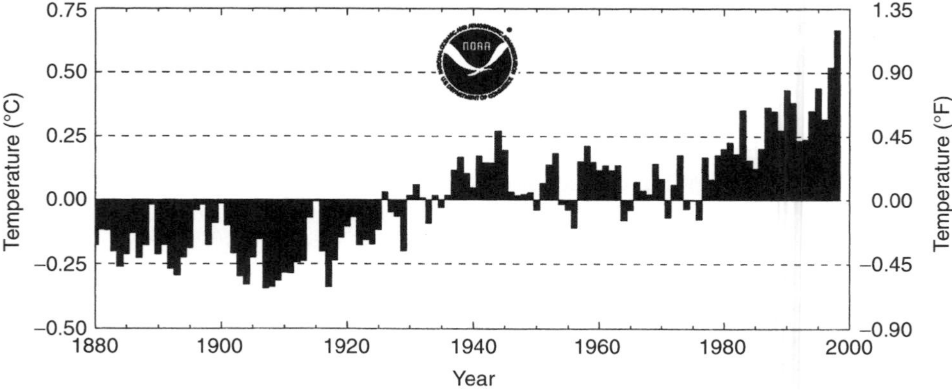

Fig. 10.1. Combined global land and ocean temperature anomalies from 1880 to 1998 relative to an 1880–1997 base period. National Climatic Data Center/NESDIS/NOAA.

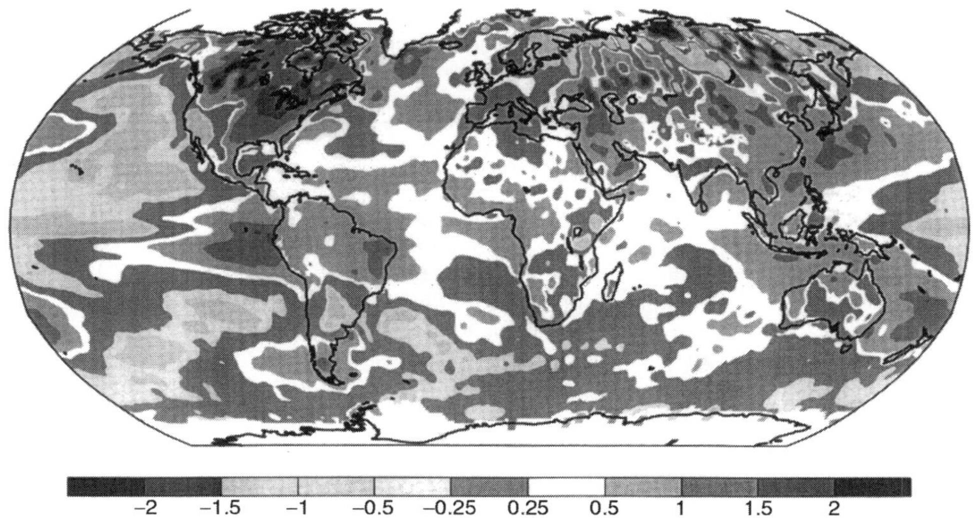

Fig. 10.2. A map of surface temperature anomalies for 1998, derived by merging both sea and land surface temperatures (*in situ* and satellite observations, 1992–1998), to show the extent of the anomalous warm temperatures. (NOAA, 1999.)

the frequency of extreme events; i.e. even if the variance of maximum daily temperatures does not change, the probability of strings of successive days with high temperatures increases substantially.

While considerable attention has been focused on the effects of higher [CO_2] on crops, less research has been directed to crop responses to predicted temperature increases. Rosenzweig and Hillel (1998) point out that high-temperature injuries commonly reduce productivity in crops grown in tropical and temperate regions, and that temperature stress is among the least well understood of all plant processes. If mean warming reaches the upper end of the predicted range (a temperature rise of approximately 4°C), developing heat-tolerant varieties of major crops will become a vital task for plant breeders.

On the other hand, potential beneficial effects of global warming include longer growing seasons, multiple cropping, better seed germination and emergence and more rapid crop growth.

Seedling germination and emergence

COOL-SEASON VEGETABLES Table 10.2 groups vegetable crops on the basis of their temperature requirements for seed germination. For cool-season vegetables, which include most leafy vegetables, brassica (cole) crops, some legumes and most root crops, the temperature range for germination is 3–17°C. Direct-seeded vegetable crops should benefit more than transplanted crops from warmer soil temperature in the spring. Since most leafy vegetables are direct-seeded and since uniform maturity (which is highly desirable) requires uniform germination, having optimal soil temperatures for germination is

Table 10.2. Minimum germination temperature (T_{min}) and heat sum (S) in degree-days for seedling emergence, and the applicable temperature (T) range for germination of various vegetables. Crops are ranked within groups by heat sum (S) in degree-days. (From Taylor, 1997.)

Group	Crop	Genus and species	T_{min} (°C)	S (degree days)	T (°C)
Leaf vegetables and brassica crops	Purslane	*Portulaca oleracea*	11.0	48	15–25
	Cress	*Lepidium sativum*	1.0	64	3–17
	Lettuce	*Lactuca sativa*	3.5	71	6–21
	Witloof, chicory	*Cichorium sativa*	5.3	85	9–25
	Endive	*Cichorium endiva*	2.2	93	3–17
	Savoy cabbage	*B. oleracea* var. *sabauda*	1.9	95	3–17
	Turnip	*B. campestris* var. *rapa*	1.4	97	3–17
	Borecole, kale	*B. oleracea* var. *acephala*	1.2	103	3–17
	Red cabbage	*B. oleracea* var. *purpurea*	1.3	104	3–17
	White cabbage	*B. oleracea* var. *capitata*	1.0	106	3–17
	Brussels sprouts	*B. oleracea* var. *gemmifera*	1.1	108	3–17
	Spinach	*Spinacea oleracea*	0.1	111	3–17
	Cauliflower	*B. oleracea* var. *botrytis*	1.3	112	3–17
	Corn salad	*Valerianella olitoria*	0.0	161	3–17
	Leek	*Allium porrum*	1.7	222	3–17
	Celery	*Apium graveolens*	4.6	237	9–17
	Parsley	*Petroselinum crispum*	0.0	268	3–17
Fruit vegetables	Tomato	*Lycopersicon esculentum*	8.7	88	13–25
	Aubergine	*Solanum melongena*	12.1	93	15–25
	Gherkin	*Cucumis sativus*	12.1	108	15–25
	Melon	*Cucumis melo*	12.2	108	15–25
	Sweet pepper	*Capsicum annuum*	10.9	182	15–25
Leguminous crops	Garden pea	*Pisum sativum*	3.2	86	3–17
	French sugar pea	*P. sativum* var. *sacharatum*	1.6	96	3–17
	Bean (French)	*Phaseolus vulgaris*	7.7	130	13–25
	Broad bean	*Vicia faba*	0.4	148	3–17
Root crops	Radish	*Raphanus sativus*	1.2	75	3–17
	Scorzonera	*Scorzonera hispanica*	2.0	90	3–17
	Beet	*Beta vulgaris*	2.1	119	3–17
	Carrot	*Daucus carota*	1.3	170	3–17
	Onion	*Allium cepa*	1.4	219	3–17

critical. At both the lower and upper ends of this range, germination is inhibited. With global warming, seed germination could be improved for spring crops, but autumn soil temperatures could become too high in some areas for good germination. For example, germination of celery (*Apium graveolens* L. var. *dulce (Mill.)* Pers.) requires a daily temperature fluctuation with night temperatures falling below 16°C, and is inhibited by temperatures above 30°C (Pressman, 1997). The rate of germination may increase more than the total percentage germination (onion data are given in Fig. 10.3), but rapid emergence makes the seedlings more competitive against diseases and insects,

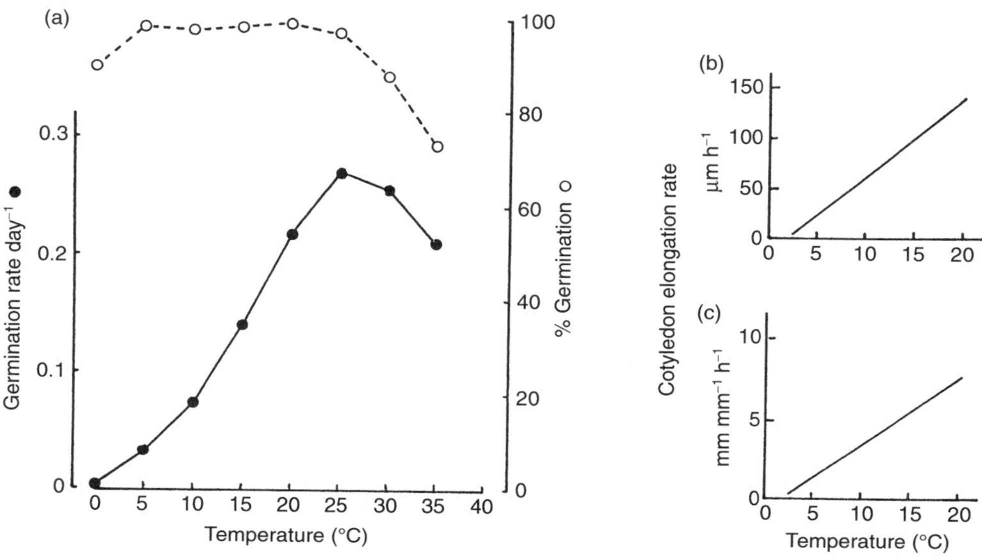

Fig. 10.3. (a) Relationship between temperature and the rate and percentage of germination of onion seeds on moist paper. Rates are reciprocals of the number of days for 50% of viable seeds to germinate (data of Harrington, 1962). (b) Relationship between temperature and rate of cotyledon elongation before hook formation for newly germinated onion seedlings cv. White Lisbon (Wheeler and Ellis, 1991). (c) Relationship between temperature and relative rate of cotyledon elongation after hook formation for the same seedlings as in (b) (Brewster, 1997).

and possibly against weeds. Thus, where both spring and autumn crops are grown, the spring crop could be seeded earlier, but planting of the autumn crop might need to be delayed. Warmer winter temperatures could allow the autumn crop to be grown farther into the winter, but higher summer temperatures could restrict the production of spring crops.

WARM-SEASON VEGETABLES. For solanaceous fruit vegetables (tomato, aubergine and pepper) and cucurbits – cucumber, squash (*Cucurbita pepo* L.) and melon (*Cucumis melo* L. Reticulatus group) – the temperature range for germination (13–25°C) is much higher than for cool-season vegetables (Table 10.2). Any soil warming would be advantageous for cucurbits, which are generally direct-seeded and have a high heat requirement. However, solanaceous fruit crops grown for the fresh market are generally seeded in heated greenhouses and not transplanted into the field until the danger of frost is past. Theoretically, these crops could be direct-seeded if soils warmed earlier, but soil temperatures are unlikely to rise sufficiently to make this practical. Also, current practices for these crops, including black plastic mulch, drip irrigation and fumigation, comprise a system that is highly efficient in terms of productivity, fruit quality and weed control. Thus, growers of fresh-market tomato and pepper are unlikely to switch to direct seeding but, because some processing tomato and pepper crops are direct-seeded, warmer soil temperatures would be beneficial. Sweet-corn, with an optimum germination temperature of 35°C (Maynard and Hochmuth, 1997), is also direct-seeded, and *sh2* supersweet lines in particular would benefit from higher soil temperatures in the field (Wolfe *et al.*, 1997).

Growth rates

For most vegetables, growth is more rapid as temperatures increase, at least up to about 25°C (Table 10.3). Figure 10.4 illustrates onion growth rate as a function of temperature. Bulbing in onions is induced by photoperiod; once induced, it occurs more rapidly at higher temperatures (Brewster, 1997). Maize growth rates increase linearly between 10 and 30°C (Wolfe *et al.*, 1997).

Even at temperatures above 25°C, plants sustain some growth through heat adaptation. In heat-adapted plants, changes in the lipid composition of chloroplast membranes raise the temperature at which the photosynthetic electron transport systems are disrupted (Fitter and Hay, 1987). Another protective mechanism in plants is the production of heat-shock proteins after sudden exposure to high temperature. These proteins may help crops to acquire tolerance to temperature stress, maintain cell integrity, prevent protein denaturation and protect the photosystem II centre. However, their exact role remains unknown (Paulsen, 1994).

Flower induction and dormancy

A greater increase in winter temperatures than in summer temperatures should reduce the potential for summer heat stress, but may lead to a lack of

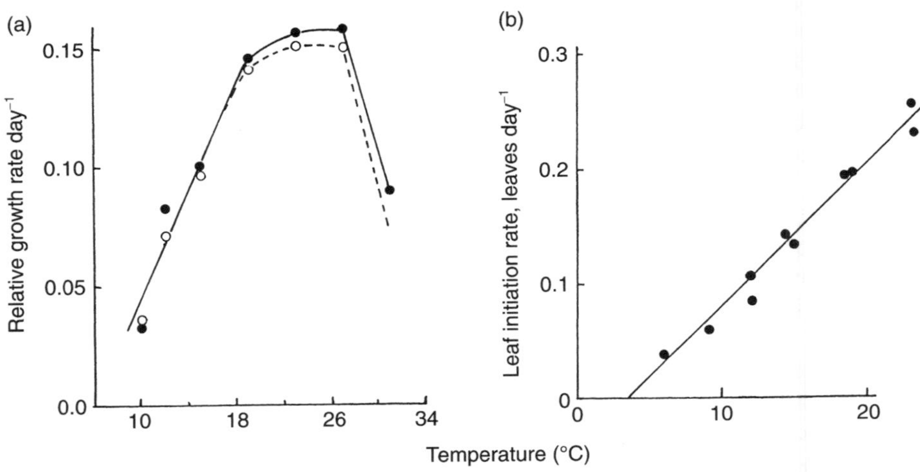

Fig. 10.4. (a) Effect of temperature on relative growth rate of whole plant dry weight (solid symbols) and of leaf area (open symbols) of onion cv. Relative growth rate (RGR) is rate of increase in dry weight per unit f existing dry weight. RGR = $1/W.\mathrm{d}W/\mathrm{d}t$, where W = dry weight and t = time. Similarly, relative growth rate of leaf area is rate of increase of leaf area per unit of existing leaf area. (b) Effect of temperature on rate of initiation of leaves by main shoot apex (i.e. not counting leaves on side shoots) of cvs Hygro, Hyton and Rijnsburger, all 'Rijnsburger types', growing in controlled environments. (From Brewster, 1997.)

vernalization (induction of flowering through low temperatures) for some crops. Flowering in brassicas, celery and onion is affected by interactions among winter temperatures, day length, and seedling age and nutritional status. Cultivar selection and planting dates are directed toward either suppressing flower initiation, in the case of celery, onion and cabbage, or delaying it, in the case of broccoli (*Brassica oleracea* L. var. *italica* Plenck.) and cauliflower (*Brassica oleracea* var. *botrytis*), until the seedling is big enough to support formation of a large head. Thus, if winters become milder, different planting dates and cultivars may be required.

In bean, high temperatures delay flowering because they enhance the short-day photoperiod requirement (Davis, 1997). In cucumber, sex expression is affected, with low temperatures leading to more female flowers (generally desirable) and high temperatures leading to production of more male flowers (Wien, 1997b).

In lettuce and spinach (*Spinacia oleracea* L.) high temperatures and long days induce flowering. Once the seed stalk starts to develop (referred to as bolting), crop quality declines significantly. Head lettuce cannot be sold at all, and leaf lettuce and spinach become tough and strong tasting. Thus, cultivars with greater resistance to bolting may need to be selected or production areas moved north as the climate warms.

Some seed production and perennial vegetable production locations may need to be moved farther north. Biennial vegetables, which include some root crops and many cole crops, require specific periods of chilling during the winter to produce a seed crop the following season. Celery requires a cold period to produce seed the following season, and high temperatures during seed development may reduce seed quality (Pressman, 1997). In perennial crops, such as chive (*Allium schoenoprasum* L.), asparagus (*Asparagus officinalis* L.), and rhubarb (*Rheum rhabarbarum* L.), low temperatures over the winter are required before new growth is initiated in the spring (Krug, 1997). Assuming dormancy requirements are fulfilled at the proper time, however, a longer growing season might increase production in perennials.

Reproductive development

High temperature affects reproductive development in two ways, both of which potentially reduce yields. Firstly, the rate of reproductive development is accelerated, which shortens the seed-filling period and the fruit maturation period. Generally, this results in lower individual seed and fruit weights and in some cases reduced concentrations of soluble solids in the fruit. In addition, in many crops the reproductive events themselves are prevented at temperatures only a few degrees above optimal. Reduced fruit set in tomato (see review by Kinet and Peet, 1997) and pepper (Wien, 1997c and references cited therein) occurs as a result of high temperature. In pepper, fruit set was reduced at 27/21°C compared with 21/16°C and no fruit set occurred at 38/32°C (Fig. 10.5). In tomato, high temperatures after pollen release decreased fruit set, yields and seed set in tomato even when pollen was produced under optimal conditions (Peet *et al.*, 1997). Overall, however, pre-anthesis stress appears to be more injurious than stress applied after pollen arrives on the

Table 10.3. Temperature demands and sensitivities of vegetable species. (From Krug, 1997.)

Temperature	Vegetable species	Frost sensitivity[a]
Hot – growth range 18–35°C; optimum range 25–27°C		
	Okra (*Abelmoschus esculentus*)	+
	Roselle (*Hibiscus sabdariffa*)	+
	Watermelon (*Citrullus lanatus* var. *vulgaris*)	+
	Melon (*Cucumis melo*)	+
	Capsicum species	+
	Sweet potato (*Ipomea batatas*)	+
Warm – growth range (10) 12–35°C; optimum 20–25°C		
	Cucumber (*Cucumis sativus*)	+
	Aubergine (*Solanum melongena*)	+
	Sweet pepper (*Capsicum annuum*)	+
	Pumpkin, squash (*Cucurbita* species)	+
	New Zealand spinach (*Tetragonia tetragonioides*)	
	Maize (*Zea mays*)	+
	Tomato (*Lycopersicon esculentum*)	+
	Phaseolus species	+
Cool – hot – growth range (5) 7–30°C; optimum 20–25°C		
	Colocasia (*Colocasia esculenta*)	
	Globe artichoke (*Cynara scolymus*)	
	Onion, shallot (*Allium cepa*)	–
	Leek (*Allium porrum*)	–
	Garlic (*Allium sativum*)	–
	Chicory (*Cichorium intybus* var. *foliosum*)	–
	Pak-choi (*Brassica chinensis*)	–
	Scorzonera (*Scorzonera hispanica*)	–
	Chives (*Allium schoenoprasum*)	–
Cool–warm–growth range (5) 7–25°C; optimum 18–25°C		
	Pea (*Pisum sativum*)	–
	Broad bean (*Vicia faba*)	–
	Cauliflower (*Brassica oleracea* convar. *botrytis* var. *botrytis*)	–
	Broccoli (*Brassica oleracea* convar. *botrytis* var. *italica*)	–
	Cabbage (*Brassica oleracea* convar. *capitata* var. *capitata*)	–
	Kohlrabi (*Brassica oleracea* convar. *caulorapa* var. *gongylodes*)	–

Table 10.3. *Continued*

Temperature	Vegetable species	Frost sensitivity[a]
	Kale (*Brassica oleracea* convar. *acephala* var. *sabellica*)	–
	Brussels sprouts (*Brassica oleracea* convar. *fruticosa* var. *gemmifera*)	–
	Turnip (*Brassica rapa* subsp. *rapa*)	–
	Rutabaga (*Brassica napus* subsp. *rapifera*)	–
	Chinese cabbage (*Brassica rapa* subsp. *pekinensis*)	–
	Parsley (*Etroselinum crispum*)	–
	Fennel (*Foeniculum vulgare*)	(+)
	Dill (*Anethum graveolens*)	(+)
	Radish (*Raphanus sativus* var. *sativus*)	(+)
	Radish (*Raphanus sativus* var. *niger*)	(+)
	Red beet (*Beta vulgaris* convar. *vulgaris*)	(–)
	Swiss chard (*Beta vulgaris* convar. *cicla*)	(–)
	Spinach (*Spinacia oleracea*)	–
	Lettuce (*Lactuca sativa* var. *capitata*)	(+)
	Endive (*Cichorium endivia*)	–
	Carrot (*Daucus carota*)	–
	Celery, celeriac (*Apium graveolens*)	(+)
	Parsnip (*Pastinaca sativa*)	–
	Potato (*Solanum tuberosum*)	+
	Lambs lettuce (*Valerianella locusta*)	–
	Rhubarb (*Rheum rhaponticum*)	–
	Asparagus (*Asparagus officinalis*)	(+)
	Horseradish (*Armoracia rusticana*)	–
	Garden cress (*Lepidium sativum*)	(+)

[a]+, sensitive to weak frost; –, relatively insensitive; (+)(–) uncertain.

stigma (Peet *et al.*, 1998). This was shown in an experiment in which heat stresses of 27 and 29°C were applied separately to male-sterile and male-fertile tomato plants of the same cultivar. Male-sterile plants receiving pollen from heat-stressed male fertiles had very reduced or no fruit set, regardless of the growth conditions of the male-sterile. Growth conditions of the male-steriles (i.e. female flowers parts and conditions after pollination) appeared to be less critical for yield (Fig. 10.6). Reduced pollen release and impaired pollen function appeared to be the main factors accounting for yield and seed set reductions. The most sensitive period was 15 to 5 days before anthesis, and a duration of 10 days was required for an effect (Sato, 1998).

Other vegetables in which reproductive development is particularly sensitive to high temperatures include bean (e.g. Konsens *et al.*, 1991; Davis, 1997) and cowpeas (*Vigna unguiculata* (L.) Walp. Subsp. *unguiculata* (L.) Walp.) (e.g. Ahmed *et al.*, 1992). In pea (*Pisum sativum* L. subsp. *sativum*),

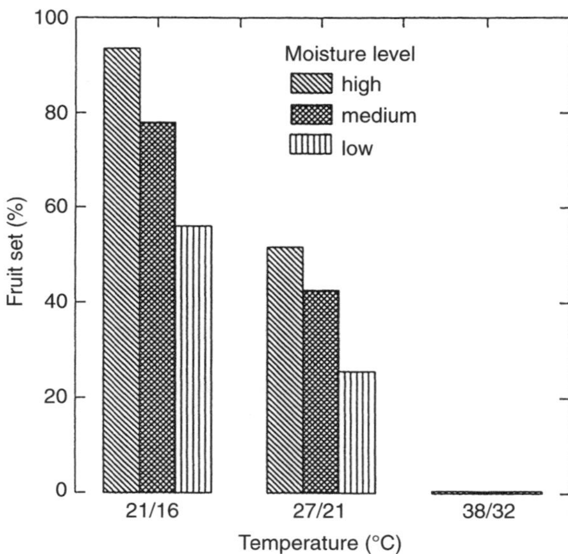

Fig. 10.5. Influence of air temperature and soil moisture on percentage of fruit set of 'World Beater' pepper grown in pots in glasshouse compartments. Fruits were removed after setting. Data are averages of 2 years' experiments, 1932–34 (Cochran, 1936). (From Wien, 1997c.)

temperatures above 25.6°C during bloom and pod set reduce flower and pod number and yields (Muehlbauer and McPhee, 1997). In maize (Wolfe *et al.*, 1997), warm temperatures can reduce seasonal productivity by accelerating developmental rates, shortening vegetative and reproductive growth phases, reducing leaf area duration and reducing ear quality (see Table 10.1 for details).

Yields

Effects of high temperatures on agronomic crops have been reviewed by Boote *et al.* (1997). Declines in harvest index, seed size and mass and seed-growth rate are rapid as temperatures increase above a critical level. Krug (1997) pointed out that increasing temperature generally increases the developmental rate, but the growth rate is not necessarily stimulated to the same degree. In an early-maturing pea cultivar, increasing the temperature from 16 to 24°C increased the differentiation rate of nodes from emergence to the appearance of the first flower, but the growth rate was reduced, which reduced total dry matter and yields. Cultivars with a proportional increase in the rate of differentiation and the rate of growth as temperature increases should be less sensitive to temperature. In onion, warmer temperatures shortened the duration of growth, accounting for a negative correlation between elevated temperatures and crop yields (Daymond *et al.*, 1997).

Predicting which vegetables will be most affected by high temperatures is difficult. It has been suggested (Hall and Allen, 1993) that indeterminate crops

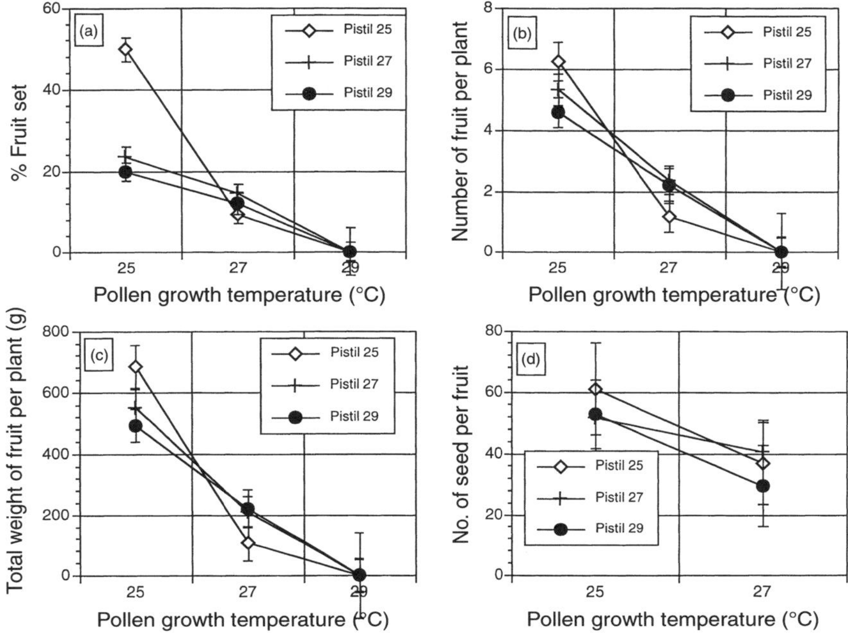

Fig. 10.6. Effect on (a) percentage fruit set, (b) fruit number, (c) fruit weight (g) and (d) seed number of exposing tomato pollen (male-fertile parent) and tomato pistils to mean daily temperatures of 25, 27 and 29°C. Pollen was collected from male-fertiles maintained in separate growth chambers and hand-applied to the stigmas of male-steriles maintained at 25°C (open triangle), 27°C (+) or 29°C (closed circle). (Peet *et al.*, 1998.)

are less sensitive to periods of heat stress because the time of flowering is extended compared with determinate crops. This is probably true for crops harvested at seed maturity. In fact, soybean, having no especially critical period in its development, is considered to have a great ability to recover from stress (Acock and Acock, 1993). Similarly, dry bean, winter squash (*Cucurbita* sp.) and pumpkin (*Cucurbita* sp.), which are harvested at maturity, should be relatively unaffected by temporary heat stress. These crops will usually keep flowering and setting fruit until a certain number of fruit develop on the vine. However, most vegetable crops that flower over an extended period, such as cucumber, snapbean, tomato and pepper, are harvested well before seed maturity. In bush snapbean and processing tomato, plants are mechanically harvested once and any pods or fruit that are too small or too large are discarded. Thus a period of heat that reduces seed set may cause considerable economic loss by making the crop less uniform in harvest maturity, even if recovery occurs later. Even in crops that are hand-harvested several times, a spread-out period of fruit set increases labour and packing costs, thus decreasing efficiency and possibly moving the crop out of its market 'window' of high demand. Thus, it could be argued that indeterminate crops also have a longer window of vulnerability to high temperatures, compared with crops such as maize, which are relatively insensitive to high temperatures except during the early stages of reproductive development (pollination and early seed development).

Cool-season crops, which are currently widely grown in the tropics, are most likely to be negatively affected by global warming. Cabbage grown at 25°C has a lower dry-matter content (and thus lower quality), a reduced growth rate, and lower water-use efficiency than cabbage grown at 20°C (Hara and Sonoda, 1982). Brussels sprouts require monthly temperatures of 17–21°C for 4 months, followed by 2 months of 12°C temperatures, during which sprouts develop. Thus not only cool weather, but also long, continuously cool weather is required (Wien and Wurr, 1997).

Since night-time temperature minima may increase more than daytime maxima (Karl *et al.*, 1991), additional night-time heat stress may be more of a factor than additional daytime heat stress. Cowpeas have been described as more sensitive to night-time than to daytime heat stress (Mutters and Hall, 1992) because they can tolerate temperatures of 30°C during the day, but not at night. Ahmed *et al.* (1992) noted premature degeneration of the tapetal layer and lack of endothecial development in cowpea, which they felt was responsible for the low pollen viability, low anther dehiscence and low pod set under 30°C night temperatures. Pepper fruit set has also been described as being more sensitive to high night temperatures than to high day temperatures (Wien, 1997c).

Went (1944) suggested that night-time heat stress most limits fruit set in tomato. In a recent re-examination of this question, Peet *et al.* (1997) reported that when heat stress was applied to male sterile tomatoes provided with pollen from non-stressed plants, mean daily temperature, rather than high night temperature, was the main factor affecting fruit and seed yield. Similarly, Peet and Bartholomew (1996) did not observe a disproportionately large heat-stress response to night temperatures ranging from 18 to 26°C when daytime temperatures were the same (26°C).

(g) Fruit quality

High temperatures may also reduce fruit quality. Reduced sugar content in fruit, such as pea, strawberry and melon produced under warm nights is often attributed to increased night-time respiration, although it may also be caused by the shorter period over which the fruit develops at high temperatures (Wien, 1997b). Other disorders associated with high or low temperatures are shown in Table 10.1.

10.2.3 Interactions

Temperature increases from global warming will occur simultaneously with [CO_2] increases, and the interaction has been suggested by a number of researchers (e.g. Idso *et al.*, 1987) to be beneficial. In a number of species, the response of vegetative growth to CO_2 enrichment increases with temperature

(Boote *et al.*, 1997 and references cited therein), presumably because the plants are better able to utilize additional carbohydrate when growth rates are more rapid and this minimizes downward acclimation to [CO_2] (Wolfe *et al.*, 1998).

The upper range for the stimulation of [CO_2] response with elevated temperature response has been reported as high as 34°C and as low as 26/20°C (Boote *et al.*, 1997 and references cited therein). In a study of four crops in the UK – wheat (*Triticum aestivum* L.), onion, carrot and cauliflower – Wheeler *et al.* (1996) found that wheat yield gains from an increase in [CO_2] to about 700 μmol mol^{-1} were offset by mean seasonal temperatures of only 1.0 and 1.8°C warmer in each of 2 years. Temperature increases of 2.8 and 6.9°C in two onion cultivars was required to offset the yield gains at [CO_2] of 560 μmol mol^{-1}. Of the four crops studied, larger yield increases in high [CO_2] were observed in crop species with higher harvest indices in current climates. Warmer temperatures reduced the duration of crop growth, and hence yield, of determinate crops such as winter wheat and onion. However, yield of carrot, which is an indeterminate crop, increased progressively with temperature over the range used in the study (Wheeler *et al.*, 1996).

In two onion cultivars (also grown in the UK), Daymond *et al.* (1996) found that an increase in bulb yield due to a rise in [CO_2] from 374 to 532 μmol mol^{-1} was offset by a temperature warming of 8.5–10.9°C in one cultivar but only a 4.0–5.8°C increase in another cultivar. They suggested that the difference in cultivar responses was due to the temperature-sensitive cultivar (i.e. requiring only a 4.0–5.8°C temperature increase to offset [CO_2] increases) being a short-season type. Thus, global climatic change may affect relative cultivar performance, and with altered [CO_2] and temperatures additional cultivar trials may be required.

Stomatal closure from high [CO_2] may reduce water losses, thereby decreasing the adverse effects of increased transpiration at higher temperatures. In addition, this closure protected photosynthesis from the effects of ozone (Reid and Fiscus, 1998) and possibly other pollutants (Allen, 1990). On the other hand, transpiration also has a cooling effect on leaf temperature. If transpirational cooling is reduced, leaves could become more vulnerable to heat stress.

Beneficial effects of elevated [CO_2] may not carry over to yields in all crops. Mitchell *et al.* (1993) found that a 4°C increase in temperature decreased wheat yield significantly, whether grown in elevated [CO_2] or not. Similar results were seen by Nair and Peet in tomato (unpublished data) and in potatoes (see section 10.3.1). In cauliflower (Wheeler *et al.*, 1995), no interaction between [CO_2] and temperature on total biomass was detected in final harvests. The total dry weight of plants grown at 531 μmol CO_2 mol^{-1} was 34% greater than from those grown at 328 μmol mol^{-1}, whereas a 1°C rise reduced dry weight by 6%. Boote *et al.* (1997) concluded that in a number of crops (including rice, wheat and soybean), high temperature damage on reproductive growth is not offset by CO_2 enrichment. This was despite predictions based on the beneficial effects of CO_2 enrichment on vegetative growth at high temperatures.

To the extent that global warming increases the frost-free period, more warm-season crops could presumably be grown in northern areas. If springtime temperatures increase, farmers may plant earlier, and/or at higher altitudes and latitudes. Such plantings will be exposed to similar risks of chilling and frost as in their current production areas. Effects of elevated [CO_2] on low-temperature tolerance are poorly understood: some reports indicate that high [CO_2] mitigates (Boese *et al.*, 1997) while others report that it increases low-temperature sensitivity (Lutze *et al.*, 1998).

10.3 Case Studies

10.3.1 Potato

Response to temperature

Short photoperiods (i.e. long nights) and cool temperatures induce tuber formation in potato (Ewing, 1997). Potato yields are particularly sensitive to high-temperature stress because tuber induction (Reynolds and Ewing, 1989; Gawronska *et al.*, 1992) and development (Krauss and Marschner, 1984) can be directly inhibited by even moderately high temperatures. There is an interaction between photoperiod and temperature: the higher the temperature, the shorter is the photoperiod required for tuberization for any given genotype (Snyder and Ewing, 1989). High temperatures can also adversely affect tuber quality by causing 'heat sprouting' which is the premature growth of stolons from immature tubers (Wolfe *et al.* 1983; Struik *et al.*, 1989).

Although photosynthesis in potato is repressed by high temperature (Ku *et al.*, 1977), it is often not as sensitive to temperature as are tuberization and partitioning of carbohydrates to the tuber reproductive sink (Reynolds *et al.*, 1990; Midmore and Prange, 1992). Therefore, moderately high temperatures can significantly reduce tuber yields even when photosynthesis and total biomass production are relatively unaffected.

In their comprehensive study, Reynolds and Ewing (1989) documented a distinction between the effects of air and soil temperature on potato physiology and yield. Cooling the soil (17–27°C) at high air temperatures (30–40°C) neither relieved any of the visible symptoms of heat stress on shoot growth nor repressed the tuberization induction signal from the leaves. This was reflected in the lack of tuberization by leaf-bud cuttings. Heating the soil (27–35°C) at cool air temperatures (17–27°C) had no apparent detrimental effect on shoot growth or induction of leaves to tuberize. However, in each case, hot soil essentially eliminated tuber development. They concluded that the induction of leaves to tuberize is affected principally by air rather than soil temperature, but that expression of the tuberization signal from the leaves can be blocked by high soil temperature.

There are various morphological responses of potato to high temperature (see review by Ewing, 1997) in addition to reduction in tuber number and size. Plants grown under high temperature are taller, with longer internodes. Leaves tend to be shorter and narrower, with smaller leaflets, and the angle of the leaf to the stem is more acute. Axillary branching at the base of the mainstem increases, more flowers are initiated, and fewer flower buds abscise. At warm temperatures, compared with cool ones, leaf and stem dry weights often increase (at the expense of tuber growth), and the leaf/stem ratio decreases.

Considerable genetic variability in response to high temperature has been reported for potato. Reynolds *et al.* (1990) observed significant differences in photosynthetic response to a 9-day heat treatment (40/30°C day/night temperature) among several accessions reported to vary in temperature sensitivity. The differences in photosynthetic rates were attributed to a number of factors, including temperature effects on leaf chlorophyll loss and senescence rate, stomatal conductance, and dark reactions of photosynthesis. Snyder and Ewing (1989) compared six cultivars and noted a tendency for the tuberization of early-maturing types to be less negatively affected by high temperatures (30/25°C day/night) than were late-maturing varieties, in which raising the temperature caused up to a 50% reduction in tuber dry weight. Reynolds and Ewing (1989) also reported genotypic variation in tuberization after exposure to high temperatures among the 319 accessions they tested.

The challenge for breeders in attempting to develop heat-tolerant potato varieties is that a genotype possessing tolerance to one aspect of heat stress may not necessarily be tolerant to other aspects. High temperature can reduce yields by affecting ability of seed tubers to sprout, photosynthetic or dark respiration rates, tuberization, partitioning of assimilates to developing tubers, and other processes, each of which may be under separate genetic control. There are also secondary reactions to stress, such as resistance to drought and increased disease pressure, that often are concomitant with high temperatures.

Response to [CO_2]

Potato possesses the C_3 photosynthetic pathway, and the tubers are a large 'sink' for carbohydrates. Typically, as much as 70–80% of total dry weight at maturity is in the tubers (Moorby, 1970; Wolfe *et al.*, 1983). Several reviews of the CO_2-enrichment literature have concluded that sustained stimulation of photosynthesis by elevated [CO_2] is most likely in C_3 plant species, such as potato, which have a large, indeterminate sink capacity for photosynthates (Stitt, 1991; Wolfe *et al.*, 1998). The experimental data for potato have not always corroborated this hypothesis, however. Collins (1976) found no significant effect of elevated [CO_2] on tuber number but did document a significant increase in tuber size and overall yield from CO_2 enrichment. Wheeler and Tibbits (1989) found that raising the [CO_2] to 1000 μmol mol^{-1} increased tuber yield by only 2% and 12% for the varieties Norland and Russet Burbank, respectively. Goudriaan and de Ruiter (1983) reported a slight reduction in yield of potato grown at elevated – compared with ambient – [CO_2], particularly when nutrients limited growth potential. They also observed mild leaf damage associated with starch accumulation (reflecting insufficient sink capacity) in plants grown in high [CO_2]. All of these studies were conducted in growth chambers or greenhouses, and the observed variation in [CO_2] response may in part be attributable to variation in pot size used, which can affect below-ground sink capacity and the magnitude of downward acclimation of photosynthesis at elevated [CO_2] (Arp, 1991; Sage 1994). In one study (Wheeler *et al.*, 1991) in which a relatively large pot size (19 L) was used and plants were grown at [CO_2] of 1000 compared with 350 μmol mol^{-1}, the varieties Norland, Russet Burbank and Denali had yield increases of 23, 35 and 40%, respectively.

A comprehensive evaluation of genotypic variation in potato yield response to [CO_2] under field conditions and with unrestricted rooting volume is needed. We have conducted experiments for one growing season with potato (var. Katahdin) in the field (Arkport fine sandy loam soil) in canopy chambers and found a yield benefit of about 70% at 700 compared with 350 μmol CO_2 mol^{-1} at ambient (non-stress) temperature (Fig. 10.7).

Temperature × atmospheric [CO_2] interaction

There have not been many studies with potato in which both [CO_2] and temperature were manipulated and the interaction evaluated. Cao et al. (1994) reported a greater tuber yield benefit from CO_2 enrichment (1000 compared with 500 μmol mol^{-1}) at a constant 20°C temperature compared with 16°C. This may reflect one of the most important and most frequently observed [CO_2] × temperature interactions: the stimulation of C_3 photosynthesis by elevated [CO_2] increases as temperatures increase within the non-stress temperature range (e.g. 15–30°C). A primary reason for this is that as temperatures increase, oxygenation by the key photosynthetic enzyme, Rubisco, increases relative to carboxylation, thereby increasing the benefit from CO_2 enrichment (Jordan and Ogren, 1984).

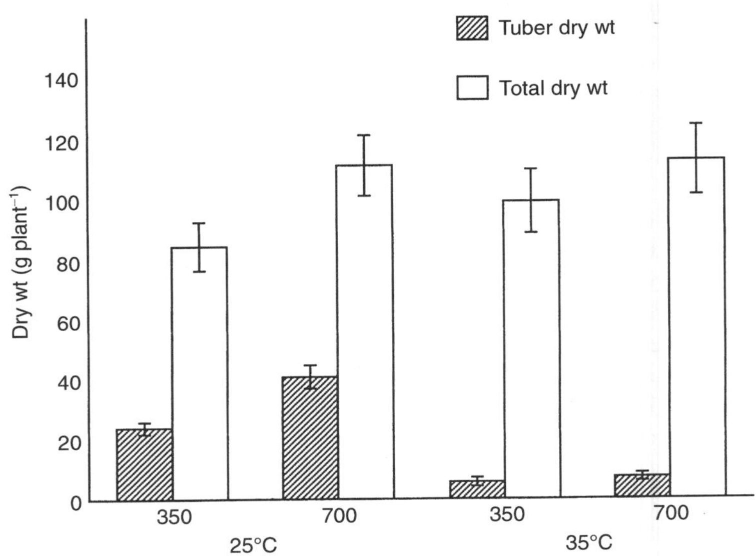

Fig. 10.7. Effect of CO_2 enrichment to 350 and 700 μmol mol^{-1} at growth temperatures of 25 and 35°C on total and tuber dry weight in potato. (Wolfe and Boese, unpublished data.)

The effect of elevated [CO_2] on yield response to heat stress will be particularly important to potato farmers in regions where an increase in the frequency of high-temperature stress events occurs concomitantly with an increase in [CO_2]. Wolfe and Boese (unpublished) compared the yield response to a [CO_2] doubling in field canopy chambers maintained at non-stress temperatures (daytime maximum temperature near 25°C) with the yield response in chambers allowed to develop moderate heat stress (daytime maximum temperatures near 35°C). Averaged across [CO_2] treatments, tuber yields were reduced by 85% in the high-temperature stress treatment, while total biomass was relatively unaffected by temperature (Fig. 10.7). This corroborates earlier findings reviewed above: potato tuberization and tuber development are more sensitive to heat stress than is either photosynthesis or total biomass accumulation. Tuber yield was increased by 71.5% (statistically significant at $P < 0.05$) by a [CO_2] doubling in the non-stress temperature treatment. Tuber yields in plants exposed to high temperatures were extremely low, regardless of [CO_2] treatment, and there was no statistically significant [CO_2] effect. These results should be viewed as preliminary since they are based on a single growing season and one variety (Katahdin), but they suggest that elevated [CO_2] will not mitigate the negative effects of high-temperature stress on tuberization and yield.

10.3.2 Lettuce

Lettuce germination and early growth rate are largely determined by temperature. The apical meristem of young lettuce plants is near the soil surface, so soil temperature is often more closely correlated with early plant growth rate than is air temperature (Wurr *et al.*, 1981). Production of high-quality lettuce generally requires a cool, mild climate. Maynard and Hochmuth (1997) suggest that the optimum average temperature for lettuce growth is 18°C. Kimball *et al.* (1967) found that most successful commercial lettuce production in the western USA occurs during periods of the year when there are at least 2 months with maximum daytime temperatures between 17 and 28°C concomitant with night-time temperatures that do not exceed 3–12°C.

In a study using temperature-gradient polyethylene tunnels in the field, Wheeler *et al.* (1993a) found that an increase in average temperature from 16.3 to 21.1°C increased early-season growth rate but shortened crop duration, thereby decreasing final yield by 17%. This is characteristic of determinate crops. Others (Wurr and Fellows, 1991) have also reported a negative relationship between temperature and head weight and density of crisphead lettuce.

Temperatures that exceed specific warm temperature thresholds can cause premature 'bolting' (elongation of internodes of the main stem to form a seed stalk) and lead to severe reductions in marketable yield in lettuce (Wien, 1997d). Some research suggests that night temperature is more critical than day temperature with regard to flower induction and bolting. Rappaport and Wittwer (1956) reported that flowering of the 'Great Lakes' variety of head lettuce occurred 21 days earlier with night temperatures of 21°C compared with 16°C. There is genotypic variation in the specific temperature for flower induction (Wurr and Fellows, 1984), and there has been some success at developing varieties resistant to bolting at night temperatures that exceed the temperature range (3–12°C) identified by Kimball *et al.* (1967).

Tipburn, a physiological disorder of lettuce often associated with rapid growth rates at warm temperatures (Wien, 1997d), is characterized by necrosis of the edges of young, rapidly expanding leaves. Only a few days of high temperature can have a sufficient negative effect on the visual quality of crisphead lettuce to lead to a complete crop failure (Misaghi *et al.*, 1992). The disorder has been well studied and is known to be associated with localized calcium deficiency, which results when leaf expansion is faster than the mobility of calcium in the plant. Improving soil calcium availability does not alleviate the problem. While foliar application of calcium to susceptible, rapidly expanding leaves is sometimes partially effective in loose-leaf types, it seldom is a practical solution.

Carbon dioxide enrichment (typically to levels of about 1000–1200 μmol mol^{-1}) has been used commercially to increase yields of greenhouse-grown lettuce (Wittwer, 1986). Quantitative scientific assessments of lettuce response to [CO_2] (Hunt *et al.*, 1984; Mortensen, 1985) indicate about a 50% increase in total dry weight when plants are grown at 1000–1200 compared with 330–350 μmol CO_2 mol^{-1}.

Wheeler *et al.* (1993b), utilizing the [CO_2] response data of Hunt *et al.* (1984) and Mortensen (1985), assumed a linear relationship between 350 and 1200 μmol CO_2 mol^{-1} to estimate the yield benefit for lettuce from a [CO_2] doubling above pre-industrial levels (e.g. 560 μmol mol^{-1}). They linked this estimate of the yield response to [CO_2] to an estimate of the effect of increasing temperature on yield based on their data (Wheeler *et al.*, 1993a) to predict the effects of various climatic change scenarios. They used temperature output from three equilibrium-doubled [CO_2] scenarios derived from three different general circulation models (from the Geophysical Fluid Dynamics Laboratory, the UK Meteorological Office and the Goddard Institute of Space Studies). In their analysis, the negative effects on yield from climatic change were compensated for to some extent by the beneficial effects from a [CO_2] doubling. The predicted maximum fresh-weight yields with climatic change were reduced by only about 5% compared with the baseline scenario (no climatic change or increase in [CO_2]). All three climatic change scenarios significantly reduced crop duration. These results suggest that in those cases where temperatures do not reach thresholds that cause bolting, tipburn, or other problems that lead to a serious loss of marketable yield, the yield reductions in a future high-[CO_2] world might be small; in some situations the shorter crop duration may allow an additional planting in a single season.

10.3.3 Carrot

Carrot, like most other vegetable root crops, is best adapted to cool temperatures. Barnes (1936) found that the optimum temperature for carrot root growth is in the range of 16–21°C. In general, this observation has been confirmed by subsequent observations of reduced yields or quality with temperature variation much beyond this range (Benjamin *et al.*, 1998). Olymbios (1973) found that temperature effects may be more complex than this, however. Increasing the temperature from 15 to 25°C increased dry weights of shoot and roots by 2.6-fold and 1.5-fold, respectively. Increasing shoot temperature from 15 to 25°C while holding the root at 15°C increased shoot weight by 36% while decreasing root weight slightly. Reversing this, increasing root temperature from 15 to 25°C while holding shoot temperature at 15°C reduced root weights even more (by 47%). Olymbios (1973) suggested that 25°C may be optimal for shoot growth while supraoptimal for root growth.

Because storage roots can be strong sinks for photosynthate, it is hypothesized that yield of crops such as carrot may be particularly responsive to increasing [CO_2]. Indeed yield increases from a [CO_2] doubling of as much as 110% have been reported for carrot (Poorter, 1993), but these results should not be viewed as typical.

Wheeler *et al.* (1994) conducted a [CO_2] study with carrot in temperature-gradient tunnels in the field so that temperature and the temperature × [CO_2] interaction could be evaluated. Temperature in the tunnels ranged from a mean of 7.5 to 10.9°C. Averaging across temperature treatments, they observed a 31% root weight increase at 551 compared with 348 μmol CO_2 mol^{-1} and a small, but consistent, increase of about 5% in the root/total dry weight ratio for plants grown at elevated [CO_2]. Cooler temperatures reduced the root total dry weight ratio at 348 μmol CO_2 mol^{-1}, but temperature had no effect on this ratio at 551 μmol mol^{-1}. Averaging across [CO_2] treatments, root yields increased about 34% for each 1°C increase in temperature. Most of the increase in yield due to temperature was attributable to faster growth rate and development at warmer temperatures. In contrast, CO_2-enrichment increased plant weight at similar developmental stages (i.e. at a given leaf number).

The study by Wheeler *et al.* (1994) found no significant interaction between [CO_2] and temperature on root yield or on total biomass when comparisons were made at either the same time after planting or at similar developmental stages. In other words, the yield benefits from increasing both [CO_2] and temperature, within the cool range of their study, were consistent across [CO_2] and temperature treatments. In contrast, Idso and Kimball (1989) reported a significant [CO_2] × temperature interaction for carrot yields, with more than twofold yield increases at relatively warm temperatures and little benefit at temperatures below 12°C. This discrepancy between studies demands further study. It may be due in part to the much wider range of temperatures they examined. Also, the Wheeler *et al.* (1994) data were analysed based on average temperatures and yield for the entire season. The Idso and Kimball (1989) study evaluated each sequential period between harvests and the corresponding temperature environment separately. More [CO_2] research at warmer temperatures approaching a stress level for carrot would also be of value in attempting to anticipate the effects of climatic change on this crop species.

10.4 Summary

Increasing [CO_2] will enhance photosynthesis and improve water-use efficiency, thus increasing yield in most crops. Relative benefits from increased [CO_2] can often be maintained with modest water and N deficiency, but yield benefits on an absolute basis are reduced when water or N limit growth. The impact of increasing temperatures is more difficult to predict. Seed germination will probably be improved for most vegetables, as will vegetative growth in regions where mean daily temperatures during the growing season remain under 25°C, assuming adequate water is available. Reproductive growth is extremely vulnerable to periods of heat stress in many important vegetable fruiting crops, such as tomato, pepper, bean and sweet-corn, and yield reductions will probably occur unless production is shifted to cooler portions of the year or to cooler production regions. This vulnerability results from the shortened duration of grain, storage tissue, or fruit-filling and from failure of various reproductive events, especially the production and release of viable pollen. Processing crops, which are sometimes direct-seeded and are more frequently grown in cool-summer areas, are more likely than fresh-market crops to benefit from higher temperatures. In general, crops with a high harvest index, high sink demand, indeterminate growth and long growth seasons are considered most likely to respond positively to the combination of higher [CO_2] and temperature. Relatively few crops have been studied, however and cultivars within a crop often differ in their responses, thus making generalizations difficult.

In many crops, high temperatures may decrease quality parameters, such as size, soluble solids and tenderness. For fresh-market vegetable producers, even minor quality flaws can make their crops completely unsaleable in some markets. Reduced or more irregular precipitation will also decrease vegetable yields and quality, although soluble solids and specific weight may increase in some crops. Leafy greens and most cole crops are generally considered to be cool-season crops, so heat stress during the growing season would be detrimental to these species. High-temperature effects on lettuce and spinach and low-temperature effects on cole crops include induction of flowering and elongation of the seedstalk. Perennial crops also require an overwinter cool period. Thus, planting dates, production areas and cultivars may need to be adjusted if temperatures change.

10.5 Impacts on the Vegetable Industry

It is not clear whether the overall impact of global warming will be positive or negative relative to production of vegetable crops. In areas where mean daily temperatures do not currently exceed 25°C during the growing season, overall effects should be beneficial, while they may be negative where growing-season temperatures are higher. Higher temperatures should also be more beneficial in areas that currently have short growing seasons and most production in the summer, compared with areas with longer frost-free periods and most production in the winter.

Fresh-market crops in the north-central and northeastern USA and processing crops (which are grown on significant acreages in such northern states as Minnesota, Wisconsin and New York and are usually direct-seeded) will be likely to experience more beneficial than adverse effects. For the bulk of the US fresh-market vegetable production, the impact is less clear. The most important fresh-market vegetable production states in the USA in 1995 in terms of acreage were California, Florida, Georgia, Texas and Arizona. In terms of crop value, California also produces 62% of the vegetables grown for processing (Maynard and Hochmuth, 1997) and in 1993 Florida produced over half the fresh-market tomatoes. In the states which are currently most important in fresh-market vegetable production, gains in frost-free periods might be offset by reduced ability to grow heat-sensitive vegetables (such as tomato, bean, sweet-corn, pepper, carrot, lettuce and cole crops) during the summer and by reduced fruit quality.

To the extent that planting and harvest dates and production areas change, market disruptions will occur. Currently, profitability in vegetable crops is highly dependent on having a 'market window' where prices are high. These market windows will become increasingly hard to predict if climatic changes significantly alter planting and harvest dates. Market disruptions could also result in temporary shortages and/or oversupplies for the consumer, and possible price changes for both consumers and producers.

It will be more difficult for small growers to adapt to changing market opportunities, because they have less ability to survive a temporary market 'glut'. It will also be harder for small growers to switch crops, because transplanting, seeding, cultivating and harvesting equipment differs greatly for the different types of vegetable crops. Additional cultivar trials may need to be conducted, which will be difficult for small growers. Even for large corporate farms, which already have multi-state operations, additional costs will be incurred if the infrastructure supporting vegetable production (packing houses, cooling facilities, housing for labourers) needs to be relocated.

References

Acock, B. and Acock, M.C. (1993) Modeling approaches for predicting crop ecosystem responses to climate change. In: Buxton, D.R., Shibles, R., Forsberg, R.A., Blad, B.L., Asay, K.H., Paulsen, G.M. and Wilson, R.F. (eds) *International Crop Science I.* Crop Science Society of America, Madison, Wisconsin, pp. 41–55.

Acock, B. and Allen, L.H. Jr (1985) Crop responses to elevated carbon dioxide concentrations. In: Strain, B.R. and Cure, J.D. (eds) *Direct Effects of Increasing Carbon Dioxide on Vegetation.* DOE/ER-0238, Office of Energy Research, US Department of Energy, Washington, DC, pp. 53–97.

Ahmed, F.E., Hall, A.E. and DeMason, D.A. (1992) Heat injury during floral development in cowpea (*Vigna unguiculata*, Fabaceae). *American Journal of Botany* 79, 784–791.

Allen, L.H. Jr (1990) Plant responses to rising carbon dioxide and potential interaction with air pollutants. *Journal of Environmental Quality* 19, 15–34.

Allen, L.H. Jr, Kirkham, M.B., Olszyk, D.M. and Whitman, C.E. (1997) *Advances in Carbon Dioxide Effects Research.* ASA Special Publication Number 61, American Society of Agronomy, Madison, Wisconsin, 228 pp.

Amthor, J.S. (1988) Growth and maintenance respiration in leaves of bean (*Phaseolus vulgaris* L.) exposed to ozone in open-top chambers in the field. *New Phytologist* 110, 319–325.

Amthor, J.S. (1997) Plant respiratory responses to elevated carbon dioxide partial pressure. In: Allen, L.H. Jr, Kirkham, M.B., Olszyk, D.M. and Whitman, C.E. (eds) *Advances in Carbon Dioxide Effects Research.* ASA Special Publication Number 61, American Society of Agronomy, Madison, Wisconsin, pp. 35–78.

Arp, W.J. (1991) Effects of source–sink relations on photosynthetic acclimation to elevated CO_2. *Plant, Cell and Environment* 14, 869–875.

Ball, M.C. and Munns, R. (1992) Plant responses to salinity under elevated atmospheric concentrations of CO_2. *Australian Journal of Botany* 40, 515–525.

Barnes, W.C. (1936) Effects of some environmental factors on growth and colour of carrots. *Memoirs of Cornell Agricultural Experimental Station* 186, 1–36.

Benjamin, L.R., McGarry, A. and Gray, A. (1998) The root vegetables: beet, carrot, parsnip and turnip. In: Wien, H.C. (ed.) *The Physiology of Vegetable Crops.* CAB International, Wallingford, UK. pp. 553–580.

Bhattacharya, S., Bhattacharya, N.C. and Strain, B.R. (1985) Rooting of sweet potato stem cuttings under CO_2-enriched environment and with IAA treatment. *HortScience* 20, 1109–1110.

Boese, S.R., Wolfe, D.W. and Melkonian, J.J. (1997) Elevated CO_2 mitigates chilling-induced water stress and photosynthetic reduction during chilling. *Plant, Cell and Environment* 20, 625–632.

Boote, K.J., Pickering, N.C. and Allen, L.H. Jr (1997). Plant modeling: advances and gaps in our capability to project future crop growth and yield in response to global climate change. In: Allen, L.H. Jr, Kirkham, M.B., Olszyk, D.M. and Whitman, C.E. (eds) *Advances in Carbon Dioxide Effects Research.* ASA Special Publication Number 61. American Society of Agronomy, Madison, Wisconsin, pp. 179–228.

Brewster, J.L. (1997) Onions and garlic. In: Wien, H.C. (ed.) *The Physiology of Vegetable Crops.* CAB International, Wallingford, UK, pp. 581–620.

Bunce, J.A. (1995) Effects of elevated carbon dioxide concentration in the dark on the growth of soybean seedlings. *Annals of Botany* 75, 365–368.

Cao, W., Tibbits, T.W. and Wheeler, R.M. (1994) Carbon dioxide interactions with irradiance and temperature in potatoes. *Advances in Space Research* 14, 243–250.

Cochran, H.L. (1936) Some factors influencing growth and fruit-setting in the pepper (*Capiscum frutescens* L.). *Cornell Agricultural Experiment Station Memoir* 190, 1–39.

Collins, W.B. (1976) Effect of carbon dioxide enrichment on growth of the potato plant. *HortScience* 11, 467–469.

Conroy, J.P. (1992) Influence of elevated atmospheric CO_2 concentrations on plant nutrition. *Australian Journal of Botany* 40, 445–456.

Crawford, R.M.M. and Wolfe, D.W. (1999) Temperature: cellular to whole plant and population responses. In: Luo, Y. and Mooney, H. (eds) *Carbon Dioxide and Environmental Stresses.* Academic Press, San Diego, pp. 61–106.

Davis, J.H.C. (1997) *Phaseolus* beans. In: Wien, H.C. (ed.) *The Physiology of Vegetable Crops.* CAB International, Wallingford, UK, pp. 409–428.

Davis, T.D. and Potter, J.R. (1989) Relations between carbohydrate, water status and adventitious root-formation in leafy pea cuttings rooted under various levels of atmospheric CO_2 and relative-humidity. *Physiologia Plantarum* 77,185–190.

Daymond, A.J., Wheeler, T.R., Hadley, P., Ellis, R.H. and Morison, J.I.L. (1997) The growth, development and yield of onion (*Allium cepa* L.) in response to temperature and CO_2. *Journal of Horticultural Science* 72, 135–145.

Drake, B.G, Gonzàlez-Meler, M.A. and Long, S.P. (1997) More efficient plants: a consequence of rising atmospheric CO_2. *Annual Review of Plant Physiology and Plant Molecular Biology* 48, 609–639.

Ewing, E.E. (1997) Potato. In: Wien, H.C. (ed.) *The Physiology of Vegetable Crops.* CAB International, Wallingford, UK, pp. 295–344.

FAOSTAT (1997) http://www.FAO.org/ Select: Statistical Databases/Agriculture/Crops Primary.

Fitter, A.H. and Hay, R.K.M. (1987) *Environmental Physiology of Plants,* 2nd edn. Academic Press, London, 423 pp.

Ford, M.A. and Thorne, G.N. (1967) Effect of CO_2 concentration on growth of sugar-beet, barley, kale and maize. *Annals of Botany* 3, 629–644.

Gale, J. (1982) Evidence for essential maintenance respiration of leaves of *Xanthium strumarium* at high temperature. *Journal of Experimental Botany* 33, 471–476.

Gawronska, H., Thornton, M.K. and Dwelle, R.B. (1992) Influence of heat stress on dry matter production and photoassimilate partitioning by four potato clones. *American Potato Journal* 69, 653–665.

Goudriaan, J. and de Ruiter, H.A. (1983) Plant growth in response to CO_2 enrichment at two levels of nitrogen and phosphorus supply. I. Dry matter, leaf area and development. *Netherlands Journal of Agricultural Science* 31, 157–169.

Hall, A.E. and Allen, L.H. Jr (1993) Designing cultivars for the climatic conditions of the next century. In: Buxton, D.R., Shibles, R., Forsberg, R.A., Blad, B.L., Asay, K.H., Paulsen, G.M. and Wilson, R.F. (eds) *International Crop Science I.* Crop Science Society of America, Madison, Wisconsin, pp. 291–297.

Hara, T. and Sonoda, Y. (1982) Cabbage head development as affected by nitrogen and temperatures. *Soil Science and Plant Nutrition* 28, 109–117.

Harrington, J.F. (1962) The effect of temperature on the germination of several kinds of vegetable seeds. *Proceedings of the XVI International Horticultural Congress,* Vol. II, pp. 435–441.

Hunt, R., Wilson, J.W., Hand, D.W. and Sweeney, D.G. (1984) Integrated analysis of growth and light interception in winter lettuce. I. Analytical methods and environmental influences. *Annals of Botany* 54, 743–757.

Idso, S.B. and Kimball, B.A. (1989) Growth response of carrot and radish to atmospheric CO_2 enrichment. *Environmental and Experimental Botany* 29, 135–139.

Idso, S.B., Kimball, B.A., Anderson, M.G. and Mauney, J.R. (1987) Effects of atmospheric CO_2 enrichment on plant growth: the interactive role of air temperature. *Agricultural Ecosystems and Environment* 20, 1–10.

Imai, K., Coleman, D.F. and Yanagisawa, T. (1984) Elevated atmospheric partial pressure of carbon dioxide and dry matter production of cassava (*Manihot esculenta* Crantz). *Japanese Journal of Crop Science* 53, 479–485.

Jordan, D.B. and Ogren, W.L. (1984) The CO_2/O_2 specificity of ribulose 1,5 bisphosphate carboxylase/oxygenase. Dependence on ribulose bisphosphate concentration, pH, and temperature. *Planta* 161, 308–313.

Karl, T.R., Kukla, G., Razuvayev, V.N., Changery, M.J., Quayle, R.G., Heim, R.R. Jr, Easterling, D.R. and Fu, C.B. (1991) Global warming: evidence for asymmetric diurnal temperature change. *Geophysical Research Letters* 18, 2253–2256.

Kimball, B.A. (ed.) (1990) *Impact of Carbon Dioxide, Trace Gases, and Climate Change on Global Agriculture: proceedings of a symposium sponsored by the American Society of Agronomy, Crop Science Society of America, and Soil Science Society of America.* ASA Special Publication No. 53, American Society of Agronomy, Crop Science Society of America, Soil Science Society of America, Madison, Wisconsin, 133 pp.

Kimball, B.A., Sims, W.L. and Welch, J.E. (1967) Plant climate analysis for lettuce. *California Agriculture* 21, 2–4.

Kinet, J.M. and Peet, M.M. (1997) Tomato. In: Wien, H.C. (ed.) *The Physiology of Vegetable Crops.* CAB International, Wallingford, UK, pp. 207–259.

Konsens, I., Ofir, M. and Kigel, J. (1991) The effect of temperature on the production and abscission of flowers and pods in snap bean (*Phaseolus vulgaris* L.). *Annals of Botany* 67, 391–399.

Krauss, A. and Marschner, H. (1984) Growth rate and carbohydrate metabolism of potato tubers exposed to high temperatures. *Potato Research* 27, 297–303.

Krug, H. (1991) *Gemüseproduktion*, 2nd edn. Verlag Paul Pavey, Berlin.

Krug, H. (1997) Environmental influences on development, growth and yield. In: Wien, H.C. (ed.) *The Physiology of Vegetable Crops.* CAB International, Wallingford, UK, pp. 101–180.

Ku, G., Edwards, E. and Tanner, C.B. (1977) Effects of light, carbon dioxide, and temperature on photosynthesis, oxygen inhibition of photosynthesis, and transpiration in *Solanum tuberosum. Plant Physiology* 59, 868–872.

Kuehny, J.S., Peet, M.M., Nelson, P.V. and Willits, D.H. (1991) Nutrient dilution by starch in CO_2-enriched chrysanthemum. *Journal of Experimental Botany* 42, 711–716.

Lutze, J.L., Roden, J.S., Holly, C.J., Wolfe, J., Egerton, J.J.G. and Ball, M.C. (1998) Elevated atmospheric CO_2 promotes frost damage in evergreen tree seedlings. *Plant, Cell and Environment* 21, 631–635.

Maynard, D.N. and Hochmuth, G.J. (1997) *Knott's Handbook for Vegetable Growers.* John Wiley & Sons, New York, 582 pp.

Mearns, L.O., Katz, R.W. and Schneider, S.H. (1984) Extreme high temperature events: changes in their probabilities with changes in mean temperature. *Journal of Climate and Applied Meteorology* 23, 1601–1613.

Midmore, D.J. and Prange, R.K. (1992) Growth responses of two *Solanum* species to contrasting temperatures and irradiance levels: relations to photosynthesis, dark respiration and chlorophyll fluorescence. *Annals of Botany* 69, 13–20.

Misaghi, I.J., Oebker, N.F. and Hine, R.B. (1992) Prevention of tipburn in iceberg lettuce during postharvest storage. *Plant Disease* 76, 1169–1171.

Mitchell, R.A.C., Mitchell,V.J., Driscoll, S.P., Franklin, J. and Lawlor, D.W. (1993) Effects of increased CO_2 concentration and temperature on growth and yield of winter wheat at two levels of nitrogen application. *Plant, Cell and Environment* 16, 521–529.

Moorby, J. (1970) The production, storage, and translocation of carbohydrates in developing potato plants. *Annals of Botany* 34, 297–308.

Mortensen, L.M. (1985) Nitrogen oxides produced during CO_2 enrichment. II. Effects on different tomato and lettuce cultivars. *New Phytologist* 101, 411–415.

Muehlbauer, F.J. and McPhee, K.E. (1997) Peas. In: Wien, H.C. (ed.) *The Physiology of Vegetable Crops.* CAB International, Wallingford, UK, pp. 429–460.

Mutters, R.G. and Hall, A.E. (1992) Reproductive responses of cowpea to high temperature during different night periods. *Crop Science* 32, 202–206.

Nederhoff, E.M. (1994) Effects of CO_2 concentration on photosynthesis, transpiration and production of greenhouse fruit vegetable crops. PhD Dissertation, Wageningen, the Netherlands, 213 pp.

NOAA (1999) *Climate of 1998.* National Oceanic and Atmospheric Administration. Annual National Climatic Data Center, January 12, 1999. http://www.ncdc.noaa.gov/ol/climate/research/1998/ann/ann98.html.

Olymbios, C. (1973) Physiological studies on the growth and development of the carrot, *Daucus carota* L. PhD Thesis, University of London, London.

Overdieck, D. (1993) Elevated CO_2 and the mineral content of herbaceous and woody plants. *Vegetatio* 104/105, 403–411.

Paulsen, G.M. (1994) High temperature responses of crop plants. In: Boote, K.J., Bennett, J.M., Sinclair, T.R. and Paulsen, G.M. (eds) *Physiology and Determination of Crop Yield.* American Society of Agronomy, Madison, Wisconsin, pp. 365–389.

Peet, M.M. and Bartholomew, M. (1996) Effect of night temperature on pollen characteristics, growth and fruitset in tomato (*Lycopersicon esculentum* Mill). *Journal of the American Society for Horticultural Science* 121, 514–519.

Peet, M.M., Willits, D.H. and Gardner, R. (1997) Response of ovule development and post-pollen production processes in male-sterile tomatoes to chronic, sub-acute high temperature stress. *Journal of Experimental Botany* 48, 101–112.

Peet, M.M., Sato, S. and Gardner, R. (1998) Comparing heat stress on male-fertile and male-sterile tomatoes. *Plant, Cell and Environment* 21, 225–231.

Poorter, H. (1993) Interspecific variation in the growth response of plants to an elevated ambient CO_2 concentration. *Vegetatio* 104/105, 77–97.

Pressman, E. (1997) Celery. In: Wien, H.C. (ed.) *The Physiology of Vegetable Crops.* CAB International, Wallingford, UK, pp 387–408.

Radoglou, K.M. and Jarvis, P.G. (1992) The effects of CO_2 enrichment and nutrient supply on growth morphology and anatomy of *Phaseolus vulgaris* L. seedlings. *Annals of Botany* 70, 245–256.

Rappaport, L. and Wittwer, S.H. (1956) Flowering in head lettuce as influenced by seed vernalization, temperature and photoperiod. *Proceedings of the American Society of Horticultural Science* 67, 429–437.

Reid, C.D. and Fiscus, E.L. (1998) Effects of elevated [CO_2] and/or ozone on limitations to CO_2 assimilation in soybean (*Glycine max*). *Journal of Experimental Botany* 49, 885–895.

Reynolds, M.P. and Ewing, E.E. (1989) Effects of high air and soil temperature stress on growth and tuberization of *Solanum tuberosum. Annals of Botany* 64, 241–247.

Reynolds, M.P., Ewing, E.E. and Owens, T.G. (1990) Photosynthesis at high temperature in tuber bearing *Solanum* species. *Plant Physiology* 93, 791–797.

Rogers, H.H. and Dahlman, R.C. (1993) Crop responses to CO_2 enrichment. *Vegetatio* 104/105, 117–131.

Rogers, H.H., Runion, G.B. and Krupa, S.V. (1994) Plant responses to atmospheric CO_2 enrichment with emphasis on roots and the rhizosphere. *Environmental Pollutants* 83, 155–167.

Rogers, H.H., Runion, G.B., Krupa, S.V. and Prior, S.A. (1997) Plant responses to atmospheric carbon dioxide enrichment: implications in root–soil microbe interactions. In: *Advances in Carbon Dioxide Effects Research.* ASA Special Publication No. 61, American Society of Agronomy, Madison, Wisconsin. pp. 1–34.

Rosenzweig, C. (1995) *Climate Change and Agriculture: Analysis of Potential International Impacts.* Proceedings of a symposium sponsored by the American

Society of Agronomy in Minneapolis, Minnesota, 4–5 Nov. 1992; organized by Division A-3 (Agroclimatology and Agronomic Modeling) and Division A-6 (International Agronomy). ASA Special Publication No. 59, American Society of Agronomy, Madison, Wis., USA. 382 pp.

Rosenzweig, C. and Hillel, D. (1998). *Climate Change and the Global Harvest Potential Impacts of the Greenhouse Effect on Agriculture.* Oxford University Press, New York, 336 pp.

Sage, R.F. (1994) Acclimation of photosynthesis to increasing atmospheric CO_2: the gas exchange perspective. *Photosynthesis Research* 39, 351–368.

Sato, S. (1998) Effects of chronic high temperature stress on the development and function of the reproductive structures of tomato (*Lycopersicon esculentum* Mill.) PhD Thesis, North Carolina State University, Raleigh.

Snyder, R.G. and Ewing, E.E. (1989) Interactive effects of temperature, photoperiod, and cultivar on tuberization of potato cuttings. *HortScience* 24, 336–338.

Sritharan, R., Caspari, H. and Lenz, F. (1992) Influence of CO_2 enrichment and phosphorus supply on growth, carbohydrates and nitrate utilization of kohlrabi plants. *Gartenbauwissenschaft* 57, 246–251.

Stitt, M. (1991) Rising CO_2 levels and their potential significance for carbon flow in photosynthetic cells. *Plant, Cell and Environment* 14, 741–762.

Struik, P.C., Geertsema, J. and Custers, C.H.M.G. (1989) Effects of shoot, root and stolon temperature on the development of the potato plant. III. Development of tubers. *Potato Research* 32, 151–158.

Taylor, A.G. (1997) Seed storage, germintion and quality. In: Wien, H.C. (ed.) *The Physiology of Vegetable Crops.* CAB International, Wallingford, UK, pp. 1–36.

Went, F.W. (1944) Plant growth under controlled conditions. II. Thermoperiodicity in growth and fruiting of the tomato. *American Journal of Botany* 31, 135–150.

Wheeler, T.R. and Ellis, R.H. (1991) Seed quality, cotyledon elongation at suboptimal temperatures, and the yield of onion. Seed. *Seed Science Research* 1, 57–67.

Wheeler, R.M. and Tibbitts, T.W. (1989) Utilization of potatoes for life support systems in space. IV. Effect of CO_2 enrichment. *American Potato Journal* 66, 25–34.

Wheeler, R.M., Tibbitts, T.W. and Fitzpatrick, A.H. (1991) Carbon dioxide effects on potato growth under different photoperiods and irradiance. *Crop Science* 31, 1209–1213.

Wheeler, T.R., Hadley, P., Ellis, R.H. and Morison, J.I.L. (1993a) Changes in growth and radiation use by lettuce crops in relation to temperature and ontogeny. *Agricultural and Forest Meteorology* 66, 173–186.

Wheeler, T.R., Hadley, P, Morison, J.I.L. and Ellis, R.H. (1993b) Effects of temperature on the growth of lettuce (*Lactuca sativa* L.) and the implications for assessing the impacts of potential climate change. *European Journal of Agronomy* 2, 305–311.

Wheeler, R.M., Morison, J.I.L., Ellis, R.H. and Hadley, P. (1994) The effects of CO_2, temperature, and their interaction on the growth and yield of carrot (*Daucus carota* L.) *Plant, Cell and Environment* 17, 1275–1284.

Wheeler, T.R., Ellis, R.H., Hadley, P. and Morison, J.I.L. (1995) Effects of CO_2, temperatures and their interaction on the growth, development and yield of cauliflower (*Brassica oleracea* L. botrytis). *Scientia Horticulturae* 60, 181–197.

Wheeler, T.R., Ellis, R.H., Hadley, P., Morison, J.I.L., Batts, G.R. and Daymond, A.J. (1996) Assessing the effects of climate change on field crop production. *Aspects of Applied Biology* 45, 49–54.

Wien, H.C. (1997a) *The Physiology of Vegetable Crops.* CAB International, Wallingford, UK, 662 pp.

Wien, H.C. (1997b) The cucurbits: cucumber, melon, squash and pumpkin. In: Wien, H.C. (ed.) *The Physiology of Vegetable Crops.* CAB International, Wallingford, UK, pp. 345–386.

Wien, H.C. (1997c) Pepper. In: Wien, H.C. (ed.) *The Physiology of Vegetable Crops.* CAB International, Wallingford, UK, pp. 259–293.

Wien, H.C. (1997d) Lettuce. In: Wien, H.C. (ed.) *The Physiology of Vegetable Crops.* CAB International, Wallingford, UK, pp. 479–509.

Wien, H.C. and Wurr, D.C.E. (1997) Cauliflower, broccoli, cabbage and Brussels sprouts. In: Wien, H.C. (ed.) *The Physiology of Vegetable Crops.* CAB International, Wallingford, UK, pp 511–552.

Willits, D.H. and Peet, M.M. (1989) Predicting yield responses to different greenhouse CO_2 enrichment schemes: cucumbers and tomatoes. *Agricultural and Forest Meteorology* 44, 275–293.

Wittwer, S.H. (1986) Worldwide status and history of CO_2 enrichment: an overview. In: Enoch, H.Z. and Kimball, B.A. (eds) *Carbon Dioxide Enrichment of Greenhouse Crops,* Vol. 1. CRC Press, Boca Raton, Florida, pp. 3–15.

Wolfe, D.W. (1994) Physiological and growth responses to atmospheric carbon dioxide concentration. In: Pessarakli, M. (ed.) *Handbook of Plant and Crop Physiology.* Marcel Dekker, New York, pp. 223–242.

Wolfe, D.W. and Erickson, J.D. (1993) Carbon dioxide effects on plants: uncertainties and implication for modeling crop response to climate change. In: Kaiser, H.M. and Drennen, T.E. (eds) *Agricultural Dimension of Global Climate Change.* St Lucie Press, Delray Beach, Florida.

Wolfe, D.W., Fereres, E. and Voss, R.E. (1983) Growth and yield response of two potato cultivars to various levels of applied water. *Irrigation Science* 3, 211–222.

Wolfe, D.W., Azanza, F. and Juvik, J.A. (1997) Sweet corn. In: Wien, H.C. (ed.) *The Physiology of Vegetable Crops.* CAB International, Wallingford, UK, pp. 461–478.

Wolfe, D.W., Gifford, R.M., Hilbert, D. and Luo, Y. (1998) Integration of photosynthetic acclimation to CO_2 at the whole-plant level. *Global Change Biology* 4, 879–893.

Wurr, D.C.E. and Fellows, J.R. (1984) The growth of three crisp lettuce varieties from different sowing dates. *Journal of Agricultural Science* 102, 733–745.

Wurr, D.C.E. and Fellows, J.R. (1991) The influence of solar radiation and temperature on the head weight of crisp lettuce. *Journal of Horticultural Science* 66, 183–190.

Wurr, D.C.E., Fellows, J.R. and Morris, G.E.L. (1981) Studies of the hearting of butterhead lettuce: temperature effects. *Journal of Horticultural Science* 56, 211–218.

11 Crop Ecosystem Responses to Climatic Change: Tree Crops

IVAN A. JANSSENS[1], MARIANNE MOUSSEAU[2] AND REINHART CEULEMANS[1]

[1]Department of Biology, University of Antwerp, Universiteitsplein 1, B-2610 Wilrijk, Belgium; [2]Laboratoire d'Ecologie Végétale, Université Paris-Sud XI, F-91405 Orsay Cedex, France

11.1 Introduction

Forest ecosystems cover little more than one-quarter of the earth's land surface (3.5 billion ha; FAO, 1995). However, approximately 85% of plant carbon (C) and 35% of soil C is found in forests (Kirschbaum and Fischlin, 1996). As such, they contain over 60% of the C stored in the terrestrial biosphere. Thus, forest ecosystems constitute a major terrestrial C reservoir, which may exert a significant impact on the atmospheric CO_2 concentrations ($[CO_2]$). Within the framework of the changing global C balance, there is a large potential for forests to act as a sink for atmospheric CO_2. This could happen either through reforestation or through enhanced tree growth rates. Enhanced tree growth rates have been attributed to the net CO_2 fertilization effect, to increased nitrogen (N) deposition rates (Spiecker *et al.*, 1996) and to a prolongation of the growing season (Myneni *et al.*, 1997). However, forests may also act as a net source and increase the atmospheric $[CO_2]$ through changes in land use such as, deforestation for timber products or the creation of pasture and rangeland.

Although most of the terrestrial biomass resides in forests, we live in a world where wood and woody products are increasingly scarce. Natural forests have been reduced from occupying nearly 46% of the earth's terrestrial ecosystems in pre-industrial times to only 27% today (Winjum and Schroeder, 1997). In recent years the need to intensify forestry has focused worldwide attention on the economic importance of tree plantations, and in particular fast-growing tree plantations. Forest plantations in the world now total approximately 130 Mha, with annual rates of establishment of about 10.5 Mha. Plantations of fast-growing trees are usually managed as short-rotation plantations (with growing cycles of less than 15 years) or as agro-forestry systems, often for the production of industrial raw materials or biomass for energy. Under the designation of fast-growing tree plantations, or short-rotation silviculture, one may find ecosystems managed for different economic objectives, with different intensities of cultural management and different levels of productivity. They may include any of a wide range of species grown under various environmental conditions. A common factor is the greater possibility that exists, relative to conventional forestry, for manipulation of both the environment and the genetics of the trees. Another obvious distinction between natural forests and agro-forestry plantations is in the processes and adaptive strategies available to managers of managed and unmanaged ecosystems.

A crop could be defined as a biological system tailored to give certain products. While planted forest trees meet the criteria of this definition, they are also economically important, whether in a managed or unmanaged stage. Global production of woody forest products is about 4 billion m^3, and that of paper and pulp is 0.5 billion tonnes (Table 11.1). Timber and pulp are not the only economic tree products. Over 33 Mha of the world's terrestrial surface is covered with fruit, palm, rubber tree, or vineyard plantations (Goudriaan, Wageningen, 1998, personal communication). Together these produce a very significant part of all human food and other consumer products (Table 11.2).

Modern silviculture is designed to handle forest stands in such a manner that the saleable yield is maximized. Productivity and yield of cultivated, planted forests (whether short-term or medium-term) are influenced by genetic and physiological factors (Tigerstedt *et al.*, 1985); so to maximize forest yields, both aspects have to be optimized in the silvicultural system. In the 1960s, agricultural crop physiologists realized drastic increases in, for example, grain production by improving the genotype/environment interactions. Tree physiologists and breeders are some 20–30 years behind their agricultural and horticultural colleagues, but in all parts of the world the forest tree is being turned into a cultivated plant. Basically, forest trees are not that different from other (agricultural or horticultural) crops.

Table 11.1. World production of wood and wood-derived products. (FAO, 1995.)

Product	Production
Woody materials (10^6 m^3 $year^{-1}$)	
Fuelwood and charcoal	1920
Industrial roundwood	1430
Sawnwood and sleepers	430
Wood based panels	140
Total	3920
Wood-derived products (10^6 t $year^{-1}$)	
Wood pulp	170
Other fibre pulp	20
Recovered waste paper	100
Paper and paper board	290
Total	580

Table 11.2. World production of economic non-woody tree products (in 10^6 t $year^{-1}$). (FAO, 1995.)

Product	Production
Multi-seed and stone fruits[a]	114
Citrus fruits	89
Bananas[b]	84
Coconut	45
Palm oil	15
Coffee and cacao beans	7
Coprah	6
Grapes[c]	6
Rubber	6
Tree nuts	4

[a]Includes apples, peaches, plums, pears, apricots, cherries, olives, avocados, mangoes and papayas.
[b]Includes banana fruits and plantains.
[c]Data from OIV (1998).

11.2 Trees vs. Non-woody Plants: Similarities and Differences

All trees are photosynthesizing C_3 plants and in many respects they do not differ from all other C_3 plants (Ceulemans and Saugier, 1991). The basic metabolic functions have already been studied and documented in great detail, and the reader is referred to a number of books and reviews on tree physiology (Raghavendra, 1991; Kozlowski and Pallardy, 1996; Rennenberg *et al.*, 1997).

11.2.1 Size, life span, acclimation and seasonality

There are a number of specific features and characteristics that distinguish trees from annuals, herbaceous perennials and agricultural crops. Compared with other plant types, trees are long-lived organisms. The predicted rapid changes in $[CO_2]$ and increases in temperature will significantly affect currently growing and future forests. Long-lived plants have more time to acclimate to changing environmental conditions than short-lived organisms. On a time scale, this acclimation might occur in the order of several years, and the acclimation process might be influenced by seasonal changes in environmental conditions (Kozlowski and Pallardy, 1996). Acclimation may be expressed through: (i) the reduction of growth, thereby limiting the sink for carbon in the growing tissues; (ii) the gradual filling of available sinks; and (iii) a reallocation of nutrients from the leaf.

Although trees are typically large in size and long in lifespan, studies on responses of tree species to changing environmental conditions are predominantly of short duration (less than two growing seasons) and with seedlings or juvenile plant materials (Ceulemans and Mousseau, 1994). This anomaly between the real world and the experimental approach can be explained by the expenses of construction and operation of experimental facilities for large trees.

The recurrent developmental events of phenology and seasonality also distinguish trees from annual or agricultural crops. The fact that trees live over multiple growing seasons implies that every year there is a considerable renewal cost of some organs (leaves and fine roots), and that trees are more responsive or susceptible to climatic changes. Furthermore, trees remain in a juvenile phase for several years. Intermittent (periodic, episodic, rhythmic) rather than continuous growth is the rule among woody plants. Even trees such as palms and mangroves, which are noted for continuous growth by producing a leaf or pair of leaves every few months, do not necessarily have constant root or shoot extension rates (Borchert, 1991).

Annual stem growth of trees can be determined from the width of the growth rings. Studies have already correlated tree-ring width with climate (Fritts, 1976) and tree-ring analysis of hardwood species appears to be an effective approach to revealing historical changes in soil chemistry, environmental changes, etc. (Becker *et al.*, 1995). Recent dendrochronological studies have indicated an increase in tree growth trends, especially in European forest trees (Spiecker *et al.*, 1996). This increase can be accounted for by changes in various climatic factors, such as increases in atmospheric temperature and [CO_2]. Growth is also enhanced by atmospheric nitrogen deposition, which is particularly high in western Europe.

11.2.2 Source/sink relations and fruit production

Another major difference between trees and non-woody plants is their source-to-sink ratio. A sink is defined as the region of a plant that is a net consumer of carbohydrates; a source is defined as the region of a plant that is a net producer of carbohydrates. The canopy, as a whole, is essentially viewed as source tissue; therefore, pruning of tree canopies results in a reduced source activity. However, a new and rapidly developing canopy may also be viewed as sink tissue. Whereas photosynthesis in non-woody plants may be reduced by limited sink strength, this is not the case for trees, because they have large C storage capacity in their woody tissues.

Fruits represent a major C sink in tree crops. Therefore, this important source/sink interaction has been studied in several species (Eckstein *et al.*, 1995; Berman and Dejong, 1996). When horticultural productivity is the goal, the allocation of resources toward reproductive processes must be maximized. However, the tree must also preserve its growth potential for future years; thus, a delicate C balance must be maintained between vegetative and reproductive needs. In spring, stored sugars and nutrients support actively growing shoots and inflorescences. Competition occurs between vegetative and reproductive meristems, and the fruit is growing essentially on the currently produced photosynthates. The relationship between fruit load and photosynthetic activity (Palmer *et al.*, 1997), as well as the effects of several climatic variables, have been intensively studied (Wibbe *et al.*, 1993; Buwalda and Lenz, 1995). Among all tree crops, cultivated fruit trees are the ones most adequately supplied with water and nutrients; thus, there should be few, if any, constraints to a positive CO_2 response. However, very few studies have dealt with the effects of increased CO_2 concentrations on fruit production (Goodfellow *et al.*, 1997). The best known study is that of Idso's group, who grew eight citrus trees in open-top chambers for 7 years (Idso and Kimball, 1997). Results showed that under elevated CO_2 there was an increase in growth of the vegetative tree parts and that while the number of fruits increased, their size did not.

11.3 Effects of Climatic Changes on Tree Growth and Physiology

11.3.1 Uniformity and variability in tree responses

In comparison with crop species, the response of tree species to CO_2 enrichment is slightly less well understood. This is probably due to the large buffering effect of the woody storage compartments. However, the effects of elevated [CO_2], and to a much lesser extent of temperature, on tree growth and physiology have been extensively examined and reviewed during the last decade. Reviews on above-ground responses (e.g. Eamus and Jarvis, 1989; Mousseau and Saugier, 1992; Ceulemans and Mousseau, 1994; Wullschleger *et al.*, 1997; Jarvis, 1998; Norby *et al.*, 1999) or on below-ground responses of trees to elevated [CO_2] (Rogers *et al.*, 1994, 1999; Norby *et al.*, 1995a) have been published. In order to estimate the magnitude and significance of these elevated [CO_2] effects on trees, Curtis and Wang (1998) performed a meta-analysis on 508 published responses of trees to elevated [CO_2].

In terms of general conclusions two main messages arise from these reviews. Firstly, climatic change effects are not caused by a single factor (e.g. elevated [CO_2]), but originate from complex interactions among various factors such as atmospheric [CO_2], air temperature, nutrient supply, tropospheric ozone level, UV-B radiation, drought frequency, etc. Secondly, although these reviews and complicated analyses have revealed some general response trends to elevated atmospheric [CO_2] (see below), there are also significant differences within and among tree species and genera in terms of their responses to elevated [CO_2]. Part of this variability might be explained by the fact that species differ in their C allocation patterns and growth potential, as well as in their responses to different growing conditions.

The main responses of tree growth and physiology to elevated atmospheric [CO_2] have been summarized (Table 11.3). Three main directives were employed when selecting the studies included in this review. Firstly, to avoid root restriction (pot effects: Arp, 1991), only studies with free-rooted trees or trees grown in very large containers were considered. Pots limit root growth and alter C partitioning; therefore, pot experiments using trees larger than seedlings may confound nutrient and CO_2 effects. Secondly, to avoid overlap with several of the reviews cited above, only the most recent

Table 11.3. Literature survey (1994–1998) on the influence of elevated atmospheric CO_2 on above-ground biomass, total leaf area, root/shoot ratio, and net photosynthetic capacity on a leaf area basis for free-rooted trees. All values are given as the percentage increase under elevated CO_2 as compared with the ambient treatment. Species name, plant age at the start of the treatment, total duration of treatment (minimum 6 months), and growth conditions have been indicated for each reference. For more information, see section 11.3.1 in the text.

Species name	Plant age at start of treatment	Duration of treatment	Growth conditions	Above-ground biomass	Leaf area	Root/ shoot ratio	Photosynthesis/ unit leaf area	References
Acacia smallii	Seed	13 months	W, N, large containers, N-fixing tree	87			ns	Polley *et al.*, 1997
Betula pendula	1 year	4.5 years	OTC, forest soil	58	43	ns	32	Rey and Jarvis, 1997
Castanea sativa	2 years	1 year	Mini-ecosystems – forest soil	52	42			Mousseau *et al.*, 1995
Citrus aurantium	30 cm tall	9 years	OTC, orchard	117				Idso and Kimball, 1997
Fagus sylvatica	10 years	3 years	BB, isolated trees, UPS	22	38		1st year: 93 2nd year: 64 3rd year: 19	Besford *et al.*, 1998; Lee *et al.*, 1998; Murray and Ceulemans, 1998
Fagus sylvatica	2.5 years	3 years	Microcosms, TUB	92	29	−19	1st year: 117 2nd year: 38	Besford *et al.*, 1998; Lee *et al.*, 1998; Murray and Ceulemans, 1998
Fagus sylvatica	2 years	2 years	Mini-ecosystems – forest soil	67–102	53–66	17–13		Epron *et al.*, 1996; Badeck *et al.*, 1997
Liriodendron tulipifera	Seed	2.5 years	OTC, natural site	ns	ns		58–62	Norby *et al.*, 1995b
Mangifera indica	Grafted seedlings, 1 year old	> 2 years	OTC, W, N	60	ns		20	Goodfellow *et al.*, 1997
Nothofagus fusca	1 year	1 year	OTC, natural site, W, N, spring summer				23 42	Hogan *et al.*, 1996
Picea abies	30 years	2 years	OTC, natural site	ns		35		Oplutsilova and Dvorak, 1997

Picea abies	3 years	3 years	Large containers, forest soil,					Hättenschwiler and
			GC, n	ns	−28	ns		Körner, 1998
			N (30 kg)	ns	−20	ns		
			N (90 kg)	ns	−24	ns		
Picea abies	3 years	2 years	OTC, natural site, W				ns	Dixon *et al.*, 1995
			w				80	
Picea sitchensis	Seed	3 years	Solar domes, free rooted bags					Townend, 1995
			WN	52				
			Wn	ns				
			wN	44				
			wn	49				
Picea sitchensis	3.5 years	2 years	OTC, UEDIN + FC N	18	7	−13		Lee *et al.*, 1998;
			n	26	16	11		Murray and
								Ceulemans, 1998
Picea sitchensis	17–21 years	4 years	BB, natural site, UEDIN	15	9		1st year: 97	Besford *et al.*, 1998;
							2nd year: 161	Lee *et al.*, 1998;
							3rd year: 47	Murray and
								Ceulemans, 1998
Pinus palustris	1 month	1.5 years	Large containers, OTC, natural					Prior *et al.*, 1997;
			site, W	ns		−16	−30	Runion *et al.*, 1997
			w	83		39	−48	
			N	30		ns		
			n	ns		ns		
Pinus ponderosa	Seed	1.5 years	OTC, W, N	86		38	86	Johnson *et al.*, 1995
			n	100		95	100	
Pinus radiata	Seed	1 year	OTC, natural site,					Hogan *et al.*, 1996
			W, N, spring				ns	
			summer				43	
Pinus sylvestris	25–30 years	2–3 years	BB, N				24	Kellomäki and Wang,
			n				31	1997a,b

Table 11.3. *Continued*

Species name	Plant age at start of treatment	Duration of treatment	Growth conditions	Above-ground biomass	Leaf area	Root/ shoot Ratio	Photosynthesis/ unit leaf area	References
Pinus sylvestris	Mature trees	4 years	GH, nat site, elevated temperature				133	Beerling, 1997
Pinus sylvestris	20–25 years	2 years	OTC, natural site, W				1st year needle: 119 2nd year needle: 109	Wang *et al.*, 1995
Pinus taeda	Seed	11 months	OTC, natural site	233	103		33	Tissue *et al.*, 1996
		19 months		111	100	–60	90	
Pinus taeda	9 years	11 months	BB, N or n,W or w				83–91	Murthy *et al.*, 1996
Pinus taeda	21 years	3 years	BB	146 (diam.)	1st flush: 43 2nd flush: ns		120	Teskey, 1995
Populus hybrids	Cuttings	1 and 2						
Beaupre		years	OTC	55–43	18– 48		66	Ceulemans *et al.*, 1996
Robusta			First year–second year	38–58	8– 8		58	
Quercus alba	Seed	4 years	OTC, natural site	135	90	–21		Norby *et al.*, 1995c
Quercus ilex	Mature trees	3 years	OTC, natural mediterranean site (macchia), UVT	49	ns	ns	122	Besford *et al.*, 1998; Lee *et al.*, 1998; Murray and Ceulemans, 1998
Quercus rubra	5 years	2 years	OTC, natural site, W				75	Dixon *et al.*, 1995
			w				ns	

BB, branch bag; GC, growth chamber; GH, greenhouse; N, fertilized; n, non-fertilized; OTC, open-top chamber; W, irrigated; w, non-irrigated; FC, Forestry Commission, UK; TUB, Technical University Berlin, Germany; UEDIN, Edinburgh University, UK; UPS, University Paris-Sud, France; UVT, Viterbo University, Italy.

publications (i.e. papers published in the period 1994–1998) were reviewed. Finally, only studies lasting more than 6 months were considered. These directives were not employed for the mycorrhizae and decomposition tables (Tables 11.4 and 11.5) in this chapter.

11.3.2 Above-ground tree responses

The principal observations that become evident from Table 11.3 are that: (i) biomass increases under elevated [CO_2] in all studies and in all tree species; (ii) photosynthesis per unit leaf area is significantly stimulated, with some exceptions; (iii) there is considerable variability among different tree genera and species in the response reactions; and (iv) there are still very few long-term studies on adult or nearly adult trees. Results of studies of less than one growing season may not represent an acclimated or long-term response, and may exclude the influence of seasonality and ontology. Size constraints are often countered by the use of juvenile trees, yet large morphological and physiological differences often exist between juvenile and mature trees (Kozlowski and Pallardy, 1996). However, a number of CO_2 enrichment studies have used adult trees equipped with branch bags (also included in Table 11.3), which allow a compromise between large tree size and tree maturity (Dufrêne *et al.*, 1993). In the following discussion, some response processes of trees to CO_2 enrichment are briefly reviewed.

Photosynthetic responses and acclimation

The response of photosynthesis in trees to CO_2 enrichment has been the subject of a number of recent reviews (Ceulemans and Mousseau, 1994; Gunderson and Wullschleger, 1994). In the majority of short-term CO_2 enrichment studies on trees, photosynthesis is enhanced. The magnitude of the response varies widely from 0 to +216% (Table 11.3). This variation could be due to inter- and intraspecific differences, to variations in experimental growth conditions (such as nutrient and water treatments), or to the period of exposure to CO_2 enrichment. Also, the variations in photosynthesis between juvenile plants and mature plants are particularly important in trees and may contribute to the large variation in the response of photosynthesis to elevated [CO_2].

Acclimation is defined by a changed photosynthetic CO_2 uptake when CO_2-enriched plants are transferred back to the ambient [CO_2]. A considerable number of papers have described photosynthetic acclimation in potted-tree experiments (Ceulemans and Mousseau, 1994), but free-rooted trees in open-top chambers might also show this phenomenon. Nevertheless, the studies summarized in Table 11.3 show that the overall effect of elevated [CO_2] on photosynthesis in soil-grown trees is strongly positive, although this stimulation decreases after several years of treatment (Ceulemans *et al.*, 1997).

Table 11.4. Effects of elevated CO_2 on tree–mycorrhiza associations. Treatment conditions and duration, response of the percentage of mycorrhizal infection (% Inf), response of total amount of mycorrhizae (Total), fine-root responses and remarkable mycorrhizal features are reported.

Species	Plant growth conditions	% Inf	Total	Other effects on mycorrhizae	Fine-root response	Reference
Betula alleghaniensis	7 months, mesocosms (700 μl l^{-1}), in competition with *Betula papyrifera*	↑			Mass ↑, length ↑	Berntson and Bazzaz, 1998
Betula papyrifera	7 months, mesocosms (700 μl l^{-1}), in competition with *Betula alleghaniensis*	↑			Mass =, length =	Berntson and Bazzaz, 1998
Betula papyrifera	7–9 months, pots in GC (700 μl l^{-1})	EM ↑	↑	Altered morphotype assemblages, extraradical hyphal length ↑	Mass ↑	Godbold and Berntson, 1997
Betula pendula	4 years, OTC (700 μl l^{-1}), no fertilizer	↑		Altered species composition	Mass ↑	Rey *et al*., 1997
Liriodendron tulipifera	6 months, pots in GC (+150 and +300 μl l^{-1})	AM =			Mass ↑	O'Neill and Norby, 1991
Liriodendron tulipifera	2.5 years, OTC (+150 and +300 μl l^{-1})	AM =			Mass ↑	O'Neill, 1994
Pinus caribaea	1 year, pots in GC (660 μl l^{-1})	=				Conroy *et al*., 1990
Pinus echinata	10 months, pots in GC (double CO_2), no fertilizer	=		Signif. ↑ in % Inf. at 34 weeks, not at final harvest	Mass ↑	Norby *et al*., 1987
Pinus echinata	6 months, pots in GC (double CO_2)	=		Signif. ↑ in % Inf. at 6 weeks, not at final harvest	Mass ↑	O'Neill *et al*., 1987
Pinus palustris	20 months, pots in OTC (720 μl l^{-1})	EM ↑	EM ↑	No changes in morphotype assemblages, effect larger at low N and adequate water	Length ↑	Runion *et al*., 1997

Pinus ponderosa	2.5 years, OTC (+175 and +350 μl l^{-1})		EM ↑	Extraradical fungal hyphae ↑, mycorrhizal turnover ↑	Area density ↑	Tingey *et al.*, 1996; Rygiewicz *et al.*, 1997
Pinus ponderosa	4 months, pots in GC (700 μl l^{-1})	↑	↑		Density ↑	DeLucia *et al.*, 1997
Pinus radiata	1 year, pots in GC (660 μl l^{-1})	=				Conroy *et al.*, 1990
Pinus strobus	7–9 weeks, pots in GC (700 μl l^{-1})	EM ↑		Altered morphotype assemblages	Mass ↑	Godbold *et al.*, 1997
Pinus sylvestris	3 months, Petri dishes in GC (600 μl l^{-1})		EM ↑	Extraradical mycelium ↑	Mass ↑, length =	Ineichen *et al.*, 1995
Pinus sylvestris	7–9 months, pots in GC (double CO_2)	EM =		No effect on total fungal mass	Mass =	Markkola *et al.*, 1996
Pinus sylvestris	3 months, pots in GC (700 μl l^{-1})	=		Elevated CO_2 did not alleviate negative effects of NH_3 and O_3	Mass =	Pérez-Soba *et al.*, 1995
Pinus taeda	4 months, pots in GC (double CO_2)	=				Lewis *et al.*, 1994
Populus tremuloides	14 months, open-bottom pots in OTC (700 μl l^{-1})	=		Extraradical mycorrhizal hyphal length ↑ under N-poor conditions, ↓ under N-rich conditions	Mass ↑	Klironomos *et al.*, 1997; Pregitzer *et al.*, 1995
Populus hybrids	2 years, OTC (+350 μl l^{-1})	AM =			Mass ↑	Ceulemans and Godbold, unpublished
Quercus alba	6 months, pots in GC (double CO_2	EM ↑		Increased mycorrhizal infection before increase in root mass, alterations in species abundance	Mass ↑	O'Neill *et al.*, 1987
Quercus alba	4 years, OTC (+150 and +300 μl l^{-1})	↑			Mass ↑	O'Neill, 1994; Norby, 1994
Tsuga canadensis	7–9 weeks, pots in GC (700 μl l^{-1})	EM = AM ↑			Mass ↑	Godbold *et al.*, 1997

OTC, open top chambers; GC, growth chambers; EM, ectomycorrhizae; AM, arbuscular mycorrhizae; =, no significant changes ($P < 0.05$); ↑, significantly enhanced ($P < 0.05$).

Table 11.5. Effects of elevated CO_2 on tree litter decomposition. Tree species and tissue, fumigation conditions and duration, decomposition conditions and duration, and changes in litter chemistry and decomposition rates are reported for every study.

Tree species and tissue	Plant growth conditions	Decomposition conditions	Response of litter quality	Response of decomposition	Reference
Acer pseudoplatanus, senesced leaves	1 season, 600 $\mu l\ l^{-1}$, pots in solar domes	8 months, chambers	C/N ↑, lignin/N ↑	=	Cotrufo *et al.*, 1994
Betula pubescens, senesced leaves	1 season, 600 $\mu l\ l^{-1}$, pots in solar domes	5 months, chambers	C/N ↑, lignin/N ↑	↓	Cotrufo *et al.*, 1994
Betula pubescens, live roots < 2 mm	1 season, 600 $\mu l\ l^{-1}$, pots in solar domes	3 months, chambers with soil			
	Fertilized		C/N =	=	Cotrufo and Ineson, 1995
	Non-fertilized		C/N ↑	=	
Castanea sativa, senesced leaves	2 years, 700 $\mu l\ l^{-1}$, pots in GC	6 months, chambers			
		Incomplete decomposer community	C/N ↑, lignin/N =	↓	Coûteaux *et al.*, 1991
		Complex decomposer community	C/N ↑, lignin/N =	↑	
Fraximus excelsior, senesced leaves	1 season, 600 $\mu l\ l^{-1}$, pots in solar domes	6 months, chambers	C/N ↑, lignin/N ↑	=	Cotrufo *et al.*, 1994
Liriodendron tulipifera, senesced leaves	1 season, + 300 $\mu l\ l^{-1}$, pots in GC, exposed to ozone	1 year, litterbags in forest floor	N ↓, lignin ↑	↓	Boerner and Rebbeck, 1995
Liriodendron tulipifera, senesced leaves	2 years, + 150 and + 300 $\mu l\ l^{-1}$, OTC	2 years, litterbags in forest floor	C/N =, lignin/N =	=	O'Neill and Norby, 1995b
Picea sitchensis, senesced needles	1 season, 600 $\mu l\ l^{-1}$, pots in solar domes	5 months, chambers	C/N =, lignin/N ↑	↓	Cotrufo *et al.*, 1994
Picea sitchensis, live roots < 2 mm	1 season, 600 $\mu l\ l^{-1}$, pots in solar domes	3 months, chamber with soils			
	Fertilized		C/N ↓	=	Cotrufo and Ineson, 1995
	Non-fertilized		C/N ↑	=	

OTC, open-top chambers; GC, growth chambers; =, no significant changes ($P < 0.05$); ↑, significantly enhanced ($P < 0.05$); ↓, significantly decreased ($P < 0.05$).

Altered source/sink relationships and canopy processes

As a result of the increased CO_2 assimilation rates, growth rates are enhanced and patterns of biomass allocation are altered (Eamus and Jarvis, 1989; Conroy *et al.*, 1990). Hence, changes in the functional relationship between plant parts are frequently observed. The balance between sources and sinks changes during the growing season. It is also affected by other environmental variables, e.g. local climate, soil and plant nutritional levels, competition and solar radiation (Bazzaz and Miao, 1993). Unlike the pronounced seasonal allocation patterns of temperate tree species, many tropical tree species accumulate dry matter throughout the year if sufficient water is available (Goodfellow *et al.*, 1997).

Although the major role of a leaf is to be a source, the production of new leaves may be considered as a sink. Very often an increase in total leaf area is found under elevated [CO_2] (Table 11.3). This can be caused by enhanced new leaf production (Guehl *et al.*, 1994; Ceulemans *et al.*, 1995), increased individual and total tree leaf area (Guehl *et al.*, 1994) or increased flush length and number of fascicles (Kellomäki and Wang, 1997b). This increased total leaf area may be displayed in various ways, which affect tree structure and architecture (Tissue *et al.*, 1996; Ceulemans *et al.*, 1995). Greater C assimilation in response to elevated [CO_2] also affects canopy architecture through increased branch production, especially secondary branching (Idso *et al.*, 1991; Ceulemans *et al.*, 1995), increased shoot length (Teskey, 1995), or increased number of growth flushes produced during the growing season (El Kohen *et al.*, 1993; Guehl *et al.*, 1994). It has also been demonstrated that phenological processes can be affected by changes in atmospheric [CO_2] and/or air temperature (Murray and Ceulemans, 1998). Remotely sensed observations indicate a lengthening of effective leaf area duration, likely leading to an accelerated growth of forest trees (Myneni *et al.*, 1997).

Larger leaf area and/or altered crown architecture will result in earlier canopy closure, and thus enhanced competition (Kellomäki and Wang, 1997b). After 2 years of elevated [CO_2] treatment, Scots pine showed a reduction of the initial growth stimulation. This was most likely due to earlier canopy closure (Jach and Ceulemans, 1999). Some studies suggest that under elevated [CO_2] the altered vertical leaf display and crown structure might alter the red/far-red ratio of understorey tree seedlings, thereby affecting their growth pattern (Arnone and Körner, 1993). Increased understanding of branch morphological and crown characteristics should equate results of physiological studies to the tree or stand level, since whole-canopy function is an integration of both physiological processes and morphological characteristics at smaller scales.

Water relations and water-use efficiency

A reduction of stomatal conductance under elevated [CO_2] might have a significant effect on water transport in trees, since the latter is roughly proportional to stomatal conductance. Hydraulic conductivity was reported to decrease with elevated [CO_2] (Tognetti *et al.*, 1996) but this effect is very species-specific (Picon *et al.*, 1996; Heath *et al.*, 1997). In the review by Scarascia-Mugnozza and de Angelis (1998), the reduction of stomatal conductance ranged from 20 to 90%. A decrease in stomatal conductance in response to CO_2 enrichment is commonly observed in both short-term studies and experiments using tree seedlings grown in pots (Mousseau and Saugier, 1992; Berryman *et al.*, 1994). Rooting restriction as a result of growing trees in pots has been suggested to result in reduced stomatal conductance either directly through drought stress, or indirectly through a feedback limitation of net photosynthesis that results from reduced root sink strength and increased internal [CO_2] (Sage, 1994). However, stomatal conductance has also been found to decrease with CO_2 enrichment when mature tree species are grown directly in the ground for more than one growing season (Kellomäki and Wang, 1997a).

The distinct structure and architecture of trees has led to the development of specific methods for measuring water transport in whole trees or branches (Granier *et al.*, 1996). These methods are based on the relationship between water flux rates through the xylem and the dissipation of heat applied to the stem. Direct measurements of water-use efficiency (WUE) – defined as the ratio of assimilated C to the amount of water transpired – are now possible through the study of the stable C isotope composition of tree rings. This is thanks to the work of Farquhar *et al.* (1982), who have shown that plant discrimination of ^{13}C ($\Delta^{13}C$) decreases as WUE increases, since both WUE and $\Delta^{13}C$ are related to the difference in internal to external CO_2 partial pressure ratio. Under elevated [CO_2], the increase in WUE is usually greater than the reduction of stomatal conductance, especially under drought conditions (Scarascia-Mugnozza and de Angelis, 1998). Because instantaneous WUE is invariably enhanced with elevated [CO_2], it is often thought that elevated [CO_2] will increase drought tolerance (Tyree and Alexander, 1993). Increased [CO_2] may thus alleviate moderate drought stress and might allow some extension of forests into drier areas. However, the hypothesized increase in drought tolerance may not always be the case in practice (Beerling *et al.*, 1996): resistance to drought may also depend on a number of factors affecting the evaporative demand or the ability of stem and root systems to transport water. In trees, as well as in some agricultural crops, drought may induce cavitation, and the risk of cavitation is greater in tree species with large-diameter xylem vessels (Cochard *et al.*, 1996).

11.3.3 Below-ground tree responses to elevated [CO_2]

Effects on nutrient uptake capacity

Whether or not the potential increase in plant growth under elevated [CO_2] is realized and will be sustained is a function of the nutrient and water availability and of the ability of plants to compete with soil biota for these resources. In elevated [CO_2], plants generally increase their nutrient uptake capacity by allocating more C to their roots, resulting in higher fine-root densities and mycorrhizal colonization (Rogers *et al.*, 1998). The subsequent increased below-ground C losses from root respiration and rhizodeposition result in significantly higher soil CO_2 efflux rates (Le Dantec *et al*., 1997; Janssens *et al*., 1998). This could explain the absence of significant growth responses to elevated [CO_2] encountered in some studies (Norby *et al*., 1992).

In a review by Rogers *et al.* (1994), 87% of 150 species showed increased fine-root biomass in response to elevated [CO_2]. Among all plant organs, fine roots generally show the greatest response to elevated [CO_2]. In addition to these increases in fine-root density, trees may enhance their nutrient uptake capacity through alterations in root morphology and architecture. Trees grown under elevated [CO_2] initiate more lateral root primordia, leading to increased root branching and a more thorough exploration of the soil (Rogers *et al*., 1999). In addition to changes in fine-root density, morphology and structure, alterations in root functioning are also frequently observed. These physiological changes include enhanced fine-root production and loss (Pregitzer *et al*., 1995; Rygiewicz *et al*., 1997), alterations in nutrient uptake kinetics (DeLucia *et al*., 1997) and, possibly, quantitative or qualitative changes in root exudation (Norby *et al*., 1995a).

As in most other plants, tree roots often have symbiotic associations with mycorrhizal soil fungi. In most natural ecosystems, mycorrhizae serve even more frequent as nutrient absorbing organs than do uninfected roots (Harley, 1984). Trees allocate 10–30% of their photoassimilates to their mycorrhizal symbionts (Markkola *et al*., 1996), which makes these one of the largest net primary productivity consumers in forest ecosystems. Under natural conditions, where water and nutrient deficiencies may occur, mycorrhizae are beneficial to their host. With their extraradical-hyphal network they explore the soil thoroughly and transfer the absorbed nutrients and water to their host (Guehl *et al*., 1990). It has been hypothesized that trees growing in elevated [CO_2] allocate more C to symbiotic N-fixers and mycorrhizae, which would result in enhanced nutrient acquisition by trees (Luxmoore, 1981). Since elevated [CO_2] has the largest effect on biomass production under high nutrient conditions, increased mycorrhizal exploration of the soil in elevated [CO_2] will be of great importance in sustaining enhanced tree growth rates, especially under low nutrient conditions (Hodge, 1996). Only half the studies on the effect of elevated [CO_2] on tree–mycorrhizal associations report increased percentages of infected root tips (Table 11.4). However, almost every study cited in Table 11.4 reported increased fine-root mass or length. Mycorrhizal colonization is, therefore, a poor parameter for representing mycorrhizal responses to CO_2 enrichment (O'Neill, 1994; Hodge, 1996). Elevated [CO_2] has also been found to accelerate colonization of newly formed root tips and to expand the extraradical mycelia (Table 11.4). Expanded extraradical mycelia could be of utmost importance, and is more significant to the tree than the degree of root infection for soil exploration and subsequent nutrient translocation to the host. With elevated [CO_2], nutrient acquisition will also be altered due to the observed changes in mycorrhizal species composition (Table 11.4), since mycorrhizal species differ in their nutrient uptake capacity. Therefore, ecological studies on root–mycorrhiza relations should concentrate not only on mycorrhizal biomass or colonization, but also on species composition and diversity.

Effects on decomposition and nutrient availability

In contrast to most non-woody crops, not all tree plantations are fertilized, and their long-term nutrient availability largely depends on litter decomposition and mineralization rates. As elevated [CO_2] stimulates biomass production, litterfall and rhizodeposition will also increase. This increased delivery of labile organic matter to the soil could influence soil microbial communities and subsequent decomposition rates, nutrient availability and C storage in soils (Curtis *et al.*, 1994). There are, however, contradictory hypotheses about the direction in which nutrient availability will be affected. Zak *et al.* (1993) suggested that the increased input of labile C would stimulate microbial biomass and mineralization, while Diaz *et al.* (1993) hypothesized that more nutrients would be immobilized in the larger microbial biomass. Reported effects of elevated [CO_2] on soil microbial biomass and community structure remain highly inconsistent (Sadowsky and Schortemeyer, 1997). More research is definitely needed to clarify how soil biota and food chains respond to elevated [CO_2], and how this will affect nutrient mineralization and immobilization.

In addition to the quantitative changes in litter production, qualitative alterations in plant tissues grown in elevated [CO_2] have also been observed. The main change in chemical composition of tissues produced under elevated [CO_2] is accumulation of total non-structural carbon (TNC) (Poorter *et al.*, 1997). Besides TNC accumulation, a reduction of the N concentration is commonly observed (Cotrufo *et al.*, 1998), even on a TNC-free basis (Poorter *et al.*, 1997). It has been hypothesized that the combination of reduced nutrient concentrations and accumulation of TNC would result in an increased synthesis of C-based secondary compounds such as lignins and tannins (Bryant *et al.*, 1983). To date, however, there is no conclusive evidence that elevated [CO_2] stimulates the production of these compounds (Penuelas and Estiarte, 1998).

Along with soil climate and structure of the soil microbial communities, substrate quality is a key factor regulating decomposition patterns and rates. Whereas initial mass loss rates are generally assumed to be stimulated by high nutrient concentrations, long-term decomposition rates are assumed to depend on lignin concentrations or lignin/N ratio (Berg, 1996). In the limited number of studies on the decomposition of tree tissues produced under elevated [CO_2] (Table 11.5), lower N concentrations did not induce the predicted decreased decomposition rates. This agrees with the review by Fog (1988), where no evidence was found for a stimulation of mass loss by higher N concentrations. Also, because there is no clear effect of elevated [CO_2] on lignin concentrations (Table 11.3), it is unlikely that long-term decomposition rates will be strongly affected. However, this should be tested. Almost every study in Table 11.5 was very short-term ($\leq$ 1 year); there is an imminent need for long-term *in situ* decomposition studies on tissues grown in elevated [CO_2].

11.3.4 Interactions with biotic components (phytophages)

Both the reduced N concentrations and the hypothesized increases in C-based secondary compounds could potentially alter herbivore feeding behaviour (Lindroth, 1995). Herbivores might need to consume more of the nutrient-poorer foliage, which could have detrimental effects on trees (Williams *et al.*, 1997). Results of several experiments indicate that the magnitude and direction of the herbivore response depend on how the concentrations of specific compounds determining insect fitness will be affected. For example, increased levels of some of these compounds could reduce the palatability of the leaves and subsequent herbivore consumption rates (Hunter and Lechowitcz, 1992). More studies on the characteristics of the biology of the herbivore population and on the interaction between trees and herbivores under elevated [CO_2] are needed before general conclusions can be drawn on the final effects on forest ecosystems.

11.3.5 Interactions with abiotic components

Tropospheric ozone (O_3) levels are increasing in parallel with [CO_2], especially around urban areas. Thus, in the future, trees will be simultaneously experiencing elevated [CO_2] and O_3 levels (see Chapter 3, this volume). Ozone enters the leaf through the stomata and diffuses within the apoplast, where it rapidly decomposes to reactive oxygen species. Responses to ozone differ among tree species or genotypes, as well as between trees of different size within species (Samuelson and Kelly, 1997). The significant impact of ozone on growth is primarily caused by changes in photosynthetic physiology, decreased stomatal conductance and accelerated senescence. Elevated atmospheric [CO_2] may reduce some of these negative effects through stomatal closure (Kellomäki and Wang, 1997a) but the opposite interaction has also been observed (Kull *et al.*, 1996).

Since both atmospheric CO_2 enrichment and increasing UV-B radiation represent components of global climatic change, the interactions of UV-B with elevated [CO_2] are particularly relevant. Only a few pertinent studies have been made, all of them under glasshouse conditions. Overall, it seems unlikely that UV-B radiation would substantially alter the CO_2 response of trees. In general, trees respond positively to elevated [CO_2] and negatively to enhanced UV-B radiation. Significant interactions between both components have been reported for some species, but not for others (e.g. Yakimchuk and Hoddinot, 1994).

11.4 Conclusions and Future Directions

Although most studies of the impact of elevated [CO_2] on trees have been made on young seedlings or small trees in various enclosures (see Table 11.3), our interest is with 'real' forests. In real forests, trees experience severe competition; therefore, competition should be included in future experimental designs. The question also arises whether or not the knowledge gained at the individual tree level can be applied to the stand level, and, if so, under which conditions? This issue is especially important with regard to leaf area. Most studies reviewed in this chapter have found that leaf area is enhanced under elevated CO_2 conditions, provided nutrient supply is not a strong limiting factor. It is questionable whether or not this stimulation of leaf area would still occur under closed canopy conditions. If there is a stimulating effect of elevated [CO_2] in a closed canopy, then the increase in leaf area index might, or might not, result in a significant increase in canopy photosynthesis. Therefore, careful attention should be paid to the proper up-scaling from the leaf to the tree, and to the ecosystem levels.

'Real' forests include not only natural or native forest ecosystems, but also planted and managed forest plantations. As these managed forest plantations (as well as agro-forestry systems) are crops of major global significance, the effects of global climatic changes on their yield or productivity are as important as, or even more important than, the effects on native or natural forests. Drought and nutrient limitation are not likely to be a major concern in plantations where intensive management practices such as irrigation and fertilization are already used to optimize yield in fast-growing stands destined for harvest on 10–12-year rotations. Likewise, seed germination and early plant competition, as key elements of forest regeneration, are less likely to be a concern of forest geneticists who now have at their disposal advanced pedigrees of loblolly pine, poplar, eucalyptus, etc. for the mechanized establishment of forest plantations. Challenging questions for the future remain. How will fast-growing trees in managed plantations respond to global change? What are the unique aspects of short-rotation forests that might cause them to respond differently and perhaps more unpredictably than natural forest ecosystems? The answers to these questions might have important implications for the kind of management options that need to be taken by forest managers either to overcome or to take advantage of changes in future climate.

This chapter has demonstrated that a great deal can be learned from detailed studies of the existing literature, since there are large amounts of available data. However, long-term field-based studies are necessary and should be continued into the future to understand the feedback processes and to predict future forest production from CO_2 enrichment. Several free-air CO_2 enrichment (FACE) studies that have been started over the last 4–5 years have the main advantages of focusing on entire ecosystem responses to elevated atmospheric [CO_2] and of being long-term experiments. Interesting and relevant results will come out of these FACE studies in the near future, but some patience is needed before long-term complex ecosystem responses (both above and below ground) can be properly evaluated.

In order to predict and evaluate the responses of forests to global climatic changes, a better understanding of the C and nutrient cycles in managed or natural forest ecosystems is badly needed. It should also be evident from this chapter that below-ground processes and components play a crucial role in the response of trees and forest ecosystems to global climatic changes, particularly their pronounced responses and their feedback control systems. Finally, whether or not tree or forest productivity will increase in the future also depends on societal attitudes toward forests. Good management and decreased deforestation will, without any doubt, result in improved worldwide forest productivity. Geo-political and geo-economic changes will play as big a role as global climatic changes.

References

Arnone, J.A. III and Körner, C. (1993) Influence of elevated CO_2 on canopy development and red:far red ratios in two-storied stands of *Ricinus communis*. *Oecologia* 94, 510–515.

Arp, W.J. (1991) Effects of source–sink relations on photosynthetic acclimation to elevated CO_2. *Plant, Cell and Environment* 14, 869–875.

Badeck, F.W., Dufrêne, E., Epron, D., Le Dantec, V., Liozon, R., Mousseau, M., Pontailler, J.Y. and Saugier, B. (1997) Sweet chestnut and beech saplings under elevated CO_2. In: Mohren, G.M.J., Kramer, K. and Sabaté, S. (eds) *Impacts of Global Change on Tree Physiology and Forest Ecosystems*. Kluwer Academic Publishers, Dordrecht, pp. 15–25.

Bazzaz, F.A. and Miao, S.L. (1993) Successional status, seed size, and responses of seedlings to CO_2, light, and nutrients. *Ecology* 74, 104–112.

Becker, M., Bert, G.D., Bouchon, J., Dupouey, J.L., Picard, J.F. and Ulrich, E. (1995) Long term changes in forest productivity in northern France: the dendrochronological approach. In: Landmann, G. and Bonneau, M. (eds) *Forest Decline and Atmospheric Deposition Effects in the French Mountains*. Springer-Verlag, Berlin, pp. 143–156.

Beerling, D.J. (1997) Carbon isotope discrimination and stomatal response of mature *Pinus sylvestris* L. exposed *in situ* for three years to elevated CO_2 and temperature. *Acta Oecologica* 18, 697–712.

Beerling, D.J., Heath, J., Woodward, F.I. and Mansfield, T.A. (1996) Drought–CO_2 interactions in trees: observations and mechanisms. *New Phytologist* 134, 235–242.

Berg, B. (1996) Climate change: does it influence the decomposition processes of plant litter and soil organic matter? *Atti della Società Italiana di Ecologia* 17, 19–22.

Berman, M.E. and Dejong, T.M. (1996) Water stress and crop load effects on fruit fresh and dry weights in peach (*Prunus persica*). *Tree Physiology* 16, 859–864.

Berntson, G.M. and Bazzaz, F.A. (1998) Regenerating temperate forest mesocosms in elevated CO_2: below-ground growth and nitrogen cycling. *Oecologia* 113, 115–125.

Berryman, C.A., Eamus, D. and Duff, G.A. (1994) Stomatal responses to a range of variables in two tropical tree species grown with CO_2 enrichment. *Journal of Experimental Botany* 45, 539–546.

Besford, R.T., Mousseau, M. and Matteucci, G. (1998) Biochemistry, physiology and biophysics of photosynthesis. In: Jarvis, P.G. (ed.) *European Forests and Global Change: the Likely Impacts of Rising CO_2 and Temperature*. Cambridge University Press, Cambridge, UK, pp. 29–78.

Boerner, R.E.J. and Rebbeck, J. (1995) Decomposition and nitrogen release from leaves of three hardwood species grown under elevated O_3 and/or CO_2. *Plant and Soil* 170, 149–157.

Borchert, R. (1991) Growth periodicity and dormancy. In: Raghavendra, A.S. (ed.) *Physiology of Trees.* John Wiley & Sons, New York, pp. 221–245.

Bryant, J.P., Chapin, F.S. III. and Klein, D.R. (1983) Carbon/nutrient balance of boreal plants in relation to vertebrate herbivory. *Oikos* 40, 357–368.

Buwalda, J.G. and Lenz, F. (1995) Water use by European pear trees growing in drainage lysimeters. *Journal of Horticultural Science* 70, 531–540.

Ceulemans, R. and Mousseau, M. (1994) Effects of elevated atmospheric CO_2 on woody plants. *New Phytologist* 127, 425–446.

Ceulemans, R. and Saugier, B. (1991) Photosynthesis. In: Raghavendra, A.S. (ed.) *Physiology of Trees.* John Wiley & Sons, New York, pp. 21–49.

Ceulemans, R., Jiang, X.N. and Shao, B.Y. (1995) Effects of elevated atmospheric CO_2 on growth, biomass production and nitrogen allocation in two *Populus* clones. *Journal of Biogeography* 22, 261–268.

Ceulemans, R., Shao, B.Y., Jiang, X.N. and Kalina, J. (1996) First- and second-year aboveground growth and productivity of two *Populus* hybrids grown at ambient and elevated CO_2. *Tree Physiology* 16, 61–68.

Ceulemans, R., Taylor, G., Bosac, C., Wilkins, D. and Besford, R.T. (1997) Photosynthetic acclimation to elevated CO_2 in poplar grown in glasshouse cabinets or in open top chambers depends on duration of exposure. *Journal of Experimental Botany* 48, 539–546.

Cochard, H., Bréda, N. and Granier, A. (1996) Whole tree hydraulic conductance and water loss regulation in *Quercus* during drought: evidence for stomatal control of embolism. *Annales des Sciences Forestières* 53, 197–206.

Conroy, J.P., Milham, P.J., Reed, M.L. and Barlow, E.W. (1990) Increases in phosphorous requirements of CO_2-enriched pine species. *Plant Physiology* 92, 977–982.

Cotrufo, M.F. and Ineson, P. (1995) Effects of enhanced atmospheric CO_2 and nutrient supply on the quality and subsequent decomposition of fine roots of *Betula pendula* Roth. and *Picea sitchensis* (Bong.) Carr. *Plant and Soil* 170, 267–277.

Cotrufo, M.F., Ineson, P. and Rowland, A.P. (1994) Decomposition of tree leaf litters grown under elevated CO_2: effect of litter quality. *Plant and Soil* 163, 121–130.

Cotrufo, M.F., Ineson, P. and Scott, A. (1998) Elevated CO_2 reduces the nitrogen concentration of plant tissues. *Global Change Biology* 4, 43–54.

Coûteaux, M.-M., Mousseau, M., Célérier, M.-L. and Bottner, P. (1991) Increased atmospheric CO_2 and litter quality: decomposition of sweet chestnut leaf litter with animal food webs of different complexities. *Oikos* 61, 54–64.

Curtis, P.S. and Wang, X. (1998) A meta-analysis of elevated CO_2 effects on woody plant mass, form, and physiology. *Oecologia* 113, 299–313.

Curtis, P.S., Zak, D.R., Pregitzer, K.S. and Teeri, J.A. (1994) Above- and belowground response of *Populus grandidentata* to elevated atmospheric CO_2 and soil N availability. *Plant and Soil* 165, 45–51.

DeLucia, E.H., Callaway, R.M., Thomas, E.M. and Schlesinger, W.H. (1997) Mechanisms of phosphorous acquisition for Ponderosa pine seedlings under high CO_2 and temperature. *Annals of Botany* 79, 111–120.

Diaz, S., Grime, J.P., Harris, J. and McPherson, E. (1993) Evidence of a feedback mechanism limiting plant response to elevated carbon dioxide. *Nature* 364, 616–617.

Dixon, M., Le Thiec, D. and Garrec, J.P. (1995) The growth and gas exchange response of soil planted Norway spruce (*Picea abies* Karst) and red oak (*Quercus rubra* L.) exposed to elevated CO_2 and naturally occurring drought. *New Phytologist* 129, 265–273.

Dufrêne, E., Pontailler, J.Y. and Saugier, B. (1993) A branch bag technique for simultaneous CO_2 enrichment and assimilation in beech (*Fagus sylvatica* L.). *Plant, Cell and Environment* 16, 1131–1138.

Eamus, D. and Jarvis, P.G. (1989) The direct effects of the increases in the global atmospheric concentration on natural and commercial temperate trees and forests. *Advances in Ecological Research* 19, 1–55.

Eckstein, K., Robinson, J.C. and Davies, S.J. (1995) Physiological responses of banana (*Musa* AAA; cavendish sub group) in the subtropics. III. Gas exchange, growth analysis and source sink interaction over a complete crop cycle. *Journal of Horticultural Science* 70, 169–180.

El Kohen, A., Venet, L. and Mousseau, M. (1993) Growth and photosynthesis of two deciduous forest species at elevated carbon dioxide. *Functional Ecology* 7, 480–486.

Epron, D., Liozon, R. and Mousseau, M. (1996) Effects of elevated CO_2 concentration on leaf characteristics and photosynthetic capacity of beech (*Fagus sylvatica*) during the growing season. *Tree Physiology* 16, 425–432.

FAO (1995) *Forest products 1991–1995*. FAO Forestry series No. 30, FAO statistic series No.137, FAO, Rome, 57 pp.

Farquhar, G.D., O'Leary, M.H. and Berry, J.A. (1982) On the relationship between carbon isotope discrimination and the intracellular carbon dioxide concentration in leaves. *Australian Journal of Plant Physiology* 9, 121–137.

Fog, K. (1988) The effect of added nitrogen on the rate of decomposition of organic matter. *Biological Reviews* 63, 433–462.

Fritts, H.C. (1976) *Tree Rings and Climate*. Academic Press, London, 567 pp.

Godbold, D.L. and Berntson, G.M. (1997) Elevated atmospheric CO_2 concentration changes ectomycorrhizal morphotype assemblages in *Betula papyrifera*. *Tree Physiology* 17, 347–350.

Godbold, D.L., Berntson, G.M. and Bazzaz, F.A. (1997) Growth and mycorrhizal colonization of three North American tree species under elevated atmospheric CO_2. *New Phytologist* 137, 433–440.

Goodfellow, J., Eamus, D. and Duff, G. (1997) Diurnal and seasonal changes in the impact of CO_2 enrichment on assimilation, stomatal conductance and growth in a long-term study of *Mangifera indica* in the wet–dry tropics of Australia. *Tree Physiology* 17, 291–299.

Granier, A., Biron, P., Bréda, N., Pontaillier, J.Y. and Saugier, B. (1996) Transpiration of trees and forest stands: short term and long term monitoring using sap flow methods. *Global Change Biology* 2, 265–274.

Guehl, J.M., Moussain, D., Falconnet, B. and Gruez, J. (1990) Growth, carbon dioxide assimilation capacity and water-use efficiency of *Pinus pinea* L. seedlings inoculated with different ectomycorrhizal fungi. *Annales des Sciences Forestières* 47, 91–100.

Guehl, J.M., Picon, C., Aussenac, G. and Gross P. (1994) Interactive effects of elevated CO_2 and soil drought on growth and transpiration efficiency and its determinants in two European forest tree species. *Tree Physiology* 14, 707–724.

Gunderson, C.A. and Wullschleger, S.D. (1994) Photosynthetic acclimation of forest trees to a doubling of atmospheric CO_2: a broader perspective. *Photosynthesis Research* 39, 369–388.

Harley, J.L. (1984) The mycorrhizal associations. In: Linskens, H.F. and Heslop-Harrison, J. (eds) *Encyclopedia of Plant Physiology*, Vol. 17. Springer-Verlag, Berlin, pp. 148–186.

Hättenschwiler, S. and Körner, C. (1998) Biomass allocation and canopy development in spruce model ecosystems under elevated CO_2 and increased N deposition. *Oecologia* 113, 104–114.

Heath, H., Kerstiens, G. and Tyree, M.T. (1997) Stem hydraulic conductance of European beech (*Fagus sylvatica* L.) and pedunculate oak (*Quercus robur* L.) grown in elevated CO_2. *Journal of Experimental Botany* 48, 1487–1489.

Hodge, A. (1996) Impact of elevated CO_2 on mycorrhizal associations and implications for plant growth. *Biology and Fertility of Soils* 23, 388–398.

Hogan, K.P., Whitehead, D., Kallarackal, J., Buwalda, J. and Rogers, G.N.D. (1996) Photosynthetic activity of leaves of *Pinus radiata* and *Nothofagus fusca* after 1 year of growth at elevated CO_2. *Australian Journal of Plant Physiology* 23, 623–630.

Hunter, A.F. and Lechowitcz, M.J. (1992) Foliage quality changes during canopy development of some northern hardwood trees. *Oecologia* 89, 316–323.

Idso, S.B. and Kimball B.A. (1997) Effects of long term atmospheric CO_2 enrichment on the growth and fruit production of sour orange trees. *Global Change Biology* 3, 89–97.

Idso, S.B., Kimball, B.A. and Allen, S.G. (1991) CO_2 enrichment of sour orange trees: 2.5 years into a long-term experiment. *Plant, Cell and Environment* 14, 351–353.

Ineichen, K., Wiemken, V. and Wiemken, A. (1995) Shoots, roots and ectomycorrhiza formation of pine seedlings at elevated atmospheric carbon dioxide. *Plant, Cell and Environment* 18, 703–707.

Jach, M.E. and Ceulemans, R. (1999) Effects of atmospheric CO_2 on phenology, growth and crown structure of Scots pine (*Pinus sylvestris*) seedlings after two years of exposure in the field. *Tree Physiology* 19, 289–300.

Janssens, I.A., Crookshanks, M., Taylor, G. and Ceulemans, R. (1998) Elevated atmospheric CO_2 increases fine root production, respiration, rhizosphere respiration and soil CO_2 efflux in young Scots pine seedlings. *Global Change Biology* 4, 871–878.

Jarvis, P.G. (1998) *European Forests and Global Change: the Likely Impacts of Rising CO_2 and Temperature*. Cambridge University Press, Cambridge, UK, 380 pp.

Johnson, D.W., Henderson, P.H., Ball, J.T. and Walker, R.F. (1995) Effects of CO_2 and N on growth and N dynamics in Ponderosa pine: results from the first two growing seasons. In: Koch, G.W. and Mooney, H.A. (eds) *Carbon Dioxide and Terrestrial Ecosystems*, Academic Press, San Diego, pp. 23–40.

Kellomäki, S. and Wang, K.Y. (1997a) Effects of elevated O_3 and CO_2 concentrations on photosynthesis and stomatal conductance in Scots pine. *Plant, Cell and Environment* 20, 995–1006.

Kellomäki, S. and Wang, K.Y. (1997b) Effects of long-term CO_2 and temperature elevation on crown nitrogen distribution and daily photosynthetic performance of Scots pine. *Forest Ecology and Management* 99, 309–326.

Kirschbaum, M.U.F. and Fischlin, A. (1996) Climate change impacts on forests. In: Watson, R.T., Zinyowera, M.C. and Ross, R.H. (eds) *Climate Change 1995 Impacts, Adaptations and Mitigation of Climate Change: Scientific–Technical Analysis.* Cambridge University Press, Cambridge, UK, pp. 99–129.

Klironomos, J.N., Rillig, M.C., Allen, M.F., Zak, D.R., Kubiske, M. and Pregitzer, K.S. (1997) Soil fungal–arthropod responses to *Populus tremuloides* grown under enriched atmospheric CO_2 under field conditions. *Global Change Biology* 3, 473–478.

Kozlowski, T.T. and Pallardy, S.G. (1996) *Physiology of Woody Plants.* Academic Press, New York, 400 pp.

Kull, O., Sober, A., Coleman, M.D., Dickson, R.E., Isebrands, J.G., Gagnon, Z. and Karnosky, D.F. (1996) Photosynthetic response of aspen clones to simultaneous exposures to ozone and CO_2. *Canadian Journal of Forest Research* 26, 639–648.

Le Dantec, V., Dufrêne, E. and Saugier, B. (1997) Soil respiration in mini-beech stands under elevated CO_2. In: Mohren, G.M.J., Kramer, K. and Sabaté, S. (eds) *Impacts of Global Change on Tree Physiology and Forest Ecosystems.* Kluwer Academic Publishers, Dordrecht, pp. 179–186.

Lee, H., Overdieck, D. and Jarvis, P.G. (1998) Biomass, growth and carbon allocation. In: Jarvis, P.G. (ed.) *European Forests and Global Change: the Likely Impacts of Rising CO_2 and Temperature.* Cambridge University Press, Cambridge, UK, pp. 126–191.

Lewis, J.D., Thomas, R.B. and Strain, B.R. (1994) Effect of elevated CO_2 on mycorrhizal colonization of loblolly pine (*Pinus taeda* L.) seedlings. *Plant and Soil* 165, 81–88.

Lindroth, R.L. (1995) CO_2 mediated changes in tree chemistry and tree–*Lepidoptera* interactions. In: Koch, G.W. and Mooney, H.A. (eds) *Carbon Dioxide and Terrestrial Ecosystems.* Academic Press, San Diego, pp. 105–118.

Luxmoore, R.J. (1981) CO_2 and phytomass. *Bioscience* 31, 626.

Markkola, A.-M., Ohtonen, A., Ahonen-Jonnarth, U. and Ohtonen, R. (1996) Scots pine responses to CO_2 enrichment. I. Ectomycorrhizal fungi and soil fauna. *Environmental Pollution* 94, 309–316.

Mousseau, M. and Saugier, B. (1992) The direct effect of increased CO_2 on gas exchange and growth of forest tree species. *Journal of Experimental Botany* 43, 1121–1130.

Mousseau, M., Dufrêne, E., El Kohen, A., Epron, D., Godard, D., Liozon, R., Pontailler, J.Y. and Saugier, B. (1995) Growth strategy and tree response to elevated CO_2 : a comparison of beech (*Fagus sylvatica)* and sweet chestnut (*Castanea sativa* Mill.). In: Koch, G.W. and Mooney, H.A. (eds) *Carbon Dioxide and Terrestrial Ecosystems.* Academic Press, San Diego, pp. 71–86.

Murray, M.B. and Ceulemans, R. (1998) Will tree foliage be larger and live longer? In: Jarvis, P.G. (ed.) *European Forests and Global Change: the Likely Impacts of Rising CO_2 and Temperature.* Cambridge University Press, Cambridge, UK, pp. 94–125.

Murthy, R., Dougherty, P.M., Zarnoch, S.J. and Allen, H.L. (1996) Effects of carbon dioxide, fertilization and irrigation on photosynthetic capacity of loblolly pine trees. *Tree Physiology* 16, 537–546.

Myneni, R.B., Keeling, C.D., Tucker, C.J., Asrar, G. and Nemani, R.R. (1997) Increased plant growth in the northern latitudes from 1981 to 1991. *Nature* 386, 698–702.

Norby, R.J. (1994) Issues and perspectives for investigating root responses to elevated atmospheric carbon dioxide. *Plant and Soil* 165, 9–20.

Norby, R.J., O'Neill, E.G., Gregory Hood, W. and Luxmoore, R.J. (1987) Carbon allocation, root exudation and mycorrhizal colonization of *Pinus echinata* seedlings grown under CO_2 enrichment. *Tree Physiology* 3, 203–210.

Norby, R.J., Gunderson, C.A., Wullschleger, S.D., O'Neill, E.G. and McCracken, M.K. (1992) Productivity and compensatory responses of yellow-poplar trees in elevated CO_2. *Nature* 357, 322–324.

Norby, R.J., O'Neill, E.G. and Wullschleger, S.D. (1995a) Belowground responses to atmospheric carbon dioxide in forests. In: McFee, W.W. and Kelly, J.M. (eds) *Carbon Forms and Functions in Forest Soils.* Soil Science Society of America, Madison, Wisconsin, pp. 397–418.

Norby, R.J., Wullschleger, S.D., Gunderson, C.A. and Nietch, C.T. (1995b) Increased growth efficiency of *Quercus alba* trees in a CO_2-enriched atmosphere. *New Phytologist* 131, 91–97.

Norby, R.J., Wullschleger, S.D. and Gunderson, C.A. (1995c) Tree responses to elevated CO_2 and the implications for forests. In: Koch, G.W. and Mooney, H.A. (eds) *Carbon Dioxide and Terrestrial Ecosystems.* Academic Press, San Diego, pp. 1–21.

Norby, R.J., Wullschleger, S.D., Gunderson, C.A., Johnson, D.W. and Ceulmans, R. (1999) Tree responses to rising CO_2 in field experiments: implications for the future forest. *Plant, Cell and Environment* 22, 683–714.

OIV (1998) *The state of viticulture in the world, statistical information in 1996*, Bulletin de l'Office International de la Vigne et du Vin, Paris, (Supplement) 803–804.

O'Neill, E.G. (1994) Responses of soil biota to elevated atmospheric carbon dioxide. *Plant and Soil* 165, 55–65.

O'Neill, E.G. and Norby, R.J. (1991) First-year decomposition dynamics of yellow-poplar leaves produced under CO_2 enrichment. *Bulletin of the Ecological Society of America* 72 (Suppl.), 208.

O'Neill, E.G. and Norby, R.J. (1996) Litter quality and decomposition rates of foliar litter produced under CO_2 enrichment. In: Koch, G.W. and Mooney, H.A. (eds) *Carbon Dioxide and Terrestrial Ecosystems.* Academic Press, San Diego, pp. 87–103.

O'Neill, E.G., Luxmoore, R.J. and Norby, R.J. (1987) Increases in mycorrhizal colonization and seedling growth in *Pinus echinata* and *Quercus alba* in an enriched CO_2 atmosphere. *Canadian Journal of Forest Research* 17, 878–883.

Oplustilova, M. and Dvorak, V. (1997) Growth processes of Norway spruce in elevated CO_2 concentration. In: Mohren, G.M.J., Kramer, K. and Sabaté, S. (eds) *Impacts of Global Change on Tree Physiology and Forest Ecosystems.* Kluwer Academic Publishers, Dordrecht, pp. 53–58.

Palmer, J.W., Giuliani, R. and Adams, H.M. (1997) Effect of crop load on fruiting and leaf photosynthesis of 'Braeburn/M26' apple trees. *Tree Physiology* 17, 741–746.

Penuelas, J. and Estiarte, M. (1998) Can elevated CO_2 affect secondary metabolism and ecosystem function? *Trends in Ecology and Evolution* 13, 20–24.

Pérez-Soba, M., Dueck, T.A., Puppi, G. and Kuiper, P.J.C. (1995) Interactions of elevated CO_2, NH_3 and O_3 on mycorrhizal infection, gas exchange and N metabolism in saplings of Scots pine. *Plant and Soil* 176, 107–116.

Picon, C., Guehl, J.M. and Fehri, A. (1996) Leaf gas exchange and carbon isotope composition responses to drought in a drought-avoiding (*Pinus pinaster*) and a drought-tolerant (*Quercus petraea*) species under present and elevated atmospheric CO_2 concentrations. *Plant, Cell and Environment* 19, 182–190.

Polley, H.W., Johnson, H.B. and Mayeux, H.S. (1997) Leaf physiology, production, water use, and nitrogen dynamics of the grassland invader *Acacia smallii* at elevated CO_2 concentrations. *Tree Physiology* 17, 89–96.

Poorter, H., Van Berkel, Y., Baxter, R., den Hertog, J., Dijkstra, P., Gifford, R.M., Griffin, K.L., Roumet, C., Roy, J. and Wong, S.C. (1997) The effect of elevated CO_2 on the chemical composition and construction costs of leaves of 27 C_3 species. *Plant, Cell and Environment* 20, 472–482.

Pregitzer, K.S., Zak, D.R., Curtis, P.S., Kubiske, M.E., Teeri, J.A. and Vogel, C.S. (1995) Atmospheric CO_2, soil nitrogen and fine root turnover. *New Phytologist* 129, 579–585.

Prior, S.A., Runion, G.B., Mitchell, R.J., Rogers, H.H. and Amthor, J.S. (1997) Effects of atmospheric CO_2 on longleaf pine: productivity and allocation as influenced by nitrogen and water. *Tree Physiology* 17, 397–405.

Raghavendra, A.S. (1991) *Physiology of Trees.* John Wiley & Sons, New York, 509 pp.

Rennenberg, H., Eschrich, W. and Ziegler, H. (1997) *Trees – Contributions to Modern Tree Physiology.* Backhuys Publishers, Leiden, 565 pp.

Rey, A. and Jarvis, P.G. (1997) Long term photosynthetic acclimation to elevated atmospheric CO_2 concentration in birch (*Betula pendula* Roth.). In: Mohren, G.M.J., Kramer, K. and Sabaté, S. (eds) *Impacts of Global Change on Tree Physiology and Forest Ecosystems.* Kluwer Academic Publishers, Dordrecht, pp. 87–92.

Rey, A., Barton, C.V.M. and Jarvis, P.G. (1997) Belowground responses to increased atmospheric CO_2 in birch (*Betula pendula* Roth.). In: Mohren, G.M.J., Kramer, K. and Sabaté, S. (eds) *Impacts of Global Change on Tree Physiology and Forest Ecosystems.* Kluwer Academic Publishers, Dordrecht, pp. 207–212.

Rogers, H.H., Runion, G.B. and Krupa, S.V. (1994) Plant responses to atmospheric CO_2 enrichment with emphasis on roots and the rhizosphere. *Environmental Pollution* 83, 155–189.

Rogers, H.H., Runion, G.B., Prior, S.A. and Torbert, H.A. (1999) Response of plants to elevated atmospheric CO_2: root growth, mineral nutrition and soil carbon. In: Luo, Y. and Mooney, H. (eds) *Carbon Dioxide and Environmental Stress.* Academic Press, New York, pp. 215–244.

Runion, G.B., Mitchell, R.J., Rogers, H.H., Prior, S.A. and Counts, T.K. (1997) Effects of nitrogen and water limitation and elevated atmospheric CO_2 on ectomycorrhiza of longleaf pine. *New Phytologist* 137, 681–689.

Rygiewicz, P.T., Johnson, M.G., Ganio, L.M., Tingey, D.T. and Storm, M.J. (1997) Lifetime and temporal occurrence of ectomycorrhizae on ponderosa pine (*Pinus ponderosa* Laws.) seedlings grown under varied atmospheric CO_2 and nitrogen levels. *Plant and Soil* 189, 275–287.

Sadowsky, M.J. and Schortemeyer, M. (1997) Soil microbial responses to increased concentrations of atmospheric CO_2. *Global Change Biology* 3, 217–224.

Sage, R.F. (1994) Acclimation of photosynthesis to increasing atmospheric CO_2: the gas exchange perspective. *Photosynthesis Research* 39, 351–368.

Samuelson, L.J. and Kelly, J.M. (1997) Ozone uptake in *Prunus serotina, Acer rubrum* and *Quercus rubra* forest trees of different sizes. *New Phytologist* 136, 255–264

Scarascia-Mugnozza, G. and de Angelis, P. (1998) Is water used more efficiently? In: Jarvis, P.G. (ed.) *European Forests and Global Change: the Likely Impacts of Rising CO_2 and Temperature.* Cambridge University Press, Cambridge, UK, 192–214.

Spiecker, H., Mielikäinen, K., Köhl, M. and Skovsgaard, J.P. (1996) *Growth Trends in European Forests: Studies from 12 Countries.* Springer-Verlag, Berlin, 372 pp.

Teskey, R.O. (1995) A field study of the effects of elevated CO_2 on carbon assimilation, stomatal conductance and leaf and branch growth of *Pinus taeda* trees. *Plant, Cell and Environment* 18, 565–573.

Tigerstedt, P.M.A., Puttonen, P. and Koski, V. (1985) *Crop Physiology of Forest Trees.* Proceedings of an International Conference on Managing Forest Trees as Cultivated Plants, held in Finland, 23–28 July, 1984, University of Helsinki, Finland.

Tingey, D.T., Johnson, M.G., Phillips, D.L., Johnson, D.W. and Ball, J.T. (1996) Effects of elevated CO_2 and nitrogen on the synchrony of shoot and root growth in ponderosa pine. *Tree Physiology* 16, 905–914.

Tissue, D.T., Thomas, R.B. and Strain, B.R. (1996) Growth and photosynthesis of loblolly pine (*Pinus taeda*) after exposure to elevated CO_2 for 19 months in the field. *Tree Physiology* 16, 49–59.

Tognetti, R., Giovannelli, A., Longobucco, A., Miglietta, F. and Raschi, A. (1996) Water relations of oak species growing in the natural CO_2 spring of Rapolano (central Italy). *Annales des Sciences Forestières* 53, 475–485.

Townend, J. (1995) Effects of elevated CO_2, water relations and nutrients on *Picea sitchensis* (Bong.) Carr. seedlings. *New Phytologist* 130, 193–206.

Tyree, M.T. and Alexander, J.D. (1993) Plant water relations and the effects of elevated CO_2: a review and suggestions for future research. *Vegetatio* 104/105, 47–62.

Wang, K., Kellomäki, S. and Laitinen, K. (1995) Effects of needle age, long-term temperature and CO_2 treatments on the photosynthesis of Scots pine. *Tree Physiology* 15, 211–218.

Wibbe, M.L., Blanke, M.M. and Lenz, F. (1993) Effect of fruiting on carbon budgets of apple tree canopies. *Trees* 8, 56–60.

Williams, R.S., Lincoln D.E. and Thomas, R.B. (1997) Effects of elevated CO_2-grown loblolly pine needles on the growth, consumption, development and pupal weight of red-headed pine sawfly larvae reared in open-top chambers. *Global Change Biology* 3, 501–511.

Winjum, J.K. and Schroeder, P.E. (1997) Forest plantations of the world: their extent, ecological attributes, and carbon storage. *Agriculture and Forest Meteorology* 84, 153–167.

Wullschleger, S.D., Norby, R.J. and Gunderson, C.A. (1997) Forest trees and their response to atmospheric CO_2 enrichment: a compilation of results. In: Allen, L.H. Jr, Kirkham, M.B., Olszyck, D.M. and Williams, C.E. (eds) *Advances in Carbon Dioxide Effects Research*. ASA Special Publication No.61, American Society of Agronomy, Madison, Wisconsin, pp. 79–100.

Yakimchuk, R. and Hoddinot, J. (1994) The influence of ultraviolet-B light and carbon dioxide enrichment on the growth and physiology of seedlings of three conifer species. *Canadian Journal of Forest Research* 24, 1–8.

Zak, D.R., Pregitzer, K.S., Curtis, P.S., Teeri, J.A., Fogel, R. and Randlett, D.L. (1993) Elevated atmospheric CO_2 and feedback between carbon and nitrogen cycles. *Plant and Soil* 151, 105–117.

12 Crop Ecosystem Responses to Climatic Change: Productive Grasslands

JOSEF NÖSBERGER, HERBERT BLUM AND JÜRG FUHRER

Institute of Plant Sciences, Swiss Federal Institute of Technology (ETH), CH-8092 Zurich, Switzerland

12.1 Introduction

The term 'grassland' refers to a plant community in which grasses are dominant, and shrubs and trees are rare. Grassland is a major agricultural resource and is of considerable importance in many temperate and tropical regions. The major grasslands are not areas of natural climax vegetation but are the result of past human activities such as forest clearance and the cultivation of arable crops. Grassland has also developed as a result of herding cattle or sheep on pasture, of cutting vegetation, or, less frequently, of sowing grass seed. Sown grassland is generally less complex botanically, and the species sown are chosen for their productivity and nutritional value, as well as for their suitability for the local environment. In many regions, temporary and permanent grassland in lowlands has considerably exceeded the total arable area. In regions unsuitable for competitive production of arable crops, grassland accounts for the major portion of the effective agricultural area. Grassland farming involves a wide range of species, soils and climatic conditions, as well as their interaction with management practices; it gives rise to many types of vegetation.

In terms of species, diversity and ecosystem type, grassland takes on special importance in mid-latitude temperate zones. This chapter will focus on the grass species perennial ryegrass (*Lolium perenne*), and the important legume, white clover (*Trifolium repens*). In humid temperate climates, these two species are the backbone of productive, high-quality grassland. Emphasis has been placed on the importance of legumes in counteracting climatic change. In addition to their primary significant role of increasing the net primary productivity of forage, legumes also improve its nutritional quality.

Grassland that receives and transforms the sun's radiant energy into biomass shows some distinct ecophysiological differences compared with arable crops. It covers the ground almost completely, and light energy is fully intercepted throughout the year. Therefore, it has a greater potential to fix CO_2 than do arable crops. In addition, because the crop is perennial, grassland soil is much less disturbed, and litter of leaves, stolons, stubble and roots and animal manure accumulates, giving rise to levels of soil organic matter that are as high as in former forest soil. Grassland accounts for about 20% of the terrestrial CO_2 fluxes of the global carbon cycle and contributes a similar share in global soil organic C. It plays a surprisingly large part in terms of C storage. This is due entirely to the large amount of organic matter that is maintained in grassland soils, because of a combination of the considerable partitioning of dry matter to below-ground plant parts and a relatively slow rate of decay of soil organic matter.

Although there is no doubt that concentrations of atmospheric CO_2 ([CO_2]) are rising rapidly and will continue to do so (Houghton *et al.*, 1996), there is considerable uncertainty about the consequences for managed grassland ecosystems. Difficulties in predicting the impact of elevated [CO_2] were first encountered in the laboratory and the growth chamber; these increase potentially as we consider processes operating on a large scale under different types of field management and over extended periods of time. Laboratory studies showed that grassland plants differed in their responses to elevated [CO_2] (Newton, 1991). The pattern of response found under controlled conditions will be subject to the modifying effects of other environmental factors in the field, some of which (e.g. temperature, rainfall and ozone) play a significant role in global environmental change. Insight into the physiological and ecological processes must be gained in order to predict the response of grassland to elevated [CO_2].

12.2 Photosynthesis and Respiration

12.2.1 Carbon dioxide

It has been known for a long time that the photosynthesis of single leaves of C_3 species generally increases under elevated [CO_2]. This increase depends not only on [CO_2], but also on light, temperature, nutrients, water, leaf age, etc. The daily average photosynthesis of young leaves increased by 30–80% at 60 Pa (600 ppm) compared with ambient (35 Pa) CO_2 partial pressure in perennial ryegrass swards (Rogers *et al.*, 1998). Drake *et al.* (1997) determined an average increase of 58% in photosynthesis from 60 experiments with elevated [CO_2].

Acclimation of photosynthesis occurs during the long-term exposure of plants to elevated [CO_2], when the initial strong increase of photosynthesis declines. Leaves grown under elevated [CO_2] show reduced photosynthesis compared with leaves grown at ambient [CO_2], when both are measured at ambient [CO_2]. Processes that affect the acclimation of photosynthesis have been related to situations in which the sink limits growth under elevated [CO_2] (Drake *et al.*, 1997). During acclimation, concentrations of carbohydrates are increased and concentrations of soluble proteins and Rubisco are decreased. The lower concentrations of protein, and especially of Rubisco, are due to reduced gene expression mediated by the increase in carbohydrate concentrations. Thus, acclimation has been interpreted as a mechanism by which plants reduce leaf N content in order to increase sink growth. Rogers *et al.* (1998) studied acclimation of leaf photosynthesis before and after a cut in perennial ryegrass swards fertilized with low or high amounts of N. Before the cut, acclimation of photosynthesis occurred at elevated [CO_2] and low N supply. After the cut, when the small leaf area and the high demand of growing leaves for carbohydrates drastically reduced the source/sink ratio, acclimation disappeared. No significant acclimation was found in ryegrass plants fertilized with high amounts of N. Acclimation of photosynthesis was also found for the canopy (Casella and Soussana, 1997) but it was relatively small (–8 to –13%) and independent of N.

Enhanced leaf photosynthesis under elevated [CO_2] should be reflected by an increase in canopy photosynthesis, provided that neither the CO_2 assimilation of the leaves in the lower canopy (where photosynthesis is not saturated) nor the leaf area index (LAI) are reduced, and they can compensate for the increased CO_2 uptake of the upper canopy. In perennial ryegrass swards grown in large containers in highly ventilated tunnels with low or high N supply, there was an average increase of 29% and 36%, respectively, in net canopy photosynthesis at double ambient [CO_2] (Casella and Sousanna, 1997). Schapendonk *et al.* (1997) reported an annual increase in net canopy assimilation of 29 and 43% in 2 subsequent years in perennial ryegrass swards grown at 70 Pa CO_2 and with ample supplies of water and nutrients. The annual harvested biomass increased by 20 and 25%. Thus, canopy photosynthesis increased much more than biomass in the second year compared with the first year. This indicates that canopy C assimilation and harvestable biomass are not closely correlated and that partitioning of carbohydrates to shoots and roots changed.

Drake *et al.* (1997) found that dark respiration of leaves is reduced by about 20% at double ambient [CO_2]. This reduction is due to an inhibition of the activity of two key enzymes of the mitochondrial electron transport chain. Due to the fundamental nature of this mechanism, it is assumed to occur in all respiring tissues (Drake *et al.*, 1997). Schapendonk *et al.* (1997) also reported reduced specific dark respiration in perennial ryegrass swards. They found no CO_2-related change in canopy dark respiration even though harvested shoot biomass (+20%) and LAI increased considerably in 1 year. In the second year, however, canopy dark respiration increased by 28% and harvested shoot biomass by 25%. Below-ground respiration (soil and roots) decreased by 10%, though root biomass increased by 29 and 86% under double ambient [CO_2] by the end of the first and second growing seasons, respectively. Casella and Soussana (1997) measured an increase of canopy dark respiration proportional to shoot biomass. They found an increased below-ground respiration of 33 and 36% at low and high levels of N fertilizer, respectively.

12.2.2 Solar radiation

Light-limited photosynthesis may account for half of the C gain of the canopy (Drake *et al.*, 1997). Under light-limiting conditions, leaf photosynthesis is mainly determined by photosynthetic light-use efficiency, i.e. by the slope of the photosynthesis–light curve and by the light compensation point. Most studies describe the photosynthesis of young leaves adapted to full light. The few measurements of photosynthesis of shaded leaves, however, clearly indicate that an increase in CO_2 enhances the light-use efficiency and decreases the light compensation point. The increased light-use efficiency can be expected due to reduced photorespiration (Long *et al.*, 1993) and the light compensation point is decreased due to reduced dark respiration. In perennial ryegrass swards, canopy photosynthesis and light-use efficiency (of incident global irradiance) increased under double ambient [CO_2] at all levels of irradiance (Schapendonk *et al.*, 1997). Since temperature may increase with irradiance, it is difficult to distinguish between effects of these two environmental factors.

12.2.3 Temperature

At low temperature, photosynthesis is saturated at low irradiance; saturation irradiance and light-saturated photosynthesis (A_{max}) increased with temperature (Nijs and Impens, 1996). Rising temperature increases the stimulation of leaf photosynthesis resulting from elevated [CO_2] (Nijs and Impens, 1996) mainly because photorespiration strongly increases with temperature at ambient [CO_2] but is strongly reduced at elevated [CO_2]. At 10° C, doubling the [CO_2] increased the A_{max} by only 4%, but it increased by 35% at 30°C (Drake *et al.*, 1997). As a result, the optimum temperature of A_{max} increased with CO_2 as follows: 2°C at 45 Pa CO_2 and 6°C at 65 Pa CO_2. The maximum temperature at which photosynthesis is still positive also increased. These changes are particularly important with respect to the predicted global warming.

In the canopy, increased leaf photosynthesis may be partially offset by temperature-induced changes in LAI and plant water status. Therefore, it is difficult to predict the response of canopy photosynthesis to increased temperature and [CO_2]. Canopy photosynthesis at elevated [CO_2] increased more in summer (49%) than in spring or autumn (26%) in perennial ryegrass swards (Casella and Soussana, 1997). This increase was due in part (only in high-N swards) to an increased LAI. Increased assimilation in summer is assumed to be due to a combination of the effects of higher air temperature and higher water-use efficiency (WUE). When grown at elevated [CO_2], a temperature increase of 3°C above ambient temperature increased canopy photosynthesis, particularly in early spring and autumn, but it caused a decrease in summer. This was more pronounced in highly fertilized swards. The increase in spring and autumn was due to increased growth and increased leaf photosynthesis at higher temperature. The decrease in summer was caused by a slightly reduced LAI as well as by lower soil moisture and increased vapour pressure deficit

(VPD), causing stomatal closure. Thus, elevated [CO_2] increased canopy photosynthesis to a greater extent at higher temperatures if growth temperature was below optimal. However, temperature increase above the optimal growth temperature decreased LAI, and thus counteracted a temperature-related increase in leaf photosynthesis.

The CO_2-related changes in canopy dark respiration and soil respiration were highly correlated with changes in the canopy gross assimilation (Casella and Soussana, 1997). Temperature effects on canopy and soil respiration paralleled those on gross assimilation. Soil respiration increased markedly at +3°C compared with ambient temperature, but only if soil moisture did not decrease. However, soil respiration decreased if soil moisture was reduced to a greater extent at higher temperature, i.e. in summer and in high-N swards.

12.2.4 Water deficits

Stomatal closure under water-limiting conditions reduces conductivity of water and CO_2, thus reducing transpiration and leaf photosynthesis. This may occur for a limited period during the light period on warm days with high irradiance. The reduced stomatal conductance at elevated [CO_2] alleviated drought stress and thus maintained a high rate of photosynthesis, whereas photosynthesis was lower at ambient [CO_2] (Owensby *et al.*, 1997). At elevated temperature, transpiration increased as a result of increased VPD, leading to a greater water deficit. Consequently, leaf photosynthesis is expected to decrease with increasing temperature when water deficit limits growth. It is difficult to predict how a simultaneous increase in temperature and CO_2 will affect canopy photosynthesis under water deficit.

At elevated [CO_2], canopy photosynthesis increased to a greater extent in summer under moderate drought stress and higher temperature (Casella and Soussana, 1997). The increase was also apparent in low-N swards, where LAI did not increase. At elevated [CO_2], a temperature increase of 3°C reduced canopy photosynthesis during drought stress. Since the effects of higher temperature and drought stress were combined, the effect of the water deficit only on canopy photosynthesis is not known. Net C exchange increased at elevated [CO_2] as a result of improved plant water status in tallgrass prairie consisting mainly of C_4 species (Owensby *et al.*, 1997).

12.2.5 Nutrient deficits

Nutrient deficiency reduces sink growth and leaf photosynthesis (Drake *et al.*, 1997). Insufficient N is usually the primary mineral nutrient deficit that limits grassland production, but occasionally P or S deficiency can be limiting. Symbiotic N fixation is regarded as one of the causes of increased net primary productivity that augments sequestering of soil C. Acclimation is related to a lower Rubisco concentration and, consequently, to a lower N concentration in the leaves. Woodrow (1994) calculated the amount of Rubisco required to maintain a constant rate of photosynthesis when [CO_2] or temperature increases. With increasing temperature, the required amount of Rubisco dropped drastically as [CO_2] increased, whereas at 5°C only a small decrease was found with increasing [CO_2]. Thus, under elevated [CO_2], less N is required to maintain the same or an even higher rate of photosynthesis than under ambient [CO_2]. Consequently, N-use efficiency increases under elevated [CO_2]. This effect is probably more important in warmer climates. Drake *et al.* (1997) summarized a number of reports showing that photosynthesis increased by about 50% at elevated [CO_2] when N availability was high. At low N availability, photosynthesis increased only 25% at elevated [CO_2]. Midday leaf photosynthesis was about the same in high- and low-N swards of perennial ryegrass in the field and increased by 35% under elevated [CO_2] for both N levels (Rogers *et al.*, 1998). In this experiment, acclimation reduced Rubisco by about 25% in the low-N swards, but leaf N and protein concentrations were not reduced at elevated [CO_2] if they were related to unit leaf area.

Canopy photosynthesis at elevated [CO_2] increased 29% in low-N and 36% in high-N swards (Casella and Soussana, 1997). The stronger effect at high N was partly due to an increased LAI, but the radiation use efficiency of canopy photosynthesis increased by 22% and 11% in the low- and high-N treatments, respectively. This result implies that the N-use efficiency of the sward also increased to an even greater extent in low N than in high N.

12.3 Water Use

12.3.1 Carbon dioxide

Plants growing at elevated [CO_2] partially close their stomata, thus reducing stomatal conductivity (g_s) and leaf transpiration (Field *et al.*, 1995). The lower g_s reduced sap flow and increased xylem pressure potential, as reflected in an improved plant water status (Owensby *et al.*, 1997). The responses of different species were very different, but a recent survey suggested an average decrease in g_s of 20% (Drake *et al.*, 1997). The decrease in g_s does not limit photosynthesis more in elevated [CO_2] than in ambient [CO_2]. Since photosynthesis at elevated [CO_2] increases and transpiration decreases, their ratio, or WUE, also increases. This increase is also observed in C_4 species and is sometimes greater in these than in C_3 species. Thus, under water-limiting conditions, the productivity of C_4 species also increased at elevated CO_2 (Owensby *et al.*, 1997).

Canopy transpiration is determined by several factors. If LAI is small, an increase in LAI will also increase the intercepted radiation, and thus transpiration. If the canopy is closed, transpiration increases only slightly with increasing LAI. In dense canopies, such as those in productive grassland, the aerodynamic resistance may be greater than the stomatal resistance. In this situation, the total resistance for water transport from within the leaf to the atmosphere does not increase as much as the increase in stomatal resistance (Shaer and van Bavel, 1987). Consequently, the transpiration of dense canopies may be only slightly reduced under elevated [CO_2] (Eamus, 1991). This was confirmed by Schapendonk *et al.* (1997), who reported a slight increase in LAI but a constant rate of transpiration in highly productive perennial ryegrass swards. Transpiration, averaged over the growing season, decreased by only 2% at elevated [CO_2], but WUE increased by 17–30% (Casella *et al.*, 1996). However, at the beginning of a regrowth, when the canopy is open, gas exchange between the canopy and the atmosphere is more vigorous, and the increased stomatal resistance may reduce transpiration. In contrast, Ham *et al.* (1995) found a CO_2-related decrease of 22% in transpiration in a tallgrass prairie at an LAI of 4–5 when soil moisture did not limit growth. This system consisted of many different species, having different growth forms. Canopy resistance increased considerably at elevated [CO_2] and was greater than aerodynamic resistance, indicating that the coupling between the canopy and the atmosphere was larger than in dense grass canopies in temperate climates. The coupling would probably be even greater in the field, where wind speed is higher than in the large open-top chambers that were used. The relative importance of decreased stomatal conductivity to reduced transpiration under elevated [CO_2] in productive grassland depends on factors (such as species, growing conditions, management and water supply) that determine the aerodynamic resistance of the canopy.

If water supply does not limit growth, then the reduced water use at elevated [CO_2] increases drainage (Casella *et al.*, 1996). Soil moisture in high-N plots increased at elevated [CO_2] in the first year but did not change in the second year. In the low-N plots, soil moisture was higher under double ambient [CO_2].

12.3.2 Solar radiation

Increased solar irradiance leads to a higher radiation load of the canopy. Plants avoid the associated increase in leaf temperature by increasing transpiration (Nijs *et al.*, 1997). Water-use efficiency is much lower in summer than in the cooler seasons, due to the higher VPD of the air and frequent light-saturated photosynthesis (Schapendonk *et al.*, 1997). Elevated [CO_2] increased WUE more in summer than in spring and autumn because CO_2-related stomatal closure reduced transpiration more than it reduced photosynthesis. The effect of elevated [CO_2] on canopy WUE was strongest at high irradiance.

Stomata opening responded to changing irradiance such as sun/shade events (Owensby *et al.*, 1997). The response time of stomata opening or closing was reduced by 30% in a C_4 species at elevated [CO_2]. This effect also contributed to reduced transpiration and improved water status of this species.

12.3.3 Temperature

Temperature directly affects plant water use by its effect on VPD and it indirectly affects transpiration by altering the size of the transpiring leaf area. VPD generally increases with diurnal and seasonal cycles of temperature. Thus, transpiration and water use closely follow these temperature cycles. If the water demand of the plants cannot be met, stomata close to reduce water loss. At elevated [CO_2], leaf transpiration decreases, thus alleviating temporal water-deficit stress. A simultaneous increase in [CO_2] and temperature, which some climate models predict will occur in the future, might partly offset the beneficial effect of elevated [CO_2] on plant growth under drought stress. A temperature increase by 2–3°C will have beneficial effects on plant growth in cooler parts of the world. However, if increasing temperatures increase potential water use to an extent that cannot be met by the supply, then an increasing water deficit would certainly reduce plant growth.

When temperature was increased by 2.5°C, g_s decreased at ambient (–15%) and elevated (–9%) [CO_2], but only at the top of the canopy (Nijs *et al.*, 1997). The increased VPD at the elevated temperature, however, more than compensated for the reduced g_s, and leaf transpiration increased by 28% and 48% at ambient and elevated [CO_2], respectively. Leaf WUE decreased as temperature increased, but the temperatures of the air and the canopy were supra-optimal during these measurements.

A temperature increase of 3°C at elevated [CO_2] increased canopy transpiration by 8% (Casella *et al.*, 1996). The soil water content decreased considerably at the elevated air temperature. This was especially pronounced in summer, when plant growth was reduced at supra-optimal temperature. However, in spring and autumn (i.e. suboptimal temperature), yield and water use increased, additionally reducing soil moisture and leading to the earlier onset of water deficit in summer. These effects were stronger in high-N than in low-N plots, due to the larger canopy's greater demand for water. The beneficial effect of a 3°C temperature increase on photosynthesis was more than offset in summer by the increased drought stress. Increased temperature did not significantly affect the annual mean of canopy WUE. However, WUE showed strong seasonal variation, with a maximum in spring and autumn and a minimum in summer. When both [CO_2] and temperature increased simultaneously, the annual mean WUE increased compared with both factors at ambient [CO_2]. In summer, the simultaneous increase in [CO_2] and temperature did not change WUE compared with both factors at ambient levels.

12.3.4 Water deficits

Plant growth and g_s decrease under drought stress. Reduced growth may result in a more open canopy, and thus a lower aerodynamic resistance if the water deficit lasts for a longer period. Under such conditions, the CO_2-induced reduction of g_s will reduce transpiration and improve WUE and plant growth, compared with ambient [CO_2]. This was shown by Casella *et al.* (1996), who observed that WUE was not affected by water deficit. Under water-limiting conditions, the increased plant growth under elevated [CO_2] may compensate for the reduced water loss due to lower g_s, resulting in the same (Casella *et al.*, 1996) or a slightly reduced (Hebeisen *et al.*, unpublished observations) level of soil moisture.

12.3.5 Nutrient deficits

Nitrogen deficiency reduces both plant growth and water use, and decreases the aerodynamic resistance, similar to the effect of a water deficit. Elevated [CO_2], which reduces g_s, could, therefore, reduce drought stress and enhance plant growth. In contrast to this expectation, WUE was also not affected by N deficit (Casella *et al.*, 1996).

12.4 Growth and Yield

The enhanced C assimilation at elevated [CO_2] increases plant growth (Kimball, 1983; Newton, 1991). However, factors other than [CO_2], particularly those limiting growth, strongly modify the [CO_2] response. Grassland plants usually have a limited availability of N and other resources, and the effects of multiple limitations (e.g. high temperature and drought stress) are often confounded.

12.4.1 Carbon dioxide

Although the variability between years was large, the yield of high-N perennial ryegrass swards responded to elevated [CO_2] in almost all experiments (Table 12.1). In some cases, high defoliation frequency reduced the yield response to elevated [CO_2].

At elevated [CO_2], leaf N concentration ([N]) decreased by an average of 15 and 20% in infrequently and frequently cut perennial ryegrass swards, respectively (Zanetti *et al.*, 1997). The N yield of the harvested biomass was reduced by an average of 18% in the frequently cut swards and remained unchanged in the infrequently cut swards. Leaf [N] decreased due to a higher concentration of water-soluble carbohydrates (Fischer *et al.*, 1997), whereas the [N] per unit leaf area did not change at elevated [CO_2] (Rogers *et al.*, 1998).

The annual yield of white clover, a legume receiving unlimited N from symbiosis, increased from 11 to 20% in 3 years (Table 12.1) (Hebeisen *et al.*, 1997). In the first 2 years, the CO_2 stimulation of the yield of white clover was significantly higher than that of perennial ryegrass. Zanetti *et al.* (1996) found that increased productivity corresponded with increased symbiotic N fixation and leaf [N] decreased only slightly at elevated [CO_2], due to a slight increase in carbohydrate concentration.

Table 12.1. Effect of elevated [CO_2] on annual yield and root biomass of swards of *Lolium perenne* (L.p.), *Trifolium repens* (T.r.) and the mixture of both (L.p./T.r.).

Species	N supply (kg ha^{-1} year^{-1})	Cuts (year^{-1})	Years	Water stress	CO_2 effect on yield (%)	CO_2 effect on rootmass (%)	Reference
L.p.	800	10	2	No	20, 25	26, 86	Schapendonk *et al., 1997*
	1000	5	1	No	29	–	Soussana *et al.,* 1996
	530	5	2	No	27, 10.5	–	Casella *et al.,* 1996
	530	5	2	Summer	14, 14	52	Casella *et al.,* 1996
	160	5	2	Summer	19, 18	45	Casella *et al.,* 1996
	560	8	3	Some	8, 6, 2		Hebeisen *et al.,* 1997
	560	4	3	Some	8, 12, 27	53, 108, 45	Hebeisen *et al.,* 1997
	140	8	3	Some	0, –20, –11	(average of	Hebeisen *et al.,* 1997
	140	4	3	Some	10, –5, 12	all	Hebeisen *et al.,* 1997
						treatments)	Hebeisen *et al.,* 1997
T.r.	560/140	8/4	3	Some	20, 20, 11	27, 40, 26	Hebeisen *et al.,* 1997
L.p./T.r.	560/140	8/4	3	Some	18	–	Hebeisen *et al.,* 1997

The annual yield of white clover/ryegrass mixtures increased by an average of 18% under elevated [CO_2] (Hebeisen *et al.*, 1997), regardless of the year, N fertilizer supply and cutting regime (Table 12.1). The enhanced symbiotic N fixation added an average of 7.4 g N m^{-2} year^{-1} to the system. A significant amount of this additional N was transferred to the ryegrass (Zanetti *et al.*, 1997). This N input into the ecosystem may also affect processes related to N cycling and C sequestration (Soussana and Hartwig, 1996). Similar to the CO_2 response in the monocultures, the positive response of the mixtures was due mainly to an increase in white clover yield in the first 2 years and to an increase in ryegrass yield in the third year (Hebeisen *et al.*, 1997). Averaged over all treatments, the proportion of white clover in the annual yield increased from 30% to 42% at elevated [CO_2]. Despite the increase at elevated [CO_2], the proportion of white clover decreased in both CO_2 treatments after 3 years, indicating that other factors may have influenced the ryegrass/white clover ratio. This phenomenon clearly indicates that results of short-term experiments on the botanical composition of a sward may not be representative.

The CO_2 response of species other than perennial ryegrass and white clover in productive grassland have received less attention in previous studies. The yield responses of individual plants of 12 native species from permanent grassland growing in perennial ryegrass swards to elevated [CO_2] were studied by Lüscher *et al.* (1998). The species represented three functional groups: seven grasses, three non-legume dicots and two legumes. There were significant differences in the CO_2 response of the functional groups. Grasses showed the smallest response, and legumes the largest. No interspecific differences were found among species of the same functional group. Thus, the species proportion of these functional groups will probably change with a continuing increase in [CO_2]. The CO_2 response of the three functional groups varied over 3 years. The response of the legumes decreased and that of the non-legume dicots increased. Similar trends in the CO_2 response over 3 years were found in the ryegrass swards and in the ryegrass/white clover mixtures. This suggests that grassland will adapt within years to elevated [CO_2]. Nine to 14 genotypes of each species were tested, but there was no appropriate intraspecific variability on which evolutionary selection related to responsiveness to elevated [CO_2] could act.

The LAI of perennial ryegrass swards, grown in enclosures with an ample supply of N, increased slightly at elevated [CO_2] (8% and 16%) in 2 subsequent years (Schapendonk *et al.*, 1997). The increased leaf area was due to a higher rate of leaf elongation in the first week of regrowth and to increased tillering. However, in the FACE experiment (Hebeisen *et al.*, 2000), LAI was not affected significantly by elevated [CO_2] in frequently cut high-N ryegrass swards. In infrequently cut swards, however, LAI increased in the third year, when the yield increased strongly. Since LAI did not usually show a significant increase, the higher yield was associated with increased carbohydrate accumulation in the leaves (Fischer *et al.*, 2000) and, thus, specific leaf area (SLA) decreased (Hebeisen *et al.*, 1999). In contrast, the LAI of white clover increased at elevated [CO_2] in all cutting and fertilizer treatments except in the third year, when the yield increased by only 11%.

Root biomass of perennial ryegrass increased to a greater extent than did yield at elevated [CO_2] (Table 12.1), reflecting increased partitioning of C to the roots (Soussana *et al.*, 1996; Schapendonk *et al.*, 1997). The root biomass of white clover was only about 30% of that of ryegrass and the increase at elevated [CO_2] was smaller than in ryegrass (Table 12.1). Root growth into mesh bags in ryegrass and white clover swards increased under elevated [CO_2] and showed strong seasonal variation (Jongen *et al.*, 1995). The specific root length did not change; the greater root mass was due to increased root length. The root mass density of both species increased in all soil layers and although the relative increase was largest in the lowest layers, the largest increase was in the top layer (Hebeisen *et al.*, 2000).

At elevated [CO_2], root biomass usually increased significantly, implying that more root litter was also produced. Because the greater quantity of root litter will contain more nutrients, decomposition will probably result in the availability of additional nutrients (Soussana *et al.*, 1996). Increased above-ground litter was also found in some instances (Blum *et al.*, 1997). The amount and quality of these litter components are important characteristics for nutrient cycling (Ball, 1997) and C sequestration (Nitschelm *et al.*, 1997). The decomposition of litter is crucial in these processes, which can last days or hundreds or even thousands of years (Ball, 1997). The decomposition of litter is affected by litter quality and by soil properties such as biological activity, soil moisture and temperature. Most studies show that litter quality changed at elevated [CO_2], causing rates of decomposition to decline. However, most studies have been short-term experiments conducted in the laboratory. The increased amount of litter and the reduced litter quality both suggest that soil organic matter will increase under elevated [CO_2]. Casella and Soussana (1997) determined a strong increase in net C input in the soil. This suggests reduced N mineralization and, thus, an increase of temporarily immobilized N and/or increased C sequestration; i.e. a long-term immobilization of N (Soussana and Hartwig, 1996).

Other pathways along which soil N losses might change are denitrification (Ineson *et al.*, 1998) or N leaching (Casella *et al.*, 1996). The yield of perennial ryegrass increased in response to CO_2 for the first 3 years of a grassland FACE experiment (Table 12.1) and continued to increase for an additional 3 years (Daepp *et al.*, 2000). This may indicate that temporal soil N pools and mineralization were increasing. At the same time, an equilibrium between new C input and new C mineralization in the soil was established (van Kessel *et al.*, 2000). Simulations with the Hurley Pasture Model (an ecosystem model that includes soil processes such as N cycling and C sequestration) predicted a long-term increase in the CO_2 response of grassland (Thornley and Cannell, 1997). Interestingly, these simulations predicted that N-poor grassland will finally show a stronger response than will N-rich grassland (Cannell and Thornley, 1998). However, the steady state was reached much faster in the N-rich (5–10 years) than in the N-poor system (100 years). Do the experiments conducted thus far, particularly those with low N input, describe only the initial phase of transition to a new steady state? The results of the long-term FACE experiment support this hypothesis, though it is still too early to draw final conclusions.

12.4.2 Temperature

The temperature response of plant growth shows an optimum curve. At low temperature, elevated [CO_2] has no or only small effects, as indicated by the lack of increase in photosynthesis. At elevated [CO_2], the annual yield of ryegrass did not change in response to a 3°C increase in temperature, but the root fraction of total biomass decreased (Casella *et al.*, 1996). However, the yield increased in spring and autumn, when temperatures were suboptimal, and decreased in summer. The decline in the CO_2 effect in summer was due to increased canopy transpiration and, thus, to reduced water availability. Comparing the combined effects of doubled [CO_2] and a temperature increase of 3°C with ambient [CO_2] and ambient temperature, the annual grass yield increased (Casella *et al.*, 1996). The higher yield was mainly produced in spring.

There was no seasonal variation in the relative yield increase of white clover at elevated compared with ambient [CO_2] (Hebeisen *et al.*, 2000), but the absolute increase was proportional to the yield. This indicates that the CO_2 effect on growth was not directly affected by temperature. In this field experiment, temporal water deficits may have favoured carbohydrate allocation to the root. The higher photosynthesis expected at elevated temperature in summer increased root growth but not yield. The LAI decreased when the temperature increased by 2.5°C (Nijs *et al.*, 1997) both at ambient and elevated [CO_2]; however, the air temperature was above optimum during this experiment.

12.4.3 Water deficits

The yield of perennial ryegrass under moderate drought stress increased by an average of 19 and 14% in low- and high-N swards, respectively, at elevated [CO_2] (Table 12.1) (Casella *et al.*, 1996). In summer, when the water deficit was greater, elevated CO_2 increased yield by 48%. In spring and autumn, the yield increased by only 6%. However, the effects of water deficit and temperature were confounded. At elevated CO_2, a temperature increase of 3°C increased the water deficit in summer, due to increased transpiration; thus, grass yield was reduced. Compared with ambient temperature and [CO_2], the yield at elevated [CO_2] and temperature did not change during periods of water deficit. The yield of tallgrass-prairie swards that are dominated by C_4 grasses increased under elevated [CO_2] only in dry years, due to an improved plant water status (Owensby *et al.*, 1996).

Root biomass of ryegrass increased during periods of drought; the increase was much larger at elevated [CO_2] (Soussana *et al.*, 1996). The CO_2 effect on root mass was twice as strong in one year with severe drought compared with years with lower water deficits (Table 12.1) (Hebeisen *et al.*, 1997).

During a period of severe water deficit, white clover yield decreased much less than did perennial ryegrass yield. The soil water content decreased by the same amount under both species. The proportion of white clover in the yield of ryegrass/white clover mixtures increased by up to 80%. The different responses of the species may indicate that growth of ryegrass was reduced during drought because the availability of mineral N was limited, due to a reduced mineralization. Because white clover does not depend on mineral N, the water deficit affected the growth and carbohydrate allocation of this species much less.

12.4.4 Nutrient deficits

While plants growing under limited N supply show reduced above-ground growth compared with non-limited plants, their root mass increased (Hebeisen *et al.*, 1997) or decreased (Jongen *et al.*, 1995; Soussana *et al.*, 1996). However, the root/shoot biomass ratio increased strongly (Soussana *et al.*, 1996) i.e. a smaller proportion of total biomass was harvested. Leaf [N] and root [N] decreased, though the reduction in root [N] was much smaller. The larger proportion of root biomass implies that the roots contained a larger proportion of available N. Under N-limited conditions, yield and LAI of perennial ryegrass decreased (yield up to 20%) at elevated [CO_2] during the first 3 years (Hebeisen *et al.*, 2000). Additional carbohydrates were stored in the leaves and stubbles (Fischer *et al.*, 1997), thus reducing leaf [N] (Zanetti *et al.*, 1997). This indicates that the sink is limiting growth more in N-limited than in well-fertilized plants (Fischer *et al.*, 1997). However, root growth strongly increased at elevated [CO_2], and increased to a greater extent in deeper soil layers (Hebeisen *et al.*, 2000). In contrast, Casella *et al.* (1996) reported a 19% increase of ryegrass yield at elevated [CO_2] in low-N swards. The weak water deficit may have increased the CO_2 response of the plants growing in large containers.

12.4.5 Ozone

Ozone (O_3) is the most important air pollutant in rural areas in industrialized countries, and concentrations during the growing season are much higher than the threshold level for effects on vegetation (Fuhrer *et al.*, 1997). Visible leaf injury, an indication of short-term effects, has been observed in a range of plant species, including white clover, a representative species of productive grasslands (Becker *et al.*, 1989). Grass species found to be sensitive to ozone include orchard grass (*Dactylis glomerata*) and timothy (*Phleum pratense*) (see Mortensen, 1992). The effect of O_3 depends largely on stomatal uptake and, thus, is influenced by environmental factors (such as VPD, atmospheric turbulence and soil moisture) that affect stomatal conductance.

Intraspecific differences in stomatal conductance may also be responsible for differences in O_3 sensitivity of cultivars (Becker *et al.*, 1989) and ecotypes (Nebel and Fuhrer, 1994). Long-term effects of O_3 are caused by cumulative O_3 exposure and consist of changes in growth, yield and reproduction (Fuhrer *et al.*, 1997). A screening experiment revealed that growth reductions are least likely to occur in species characterized by a stress-tolerator strategy (i.e. S-strategy, *sensu* Grime, 1979) (Bungener *et al.*, 1999). Physiological effects on individual leaves include reduced photosynthetic capacity, increased dark respiration and a reduced leaf duration. These effects can be associated with changes in biomass allocation of the whole plant. In graminoids, biomass is preferentially allocated to shoots at the expense of allocation to roots and seeds. In clover, biomass allocation to stolons and roots is reduced, which may in turn affect regrowth, winter survival (Blum *et al.*, 1983) and N fixation (Montes *et al.*, 1983). Rebbeck *et al.* (1988) observed a reduction in the starch content of roots but not of shoots of ladino clover with increasing O_3; soluble-sugar contents were unaffected in both roots and shoots. In a 0.09 ppm O_3 treatment, the root/shoot ratio was reduced by 24% in crimson clover and by 22% in annual ryegrass compared with an O_3-free treatment (Bennett and Runeckles, 1977). In both species SLA and leaf area ratio (LAR) were lowered by O_3, but increased net assimilation rates did not affect relative growth rate (RGR).

During growth and regrowth of pot-grown tall fescue, shoot dry weight was not affected by increasing O_3 levels. However, under the same conditions, regrowth of ladino clover declined whether grown alone or in combination with fescue (Montes *et al.*, 1982). In the same study, the total forage yield of the mixture was reduced, and the clover/fescue ratio declined. There is evidence that effects of O_3 differ depending on whether plants are grown in monoculture or in mixtures. For instance, the total dry weight of crimson clover and annual ryegrass decreased less in a mixture. In a mixture of white clover and perennial ryegrass, constant elevation of O_3 caused a decline in the clover fraction, which was partly compensated for by increased growth of ryegrass (Nussbaum *et al.*, 1995). When the same total dose of O_3 was applied episodically, growth of ryegrass and total forage yield declined. This indicates the importance of the effects of intermittent peak concentrations of O_3 on the relatively O_3-tolerant ryegrass.

The different effect of O_3 on grasses and clover grown in potted mixtures was confirmed by experiments with field-grown plants. Two-year studies with timothy and red clover (Kohut *et al.*, 1988) or tall fescue and ladino clover (Heagle *et al.*, 1989) pastures showed clear increases in the grass/clover ratio, with increasing O_3 causing a reduction in the clover fraction combined with a stable, or even larger, fraction of the grass. In the latter study, total forage yield was reduced by 10% at ambient levels of O_3; there was no significant effect of reduced irrigation on the O_3 response. In a 2-year experiment with a fescue and clover pasture, there was a negative effect of O_3 on the fractional clover yield. The effect was much larger during the second year compared with the first year (Blum *et al.*, 1983). In a 1-year experiment with two cutting frequencies, the effect of O_3 on the ratio of ryegrass and the more sensitive clover varied from harvest to harvest, probably because of the time of cutting relative to the O_3 episodes and differences in the phenology of the species (Wilbourn *et al.*, 1995).

Two recent studies on the response of grass/clover mixtures to O_3 have produced contrasting results. The study by Fuhrer *et al.* (1994), using a standard seed mixture, gave results very similar to those found by Heagle *et al.* (1989), i.e. no effect of O_3 on the total 2-year yield of the grasses but a substantial decline in clover yield. In contrast, a Swedish study showed a linear decrease in forage yield with increasing O_3, but no effect on the clover fraction (Pleijel *et al.*, 1996). This suggests that the effect of O_3 on the competitive balance depends on the ratio of the species and cultivar sensitivities. Overall, the available data suggest that the clover decline in grass and clover mixtures is accelerated by O_3, but that the extent of the effect depends on the species and cultivars used (Fig. 12.1).

In ambient air, periods of elevated O_3 occur at irregular intervals. Their effect on the competitive balance in plant mixtures depends not only on the impact during the O_3 episodes, but also on the capacity of the species to recover between episodes. Wilbourn *et al.* (1995) observed quick recovery of a clover canopy after a period of elevated O_3, but there was a persistent effect on stolon density, which may affect the long-term performance of clover. Experiments with ryegrass and clover mixtures confirmed that the overall effect on the clover proportion may be related to the cumulative exposure to O_3 in all preceding growth periods (Nussbaum *et al.*, 1995). In a frequently cut pasture, clover declined rapidly in plots exposed to elevated O_3 during the first season but recovered during the second season when the same plots were exposed to a reduced concentration of O_3 (Fuhrer *et al.*, 1994). Therefore, the extent of negative effects of O_3 on clover persistence in productive pastures could be counteracted by introducing tolerant lines of clover.

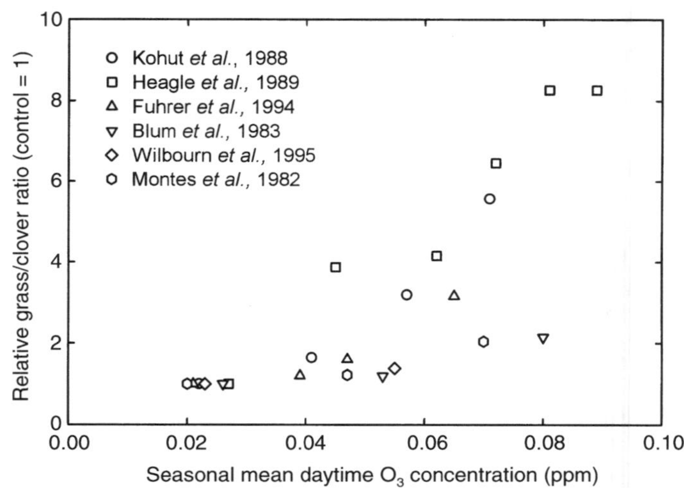

Fig. 12.1. Effect of O_3 on grass/clover ratio, relative to the value for the control treatment.

12.5 Management and Breeding Strategies for Future Climates

Temperate grasslands range from intensively managed monocultures of sown species to species-rich natural and semi-natural communities. Local distributions and productivity of these communities are controlled by variations in the interaction of climate, soil, germplasm, interspecies competition and management. The key climatic factors are rainfall, temperature, and available solar radiation. Topography and soil properties help to determine available soil moisture, while nutrient supply, utilization regime and [CO_2] may also be of great importance. There are complex interactions among growth-limiting environmental factors, CO_2 and management. Many questions remain to be answered by studying grassland ecosystems with diverse characteristics on a scale where the relationships between biotic systems and the environment can be manipulated and quantified. The lack of actual knowledge is greatly limiting the assessment of the outcome of climatic changes which could result in the adaptation of management practices and changes in the priorities of plant-breeding objectives.

Different species can show marked differences in their responses to increasing [CO_2], rising temperature and drought stress. This will probably result in major alterations in the botanical composition of temperate grasslands in the future. Effective management of grasslands depends on understanding the ecological and physiological processes that occur within them. From the perspective of management, it is important to identify when each of the potential factors causing changes in grassland may be most influential. In the Swiss FACE experiment, for example, the proportion of white clover in a binary mixture with ryegrass was more strongly affected by N fertilization than by elevated [CO_2] (Hebeisen *et al.*, 1997). Management can drastically alter the magnitude and the sign of the response of grasslands to climatic change; the response can be highly site-specific (Thornley and Cannell, 1997).

Water is the main factor limiting plant productivity in many regions; the interaction of water availability and rising [CO_2] is a major concern with respect to its effects on plant growth. Temperature increase from rising CO_2 might cause a larger evaporative demand, leading to increased evapotranspiration and reduced soil water. This will probably have very important impacts on agriculture and specifically on productive grassland with a high water demand (Rogers and Dahlman, 1993; Idso and Idso, 1994). In the future, grassland might be at great disadvantage in regions with an increased frequency and intensity of drought. Small changes in drought tolerance can lead to long-term changes in the botanical composition. As a consequence, the actual agricultural importance of a species may alter with changing climatic conditions.

The fescue (*Festuca*) and ryegrass (*Lolium*) genera include the most important temperate grasses that are sown widely for forage. In general, the ryegrasses grow faster and are more palatable and digestible, while the fescues are more robust and resistant to stresses, such as, drought and cold. Drought encompasses not only reduced availability of soil moisture, but also high-evaporation load, supra-optimal temperatures, potentially damaging insulation and reduced availability of mineral nutrients; it also increases soil hardness.

Interspecific hybridization leads to new ways of breeding forage grasses for drought tolerance. It seems that ryegrass × fescue hybrids not only have valuable agronomic potential, but also provide experimental material for a better understanding of tolerance at the physiological and genetic levels. Humphreys *et al.* (1997, 1998) demonstrated that such hybrids provide germplasm that is acclimated to both drought and freezing stresses, in some cases to a greater extent than is the fescue parent. Long-term responses of perennial forage grasses to severe drought have been related to 'dormancy' or 'quiescence' mechanisms that develop in basal buds and enable the survival of species, such as *Phalaris aquatica* (Oram,1983). The main difference between the survival of cultivars of ryegrass and cocksfoot after severe drought was associated with carbohydrate metabolism and, in particular, with a high content of fructans, having a high degree of polymerization (Voltaire *et al.*, 1998).

12.6 Conclusions and Future Research Directions

The physiological effects of climatic change on plant growth under laboratory conditions are well known. Future CO_2 research must be conducted within the context of a managed ecosystem, where other powerful forces such as fertilization, as well as type and frequency of defoliation, act directly or via the soil on the vegetation. Assessing the effects of climatic change impinges on one of the central challenges of grassland ecology: that of scaling issues and identification of the interaction of processes of different temporal and spatial scales. In terms of their role in global energy and chemical fluxes, grasslands may play a more important role than is generally appreciated and reflected in global carbon models and budgets.

Interactions among the effects of growth factors and raising [CO_2] are manifold and difficult to predict. A stimulatory effect of elevated [CO_2] on plant growth and yield was found in many experiments under conditions of unlimited water and nutrient availability as well as under water and nutrient stress. Only at low temperatures is there little, if any, effect of raising [CO_2]. Indeed, cool regions of the planet should profit from higher temperatures. However, the amount by which yield will finally increase under elevated [CO_2], is not yet known, particularly in nutrient-limited situations. Long-term experiments with different management treatments that allow a realistic assessment of the long-term response of grassland to elevated [CO_2] must be conducted. The FACE technology provides a means to increase [CO_2] and quantify the response of critical processes regulating carbon exchange without modifying the microclimate as other exposure approaches do.

Species composition will change under elevated [CO_2]. Legumes are expected to benefit more than grasses, at least in the short term, and other species will benefit from the increased N input into the system due to enhanced symbiotic N fixation. To determine the effects of climatic change on species composition, long-term experiments are necessary.

Grassland soils contain large C and N pools which may increase under elevated [CO_2] and decrease with rising temperature. Increased C sequestration in grassland soils may contribute to slowing the increasing [CO_2]. It is not yet known how C sequestration will change under elevated [CO_2]; much depends on the still uncertain climatic change.

References

Ball, A.S. (1997) Microbial decomposition at elevated CO_2 levels: effect of litter quality. *Global Change Biology* 3, 379–386.

Becker, K., Saurer, M., Egger, A. and Fuhrer, J. (1989) Sensitivity of white clover to ambient ozone in Switzerland. *New Phytologist* 112, 235–243.

Bennett, J.P. and Runeckles, V.C. (1977) Effects of low levels of ozone on plant competition. *Journal of Applied Ecology* 14, 877–880.

Blum, H., Hendrey, G.R. and Nösberger, J. (1997) Effects of elevated CO_2, N fertilization and cutting regime on the production and quality of *Lolium perenne* L. shoot necromass. *Acta Oecologica* 18, 291–296.

Blum, U., Heagle, A.S., Burns, J.C. and Linthurst, R.A. (1983) The effects of ozone on fescue–clover forage: regrowth, yield and quality. *Environmental and Experimental Botany* 23, 121–132.

Bungener, P., Nussbaum, S., Grub, A. and Fuhrer, J. (1998) Growth response of grassland species to ozone in relation to soil moisture condition and plant strategy. *New Phytologist* 142, 283–293.

Cannell, M.G.R. and Thornley, J.H.M. (1998) N-poor ecosystems may respond more to elevated [CO_2] than N-rich ones in the long-term. A model analysis of grassland. *Global Change Biology* 4, 431–442.

Casella, E. and Soussana, J.F. (1997) Long-term effects of CO_2 enrichment and temperature increase on the carbon balance of a temperate grass sward. *Journal of Experimental Botany* 48, 1309–1321.

Casella, E., Soussana, J.F. and Loiseau, P. (1996) Long-term effects of CO_2 enrichment and temperature increase on a temperate grass sward. I. Productivity and water use. *Plant and Soil* 182, 83–99.

Daepp, M., Suter, D., Lüscher, A., Almeida, J.P.F., Isopp, H., Hartwig, U.A., Blum, H. and Nösberger, J. (2000) Yield response of *Lolium perenne* swards to free air CO_2 enrichment increased over six years in a high-N input system. *Global Change Biology* (in press).

Drake, B.G., Gonzàlez-Meler, M.A. and Long, S.P. (1997) More efficient plants: a consequence of rising atmospheric CO_2? *Annual Review of Plant Physiology and Plant Molecular Biology* 48, 609–639.

Eamus, D. (1991) The interaction of rising CO_2 and temperatures with water use efficiency. *Plant, Cell and Environment* 14, 843–852.

Field, C.B., Jackson, R.B. and Mooney, H.A. (1995) Stomatal responses to increased CO_2: implications from the plant to the global scale. *Plant, Cell and Environment* 18, 1214–1225.

Fischer, B.U., Frehner, M., Hebeisen, T., Zanetti, S., Lüscher, A., Hartwig, U.A., Hendrey, G.R., Blum, H. and Nösberger, J. (1997) Source–sink relations in *Lolium perenne* L. as reflected by carbohydrate concentrations in leaves and stubbles during regrowth in a free air carbon dioxide enrichment (FACE) experiment. *Plant, Cell and Environment* 20, 945–952.

Fuhrer, J., Shariat-Madari, H., Perler, R., Tschannen, W. and Grub, A. (1994) Effects of ozone on managed pasture. II. Yield, species composition, canopy structure, and forage quality. *Environmental Pollution* 86, 307–314.

Fuhrer, J., Skärby, L. and Ashmore, M. (1997) Critical levels for ozone effects on vegetation in Europe. *Environmental Pollution* 97, 91–106.

Grime, J.P. (1979) *Plant Strategies and Vegetation Processes.* John Wiley & Sons, Chichester, UK.

Ham, J.M., Owensby, C.E., Coyne, P.I. and Bremer, D.J. (1995) Fluxes of CO_2 and water vapor from a prairie ecosystem exposed to ambient and elevated CO_2. *Agricultural and Forest Meteorology* 77, 73–93.

Heagle, A.S., Rebbeck, J., Shafer, S.R., Blum, U. and Heck, W.W. (1989) Effects of long-term ozone exposure and soil moisture deficit on growth of a Ladino clover–tall fescue pasture. *Phytopathology* 79, 128–136.

Hebeisen, T., Lüscher, A., Zanetti, S., Fischer, B.U., Hartwig, U.A., Frehner, M., Hendrey, G.R., Blum, H. and Nösberger, J. (1997) The different responses of *Trifolium repens* L. and *Lolium perenne* L. grassland to free air CO_2 enrichment and management. *Global Change Biology* 3, 149–160.

Houghton, J.T., Meira Filho, L.G., Callander, B.A., Harris, N., Kattenberg, A. and Maskell, K. (eds) (1996) *Climate Change 1995. The Science of Climate Change.* Cambridge University Press, Cambridge, UK.

Humphreys, M., Thomas, H.M., Harper, J., Morgan, G., James, A., Ghamari-Zare, A. and Thomas, H. (1997) Dissecting drought- and cold-tolerance traits in the Lolium–Festuca complex by introgression mapping. *New Phytologist* 137, 55–60.

Humphreys, M.W., Pasakinskiene, I., James, A.R. and Thomas, H. (1998) Physically mapping quantitative traits for stress-resistance in the forage grasses. *Journal of Experimental Botany,* 49, 1611–1618.

Idso, K.E. and Idso, S.B. (1994) Plant-responses to atmospheric CO_2 enrichment in the face of environmental constraints – a review of the past 10 years of research. *Agricultural and Forest Meteorology* 69, 153–203.

Ineson, P., Coward, P.A. and Hartwig, U.A. (1998) Trace gas measurements in the Zürich FACE experiment. *Plant and Soil* 198, 89–95.

Jongen, M., Jones, M.B., Hebeisen, T., Blum H. and Hendrey, G.R. (1995) The effects of elevated CO_2 concentrations on the root growth of *Lolium perenne* and *Trifolium repens* grown in a FACE system. *Global Change Biology* 1, 361–371.

Van Kessel, C., Nitschelm, J., Horwath, W., Harris, D., Walley, F., Lüscher, A. and Hartwig, U. (2000) Carbon-13 input and turn-over in a pasture soil exposed to long term elevated atmospheric CO_2. *Global Change Biology* (in press).

Kimball, B.A. (1983) Carbon dioxide and agricultural yield: an assessment and analysis of 430 prior observations. *Agronomy Journal* 75, 779–88.

Kohut, R.J., Laurence, J.A. and Amundson, R.G. (1988) Effects of ozone and sulfur dioxide on yield of red clover and timothy. *Journal of Environmental Quality* 17, 580–585.

Long, S.P., Baker, N.R. and Raines, C.A. (1993) Analysing the response of photosynthetic CO_2 assimilation to long-term elevation of atmospheric CO_2 concentration. *Vegetatio* 104/105, 33–45.

Lüscher, A., Hendrey, G.R. and Nösberger, J. (1998) Long-term responsiveness to free air CO_2 enrichment of functional types, species and genotypes of permanent grassland. *Oecologia* 113, 37–45.

Montes, R.A., Blum, U. and Heagle, A.S. (1982) The effects of ozone and nitrogen fertilizer on tall fescue, ladino clover, and fescue–clover mixture. I. Growth, regrowth, and forage production. *Canadian Journal of Botany*, 60, 2745–2752.

Montes, R.U., Blum, U., Heagle, A.S. and Volk, R.J. (1983) The effects of ozone and nitrogen fertilizer on tall fescue, ladino clover, and fescue–clover mixture. II. Nitrogen content and nitrogen fixation. *Canadian Journal of Botany* 61, 2159–2168.

Mortensen, L. (1992) Effects of ozone on growth of seven grass and one clover species. *Acta Agronomica Scandinavica* (Section B), *Soil and Plant Science* 42, 235–239.

Nebel, B. and Fuhrer, J. (1994) Inter- and intraspecific differences in ozone sensitivity in semi-natural plant communities. *Angewandte Botanik* 68, 116–121.

Newton, P.C.D. (1991) Direct effects of increasing carbon dioxide on pasture plants and communities. *New Zealand Journal of Agricultural Research* 34, 1–24.

Nijs, I. and Impens, I. (1996) Effects of elevated CO_2 concentration and climate-warming on photosynthesis during winter in *Lolium perenne*. *Journal of Experimental Botany* 47, 915–924.

Nijs, I., Ferris, R., Blum, H., Hendrey, G. and Impens, I. (1997) Stomatal regulation in a changing climate: a field study using free air temperature increase (FATI) and free air CO_2 enrichment (FACE). *Plant, Cell and Environment* 20, 1041–1050.

Nitschelm, J.J., Lüscher, A., Hartwig, U.A. and van Kessel, C. (1997) Using stable isotopes to determine soil carbon input differences under ambient and elevated atmospheric CO_2 conditions. *Global Change Biology* 3, 411–416.

Nitschelm, J.J., Walley, F.L., Hartwig, U.A., Lüscher, A. and van Kessel, C. (2000) Carbon-13 turnover in a pasture soil subjected to free-air carbon dioxide enrichment. *Global Change Biology* (in press).

Nussbaum, S., Geissmann, M. and Fuhrer, J. (1995) Ozone-exposure–response relationships for mixtures of perennial ryegrass and white clover depend on ozone exposure patterns. *Atmospheric Environment* 29, 989–995.

Oram, R.N. (1983) Ecotypic differentiation for dormancy levels in oversummering buds of *Phalaris aquatica* L. *Botanical Gazette* 144, 544–551.

Owensby, C.E., Ham, J.M., Knapp, A.K., Rice, C.W., Coyne, P.I. and Auen, L.M. (1996) Ecosystem-level responses of Tallgrass Prairie to elevated CO_2. In: Koch, G. And Mooney, H. (eds) *Carbon Dioxide and Terrestrial Ecosystems.* Physiological Ecology Series, Academic Press, New York, pp. 175–193.

Owensby, C.E., Ham, J.M., Knapp, A.K., Bremer, D. and Auen, L.M. (1997) Water vapour fluxes and their impact under elevated CO_2 in a C4-tallgrass prairie. *Global Change Biology* 3, 189–195.

Pleijel, H., Karlsson, G.P., Sild, E., Danielsson, H., Skärby, L. and Selldén, G. (1996) Exposure of a grass–clover mixture to ozone in open-top chambers – effects on yield, quality and botanical composition. *Agriculture, Ecosystems and Environment* 59, 55–62.

Rebbeck, J., Blum, U. and Heagle, A.S. (1988) Effects of ozone on the regrowth and energy reserves of a Ladino clover–tall fescue pasture. *Journal of Applied Ecology* 25, 65–681.

Rogers, A., Fischer, B.U., Bryant, J., Frehner, M., Blum, H., Raines, C.A. and Long, S. (1998) Acclimation of photosynthesis to elevated CO_2 under low N nutrition is effected by the capacity for assimilate utilisation. Perennial ryegrass under free-air CO_2 enrichment (FACE). *Plant Physiology* 118, 683–689.

Rogers, H.H. and Dahlman, R.C. (1993) Crop responses to CO_2 enrichment. *Vegetatio* 104, 117–131.

Schapendonk, A.H.C.M., Dijkstra, P., Groenwold, J., Pot, C.S. and Geijn, S.C., van de (1997) Carbon balance and water use efficiency of frequently cut *Lolium perenne* L. swards at elevated carbon dioxide. *Global Change Biology* 3, 207–216.

Shaer, Y.A. and Van Bavel, C.H.M. (1987) Relative role of stomatal and aerodynamic resistances in transpiration of a tomato crop in a CO_2-enriched greenhouse. *Agricultural and Forest Meteorology* 41, 77–85.

Soussana, J.F. and Hartwig, U.A. (1996) The effect of elevated CO_2 on symbiotic N_2 fixation: a link between the carbon and nitrogen cycles. *Plant and Soil* 187, 321–332.

Soussana, J.F., Casella, E. and Loiseau, P. (1996) Long-term effects of CO_2 enrichment and temperature increase on a temperate grass sward. II. Plant nitrogen budgets and root fraction. *Plant and Soil* 182, 101–114.

Thornley, J.H.M. and Cannell, M.G.R. (1997) Temperate grassland responses to climate change: an analysis using the Hurley Pasture Model. *Annals of Botany* 80, 205–221.

Voltaire, F., Thomas, H., Bertagne, N., Bourgeois, E., Gauthier, M.F. and Lelièvre, F. (1998) Survival and recovery of perennial forage grasses under prolonged Mediterranean drought. II. Water status, solute accumulation, abscisic acid concentration and accumulation of dehydrin transcripts in bases of immature leaves. *New Phytologist* 140, 451–460.

Wilbourn, S., Davison, A.W. and Ollerenshaw, J.H. (1995) The use of an unenclosed field fumigation system to determine the effects of elevated ozone on a grass–clover mixture. *New Phytologist* 129, 23–32.

Woodrow, I.E. (1994) Control of steady-state photosynthesis in sunflowers growing in enhanced CO_2. *Plant, Cell and Environment* 17, 277–286.

Zanetti, S., Hartwig, U.A., Lüscher, A., Hebeisen, T., Frehner, M., Fischer, B.U., Hendrey, G.R., Blum, H. and Nösberger, J. (1996) Stimulation of symbiotic N_2 fixation in *Trifolium repens* L. under elevated atmospheric pCO_2 in a grassland ecosystem. *Plant Physiology* 112, 575–583.

Zanetti, S., Hartwig, U.A., van Kessel, C., Lüscher, A., Hebeisen, T., Frehner, M., Fischer, B.U., Hendrey, G.R., Blum, H. and Nösberger, J. (1997) Does nitrogen nutrition restrict the CO_2 response of fertile grassland lacking legumes? *Oecologia* 112, 17–25.

13 Crop Ecosystem Responses to Climatic Change: Rangelands

H. WAYNE POLLEY[1], JACK A. MORGAN[2], BRUCE D. CAMPBELL[3] AND MARK STAFFORD SMITH[4]

[1]*USDA-ARS, Grassland, Soil and Water Research Laboratory, 808 E. Blackland Road, Temple, TX 76502, USA;* [2]*USDA-ARS, Crops Research Laboratory, Fort Collins, CO 80526, USA;* [3]*AgResearch, Grasslands Research Centre, Private Bag 11008, Palmerston North, New Zealand;* [4]*CSIRO Division of Wildlife and Ecology, PO Box 2111, Alice Springs, NT 0871, Australia*

13.1 Introduction

Rangelands are defined as natural or semi-natural areas that produce plants grazed by wild and domesticated animals (Stoddart *et al.*, 1975). Included in this definition are unimproved grasslands, savannas, shrublands containing both grasses and woody plants, and hot and cold deserts (Fig. 13.1). These ecosystems occur on every continent except Antarctica and cover over 40%

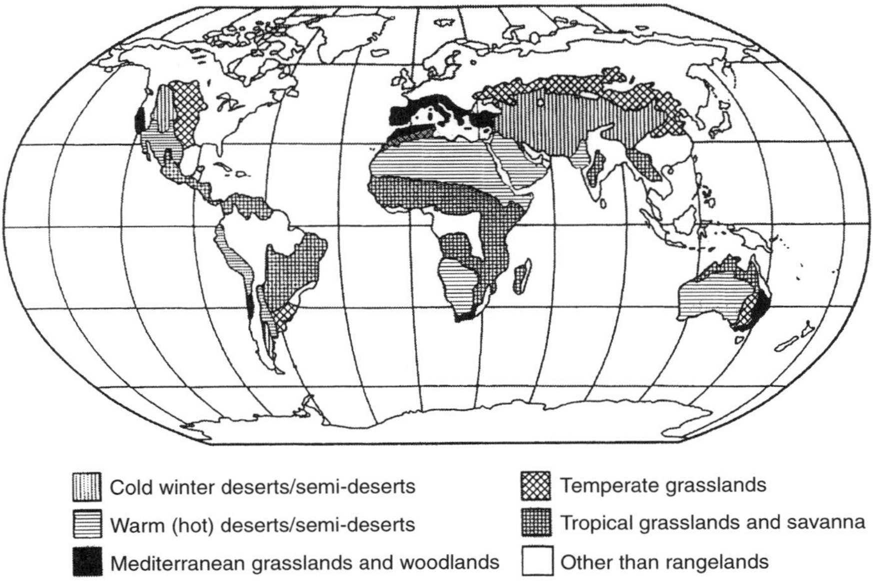

Fig. 13.1. Distribution of the world's rangelands. (Redrawn from Allen-Diaz, 1996.)

of the terrestrial land surface, with the greatest area in Africa and Asia (Allen-Diaz, 1996).

Environmentally determined rangelands are found where climate (chiefly water balance and minimum temperature) and soils interact to prevent occupation by dense stands of trees. Human activities, broadly classified as land use patterns, are another major determinant of the composition and structure of rangelands, and are an example of the alterations to earth considered here as global change. Land use changes include manipulation of fire regimes and other natural disturbances, adjustments in the intensity and duration of grazing, and fragmentation of once continuous rangelands by intensive agriculture and urbanization. Global change also includes changes in atmospheric composition, e.g. carbon dioxide concentration ([CO_2]) and resultant modifications in climate (temperature, precipitation). Some aspects of global change, like atmospheric and land use change, are well underway. Others, including shifts in climate, appear imminent.

Anticipated global changes could dramatically alter the extent and productivity of rangelands, but prediction and risk assessment are complicated by the diverse nature of these ecosystems and varied goals of managers (Campbell *et al.*, 1996; Stafford Smith, 1996).

1. Rangelands include a variety of plant species and growth forms (grasses, herbs, trees, shrubs) that respond in different ways and at different rates to the environment and to management inputs.
2. Rangelands are spatially and temporally variable. This natural variability increases the difficulty of discerning effects of management from those of the environment and environmental change (Campbell *et al.*, 1996; Stafford Smith, 1996).
3. Rangelands traditionally have been used to produce livestock, but these ecosystems provide other 'goods and services', including recreation, water and fuel wood (Fig. 13.2). Atmospheric and climatic change, combined with social, economic and demographic forces, will influence the product or combination of products for which rangelands are managed, and the intensity of future land use.

In this chapter, we review climatic and atmospheric changes expected during the 21st century and suggest some of the consequences of these changes for rangelands that are used primarily for grazing. Non-grazing uses of rangelands are briefly noted to illustrate reciprocal interactions between land use and climatic and atmospheric changes. Finally, we suggest management implications of global change.

13.2 Atmospheric and Climatic Change

The atmospheric [CO_2] has risen during the past 200 years from approximately 280 μmol mol^{-1} in pre-industrial times to 360 μmol mol^{-1} today, and is projected to double over present-day concentration during the 21st century (Alcamo *et al.*, 1996). Other trace gases (CH_4, N_2O, NO_x, CO) are also

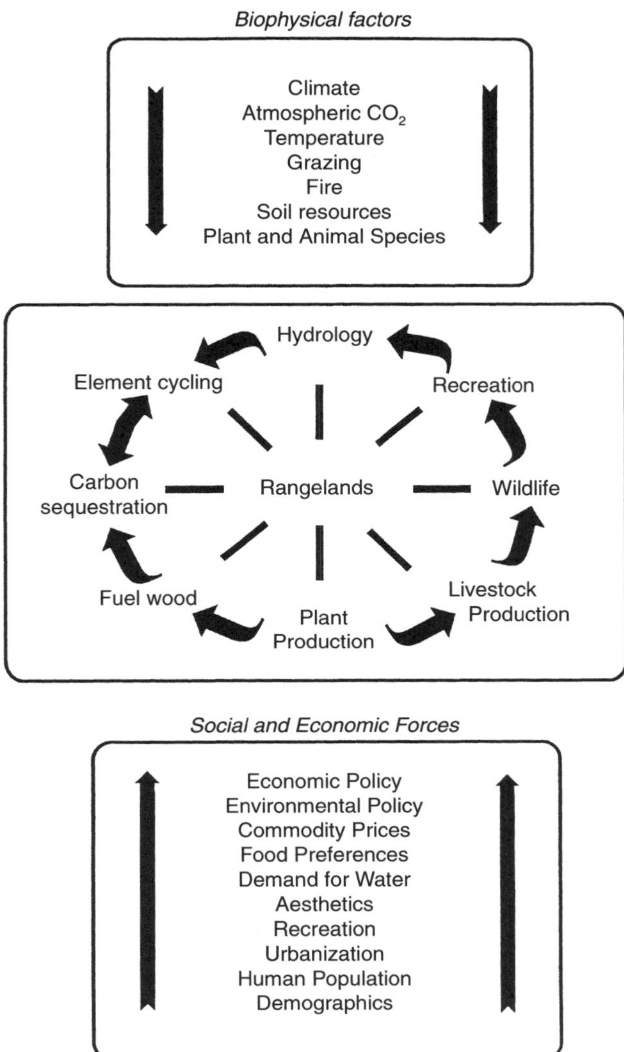

Fig. 13.2. Biophysical, social and economic forces influencing the goods and services for which rangelands are managed.

increasing rapidly, and will contribute to climate change. These increases result from human activities, especially combustion of fossil fuels, land use changes and agriculture. Increases in [CO_2] alone are predicted to warm the earth by 2–4.5°C by the middle of the 21st century, with a resultant increase in precipitation and storm intensity (Giorgi *et al.*, 1998).

Evidence from long-term climatic trends of the last century and modelled effects of [CO_2] (Giorgi *et al.*, 1998; Karl, 1998) suggests that warming will be greatest at high northern latitudes during autumn and winter. Consequently, temperate grasslands of North America and central Asia may experience relatively more warming than tropical and subtropical grasslands of Africa, Australia, and South and Central America. In high and mid-latitudes, global warming may lead to increased winter precipitation. Predictions are more variable for the tropics, but many models predict more precipitation over India and southeast Asia. Inclusion of aerosols in modelling exercises reduces the change in temperature and precipitation patterns, and sometimes gives regional results that contrast with those obtained by modelling CO_2 responses alone (Giorgi *et al.*, 1998).

13.3 Impacts on Plant Productivity

13.3.1 Atmospheric [CO_2]

Most forage species on rangelands have either the C_3 or C_4 photosynthetic pathway. Over 95% of the world's plant species, including most woody plants, utilize the C_3 pathway. Photosynthesis in C_3 plants is not CO_2-saturated at the present atmospheric concentration, so increasing [CO_2] is predicted to stimulate carbon gain and productivity in these species (Drake *et al.*, 1997). Plants with C_4 photosynthetic pathways comprise fewer than 5% of the world's species, but are an important component of tropical and subtropical grasslands (Ehleringer *et al.*, 1997). The final steps of photosynthesis in C_4 plants occur in bundle sheath cells, where a highly efficient biochemical pump maintains CO_2 at concentrations that nearly saturate photosynthesis when atmospheric [CO_2] is near the current 360 μmol mol^{-1} (Bowes, 1993). The C_4 metabolism does not, however, preclude photosynthetic and growth responses to CO_2 enrichment (Ghannoum *et al.*, 1997; LeCain and Morgan, 1998). Wand and Midgley (1998, unpublished results), for instance, measured a growth enhancement of 15% in C_4 grasses compared with 23% in C_3 grasses on doubling [CO_2] over the present level.

The photosynthetic pathway partially explains growth responses to CO_2, but CO_2 effects on transpiration and plant water-use efficiency (WUE; biomass produced per unit of transpiration) will be at least as important as photosynthetic metabolism in the future productivity of rangelands. Stomata of most plant species partially close as [CO_2] increases (Field *et al.*, 1995; Drake *et al.*, 1997). As this partial closure tends to reduce transpiration more than photosynthesis, leaf-level WUE (photosynthesis/transpiration) rises with [CO_2] (Polley *et al.*, 1996a). Reduced water loss and enhanced WUE can also be realized at the canopy level (Kirkham *et al.*, 1991; Nie *et al.*, 1992; Ham *et al.*, 1995), improving plant and soil water relations (Knapp *et al.*, 1994; Morgan *et al.*, 1994; Wilsey *et al.*, 1997), increasing plant production under water limitation (Owensby *et al.*, 1993a) and lengthening the growing season (Chiariello and Field, 1996). Water relation benefits of CO_2 enrichment largely explained the greater growth enhancement of C_4 than C_3 grasses in tallgrass prairie during dry years and the similar growth responses of C_3 and C_4 grasses under typically water-limiting conditions in shortgrass steppe (Owensby *et al.*,

1993a, 1997; Hunt *et al.*, 1996; Coughenour and Chen, 1997). Water relation benefits also apply to annual C_3 grasslands, where effects of elevated [CO_2] on plant production are more evident in drier years (Jackson *et al.*, 1995).

The ability to recover from defoliation is a major determinant of plant productivity and persistence on grazing lands. Recovery of grasses following grazing is controlled initially by re-mobilization of reserves, followed by photosynthetic gains (Caldwell *et al.*, 1981). To the extent that CO_2 enrichment increases photosynthesis and storage of reserves, it should enhance recovery from grazing. Plant response to defoliation depends on complicated interactions between grazing history and the environment (Milchunas *et al.*, 1988) and so effects of CO_2 enrichment are not likely to be simple. In a controlled environment, CO_2 enrichment had little effect on regrowth of grasses from three distinctly different grasslands (Wilsey *et al.*, 1997).

13.3.2 Temperature

Carbon dioxide enrichment and global warming are predicted to increase net primary production on most rangelands (Baker *et al.*, 1993; Parton *et al.*, 1995; Coughenour and Chen, 1997; Neilson *et al.*, 1998). Because of severe cold-temperature restrictions on growth rate and duration, warmer temperatures alone should enhance production in high- and mid-latitude and high-altitude rangelands (Baker *et al.*, 1993; Körner *et al.*, 1996; Rounsevell *et al.*, 1996). Warmer temperatures should also enhance the growth response of most C_3-dominated grasslands to rising [CO_2] (Long, 1991; Jones and Jongen, 1996; Coughenour and Chen, 1997; Drake *et al.*, 1997). This positive effect of warmer temperatures on production may be lessened, however, by an accompanying increase in evapotranspiration (ET) rate in drier systems such as the arid and semi-arid rangelands of Central and South America, Africa, the Middle East, Asia and Australia.

13.3.3 Precipitation

Current models yield widely varying estimates of future patterns in precipitation (Giorgi *et al.*, 1998), making it difficult to predict consequences of altered hydrological cycles for rangelands. Productivity on most rangelands is limited by water (Campbell *et al.*, 1997); therefore changes in the amount of precipitation will significantly impact these systems. Arid and semi-arid lands will be most sensitive to changes in precipitation, while usually wet mountain meadows will be minimally affected. Shifts in seasonal patterns of precipitation and predicted increases in storm intensity will probably have a greater impact on rangelands than shifts in precipitation amounts (Giorgi *et al.*, 1998). It is widely agreed that storm intensity will increase, resulting in greater runoff and concentration of water in smaller portions of the landscape. Such changes could reduce productivity or increase its heterogeneity (Campbell *et al.*, 1997; but see Williams *et al.*, 1998). The proportion of annual precipitation that falls during winter months is predicted to increase at high to mid-latitudes (Giorgi *et al.*, 1998). Such a change in seasonality of precipitation, combined with warming (predicted to be greatest at high northern latitudes) and increased runoff resulting from more severe storms, could increase the incidence and severity of summer droughts in semi-arid grasslands of North America and Asia. Conversely, any increase in rainfall during the growing season will help to mitigate the desiccating effects of warmer temperatures.

13.3.4 Soil feedbacks

Long-term responses of rangelands to global change ultimately depend on the soil and its ability to supply nutrients, as well as water. Carbon dioxide enrichment appears to improve the efficiency with which plants utilize nutrients for growth (Stock and Midgley, 1995; Drake *et al.*, 1997). Interactions between [CO_2] and nutrients are complicated, however, and plant responses to CO_2 enrichment may be constrained by low fertility, especially in relatively mesic environments (Sage, 1994; Stock and Midgley, 1995).

13.4 Impacts on Forage Quality

13.4.1 Plant–animal interface

Animal production on rangelands, as in other grazing systems, depends on the quality as well as the quantity of forage. Key quality parameters for rangeland forage include fibre content and concentrations of crude protein, non-structural carbohydrates, minerals and secondary toxic compounds. Ruminants require forage with about 7% crude protein (as a percentage of dietary dry matter) for maintenance, 10–14% protein for growth and 15% protein for lactation (Ulyatt *et al.*, 1980). Optimal rumen fermentation also requires a balance between ruminally available protein and energy (Dove, 1996). The rate at which digesta pass through the rumen depends on the fibre content of forage. Increasing fibre content slows passage and reduces animal intake.

13.4.2 Atmospheric [CO_2]

Based on expected vegetation changes and known environmental effects on forage protein, carbohydrate and fibre contents (e.g. Wilson, 1982; Owensby *et al.*, 1993b, 1996), both positive and negative changes in forage quality are possible as a result of atmospheric and climatic change (Table 13.1). Effects of CO_2 enrichment on crude protein content of forage, for example, are likely to be negative, for plant nitrogen concentration usually declines at elevated [CO_2] (Owensby *et al.*, 1993b; Cotrufo *et al.*, 1998). Limited evidence suggests that the decline is greater when soil nitrogen availability is low (Bowler and Press, 1996; Wilsey, 1996), implying that rising CO_2 could reduce the digestibility of

Table 13.1. Potential changes in forage quality arising from atmospheric and climatic change.

Change	Examples of positive effects on forage quality	Examples of negative effects on forage quality
Life-form distributions	Decrease in proportion of woody shrubs and increase in grasses in areas with increased fire frequency (Ryan, 1991)	Increase in proportion of woody species because of elevated CO_2, increases in rainfall event sizes and longer intervals between rainfall events (Stafford Smith *et al.*, 1995)
Species or functional group distributions	Increase in C_3 grasses relative to C_4 grasses with higher CO_2 (Johnson *et al.*, 1993)	Increase in proportion of C_4 grasses relative to C_3 grasses due to higher temperatures (Campbell *et al.*, 1996) or changes in availability of water at elevated CO_2 (Owensby *et al.*, 1997). Increase in plants poisonous to animals
Plant biochemical properties	Increase in non-structural carbohydrates at elevated CO_2 (Read *et al.*, 1997). Increase in crude protein with reduced rainfall	Decrease in tissue nitrogen contents and increased fibre contents as result of reduced photosynthetic protein contents at elevated CO_2 or higher temperatures (Sage *et al.*, 1989; Owensby *et al.*, 1993b, 1996; Soussana *et al.*, 1996; Read *et al.*, 1997). No change or decrease in crude protein in regions with more summer rainfall

forages that are already of poor quality for ruminants. Such reductions in forage quality would have pronounced negative effects on animal growth, reproduction and mortality (Owensby *et al.*, 1996) and could render livestock production unsustainable unless animal diets are supplemented with N (e.g. urea, soybean meal). Concentrations of some of the plant products that are toxic to animals may also increase in a CO_2-rich environment.

13.4.3 Botanical composition and animal selectivity

Both positive and negative effects on forage quality are possible for individual species, but the total quantity of nutrients on offer to a grazing animal is determined by the relative abundances of plant species in vegetation. Carbon dioxide enrichment initially reduced crude protein content of both species in a grass–clover mixture, but the protein content of the entire sward eventually increased at elevated [CO_2] because of a greater overall proportion of high-N clover (Schenk *et al.*, 1997). Similar effects are likely on rangelands, which contain complex mixtures of species of differing ecology and forage quality.

Ultimately, the quality of livestock diets is determined both by the quality of the forage on offer and by selectivity of animals during grazing. Research efforts have focused primarily on changes in forage quality to the near exclusion of potential changes in grazing behaviour by animals. Selective grazing is a significant feature of livestock on rangelands, where utilization is much lower than in intensively managed pastures. There is a need, therefore, to determine whether higher temperatures or other global changes will alter grazing behaviour and whether changes in grazing behaviour could compensate for a general decline in forage quality.

13.5 Impacts on Plant Species Composition

13.5.1 Importance of botanical composition

Research has emphasized global change effects on plant and ecosystem production and water balance, but changes in plant species composition could have at least as great an impact on the goods and services that rangelands provide as might changes in production. Rangelands are used primarily for grazing. For most domestic herbivores, the preferred forage is grass. Other plants, including trees, shrubs, and other broadleaf species, can lessen livestock production and profitability by reducing availability of water and other resources to grasses, making desirable plants unavailable to livestock, or physically complicating livestock management, or poisoning grazing animals (Dahl and Sosebee, 1991). The functioning of ecosystems can also be changed by addition or loss of plants which greatly affect disturbance regimes or soil resource (Vitousek, 1990). The spread of the annual grass *Bromus tectorum* (cheatgrass) through the intermountain region of western North America, for example, altered the frequency and timing of wildfires, and reduced establishment of perennial herbaceous species by pre-empting soil water early in the growing season (Young, 1991).

13.5.2 Environmental controls on species composition

The plant species composition of a region is largely determined by climate and soils, with fire regime, grazing and other land uses more locally important. The primary climatic control on the distribution and abundance of plants is water balance (Stephenson, 1990), especially on rangelands, where species composition is highly correlated with both the amount of water that plants use and its availability in time and space (Parton *et al.*, 1994).

13.5.3 Atmospheric [CO_2]

Carbon dioxide enrichment should slow canopy-level evapotranspiration (ET) (Drake, 1992; Ham *et al.*, 1995) and the rate or extent of soil water depletion (Kirkham *et al.*, 1991; Owensby *et al.*, 1993a; Jackson *et al.*, 1994; Field *et al.*,

1997), unless stomatal closure is compensated by atmospheric or other feedbacks. Plants that are less tolerant of water stress than current dominants should be favoured. Three general mechanisms may contribute to compositional changes.

1. By increasing WUE, CO_2 enrichment increases maximum leaf area and competition for light (Woodward, 1993). These changes should favour progressively taller and less drought-tolerant plants (Smith and Huston, 1989).
2. Slower ET minimizes the decline in soil moisture during periods between rainfall events. Wetter soils should enhance reproduction and survival of drought-sensitive species (Jackson *et al.*, 1995; Chiariello and Field, 1996).
3. By fostering wetter soils, CO_2 enrichment may increase deep percolation of water and favour more deeply rooting plants, like trees and shrubs, at the expense of shallow-rooting grasses (Polley *et al.*, 1997). Percolation below 1 m in California sandstone grassland increased by 20% on doubling [CO_2], despite a concomitant 20% increase in plant biomass (Jackson *et al.*, 1998).

Paradoxically, species composition may be more sensitive to a CO_2-mediated decrease in transpiration on relatively mesic rangelands (Table 13.2). Water savings are reduced when leaf area or soil evaporation increase – changes that are more likely on dry rangelands with open vegetative canopies than in mesic systems with closed vegetation.

Benefits of CO_2 enrichment to droughted plants may not be restricted to slower ET and soil water depletion. Higher [CO_2] can also extend seedling survival during drought (Polley *et al.*, 1996b). The full impact of this effect remains to be established, but density of the dominant species in California annual grassland (*Avena barbata*) was 87% greater at elevated than ambient [CO_2] during a dry year, apparently because of greater survivorship (Jackson *et al.*, 1995).

13.5.4 Precipitation regimes

A warmer climate means a more intense hydrological cycle, accompanied by an increase in the frequency of extreme events such as heavy rains and

Table 13.2. Predicted effects of CO_2 enrichment on soil water content and components of the hydrological cycle on arid and relatively mesic rangelands.

Parameter	Mesic rangelands	Arid rangelands
Transpiration/leaf area	Reduced	Reduced
Leaf area	Increased during dry periods	Increased
Total transpiration	Reduced, especially when wet	Small reduction
Soil evaporation	Little change	Possible increase
Percolation	Increased	No change
Runoff	Possible increase	No change
Soil water content		
Shallow	Increased	Small or no increase
Deep	Potentially increased	No change

droughts. Changes in the timing and intensity of rainfall will be especially important on arid rangelands, where plant community dynamics are 'event-driven' (Walker, 1993; Wiegand *et al.*, 1995) and the seasonality of precipitation determines which plant growth strategies are successful (Westoby, 1980). The timing of precipitation also affects the vertical distribution of soil water, which controls the relative abundances of plants that root at different depths (Ehleringer *et al.*, 1991; Weltzin and McPherson, 1997).

13.5.5 Temperature

Temperature influences botanical composition in several ways. Global warming, for example, may favour C_4 over C_3 grasses (Field and Forbe, 1990; Epstein *et al.*, 1997) by increasing the minimum daily temperature during the growing season (Teeri and Stowe, 1976) or by increasing photorespiration and reducing quantum yield in C_3 species (Ehleringer *et al.*, 1997). However, the effect of CO_2 enrichment on quantum yield is opposite to that of higher temperature. Doubling [CO_2] can, in fact, more than offset the decline in quantum yield of C_3 plants caused by a 2°C rise in temperature (Long, 1991) and render C_3/C_4 distributions and abundances relatively insensitive to the effects of higher temperature on quantum yield.

Extreme temperatures could become more frequent with global warming and could influence species distributions and abundances by affecting plant reproduction, competitive ability or survivorship. Plant responses to extreme temperatures appear to be species-specific, and thus are difficult to predict (Coleman *et al.*, 1991; Bassow *et al.*, 1994).

13.5.6 Interactions between temperature and [CO_2]

Many of the influences of higher temperature and [CO_2] on plants are not additive, so combined effects are not readily predictable from knowledge of individual effects (Bazzaz *et al.*, 1996). Neither do plants respond as predictably to temperature or CO_2 as to other factors or resources (e.g. water, nitrogen). Species within a given growth form responded similarly to changes in nutrient availability and light in tussock tundra, but showed no consistent response to higher temperature (Chapin *et al.*, 1995). Progress in predicting the response of vegetation to a warmer climate may require a better understanding of indirect effects of temperature on soil resources to which species respond more predictably (Chapin *et al.*, 1995). Warming in montane vegetation produced species changes apparently explained by temperature effects on soil moisture (Harte and Shaw, 1995).

13.5.7 Changes on global and regional scales

Long-term and global-scale responses of vegetation to higher [CO_2] and temperature have been estimated by modelling the equilibrium distribution of earth's vegetation from hydrological and plant physiological parameters (Woodward, 1993; Neilson, 1995; Haxeltine and Prentice, 1996). Both the MAPSS (Neilson, 1995) and BIOME3 (Haxeltine and Prentice, 1996) models predict a mean increase in leaf area index of earth's vegetation following a doubling in [CO_2] (Neilson *et al.*, 1998). The total area of grassland and shrubland in simulations either remains unchanged or increases by as much as 27%, depending on the scenario modelled.

Regional-scale estimates of climate change and associated impacts are highly uncertain. Climate models generally predict a greater-than-average increase in temperature in southern Europe and central North America, accompanied by reduced precipitation and soil moisture during summer (Giorgi *et al.*, 1998). Precipitation is particularly difficult to predict at regional scales, but there is some consensus that winter precipitation will increase at mid-latitudes. In regions like the southwestern USA, increased winter precipitation could favour large and deep-rooting woody plants over shallow-rooting warm-season grasses (Weltzin and McPherson, 1997). Wetter winters during the recent past may already have contributed to invasion of desert communities in Arizona, USA, by red brome (*Bromus rubens*), a winter annual (Betancourt, 1996) and of grasslands in the southwestern USA by shrubs (Neilson, 1986).

13.5.8 Local and short-term changes

The ability to predict vegetation change declines at lower spatial and temporal scales, i.e. as greater details of vegetation dynamics are required. There are at least two reasons. Firstly, transient responses of vegetation to global change depend on how quickly various species can disperse propagules across landscapes that are sometimes fragmented. Secondly, disturbances, biotic interactions and other local-scale processes become more important in vegetation dynamics at lower spatial and temporal scales.

Of necessity, field experiments with different CO_2 concentrations and temperatures are conducted at local scales, where variability in vegetation is high. Nevertheless, patterns of vegetation response are beginning to emerge.

1. Directional shifts in the composition of vegetation occur most consistently when global change treatments alter water availability (Owensby *et al.*, 1993a; Harte and Shaw, 1995; Chiariello and Field, 1996).

2. Carbon dioxide enrichment usually alters species abundances in multi-species communities, even when there is no net stimulation of total biomass (Körner, 1996).

3. Plant response to [CO_2] or temperature in multispecies communities is not readily predictable from the response of individually grown plants or from plant morphology or physiology (Chapin *et al.*, 1995; Körner, 1995; Leadley and Körner, 1996). This occurs because expression of the multiple direct and indirect effects of [CO_2] and temperature on plant growth and development depends on complex interactions among other environmental and biotic factors.

4. As a consequence of the above, species response to [CO_2] and temperature is often highly context-specific (Roy *et al.*, 1996).
5. Vegetation dynamics may be as sensitive to the secondary or indirect effects of atmospheric and climatic change as to direct effects of global changes on plant growth. In closed-canopy vegetation in particular (Roy *et al.*, 1996), changes in dominance may correlate better with changes in plant morphology, development and phenology than with more direct effects of [CO_2] or temperature on growth (Reekie and Bazzaz, 1989). At larger scales, effects of atmospheric and climatic change on fire frequency and intensity and on soil water and N availability will probably influence botanical composition to a much greater extent than global change effects on production.
6. Effects of CO_2 enrichment on species composition and the rate of species change will probably be greatest in disturbed or early-successional communities where nutrient and light availability are high and species change is more highly influenced by growth-related parameters (Arnone, 1996).
7. Rangeland vegetation will be influenced more by management practices (land use) than by atmospheric and climatic change. Global change effects will be superimposed on and modify those resulting from land use patterns in ways that are as yet uncertain.

Vegetation changes of greatest concern to managers are those that are essentially irreversible within the constraints of traditional management, and that fundamentally alter rangeland structure and function. Such shifts between 'alternate stable states' of vegetation usually occur when changes in soil properties, disturbance regimes or animal populations remove limitations on increasing plants or create or enforce limitations on current dominants.

Vegetation change can occur gradually, as when woody plants replace grasses following prolonged grazing, but can also occur rapidly, as when a threshold of soil loss is crossed that prevents continued dominance by current occupants of a site (Friedel, 1991). Gradual vegetation change is more common on mesic grazing lands, whereas rapid or 'episodic' change is more prevelant on arid rangelands (Walker, 1993; Wiegand *et al.*, 1995). Changes in precipitation could cause rapid shifts in vegetation on arid rangelands, but global change will more often influence the susceptibility of vegetation to other factors than directly alter the 'state' of vegetation. Unfortunately, there is no universal method for recognizing the proximity of rangelands to thresholds of vegetation change or for predicting global change effects on the susceptibility of rangelands to change.

13.6 Management Implications

13.6.1 Rangelands managed for livestock production

Global change has implications for both land managers and national or regional policy-makers. Some effects of atmospheric and climatic change can be accommodated quite easily by each group. Others will require changes in practices and policies. Anticipated changes in forage quality and quantity, for example, are not likely to be novel, and could be dealt with by feeding supplements, albeit at an economic cost. Modest shifts in primary productivity could be accommodated by adjusting stocking rates, while climatic impacts on pests and diseases might require a shift in livestock type or breed. On the other hand, innovation and changes in management practices may be required to deal with substantial changes in plant composition, including invasions by different forage species or changes to non-forage species such as trees and shrubs.

These factors could create systematic changes in enterprise profitability, with consequences for regional economies. A simulation study of a beef cattle ranch in northern Australia (Campbell *et al.*, 1997) demonstrated the relative effects of different factors on whole-enterprise profitability (Table 13.3). The precise figures should be given no significance, because they are strongly dependent on enterprise, management and markets, but they make two important points.

1. Changes in pasture composition may have a far greater effect on profitability than changes in plant productivity.
2. Modest shifts in prices will affect profitability as much as other changes.

Rangeland managers already must cope with highly variable physical and marketing environments. Any factor that reduces variability should simplify management and potentially increase profitability. A decrease in interannual variability of forage production, for instance, could benefit profitability, whether the decrease results from changes in plant growth patterns or from

Table 13.3. Simulated impact of global changes on profitability of a commercial cattle ranch in northern Australia (Campbell *et al.*, 1997, supplemented with more recent unpublished simulations). Simulations ran for 100 years with realistic weather and with constant management strategy, costs and prices. Results are the change from a baseline simulation for an enterprise with annual turnover of Aus$450,000.

Parameter impacted	Implementation	Change in annual profit (Aus$'000s)
Plant productivity	Assumes the maximal realistic effect of CO_2 on transpiration, nitrogen and radiation use efficiency (a substantial part of the benefit results from reduced interannual variability)	+37
Forage composition	20% reduction in perennial grasses	−60
Tree/grass balance	2.5% increase in tree basal area with consequent reduction in forage production	−107
Forage quality	10% decrease in liveweight gain per year (this could be compensated by supplements)	−35
Sale prices	10% increase in sale prices	+82
Transport costs	10% increase in direct fuel costs	−2

changes in climate (Campbell *et al.*, 1997). On the other hand, profitability could be reduced by climatic changes (such as a greater incidence of drought) that increase variability in forage production. Given the variety of environments on rangelands, we should expect the relative importance of various global change effects to differ among regions. Indeed, interannual variability is of greater concern on arid than mesic rangelands, while the anticipated decline in forage quality will be a greater problem in infertile than fertile environments (Table 13.4).

At the regional scale, then, it is clear that there could be systematic changes in profitability. We are, however, only beginning to acquire the information and to develop the tools necessary to predict these changes reliably. The simulation approach described above could be expanded to consider regional implications of changes for rangeland profitability. Baker *et al.* (1993) described such an effort for the western USA, while Campbell *et al.* (1996) described a general approach for assessing the management implications and economic consequences of global change. To provide relevant information, global change studies must focus on issues appropriate to different regions. In general, however, global change effects on diet quality and vegetation composition are the two areas of poorest knowledge.

13.6.2 Rangelands managed for other uses

Rangeland issues of greatest concern to society will differ among regions and may not include livestock production. Results of an informal survey of policy-makers in northern Australia and the western USA, for example, clearly demonstrate that livestock productivity is but one of a suite of concerns in a changing world (Table 13.5). In general, commercial rangelands of the world are tending towards less intensive grazing by livestock, with increased emphasis on other uses such as urbanization (USA), game ranching (South Africa, USA) and the transfer of former homelands to indigenous peoples

Table 13.4. Atmospheric change and its management implications for rangelands that continue to be used for grazing.

System description	Fertile	Infertile
Subhumid/ subtropical	Drop in forage quality, but probably easily compensated for at a cost with supplements; risk of shifting tree/grass balance	Drop in forage quality most significant. May cause marginal areas to be taken out of production
Arid	Increase in WUE and consequent reduction in variability of plant production between years significantly improves reliability and profitability	Minor reduction in variability and increase in production efficiency

(South Africa, Australia). Global change effects on species composition are likely to be more important than those on production in these rangelands (Table 13.6). Meanwhile, rangelands utilized for subsistence (such as those in the Sahel, the Thar Desert in India, and the margins of China) are being used more intensively to meet the needs of expanding human populations. Changes in plant production are the dominant concern in these areas.

Table 13.5. Issues raised by policy-makers for commercial rangelands of Australia and USA (unpublished data, Mark Stafford Smith, Jack Morgan).

Australia	USA
Woodland thickening	Aging of agricultural population
Costs and impacts of supplementary feeding	Urbanization and increasing land use conflict
Industry viability on marginal lands – grazing vs. cropping	Consumptive vs. non-consumptive uses of wildlife
Fire frequencies and degradation risks	Recreation interests
Opportunities for carbon sequestration	Water quality and supply
Changing drought frequency and tax incentives for managing it	Plant species migration
Regional conservation of biodiversity	Government fees and incentives on public and private land uses

Table 13.6. Implications of global change for regions where land use is dominated by intensification vs. land abandonment.

General land use pressure	Implications of global change
Intensification (Africa, China, Asia, etc.)	Rising population places more intensive demands on food production Effects of atmospheric changes will vary across systems (see Table 13.4) and depend greatly on regional climate change. Effects on plant production will be important, but compositional change towards woody plants is not likely to be a problem, given human needs for fuel wood
Reduction in grazing (USA, Australia, Europe, some of S. Africa, some of S. America)	Abandonment or substantial reduction in stocking rates tends to reduce the amount of rangeland used for livestock production and the intensity of land use. Both changes lead to greater game ranching, tourism, nature conservation, etc. Most direct atmospheric effects will be unimportant for primary land uses (e.g. tourism, hunting), although non-domestic grazers may suffer from diet quality reductions. Increases in trees and shrubs could decrease fire, while greater fuel production could increase fire frequency

13.6.3 Management implications – summary

On commercial rangelands, profitability may change and management may have to be adjusted to accommodate vegetation change. The general need to optimize stocking rates will continue to be essential, and flexibility in management will remain important. The rate at which management must adjust to accommodate climatic and atmospheric change will be considerably slower than adjustments required to track market fluctuations and other aspects of the normal operating environment. The situation will differ on marginal rangelands. On the productive margins, land use pressures are likely to cause substitution by attempts at cropping. On the least productive margins, changes in profitability may tilt land use decisions in favour of or against grazing, depending on local conditions. In general, however, land use decisions on marginal rangelands will be driven by factors other than global change effects on plant and animal productivity – these will simply serve to enhance or delay existing trends.

13.7 Conclusions

Of the global change effects discussed here, none is potentially more significant for rangelands than a shift in botanical composition. Yet species change is also among the most difficult of global change effects to predict, because it is highly context-specific at the spatial and temporal scales of field experiments. What is needed is a more systematic examination of the ways in which global changes interact with and modify the characteristics of plants that determine their responses to disturbances and environmental conditions. Vegetation change will probably be more closely coupled to changes in soil resources than to immediate physiological responses of plants to [CO_2] concentration or temperature. Understanding how soil resources respond to global change is thus also a priority. Predicting and adapting to these changes are among the major challenges faced by rangeland scientists, land managers and policy-makers.

Acknowledgements

This chapter benefited greatly from the comments and input of Debra Coffin, Justin Derner, Rodney Pennington, Charles Tischler, Jake Weltzin and Brian Wilsey.

References

Alcamo, J., Kreileman, G.J.J., Bollen, J.C., van den Born, G.J., Gerlagh, R., Krol, M.S., Toet, A.M.C. and de Vries, H.J.M. (1996) Baseline scenarios of global environmental change. *Global Environmental Change* 6, 261–303.

Allen-Diaz, B. (1996) Rangelands in a changing climate: impacts, adaptations, and mitigation. In: Watson, R.T., Zinyowera, M.C., Moss, R.H. and Dokken, D.J. (eds) *Climate Change 1995 – Impacts, Adaptations and Mitigation of Climate Change: Scientific–Technical Analyses.* Cambridge University Press, New York, pp. 130–158.

Arnone, J.A. III (1996) Predicting responses of tropical plant communities to elevated CO_2: lessons from experiments with model ecosystems. In: Körner, Ch. and Bazzaz, F.A. (eds) *Carbon Dioxide, Populations, and Communities.* Academic Press, San Diego, pp. 101–121.

Baker, B.B., Hanson, J.D., Bourdon, R.M. and Eckert, J.B. (1993) The potential effects of climate change on ecosystem processes and cattle production on US rangelands. *Climatic Change* 25, 97–117.

Bassow, S.L., McConnaughay, K.D.M. and Bazzaz, F.A. (1994) The response of temperate tree seedlings grown in elevated CO_2 atmospheres to extreme temperature events. *Ecological Applications* 4, 593–603.

Bazzaz, F.A., Bassow, S.L., Berntson, G.M. and Thomas, S.C. (1996) Elevated CO_2 and terrestrial vegetation: implications for and beyond the global carbon budget. In: Walker, B.H. and Steffen, W.L. (eds) *Global Change and Terrestrial Ecosystems.* Cambridge University Press, Cambridge, UK, pp. 43–76.

Betancourt, J.L. (1996) Long- and short-term climate influences on southwestern shrublands. In: Barrow, J.R., McArthur, E.D., Sosebee, R.E. and Taucsh, R.J. (compilers) *Proceedings: Shrubland Ecosystem Dynamics in a Changing Environment.* USDA Forest Service, Intermountain Research Station, Ogden, Utah, pp. 5–9.

Bowes, G. (1993) Facing the inevitable: plants and increasing atmospheric CO_2. *Annual Review of Plant Physiology and Molecular Biology* 44, 309–332.

Bowler, J.M. and Press, M.C. (1996) Effects of elevated CO_2, nitrogen form and concentration on growth and photosynthesis of a fast- and slow-growing grass. *New Phytologist* 132, 391–401.

Caldwell, M.M., Richards, J.H., Johnson, D.A., Nowak, R.S. and Dzurec, R.S. (1981) Coping with herbivory: photosynthetic capacity and resource allocation in two semiarid *Agropyron* bunchgrasses. *Oecologia* 50, 14–24.

Campbell, B.D., McKeon, G.M., Gifford, R.M., Clark, H., Stafford Smith, D.M., Newton, P.C.D. and Lutze, J.L. (1996) Impacts of atmospheric composition and climate change on temperate and tropical pastoral agriculture. In: Bouma, W.J., Pearman, G.I. and Manning, M.R. (eds) *Greenhouse: Coping with Climate Change.* CSIRO Publishing, Melbourne, pp. 171–189.

Campbell, B.D., Stafford Smith, D.M. and McKeon, G.M. (1997) Elevated CO_2 and water supply interactions in grasslands: a pastures and rangelands management perspective. *Global Change Biology* 3, 177–187.

Chapin, F.S. III, Shaver, G.R., Giblin, A.E., Nadelhoffer, K.J. and Laundre, J.A. (1995) Responses of arctic tundra to experimental and observed changes in climate. *Ecology* 76, 694–711.

Chiariello, N.R. and Field, C.B. (1996) Annual grassland responses to elevated CO_2 in multiyear community microcosms. In: Körner, C. and Bazzaz, F.A. (eds) *Carbon Dioxide, Populations, and Communities.* Academic Press, San Diego, pp. 139–157.

Coleman, J.S., Rochefort, L., Bazzaz, F.A. and Woodward, F.I. (1991) Atmospheric CO_2, plant nitrogen status and the susceptibility of plants to an acute increase in temperature. *Plant, Cell and Environment* 14, 667–674.

Cotrufo, M.F., Ineson, P. and Scott, A. (1998) Elevated CO_2 reduces the nitrogen concentration of plant tissues. *Global Change Biology* 4, 43–54.

Coughenour, M.B. and Chen, D.-X. (1997) Assessment of grassland ecosystem responses to atmospheric change using linked plant–soil process models. *Ecological Applications* 7, 802–827.

Dahl, B.E. and Sosebee, R.E. (1991) Impacts of weeds on herbage production. In: James, L.F., Evans, J.O., Ralphs, M.H. and Child, R.D. (eds) *Noxious Range Weeds.* Westview Press, Boulder, Colorado, pp. 153–164.

Dove, H. (1996) The ruminant, the rumen and the pasture resource: nutrient interactions in the grazing animal. In: Hodgson, J. and Illius, A.W. (eds) *The Ecology and Management of Grazing Systems.* CAB International, Wallingford, UK, pp. 219–246.

Drake, B.G. (1992) A field study of the effects of elevated CO_2 on ecosystem processes in a Chesapeake Bay wetland. *Australia Journal of Botany* 40, 579–595.

Drake, B.G., Gonzàlez-Meler, M.A. and Long, S.P. (1997) More efficient plants: a consequence of rising atmospheric CO_2? *Annual Review of Plant Physiology and Molecular Biology* 48, 609–639.

Ehleringer, J.R., Phillips, S.L., Schuster, W.S.F. and Sandquist, D.R. (1991) Differential utilization of summer rains by desert plants. *Oecologia* 88, 430–434.

Ehleringer, J.R., Cerling, T.E. and Helliker, B.R. (1997) C_4 photosynthesis, atmospheric CO_2, and climate. *Oecologia* 112, 285–299.

Epstein, H.E., Lauenroth, W.K., Burke, I.C. and Coffin, D.P. (1997) Productivity patterns of C_3 and C_4 functional types in the US Great Plains. *Ecology* 78, 722–731.

Field, C.B., Jackson, R.B. and Mooney, H.A. (1995) Stomatal responses to increased CO_2: implications from the plant to the global scale. *Plant, Cell and Environment* 18, 1214–1225.

Field, C.B., Lund, C.P., Chiariello, N.R. and Mortimer, B.E. (1997) CO_2 effects on the water budget of grassland microcosm communities. *Global Change Biology* 3, 197–206.

Field, T.R.O. and Forbe, M.B. (1990) Effects of climate warming on the distribution of C_4 grasses in New Zealand. *Proceedings of the New Zealand Grassland Association* 51, 47–50.

Friedel, M.H. (1991) Range condition assessment and the concept of thresholds: a viewpoint. *Journal of Range Management* 44, 422–426.

Ghannoum, O., von Caemmerer, S., Barlow, E.W.R. and Conroy, J.P. (1997) The effect of CO_2 enrichment and irradiance on the growth, morphology and gas exchange of a C_3 (*Panicum laxum*) and a C_4 (*Panicum antidotale*) grass. *Australian Journal of Plant Physiology* 24, 227–237.

Giorgi, R., Meehl, G.A., Kattenberg, A., Grassl, H., Mitchell, J.F.B., Stouffer, R.J., Tokioka, T., Weaver, A.J. and Wigley, T.M.L. (1998) Simulation of regional climate change with global coupled climate models and regional modelling techniques. In: Watson, R.T., Zinyowera, M.C., Moss, R.H. and Dokken, D.J. (eds) *The Regional Impacts of Climate Change: an Assessment of Vulnerability.* Cambridge University Press, New York, pp. 427–437.

Ham, J.M., Owensby, C.E., Coyne, P.I. and Bremer, D.J. (1995) Fluxes of CO_2 and water vapor from a prairie ecosystem exposed to ambient and elevated atmospheric CO_2. *Agricultural and Forest Meteorology* 77, 73–93.

Harte, J. and Shaw, R. (1995) Shifting dominance within a montane vegetation community: results of a climate-warming experiment. *Science* 267, 876–880.

Haxeltine, A. and Prentice, I.C. (1996) BIOME3: an equilibrium terrestrial biosphere model based on ecophysiological constraints, resource availability and competition among plant functional types. *Global Biogeochemical Cycles* 10, 693–710.

Hunt, H.W., Elliott, E.T., Detling, J.K., Morgan, J.A. and Chen, D.-X. (1996) Responses of a C_3 and a C_4 perennial grass to elevated CO_2 and temperature under different water regimes. *Global Change Biology* 2, 35–47.

Jackson, R.B., Sala, O.E., Field, C.B. and Mooney, H.A. (1994) CO_2 alters water use, carbon gain, and yield for the dominant species in a natural grassland. *Oecologia* 98, 257–262.

Jackson, R.B., Luo, Y., Cardon, Z.G., Sala, O.E., Field, C.B. and Mooney, H.A. (1995) Photosynthesis, growth and density for the dominant species in a CO_2-enriched grassland. *Journal of Biogeography* 22, 221–225.

Jackson, R.B., Sala, O.E., Paruelo, J.M. and Mooney, H.A. (1998) Ecosystem water fluxes for two grasslands in elevated CO_2: a modelling analysis. *Oecologia* 113, 537–546.

Johnson, H.B., Polley, H.W. and Mayeux, H.S. (1993) Increasing CO_2 and plant–plant interactions: effects on natural vegetation. *Vegetatio* 104/105, 157–170.

Jones, M.B. and Jongen, M. (1996) Sensitivity of temperate grassland species to elevated atmospheric CO_2 and the interaction with temperature and water stress. *Agricultural and Food Science in Finland* 5, 271–283.

Karl, T.R. (1998) Regional trends and variations of temperature and precipitation. In: Watson, R.T., Zinyowere, M.C. and Moss, R.H. (eds) *The Regional Impacts of Climate Change: an Assessment of Vulnerability*. Cambridge University Press, New York, pp. 413–425.

Kirkham, M.B., He, H., Bolger, T.P., Lawlor, D.J. and Kanemasu, E.T. (1991) Leaf photosynthesis and water use of big bluestem under elevated carbon dioxide. *Crop Science* 31, 1589–1594.

Knapp, A.K., Fahnestock, J.T. and Owensby, C.E. (1994) Elevated atmospheric CO_2 alters stomatal responses to variable sunlight in a C_4 grass. *Plant, Cell and Environment* 17, 189–195.

Körner, Ch. (1995) Biodiversity and CO_2: global change is under way. *Gaia* 4, 234–243.

Körner, Ch. (1996) The response of complex multispecies systems to elevated CO_2. In: Walker, B.H. and Steffen, W.L. (eds) *Global Change and Terrestrial Ecosystems.* Cambridge University Press, Cambridge, UK, pp. 20–42.

Körner, Ch., Diemer, M., Schäppi, B. and Zimmermann, L. (1996) Response of alpine vegetation to elevated CO_2. In: Koch, G.W. and Mooney, H.A. (eds) *Carbon Dioxide and Terrestrial Ecosystems.* Academic Press, San Diego, pp. 177–196.

Leadley, P.W. and Körner, Ch. (1996) Effects of elevated CO_2 on plant species dominance in a highly diverse calcareous grassland. In: Körner, Ch. and Bazzaz, F.A. (eds) *Carbon Dioxide, Populations, and Communities.* Academic Press, San Diego, pp. 159–175.

LeCain, D.R. and Morgan, J.A. (1998) Growth, photosynthesis, leaf nitrogen and carbohydrate concentrations in NAD-ME and NADP-ME C_4 grasses grown in elevated CO_2. *Physiologia Plantarum* 102, 297–306.

Long, S.P. (1991) Modification of the response of photosynthetic productivity to rising temperature by atmospheric CO_2 concentrations: has its importance been underestimated? *Plant, Cell and Environment* 14, 729–739.

Milchunas, D.G., Sala, O.E. and Lauenroth, W.K. (1988) A generalized model of the effects of grazing by large herbivores on grassland community structure. *American Naturalist* 132, 87–106.

Morgan, J.A., Knight, W.G., Dudley, L.M. and Hunt, H.W. (1994) Enhanced root system C-sink activity, water relations and aspects of nutrient acquisition in mycotrophic *Bouteloua gracilis* subjected to CO_2 enrichment. *Plant and Soil* 165, 139–146.

Neilson, R.P. (1986) High-resolution climatic analysis and southwest biogeography. *Science* 232, 27–34.

Neilson, R.P. (1995) A model for predicting continental-scale vegetation distribution and water balance. *Ecological Applications* 5, 362–385.

Neilson, R.P., Prentice, I.C., Smith, B., Kittel, T. and Viner, D. (1998) Simulated changes in vegetation distribution under global warming. In: Watson, R.T., Zinyowere, M.C. and Moss, R.H. (eds) *The Regional Impacts of Climate Change: an Assessment of Vulnerability*. Cambridge University Press, New York, pp. 439–456.

Nie, D., He, H., Mo, G., Kirkham, M.B. and Kanemasu, E.T. (1992) Canopy photosynthesis and evapotranspiration of rangeland plants under doubled carbon dioxide in closed-top chambers. *Agricultural and Forest Meteorology* 61, 205–217.

Owensby, C.E., Coyne, P.I., Ham, J.M., Auen, L.A. and Knapp, A.K. (1993a) Biomass production in a tallgrass prairie ecosystem exposed to ambient and elevated CO_2. *Ecological Applications* 3, 644–653.

Owensby, C.E., Coyne, P.I. and Auen, L.M. (1993b) Nitrogen and phosphorus dynamics of a tallgrass prairie ecosystem exposed to elevated carbon dioxide. *Plant, Cell and Environment* 16, 843–850.

Owensby, C.E., Cochran, R.C. and Auen, L.M. (1996) Effects of elevated carbon dioxide on forage quality for ruminants. In: Körner, Ch. and Bazzaz, F.A. (eds) *Carbon Dioxide, Populations and Communities*. Academic Press, San Diego, pp. 363–371.

Owensby, C.E., Ham, J.M., Knapp, A.K., Bremner, D. and Auen, L.M. (1997) Water vapour fluxes and their impact under elevated CO_2 in a C_4-tallgrass prairie. *Global Change Biology* 3, 189–195.

Parton, W.J., Ojima, D.S. and Schimel D.S. (1994) Environmental change in grasslands: assessment using models. *Climatic Change* 28, 111–141.

Parton, W.J., Scurlock, J.M.O., Ojima, D.S., Schimel, D.S., Hall, D.O. and Scopegram Group Members (1995) Impact of climate change on grassland production and soil carbon worldwide. *Global Change Biology* 1, 13–22.

Polley, H.W., Johnson, H.B., Mayeux, H.S., Brown, D.A. and White, J.W.C. (1996a) Leaf and plant water use efficiency of C_4 species grown at glacial to elevated CO_2 concentrations. *International Journal of Plant Sciences* 157, 164–170.

Polley, H.W., Johnson, H.B., Mayeux, H.S., Tischler, C.R. and Brown, D.A. (1996b) Carbon dioxide enrichment improves growth, water relations and survival of droughted honey mesquite (*Prosopis glandulosa*) seedlings. *Tree Physiology* 16, 817–823.

Polley, H.W., Mayeux, H.S., Johnson, H.B. and Tischler, C.R. (1997) Viewpoint: atmospheric CO_2, soil water, and shrub/grass ratios on rangelands. *Journal of Range Management* 50, 278–284.

Read, J.J., Morgan, J.A., Chatterton, N.J. and Harrison, P.A. (1997) Gas exchange and carbohydrate and nitrogen concentrations in leaves of *Pascopyrum smithii* (C_3) and *Bouteloua gracilis* (C_4) at different carbon dioxide concentrations and temperatures. *Annals of Botany* 79, 197–206.

Reekie, E.G. and Bazzaz, F.A. (1989) Competition and patterns of resource use among seedlings of five tropical trees grown at ambient and elevated CO_2. *Oecologia* 79, 212–222.

Rounsevell, M.D.A., Brignall, A.P. and Siddons, P.A. (1996) Potential climate change effects on the distribution of agricultural grassland in England and Wales. *Soil Use and Management* 12, 44–51.

Roy, J., Guillerm, J.-L., Navas, M.-L. and Dhillion, S. (1996) Responses to elevated CO_2 in Mediterranean old-field microcosms: species, community, and ecosystem components. In: Körner, Ch. and Bazzaz, F.A. (eds) *Carbon Dioxide, Populations, and Communities*. Academic Press, San Diego, pp. 123–138.

Ryan, K.C. (1991) Vegetation and wildland fire: implications of global climate change. *Environment International* 17, 169–178.

Sage, R.F. (1994) Acclimation of photosynthesis to increasing atmospheric CO_2: the gas exchange perspective. *Photosynthesis Research* 39, 351–368.

Sage, R.F., Sharkey, T.D. and Seeman, J.R. (1989) Acclimation of photosynthesis to CO_2 in five C_3 species. *Plant Physiology* 89, 590–596.

Schenk, U., Jager, H.-J. and Weigel, H.-J. (1997) The response of perennial ryegrass/white clover mini-swards to elevated atmospheric CO_2 concentrations: effects on yield and fodder quality. *Grass and Forage Science* 52, 232–241.

Smith, T. and Huston, M. (1989) A theory of the spatial and temporal dynamics of plant communities. *Vegetatio* 83, 49–69.

Soussana, J.F., Casella, E. and Loiseau, P. (1996) Long-term effects of CO_2 enrichment and temperature increase on a temperate grass sward. II. Plant nitrogen budgets and root fraction. *Plant and Soil* 182, 101–114.

Stafford Smith, M. (1996) Management of rangelands: paradigms at their limits. In: Hodgson, I. and Illius, A.W. (eds) *The Ecology and Management of Grazing Systems.* CAB International, Wallingford, UK, pp. 325–357.

Stafford Smith, M., Campbell, B.D., Archer, S. and Steffen, W. (1995) *GCTE Focus 3 – Pastures and Rangelands Network: an Implementation Plan.* Global Change and Terrestrial Ecosystems Report No. 3, CSIRO, Canberra, Australia, 59 pp.

Stephenson, N.L. (1990) Climatic control of vegetation distribution: the role of the water balance. *American Naturalist* 135, 649–670.

Stock, W.D. and Midgley, G.F. (1995) Ecosystem response to elevated CO_2: nutrient availability and nutrient cycling. In: Moreno, J.M. and Oechel, W.C. (eds) *Global Change and Mediterranean-type Ecosystems.* Ecological Studies 117. Springer-Verlag, New York, pp. 326–342.

Stoddart, L.A., Smith, A.D. and Box, T.W. (1975) *Range Management.* McGraw-Hill, New York, 532 pp.

Teeri, J.A. and Stowe, L.G. (1976) Climatic patterns and the distribution of C_4 grasses in North America. *Oecologia* 23, 1–12.

Ulyatt, M.J., Fennessy, P.F., Rattray, P.V. and Jagush, K.T. (1980) The nutritive value of supplements. In: Drew, K.R. and Fennesy, P.F. (eds) *Supplementary Feeding.* New Zealand Society of Animal Production, Mosgeil, New Zealand, pp.157–184.

Vitousek, P.M. (1990) Biological invasions and ecosystem processes: towards an integration of population biology and ecosystem studies. *Oikos* 57, 7–13.

Walker, B.H. (1993) Rangeland ecology: understanding and managing change. *Ambio* 22, 80–87.

Weltzin, J.F. and McPherson, G.R. (1997) Spatial and temporal soil moisture resource partitioning by trees and grasses in a temperate savanna, Arizona, USA. *Oecologia* 112, 156–164.

Westoby, M. (1980) Elements of a theory of vegetation dynamics in arid rangelands. *Israel Journal of Botany* 28, 169–194.

Wiegand, T., Milton, S.J. and Wissel, C. (1995) A simulation model for a shrub ecosystem in the semiarid Karoo, South Africa. *Ecology* 76, 2205–2221.

Williams, K.J., Wilsey, B.J., McNaughton, S.J. and Banyikwa, F.F. (1998) Temporally variable rainfall does not limit yields of Serengeti grasses. *Oikos* 81, 463–470.

Wilsey, B.J. (1996) Urea additions and defoliation affect plant responses to elevated CO_2 in a C_3 grass from Yellowstone National Park. *Oecologia* 108, 321–327.

Wilsey, B.J., Coleman, J.S. and McNaughton, S.J. (1997) Effects of elevated CO_2 and defoliation on grasses: a comparative ecosystem approach. *Ecological Applications* 7, 844–853.

Wilson, J.R. (1982) Environmental and nutritional factors affecting herbage quality. In: Hacker, J.B. (ed.) *Nutritional Limits to Animal Production from Pastures.* Commonwealth Agricultural Bureaux, Farnham Royal, UK, pp. 111–131.

Woodward, F.I. (1993) Leaf response to the environment and extrapolation to larger scales. In: Solomon, A.M. and Shugart, H.H. (eds) *Vegetation Dynamics and Global Change.* Chapman & Hall, New York, pp. 71–100.

Young, J.A. (1991) Cheatgrass. In: James, L.F., Evans, J.O., Ralphs, M.H. and Child, R.D. (eds) *Noxious Range Weeds.* Westview, Boulder, Colorado, pp. 408–418.

14 Crop Ecosystem Responses to Climatic Change: Crassulacean Acid Metabolism Crops

PARK S. NOBEL

Department of Biology-OBEE, University of California, Los Angeles CA 90095-1606, USA

14.1 Introduction and Background

The best known crassulacean acid metabolism (CAM) crop species is the bromeliad pineapple (*Ananas comosus*), which is cultivated for its fruit on 720,000 ha in about 40 countries (Table 14.1; Bartholomew and Rohrbach, 1993). However, another CAM species *(Opuntia ficus-indica)*, referred to as prickly pear, prickly pear cactus, cactus pear and nopal, is cultivated on just over 1 Mha in about 30 countries (Table 14.1; Russell and Felker, 1987; Nobel, 1996a; Mizrahi *et al.*, 1997). Most such cultivation is for its stem segments (termed cladodes) that are used both for forage and fodder for cattle, goats and sheep; for example, about 400,000 ha are so utilized in Brazil. Cladodes are also harvested as a vegetable for human consumption, especially in Mexico. About 10% of the area for cultivation of *O. ficus-indica* and closely related opuntias is for their fruit, which has long been an important crop in Sicily and is cultivated most extensively in Mexico. About 20 other species in six genera of cacti are also cultivated for their fruits. These include species of *Stenocereus* and other columnar cacti in Mexico (Pimienta-Barrios and Nobel, 1994; P.S. Nobel, unpublished observations) and *Hylocereus* and *Seleniocereus* in Colombia, Israel, Mexico, USA, Vietnam, and other countries (Gibson and Nobel, 1986; Mizrahi *et al.*, 1997). This market is relatively small (Table 14.1) but expanding rapidly.

In the early part of the 20th century in various countries in eastern Africa, in substantial areas in India, and in the Americas, another CAM species, *Agave sisalana*, was extensively cultivated for the fibre in its leaves (Gentry, 1982). *Agave fourcroydes* was also cultivated in Mexico, mostly in the Yucatan peninsula. Because of the advent of synthetic fibres that were cheaper and more tolerant of moisture, the cultivation of these two agave species has declined substantially. However, its cultivation is still appreciable (Table 14.1) and certain related species are being considered as new fibre crops

Table 14.1. Principal cultivated CAM species, with part harvested and worldwide cultivation area indicated. Data involve extrapolations and estimates.

Species	Harvested part or purpose	Area of cultivation (ha)	References
Agave fourcroydes, *A. sisalana*	Leaf fibre	400,000	Gentry, 1982; FAO FAOSTAT Statistics Database (http://apps.fao.org/)
Agave mapisaga, *A. salmiana*, *A. tequilana*, various other agaves	Beverages, some fodder	88,000	Gentry, 1982; Nobel, 1994
Ananas comosus	Fruit	720,000	Bartholomew and Rohrbach, 1993; FAO FAOSTAT Statistics Database
Hylocereus, *Seleniocereus* and *Stenocereus* species, various other cacti	Fruit	14,000	Mizrahi *et al.*, 1997; Pimienta-Barrios and Nobel, 1994
Opuntia ficus-indica, other opuntias	Fodder, forage, vegetable	900,000	Barbera *et al.*, 1995; Nobel, 1994
	Fruit	95,000	
	Cochineal dye, chemicals	50,000	
Total		2,267,000	

(McLaughlin, 1993; Ravetta and McLaughlin, 1996). Other agaves are currently cultivated for alcoholic beverages, especially *Agave tequilana* for tequila. In addition, about ten species of agave are cultivated for mescal (also spelled mezcal), which, like tequila, is a distilled beverage, and for pulque, a fermented beverage. Commercial production of all these beverages occurs in Mexico.

The present worldwide cultivation of agaves as crops totals nearly 500,000 ha (Table 14.1). To this number can be added those agaves used as ornamental plants or for fences and erosion control. In addition to agaves, many species of cacti are also so utilized for these purposes. Because none of these CAM crops is particularly tolerant of freezing temperatures, nearly all cultivation of agaves, cacti and pineapple occurs within 30° latitude of the Equator (Nobel, 1988; Bartholomew and Rohrbach, 1993; Bartholomew and Malézieux, 1994). If temperatures rise as predicted during global climatic change, the regions suitable for the cultivation of such CAM plants will expand (Nobel, 1996b).

The most convincing evidence indicating that agaves, cacti and pineapple are CAM plants is the substantial nocturnal CO_2 uptake by their photosynthetic organs, which occurs in all three groups (Joshi *et al.*, 1965; Neales, 1973; Sale and Neales, 1980; Nose *et al.*, 1986; Nobel, 1988; Borland and Griffiths, 1989; Medina *et al.*, 1991). Because photosynthesis cannot occur without light, the CO_2 taken up at night cannot be immediately fixed into photosynthetic products such as glucose and sucrose. Rather, the CO_2 is incorporated into phosphoenolpyruvate (PEP) by the enzyme PEP carboxylase, leading to the formation of an organic acid such as malate. Nocturnal acidification of the chlorenchyma can be easily tested to demonstrate CAM in various species, including agaves, cacti and pineapple (Neales, 1973; Friend and Lydon, 1979; Nobel, 1988; Medina *et al.*, 1991, 1993). A sophisticated but indirect method for determining whether or not a particular species uses CAM is to find the ratio of various carbon isotopes in the plant tissues. In particular, differences in the enzymes involved in the initial fixation of carbon and the subsequent biochemical processing of the fixed carbon lead to unique isotopic signatures for the three photosynthetic pathways: CAM (which is the only photosynthetic type with nocturnal stomatal opening), C_3 and C_4. Thus, isotopic analysis of carbon by mass spectroscopy of tissue samples can indicate the photosynthetic pathway. This has been done for all three groups of CAM crops (agaves, cacti and pineapple; Nobel, 1988; Medina *et al.*, 1994).

A relevant question with respect to global climatic change and its influences on productivity (the major theme of this book) is what is the effect of increasing atmospheric CO_2 concentrations ([CO_2]) on net CO_2 uptake by CAM plants? Another question about CAM plants concerns effects on their gas exchange and productivity caused by changes in other environmental factors, such as air temperature, photosynthetic photon flux (PPF, wavelengths of 400–700 nm that are absorbed by photosynthetic pigments) and soil water status (quantified by the soil water potential and reflecting the effects of rainfall). Environmental effects on net CO_2 uptake and biomass productivity among CAM plants have been studied most extensively for *O. ficus-indica*, but sufficient data exist for predictions for other commercial CAM species as well.

One way to quantify the effects of environmental factors on net CO_2 uptake is to use an environmental productivity index (EPI; Nobel, 1984, 1988, 1991b). The individual environmental factors affect net CO_2 uptake multiplicatively, not additively. For instance, if prolonged drought causes daily stomatal opening to cease, then no net CO_2 uptake will occur, regardless of whether or not light levels and temperatures are optimal for CO_2 uptake. EPI can be represented as follows:

$$\text{EPI} = \text{fraction of maximal daily net } CO_2 \text{ uptake} \quad \text{(Eqn 14.1)}$$

$$= \text{water index} \times \text{temperature index} \times \text{PPF index}$$

The water index, which like the other indices ranges from 0 when it is totally limiting for daily net CO_2 uptake to 1 when it does not limit CO_2 uptake, represents the fractional limitation due to soil water availability; hence, it progressively decreases during drought. The temperature index quantifies the limitations of temperature on daily net CO_2 uptake, and the PPF index quantifies the effects of light. Because CAM plants take up CO_2 primarily at night, the photosynthetic responses represented by the PPF index relate total daily net CO_2 uptake over 24 h periods (mol CO_2 m^{-2} day^{-1}) to the total daily PPF (mol photons m^{-2} day^{-1}). All the indices in equation 14.1 are determined over 24 h periods.

A fourth multiplicative index can be incorporated into equation 14.1 to quantify the effects of nutrients on daily net CO_2 uptake. Such a nutrient index has been determined for various species of agaves and cacti, for which five elements (N, P, K, B and Na) have major influences on daily net CO_2 uptake (Nobel, 1989). For plants maintained under various ambient CO_2 levels for prolonged periods (months), the effects of $[CO_2]$ are most readily incorporated into predictions of net CO_2 uptake by making measurements over 24 h under conditions of wet soil, optimal temperatures and saturating PPF. Increases in $[CO_2]$ can also affect the influences of other environmental factors on net CO_2 uptake by CAM plants.

14.2 Gas Exchange

Crops with C_3 and C_4 type photosynthesis, which are the principal focus of this book and currently represent over 98% of cultivated crops in terms of land area utilized, inherently have a much lower water-use efficiency (WUE) than CAM plants. The WUE equals the net CO_2 fixed by photosynthesis divided by the water lost via transpiration on an instantaneous, daily or seasonal basis. The water vapour content of saturated air, which is the case for nearly all the air spaces within the shoots of plants, increases essentially exponentially with temperature. For example, air saturated with water vapour contains 6.8 g of water vapour m^{-3} at 5°C, 17.3 g m^{-3} at 20°C, and 39.7 g m^{-3} at 35°C. The water vapour content of the air surrounding plants is generally far below the saturation value and does not change much during the course of a day unless a major change in weather occurs. For air containing 4.0 g of water vapour m^{-3} (relative humidity of 59% at 5°C, 23% at 20°C, and 10% at 35°C), the shoot-to-air difference in water vapour content is 2.8 g m^{-3}, 13.3 g m^{-3}, and 35.7 g m^{-3} at the respective temperatures. The rate of water loss depends on this shoot-to-air difference in water vapour content and the amount of stomatal opening; for the same amount of stomatal opening, transpiration is thus 13.3/2.8 or 4.8-fold higher at 20°C than at 5°C and 35.7/13.3 or 2.7-fold higher at 35°C than at 20°C. If the shoot and air temperatures are 15°C cooler during the night than during the day (which is realistic where CAM plants are grown) transpiration would be three to five times lower during the night than during the day for the same amount of stomatal opening. This underscores the importance of nocturnal stomatal opening for water conservation by CAM plants. CAM crops also tend to have a lower maximal stomatal conductance than do C_3 and C_4 crops, which further reduces water loss (Nobel, 1988, 1994).

Plants using the C_3 or the C_4 photosynthetic pathway have a net CO_2 uptake only during the day, whereas CAM plants take up CO_2 predominantly during the night. However, they can also take up CO_2 during the day, especially when soil water is not limiting (Fig. 14.1). Any evaluation of shoot gas exchange over 24 h periods, as is necessary and conventional for CAM plants (Fig. 14.1), automatically includes the contribution of respiration to net CO_2 uptake. The maximal instantaneous rates of net CO_2 uptake by *Agave mapisaga* and *O. ficus-indica* occur at night and can be greater than those

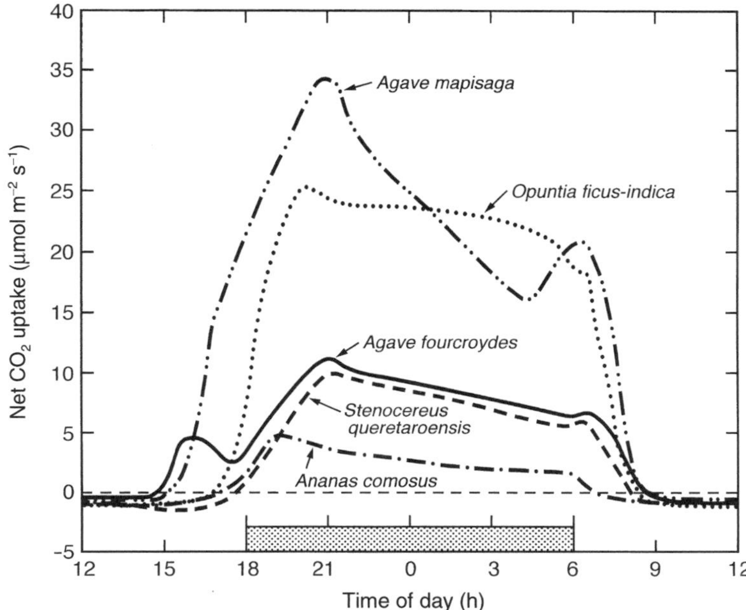

Fig. 14.1. Net CO_2 uptake over 24 h periods for various cultivated CAM species under approximately optimal conditions. Data for leaves of *Agave mapisaga* are from Nobel *et al.* (1992); for leaves of *Agave fourcroydes* from Nobel (1985); for leaves of *Ananas comosus* at a suboptimal PPF averaging 360 μmol m^{-2} s^{-1} from Medina *et al.* (1991); for stems of *Opuntia ficus-indica* from Nobel (1988) and P.S. Nobel (unpublished observations); and for stems of *Stenocereus queretaroensis* from Nobel and Pimienta-Barrios (1995) and P.S. Nobel (unpublished observations).

of most other perennials whose maximal rates occur during the day. The instantaneous net CO_2 uptake rates of these highly productive CAM species can exceed 25 μmol m^{-2} s^{-1} (Fig. 14.1), whereas nearly all ferns, shrubs and trees, as well as many C_3 and C_4 crops, have lower maximal uptake rates (Nobel, 1991b). The maximal net CO_2 uptake rates for *A. fourcroydes* and *Stenocereus queretaroensis* are about 10 μmol m^{-2} s^{-1} and are similar to those of many other perennials. Maximum instantaneous values for net CO_2 uptake by *Ananas comosus* examined over 24 h periods by various research groups are relatively low (Bartholomew and Malézieux, 1994); for example, 2.2 μmol m^{-2} s^{-1} (Borland and Griffiths, 1989), 2.4 μmol m^{-2} s^{-1} (Nose *et al.*, 1986), 3.6 μmol m^{-2} s^{-1} (Neales *et al.*, 1980) and 4.9 μmol m^{-2} s^{-1} with a suboptimal PPF (Fig. 14.1; Medina *et al.*, 1991). Indeed, net CO_2 uptake by *A. comosus* has apparently not been measured under optimal conditions. Yet because of the high WUE of all of these CAM species, they can have a relatively high productivity when cultivated under dryland (non-irrigated) conditions. In the future, the most important and relevant comparisons among crops differing in photosynthetic pathway may be based on WUE, because the available supply of groundwater is steadily decreasing, as is the supply of uncontaminated surface water available for irrigation at a reasonable cost.

14.3 Values for Component Indices of Environmental Productivity Index

To determine the component indices of EPI (equation 14.1), plants are generally placed in environmental chambers, and one environmental factor is varied while the others are maintained at optimal values. Net CO_2 uptake is then measured over 24 h periods. The only commercial CAM species for which the environmental indices relating to soil water status, temperature and PPF (equation 14.1) have been fully determined are *A. fourcroydes*, *O. ficus-indica* and *S. queretaroensis*; partial results are available for *Ananas comosus* (Neales *et al.*, 1980; Sale and Neales, 1980; Nose *et al.*, 1986) and *Agave salmiana* (Nobel *et al.*, 1996). Thus, these are the species whose net CO_2 uptake over 24 h periods can most readily be predicted in response to the environmental changes accompanying global climatic change. Extrapolations to other CAM species can be made if differences among the examined species can be rationalized.

At about 9 days of drought for *A. fourcroydes* and 13 days for *A. salmiana* the water index decreases 50% from its value of 1 under wet conditions (Fig. 14.2). The water index decreases more slowly for the stems of the two cacti used as crops because they have a much greater volume of tissue water storage per unit area across which water can be transpired. Drought of 23 days for *O. ficus-indica* and 36 days for *S. queretaroensis* reduces the water index by 50% (Fig. 14.2). The volume available for water storage per unit surface area (which indicates the average tissue depth supplying water for transpiration) is about 4 mm for leaves of *A. fourcroydes* (Nobel, 1985), 8 mm for leaves of *A. salmiana* (Nobel *et al.*, 1996), 22 mm for the stems of *O. ficus-indica* (Nobel, 1988) and 78 mm for stems of *S. queretaroensis* (Nobel and Pimienta-Barrios, 1995). A water index based on nocturnal acid accumulation (an indirect indication of total daily net CO_2 uptake) decreases 50% in 12 days for *A. salmiana*, which has an average water storage depth of 7 mm in its leaves (Nobel and Meyer, 1985), and in 7 days for *A. tequilana*, which has an average water storage depth of 3 mm in its leaves (Nobel and Valenzuela, 1987). For these five species, an empirical relation of 4.8 times the square root of the tissue depth for water storage (in millimetres) accurately predicts the drought duration in days that will cause the daily net CO_2 uptake to be halved. Unfortunately, the response of the water index to drought has not been studied for *Ananas comosus* (Bartholomew and Malézieux, 1994).

The optimal temperatures for net CO_2 uptake by the five cultivated CAM species presented in Fig. 14.3 fall within the range found for other CAM species (Nobel, 1988), but the temperatures are lower than for nearly all other cultivated crops. In particular, most net CO_2 uptake for commercial CAM species occurs at night; consequently, nocturnal temperatures are more important than are diurnal ones with respect to total daily net CO_2 uptake.

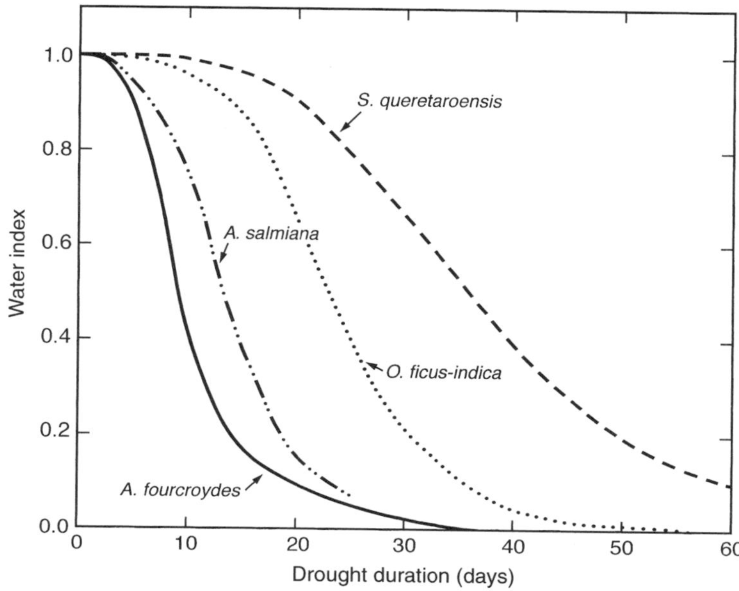

Fig. 14.2. Responses of the water index (see equation 14.1) to drought duration for *Agave fourcroydes, A. salmiana, O. ficus-indica,* and *S. queretaroensis*. Drought refers to the period when the shoot has a lower water potential than the soil just outside the roots in the centre of the root zone. Data for *A. fourcroydes* are from Nobel (1985); for *A. salmiana* from Nobel *et al.* (1996); for *O. ficus-indica* from Nobel and Hartsock (1983, 1984); and for *S. queretaroensis* from Nobel and Pimienta-Barrios (1995) and P.S. Nobel (unpublished observations).

Maximal daily net CO_2 uptake occurs at night-time temperatures of about 18°C for *A. fourcroydes*, 13°C for *A. salmiana*, 15°C for *Ananas comosus*, 15°C for *O. ficus-indica*, and 16°C for *S. queretaroensis* (Fig. 14.3). The optimal temperature for nocturnal net CO_2 uptake is about 15°C for *Agave americana* (Neales, 1973), and the optimal temperatures for nocturnal acid accumulation are about 12°C for *A. salmiana* (Nobel and Meyer, 1985; Nobel *et al.*, 1996) and 15°C for *A. tequilana* (Nobel and Valenzuela, 1987). Thus the optimal night temperature for daily net CO_2 uptake is about 15°C for all of these cultivated CAM species. This is a major consideration when determining where CAM plants will be cultivated. Temperature cannot be easily manipulated in the field, other than by location of the cultivated plots, whereas the water status can be controlled by irrigation and by light interception through spacing of plants, which affects interplant shading.

The responses of daily net CO_2 uptake to PPF are remarkably similar for the commercial CAM species examined, with 50% of maximal uptake occurring at a total daily PPF of 10 mol m^{-2} day^{-1} for *A. fourcroydes*, *Ananas comosus* and *S. queretaroensis* and at 11 mol m^{-2} day^{-1} for *O. ficus-indica* (Fig. 14.4). Moreover, the total daily PPF at which total daily net CO_2 uptake was zero was about 2 mol m^{-2} day^{-1} for all four species, with 95% of the

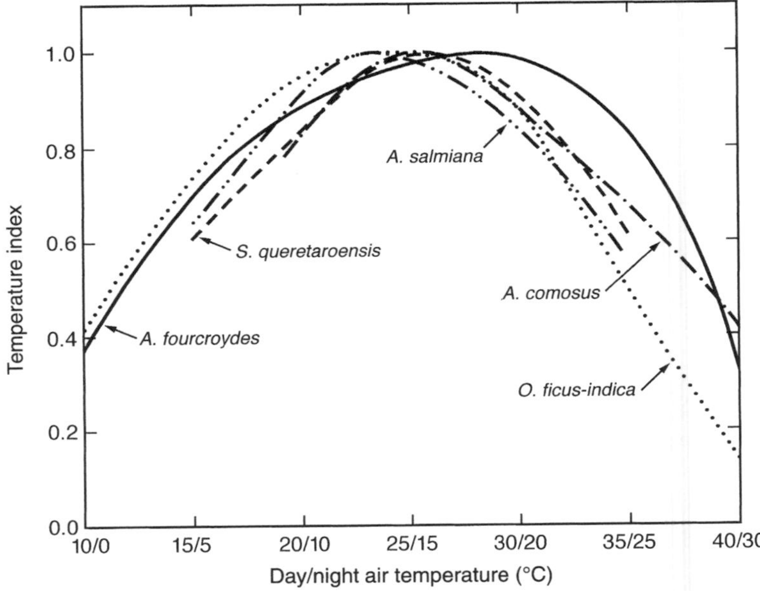

Fig. 14.3. Responses of the temperature index (see equation 14.1) to day/night air temperatures for *Agave fourcroydes, A. salmiana, Ananas comosus, O. ficus-indica* and *S. queretaroensis*. The plants were routinely kept at a particular day/night temperature regime for 10 days to allow for acclimation (Nobel, 1988). Data are for mean night temperatures for *S. queretaroensis* and for constant night temperatures for the other species. They are from the references cited in Fig. 14.2, plus Connelly (1972), Neales *et al.* (1980) and Bartholomew and Malézieux (1994) for *Ananas comosus*.

maximal net CO_2 uptake being achieved at a total daily PPF of about 25 mol m^{-2} day^{-1} (Fig. 14.4). Half of maximal nocturnal acid accumulation occurred at a total daily PPF of 9 mol m^{-2} day^{-1} for *A. salmiana* (Nobel and Meyer, 1985) and at 11 mol m^{-2} day^{-1} for *A. tequilana* (Nobel and Valenzuela, 1987); therefore the PPF responses of such cultivated CAM species are nearly identical. In this regard, the chlorenchyma for all of the agave and cactus species is a relatively thick 3–5 mm, and the amount of chlorophyll per unit surface area is a relatively high 0.7–0.9 g m^{-2} (Nobel, 1988). Thus, the light-absorbing properties of the photosynthetic organs are similar, leading to their similar responses of net CO_2 uptake to incident PPF as quantified by the PPF index (Fig. 14.4).

As mentioned above, a nutrient index can also be incorporated as a multiplicative factor in equation 14.1, although relatively little information on nutrient responses of CAM crops is available. Based on studies with four agave species and eleven cactus species, most of which are not crops, the five soil elements having the greatest effect on net CO_2 uptake, growth and biomass productivity are N, P, K, B and Na (Nobel, 1989); other macronutrients and micronutrients will likely have important effects as well and may be

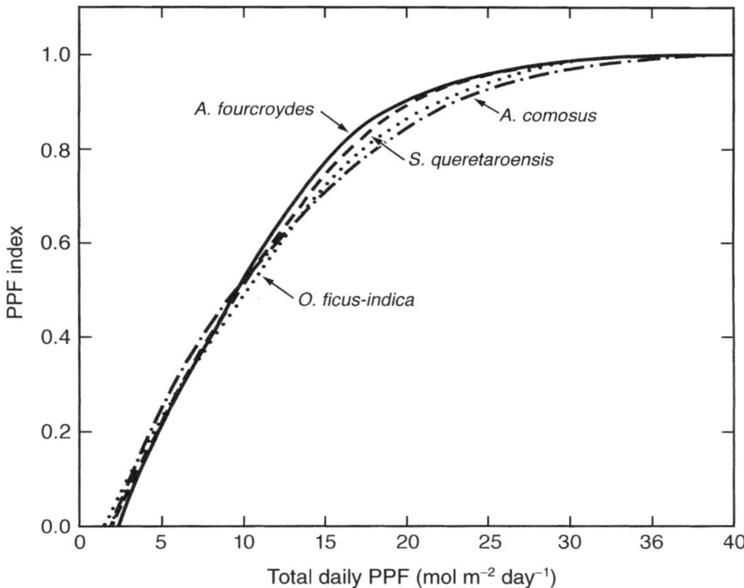

Fig. 14.4. Responses of the PPF index (see equation 14.1) to the total daily PPF for *Agave fourcroydes, Ananas comosus, O. ficus-indica,* and *S. queretaroensis.* Data for the agave and the cacti are for the PPF in the planes of the photosynthetic surfaces and are from the references cited in Fig. 14.2. For pineapple, data are for the PPF in a horizontal plane incident on the canopy and are from Sale and Neales (1980) and Nose *et al.* (1986).

particularly limiting in certain soils. Growth and biomass productivity are optimal at about 3 mg N g^{-1} (0.3% by soil mass), 60 μg P g^{-1}, 250 μg K g^{-1} and 1.0 μg B g^{-1} (Nobel, 1989). Suboptimal levels leading to half the maximal values are about 0.7 mg N g^{-1}, 5 μg P g^{-1}, 3 μg K g^{-1} and 0.04 μg B g^{-1}.

Soil Na inhibits biomass productivity for the agaves and cacti tested with 20% inhibition occurring at about 60 μg Na g^{-1} and 50% inhibition occurring at about 150 μg Na g^{-1} (Nobel, 1989). The decrease in growth resulting from increasing soil salinity is apparently less for *Ananas comosus* than for agaves and cacti, but little quantitative data are available (Bartholomew and Malézieux, 1994). In this regard, salinity can increase in arid and semi-arid regions as a consequence of poor land management practices and even global climatic change (Schlesinger *et al.*, 1990; Vitousek, 1994).

14.4 Productivity

Most productivity studies on agronomic C_3 and C_4 species were done before 1975 (Loomis and Gerakis, 1975). After that time, the focus has been on increasing their harvest index (fraction of the plant harvested), as well as on aspects of plant performance that can be affected by molecular biology/

biotechnology. Very little such attention has been devoted to CAM plants, although efforts to preserve the germplasm with field plantings in various countries have been made with prospects for increased productivity and future biotechnology in mind. CAM plants in nature are generally relatively slow-growing, but productivity for certain cultivated CAM species can be high. For example, the dry weight productivity of two cultivated CAM species, *Opuntia amyclea* and *O. ficus-indica*, equals or exceeds the highest values for all C_3 crops and trees and is exceeded by only a few C_4 crops (Table 14.2; Nobel, 1991a, 1996a). Moreover, because of the water-conserving characteristics of CAM (high WUE), the productivity of CAM plants can be substantially higher

Table 14.2. Maximal net CO_2 uptake and productivity of cultivated CAM plants, with comparisons for the C_3 and C_4 species with the highest productivities.

Species	Maximal net CO_2 uptake rate (μmol m^{-2} s^{-1})	Maximal daily net CO_2 uptake (mmol m^{-2} day^{-1})	Maximal productivity (t dry weight ha^{-1} $year^{-1}$)	References
Agave fourcroydes, A. sisalana, A. tequilana	11	360–400	about 15	Nobel, 1985; Nobel and Valenzuela, 1987; Nobel, 1988
Agave mapisaga, A. salmiana	29–34	1050–1170	38–42	Nobel *et al.*, 1992
Ananas comosus	5	240–310	20–31	Bartholomew and Kadzimin, 1977; Bartholomew, 1982; Medina *et al.*, 1991; Bartholomew and Rohrbach, 1993; Zhu *et al.*, 1997a
Opuntia amyclea, O. ficus-indica	26–28	940	45–50	Nobel *et al.*, 1992; Garcia de Cortázar and Nobel, 1992
Stenocereus queretaroensis	10	320	—	Nobel and Pimienta-Barrios, 1995
Six highest C_3 crops	av. 39	av. about 900	29–45	Loomis and Gerakis, 1975; Loomis, 1983; Nobel, 1991a
Six highest C_3 trees	–	–	36–44	Jarvis and Leverenz, 1983; Nobel, 1991a
Six highest C_4 crops	av. 46	av. about 1200	32–70	Loomis and Gerakis, 1975; Loomis, 1983; Nobel, 1991a

than that of C_3 and C_4 plants under conditions of low availability of soil water, an increasingly important aspect for crops worldwide.

Productivity per unit ground area per year, which is crucial for assessing agronomic potential, can be related to the net CO_2 uptake characteristics of plants (Nobel, 1988). In this regard, *A. mapisaga, A. salmiana, O. amyclea,* and *O. ficus-indica* have the highest maximal net CO_2 uptake rates among CAM crops, averaging 29 μmol m^{-2} s^{-1}, which is only 30% less than the average reported maximal rates for the C_3 and C_4 crops with the highest rates (Table 14.2). Net CO_2 uptake for CAM plants is generally measured and expressed for the surface most exposed to the incident PPF (one side of a leaf or a stem, which is usually opaque). For the thin leaves of C_3 and C_4 plants, net CO_2 uptake is measured for both sides of a leaf but is expressed per unit area of one side (the projected leaf area). If net CO_2 uptake by the sides of leaves or stems of CAM plants facing away from the highest PPF is also included, then the maximal rates are actually rather similar among plants of the three photosynthetic pathways. Because net CO_2 uptake by CAM plants can occur during both day and night (Fig. 14.1), compared with C_3 and C_4 plants which have net CO_2 uptake only during the day, the maximal daily net CO_2 uptake for these four highly productive CAM species is intermediate between the six C_3 crops with the highest values and the equivalent six C_4 species, even though a different area basis is conventionally used (Table 14.2). Moreover, for the CAM crops, the maximal total daily net CO_2 uptake is highly correlated with the maximal instantaneous net CO_2 uptake rate. Plant productivity per unit ground area for the various CAM crop plants considered is also highly correlated with the total daily net CO_2 uptake per unit photosynthetic surface area, although the actual productivity in the field depends on interplant spacing.

CAM plants minimize the energetically expensive release of CO_2 via photorespiration by presenting high levels of CO_2 to the pivotal enzyme, ribulose-1,5-bisphosphate carboxylase-oxygenase (Rubisco) (Nobel, 1988, 1991a, 1994). Thus photorespiration is relatively low for CAM plants, as is also the case for C_4 plants. Based on energy costs for the production of ATP and NADPH and the number of these molecules required for the fixation of CO_2 into carbohydrates, the total cost per carbon molecule fixed for CAM species is lower than that for C_3 species and only slightly higher than that for C_4 species (Nobel, 1991a, 1996a). Thus the intermediate values of total daily net CO_2 uptake and of long-term productivity by highly productive CAM species compared with similar C_3 and C_4 crops (Table 14.2) can be explained at a cellular level. This increases confidence in predictions that certain CAM crops will be more extensively cultivated in the future (Nobel, 1994, 1996a,b; Mizrahi *et al.*, 1997), because the CAM pathway can lead to high productivity.

14.5 Effects of Elevated [CO_2]

Although the long-term effects of months of elevated [CO_2] on net CO_2 uptake and productivity are relatively unstudied for commercial CAM species, such effects must be evaluated in view of the global climatic change caused by the just over 2 μmol CO_2 mol^{-1} current annual increase in [CO_2] (Houghton *et al.*, 1990; King *et al.*, 1992; Vitousek, 1994). In this regard, *Ananas comosus* growing for 4 months at 730 μmol CO_2 mol^{-1} has 23% more dry mass than plants growing at 330 μmol mol^{-1} (Zhu *et al.*, 1997b). Its nocturnal increase in leaf acidity is 42% higher for the approximately doubled [CO_2]. Studies using $\delta^{13}C$ indicate that most of the enhancement in net CO_2 uptake for *Ananas comosus* under elevated [CO_2] is at night via the CAM pathway, rather than during the day via the C_3 pathway (Zhu *et al.*, 1997b). Apparently the only species of agaves whose long-term responses to elevated [CO_2] have been studied are *A. deserti*, which is a CAM species but not a cultivated one, and *A. salmiana*. Raising the [CO_2] from 370 to 750 μmol mol^{-1} increases daily net CO_2 uptake by 49% throughout a 17-month period for *A. deserti* (Graham and Nobel, 1996). More leaves are produced per plant under the doubled [CO_2]. The combination of increased total leaf surface area and increased net CO_2 uptake per unit leaf area enhances dry mass accumulation by *A. deserti* under the doubled [CO_2] by 88% (Graham and Nobel, 1996). *Agave salmiana* grown for 4.5 months at 730 μmol CO_2 mol^{-1} has 55% more new leaves, 52% more fresh mass and a 59% higher net CO_2 uptake rate than when grown at 370 μmol mol^{-1} (Nobel *et al.*, 1996). Another experiment with *A. salmiana* and *S. queretaroensis* maintained at 720 μmol CO_2 mol^{-1} for 1 month led to a 36% higher net CO_2 uptake for both species when compared with plants grown with 360 μmol mol^{-1} (Nobel, 1996b).

Most research on the effects of elevated [CO_2] on a cultivated CAM species has been done with *O. ficus-indica*, and ranges from studies focusing on enzymes (Israel and Nobel, 1994; Nobel *et al.*, 1996; Wang and Nobel, 1996) to those focusing on productivity (Nobel, 1991c; Cui *et al.*, 1993; Nobel and Israel, 1994). Responses of plants to elevated [CO_2] depends on other factors besides CO_2, with one of the most important being soil volume (Arp, 1991; Thomas and Strain, 1991; Berntson *et al.*, 1993). When soil volume per plant is low, responses of net CO_2 uptake to a doubling of the [CO_2] are minimal, as is also the case for *O. ficus-indica* maintained in pots for 4 months (Nobel *et al.*, 1994). For *O. ficus-indica* in the field for 5 months, daily net CO_2 uptake is 35% higher at 520 μmol mol^{-1} and 49% higher at 720 μmol mol^{-1} than at 370 μmol mol^{-1}. Production of dry mass is 21% and 55% higher, respectively, at the two elevated [CO_2] levels compared with the current ambient one (Cui *et al.*, 1993). Such effects are maintained after canopy closure and for a total of 15 months; biomass productivity is then 47 t dry weight ha^{-1} $year^{-1}$ at 370 μmol CO_2 mol^{-1} and 65 t ha^{-1} $year^{-1}$ at 720 μmol mol^{-1} (Nobel and Israel, 1994). The responses of daily net CO_2 uptake by *O. ficus-indica* to other environmental factors have also been checked at the doubled [CO_2]. In particular, raising [CO_2] increases the percentage enhancement of net CO_2 uptake by increases during drought, as temperature is raised and as the PPF is lowered (Nobel and Israel, 1994). Consequently, EPI (equation 14.1) can be adjusted to reflect different responses of net CO_2 uptake to environmental factors as [CO_2] increases in the future. To summarize the results for all of the cultivated CAM species examined for months under elevated [CO_2], daily net CO_2 uptake and biomass productivity increases about 1% for each increase of 10 μmol mol^{-1} in atmospheric [CO_2].

14.6 Conclusions and Future Research Directions

Environmental influences on net CO_2 uptake and, ultimately, productivity are predictable for the CAM species that are cultivated commercially, especially agaves and cacti. Such CAM plants do not tolerate freezing temperatures well – injury occurs at −5°C for most species (Nobel, 1988) – and so cultivation is limited to low latitudes. Increases in latitudinal ranges for such species are expected to accompany future global warming (Nobel, 1996b). Net CO_2 uptake is maximal at night air temperatures near 15°C, whereas day temperatures are relatively unimportant, as stomates of commercially important CAM crops tend to be closed then. Because of similar chlorenchyma characteristics among species, the PPF responses are similar, with 50% of maximal net CO_2 uptake occurring at a modest total daily PPF of 10 mol m^{-2} incident on the photosynthetic surface. Such responses cause net CO_2 uptake per unit ground area to be maximal at a leaf or stem area index (total surface area basis) per unit ground area of about 4 (Nobel, 1991a). Decreases in availability of soil water during drought cause net CO_2 uptake by such cultivated CAM plants to rely on water stored in the shoot. The average water storage depth in the transpiring organs of CAM plants directly indicates the drought duration that can be tolerated while still maintaining substantial rates of net CO_2 uptake. The responses to elevated [CO_2] are also rather similar among the CAM crops considered and amount to a 1% increase in daily net CO_2 uptake, and hence productivity, per 10 μmol mol^{-1} increase in [CO_2] (Nobel, 1996b).

Future research will inevitably be devoted to biotechnological improvements in the yield and general performance of cultivated CAM species. Because agaves and cacti are currently important crops in Mexico and other mainly Latin American countries, information transfer from these to other regions where they can be cultivated profitably will be important (Nobel, 1994). This will also entail changes in cultural practices as new fruits, vegetables and beverages become available and new sources of forage and fodder are used for domesticated animals. The increased [CO_2] accompanying global climatic change will be advantageous for net CO_2 uptake and productivity by CAM crops. The accompanying increased temperatures will favour cultivation at higher latitudes but a somewhat lower daily net CO_2 uptake during the warmer part of the year in the hotter regions presently used for cultivation. The effects of changes in other environmental factors accompanying global climatic change (such as seasonal rainfall patterns and cloudiness) on net CO_2 uptake and productivity can also be predicted (equation 14.1). The high productivity of certain cultivated CAM species under optimal conditions (Table 14.2) and the generally high WUE of CAM plants under all conditions warrants increased agronomic exploitation of such plants in the future. These plants can be expected to have generally favourable or, in any case, predictable responses of daily net CO_2 uptake to the forecasted environmental changes.

References

Arp, W.J. (1991) Effects of source–sink relations on photosynthetic acclimation to elevated CO_2. *Plant, Cell and Environment* 14, 869–875.

Barbera, G., Inglese, P. and Pimienta-Barrios, E. (eds) (1995) *Agro-ecology, Cultivation and Uses of Cactus Pear.* FAO, Rome, 219 pp.

Bartholomew, D. (1982) Environmental control of carbon assimilation and dry matter production by pineapple. In: Ting, I.P. and Gibbs, M. (eds) *Crassulacean Acid Metabolism.* American Society of Plant Physiologists, Rockville, Maryland, pp. 278–294.

Bartholomew, D.P. and Kadzimin, S.B. (1977) Pineapple. In: Alvin, P. de T. and Kozlowski, T.T. (eds) *Ecophysiology of Tropical Crops.* Academic Press, New York, pp. 113–156.

Bartholomew, D.P. and Malézieux, E.P. (1994) Pineapple. In: Schaffer, B. and Anderson, P.C. (eds) *Handbook of Environmental Physiology of Fruit Crops,* Vol. II, *Sub-Tropical and Tropical Crops.* CRC Press, Boca Raton, Florida, pp. 243–291.

Bartholomew, D.P. and Rohrbach, K.G. (eds) (1993) *First International Pineapple Symposium. Acta Horticulturae* 334, 1–471.

Berntson, G.M., McConnaughay, K.D.M., and Bazzaz, F.A. (1993) Elevated CO_2 alters deployment of roots in 'small' growth containers. *Oecologia* 94, 558–564.

Borland, A.M. and Griffiths, H. (1989) The regulation of citric acid accumulation and carbon recycling during CAM in *Ananas comosus. Journal of Experimental Botany* 40, 53–60.

Connelly, P.R. (1972) The effects of thermoperiod on the carbon dioxide uptake and compensation point of the pineapple plant, *Ananas comosus* (L.) Merr. PhD Dissertation, University of Hawaii, Honolulu.

Cui, M., Miller, P.M., and Nobel, P.S. (1993) CO_2 exchange and growth of the crassulacean acid metabolism plant *Opuntia ficus-indica* under elevated CO_2 in open-top chambers. *Plant Physiology* 103, 519–524.

Friend, D.J.C. and Lydon, J. (1979) Effects of daylength on flowering, growth, and CAM of pineapple (*Ananas comosus* [L.] Merrill). *Botanical Gazette* 140, 280–283.

Garcia de Cortázar, V. and Nobel, P.S. (1992) Biomass and fruit production for the prickly pear cactus, *Opuntia ficus-indica. Journal of the American Society of Horticultural Science* 117, 558–562.

Gentry, H.S. (1982) *Agaves of Continental North America.* University of Arizona Press, Tucson, 670 pp.

Gibson, A.C. and Nobel, P.S. (1986) *The Cactus Primer.* Harvard University Press, Cambridge, Massachusetts, 286 pp.

Graham, E.A. and Nobel, P.S. (1996) Long-term effects of a doubled atmospheric CO_2 concentration on the CAM species *Agave deserti. Journal of Experimental Botany* 47, 61–69.

Houghton, J.T., Jenkins, G.J., and Ephraums, J. (eds) (1990) *Climate Change. The IPCC Scientific Assessment.* Cambridge University Press, Cambridge, UK, 364 pp.

Israel, A.A. and Nobel, P.S. (1994) Activities of carboxylating enzymes in the CAM species *Opuntia ficus-indica* grown under current and elevated CO_2 concentrations. *Photosynthesis Research* 40, 223–229.

Jarvis, P.G. and Leverenz, J.W. (1983) Productivity of temperate, deciduous and evergreen forests. In: Lange, O.L., Nobel, P.S., Osmond, C.B. and Ziegler, H. (eds) *Encyclopedia of Plant Physiology, New Series,* Vol. 12D. *Physiological Plant Ecology IV. Ecosystem Processes.* Springer-Verlag, Berlin, pp. 233–280.

Joshi, M., Boyer, J. and Kramer, P. (1965) Growth, carbon-dioxide exchange, transpiration, and transpiration ratio of pineapple. *Botanical Gazette* 126, 174–179.

King, A.W., Emanuel, W.R. and Post, W.M. (1992) Projecting future concentrations of atmospheric CO_2 with global carbon cycle models: the importance of simulating historical change. *Environmental Management* 16, 91–108.

Loomis, R.S. (1983) Productivity of agricultural systems. In: Lange, O.L., Nobel, P.S., Osmond, C.B. and Ziegler, H. (eds) *Encyclopedia of Plant Physiology, New Series,* Vol. 12D, *Physiological Plant Ecology IV. Ecosystem Processes.* Springer-Verlag, Berlin, pp. 151–172.

Loomis, R.S. and Gerakis, P.A. (1975) Productivity of agricultural ecosystems. In: Cooper, J.P. (ed.) *Photosynthesis and Productivity in Different Environments.* Cambridge University Press, Cambridge, UK, pp. 145–172.

McLaughlin, S.P. (1993) Development of *Hesperaloe* species (Agavaceae) as new fibre crops. In: Janick, J. and Simon, J.E. (eds) *New Crops.* John Wiley & Sons, New York, pp. 435–442.

Medina, E., Popp, M., Lüttge, U. and Ball, E. (1991) Gas exchange and acid accumulation in high and low light grown pineapple cultivars. *Photosynthetica* 25, 489–498.

Medina, E., Popp, M., Olivares, E., Jenett, H.-P. and Lüttge, U. (1993) Daily fluctuations of titratable acidity, content of organic acids (malate and citrate) and soluble sugars of varieties and wild relatives of *Ananas comosus* L. growing under natural tropical conditions. *Plant, Cell and Environment* 16, 55–63.

Medina, E., Ziegler, H., Lüttge, U., Trimborn, P. and Francisco, M. (1994) Light conditions during growth as revealed by $\delta^{13}C$ values of leaves of primitive cultivars of *Ananas comosus,* an obligate CAM species. *Functional Ecology* 8, 298–305.

Mizrahi, Y., Nerd, A. and Nobel, P.S. (1997) Cacti as crops. *Horticultural Reviews* 18, 291–319.

Neales, T.F. (1973) The effect of night temperature on CO_2 assimilation, transpiration, and water use efficiency in *Agave americana* L. *Australian Journal of Biological Sciences* 26, 705–714.

Neales, T.F., Sale, P.J.M. and Meyer, C.P. (1980) Carbon dioxide assimilation by pineapple plants, *Ananas comosus* (L.) Merr. II. Effects of variation of day/night temperature regime. *Australian Journal of Plant Physiology* 7, 375–385.

Nobel, P.S. (1984) Productivity of *Agave deserti*: measurement by dry weight and monthly prediction using physiological responses to environmental parameters. *Oecologia* 64, 1–7.

Nobel, P.S. (1985) PAR, water, and temperature limitations on the productivity of cultivated *Agave fourcroydes* (henequen). *Journal of Applied Ecology* 22, 157–173.

Nobel, P.S. (1988) *Environmental Biology of Agaves and Cacti.* Cambridge University Press, New York, 270 pp.

Nobel, P.S. (1989) A nutrient index quantifying productivity of agaves and cacti. *Journal of Applied Ecology* 26, 635–645.

Nobel, P.S. (1991a) Tansley Review No. 32. Achievable productivities of certain CAM plants: basis for high values compared with C_3 and C_4 plants. *New Phytologist* 119, 183–205.

Nobel, P.S. (1991b) *Physicochemical and Environmental Plant Physiology.* Academic Press, San Diego, 635 pp.

Nobel, P.S. (1991c) Environmental productivity indices and productivity for *Opuntia ficus-indica* under current and elevated atmospheric CO_2 levels. *Plant, Cell and Environment* 14, 637–646.

Nobel, P.S. (1994) *Remarkable Agaves and Cacti.* Oxford University Press, New York, 166 pp.

Nobel, P.S. (1996a) High productivity of certain agronomic CAM species. In: Winter, K. and Smith, J.A.C. (eds) *Crassulacean Acid Metabolism: Biochemistry, Ecophysiology and Evolution.* Springer, Berlin, pp. 255–265.

Nobel, P.S. (1996b) Responses of some North American CAM plants to freezing temperatures and doubled CO_2 concentrations: implications of global climate change for extending cultivation. *Journal of Arid Environments* 34, 187–196.

Nobel, P.S. and Hartsock, T.L. (1983) Relationships between photosynthetically active radiation, nocturnal acid accumulation, and CO_2 uptake for a crassulacean acid metabolism plant, *Opuntia ficus-indica. Plant Physiology* 71, 71–75.

Nobel, P.S. and Hartsock, T.L. (1984) Physiological responses of *Opuntia ficus-indica* to growth temperature. *Physiologia Plantarum* 60, 98–105.

Nobel, P.S. and Israel, A.A. (1994) Cladode development, environmental responses of CO_2 uptake, and productivity for *Opuntia ficus-indica* under elevated CO_2. *Journal of Experimental Botany* 45, 295–303.

Nobel, P.S. and Meyer, S.E. (1985) Field productivity of a CAM plant, *Agave salmiana*, estimated using daily acidity changes under various environmental conditions. *Physiologia Plantarum* 65, 397–404.

Nobel, P.S. and Pimienta-Barrios, E. (1995) Monthly stem elongation for *Stenocereus queretaroensis*: relationships to environmental conditions, net CO_2 uptake, and seasonal variations in sugar content. *Environmental and Experimental Botany* 35, 17–24.

Nobel, P.S. and Valenzuela, A.G. (1987) Environmental responses and productivity of the CAM plant, *Agave tequilana. Agricultural and Forest Meteorology* 39, 319–334.

Nobel, P.S., Garcia-Moya, E. and Quero, E. (1992) High annual productivity of certain agaves and cacti under cultivation. *Plant, Cell and Environment* 15, 329–335.

Nobel, P.S., Cui, M., Miller, P.M. and Luo, Y. (1994) Influences of soil volume and an elevated CO_2 level on growth and CO_2 exchange for the crassulacean acid metabolism plant *Opuntia ficus-indica. Physiologia Plantarum* 90, 173–180.

Nobel, P.S., Israel, A.A. and Wang, N. (1996) Growth, CO_2 uptake, and responses of the carboxylating enzymes to inorganic carbon in two highly productive CAM species at current and doubled CO_2 concentrations. *Plant, Cell and Environment* 19, 585–592.

Nose, A., Heima, K., Miyazato, K. and Murayama, S. (1986) Effects of day-length on CAM type CO_2 and water vapor exchange in pineapple plants. *Photosynthetica* 20, 20–28.

Pimienta-Barrios, E. and Nobel, P.S. (1994) Pitaya (*Stenocereus* spp., Cactaceae): an ancient and modern fruit crop of Mexico. *Economic Botany* 48, 76–83.

Ravetta, D.A. and McLaughlin, S.P. (1996) Ecophysiological studies in *Hesperaloe funifera* (Agavaceae): a potential new CAM crop. Seasonal patterns in photosynthesis. *Journal of Arid Environments* 33, 211–223.

Russell, C.E. and Felker, P. (1987) The prickly pears (*Opuntia* spp., Cactaceae): a source of human and animal food in semiarid regions. *Economic Botany* 41, 433–445.

Sale, P.J.M. and Neales, T.F. (1980) Carbon dioxide assimilation by pineapple plants, *Ananas comosus* (L.) Merr. I. Effects of daily irradiance. *Australian Journal of Plant Physiology* 7, 363–373.

Schlesinger, W.H., Reynolds, J.F., Cunningham, G.L., Huenneke, L.F., Jarrell, W.M., Virginia, R.A. and Whitford, W.G. (1990) Biological feedbacks in global desertification. *Science* 243, 1043–1048.

Thomas, R.B. and Strain, B.R. (1991) Root restriction as a factor in photosynthetic acclimation of cotton seedlings grown in elevated carbon dioxide. *Plant Physiology* 96, 627–634.

Vitousek, P.M. (1994) Beyond global warming: ecology and global change. *Ecology* 75, 1861–1876.

Wang, N. and Nobel, P.S. (1996) Doubling the CO_2 concentration enhanced the activity of carbohydrate-metabolism enzymes, source carbohydrate production, photoassimilate transport, and sink strength for *Opuntia ficus-indica. Plant Physiology* 110, 893–902.

Zhu, J., Bartholomew, D.P., and Goldstein, G. (1997a) Effect of temperature, CO_2, and water stress on leaf gas exchange and biomass accumulation of pineapple. In: Martin-Prével, P. (ed.) *Second International Pineapple Symposium. Acta Horticulturae* 425, 297–308.

Zhu, J., Bartholomew, D.P., and Goldstein, G. (1997b) Effect of elevated carbon dioxide on the growth and physiological responses of pineapple, a species with crassulacean acid metabolism. *Journal of the American Society of Horticultural Science* 122, 233–237.

15 Crop Ecosystem Responses to Climatic Change: Crop/Weed Interactions

JAMES A. BUNCE AND LEWIS H. ZISKA

Climate Stress Laboratory, USDA-ARS-BARC, B-046A, 10300 Baltimore Avenue, Beltsville, MD 20705-2350, USA

15.1 Introduction

15.1.1 The impact of weeds

Plants have managed to occupy nearly all terrestrial habitats. Wherever humans attempt to reserve areas of land for the growth of directly useful species (crops), other species (weeds) grow there also. Because all plants use the same basic resources of light, minerals, water and carbon dioxide, and have the ability to disperse, the growing of crop plants involves a constant struggle against weeds.

Weeds reduce the production of crops primarily by competition for resources but sometimes also by chemical interference. Weed control sufficient to prevent any loss of crop production is usually not economically practical. In the USA, where labour is expensive and herbicide use has become widespread (Bridges, 1992), production losses due to weeds averaged about 7% under the best management practices for nine principal crops (Table 15.1). However, if no herbicide was applied, production losses due to weeds averaged 35% (Table 15.1). These estimates seem consistent with earlier estimates of production losses of 12% for the world as a whole, with 25% losses in traditional agriculture systems (Parker and Fryer, 1975). These substantial crop losses due to weeds are often similar in magnitude to those caused by insects, diseases and unfavourable weather.

It is difficult to estimate the costs of controlling weeds. Two obvious costs are the direct expenses in labour, herbicides, and fuel and machinery to suppress weeds, and the indirect costs of environmental impacts such as soil erosion and pollution. In the USA, farm expenses related to weed control, fuel, labour, pesticides and fertilizer are roughly 20% of the total farm production expenses (USDA, 1988), although the proportion of these directly related to overcoming weeds is unclear. The cost of herbicides alone for the USA

Table 15.1. Estimated production losses due to weeds for selected C_3 and C_4 crops in the USA. Production loss percentages were calculated from data in Bridges (1992) and applied to 1997 production figures. Average crop losses due to weeds were 7% for Best Management Practices and 35% for Best Management Practices Without Herbicide. Production values for sugarcane refer to bulk cane.

Crop	Photosynthetic pathway	Best management practices ('000 Mt)	Best management practices without herbicides ('000 Mt)
Barley	C_3	408 (5)	1,663 (20)
Maize	C_4	14,665 (6)	71,431 (30)
Cotton	C_3	832 (8)	5,102 (49)
Potato	C_3	1,204 (6)	6,149 (29)
Rice, paddy	C_3	553 (7)	4,378 (54)
Sorghum	C_4	1,333 (8)	5,925 (35)
Soybean	C_3	5,778 (8)	28,446 (38)
Sugarcane	C_4	2,503 (9)	9,957 (37)
Wheat	C_3	3,919 (6)	13,165 (20)
Total		31,195	146,216

(US$3 billion) is approximately equal to the value (US$4 billion) of the crop production lost due to weeds under the 'best management practices' (Bridges, 1992), for a minimum cost of US$7 billion. The technologies employed in weed control and the associated costs vary widely in different parts of the world, therefore a global estimate of the direct costs of weed control would be highly complex. The environmental costs are even more difficult to quantify. For example, a shift from conventional tillage to a 'no till' system would likely reduce soil erosion and increase the amount of carbon sequestered in the soil, but might also increase the risk of pollution from herbicide use and the development of herbicide resistance.

While the costs of weeds in terms of control and production losses are not always financially quantifiable, they are widespread and substantial. Any changes in weed/crop interactions which might result from global climatic change could be important in terms of crop production, economics and sustainability. The following sections review the experimental evidence concerning the relative responses of weeds and crops to increases in the atmospheric concentration of carbon dioxide ([CO_2]) and changes in climate. Possible impacts of environmental changes on weed management are discussed and directions for future research are suggested. The approach taken here will of necessity be somewhat theoretical, since there is a dearth of field studies of crop/weed interactions under conditions of global change.

15.1.2 Global climatic change

Some of the anticipated global changes in environment resulting from human activities, such as an increase in atmospheric [CO_2], warming, increased variability of weather and increased air pollutants and ultraviolet radiation, have been discussed in earlier chapters. In terms of weed/crop interactions, two aspects of global climatic change have received the most experimental attention: increasing atmospheric [CO_2] and potential changes in temperature.

Atmospheric [CO_2] has increased from about 280 μmol mol^{-1} before industrialization to about 365 μmol mol^{-1} currently, and could reach 600 μmol mol^{-1} by the end of the next century (Houghton *et al.*, 1996). Increases in [CO_2] within this range usually stimulate net photosynthesis and growth in plants with the C_3 photosynthetic pathway by increasing the [CO_2] in the leaf interior, which stimulates the carboxylation reaction of the ribulose bisphosphate carboxylase/oxygenase enzyme and decreases photorespiration by competitively inhibiting the oxygenation function (Long, 1991). In contrast, plants with the C_4 photosynthetic pathway have an internal biochemical pump that concentrates CO_2 at the site of ribulose bisphosphate carboxylase-oxygenase, and photosynthesis and growth are often saturated at or even below the current atmospheric [CO_2] (e.g. Dippery *et al.*, 1995).

There is increasing agreement among climatic modellers that the projected increases in greenhouse gases will be accompanied by an increase of 1.5–4.5°C in mean surface air temperature by the end of the 21st century (Houghton *et al.*, 1996). The projected increase in mean surface temperature may be accompanied by changes in the frequency and distribution of precipitation and increased evapotranspiration. Changes in temperature and precipitation and increasing [CO_2] all have potentially important consequences for crop/weed interactions, which is evident from a consideration of the basic biology of weeds and crops.

15.1.3 Weed and crop biology

Many crops were derived from species ecologically and genetically similar to weeds. This is reflected by the frequency of congeneric weeds and crops. For example, of the major weeds listed by Patterson (1985), about a quarter of the 49 genera also contain crop species. The relatedness of weeds and crops is also evidenced by several important weeds serving as crop plants in some locations (cf. Holm *et al.*, 1977). Basic features such as potential growth rate, resource use, stress tolerance and reproductive effort (Grime, 1979; Patterson, 1985) tend to be similar for herbaceous crops and weeds.

Yet there seems to be a striking difference: for the most important species on a worldwide basis, crops have predominantly C_3 photosynthetic metabolism and weeds disproportionally have C_4 metabolism. For example, among the 18 most troublesome weeds in the world (Holm *et al.*, 1977), 14 are C_4, whereas of the 86 plant species that supply most of the world's food, only five are C_4 species (Patterson, 1995a). The prevalence of C_4 metabolism among weed species is striking because probably fewer than 5% of angiosperm species are C_4 (cf. Patterson, 1985). This high frequency of C_4 metabolism in weeds has potentially important implications for responses of agricultural systems to global environmental changes. Many species with the C_4 photosynthetic pathway are adapted to hot, dry climates (Ehleringer *et al.*, 1997), and could be favoured by global warming, although (as discussed later) the issue is more complex. Because of the greater responsiveness to increasing $[CO_2]$ in C_3 species, many experiments and most reviews concerned with weeds and global change (e.g. Patterson and Flint, 1990; Patterson, 1993, 1995a; Froud-Williams, 1996) have focused on C_3 crop/C_4 weed interactions.

This generalization – that crops are C_3 and weeds are C_4 – should not limit discussion concerning crop/weed interactions, as there are certainly C_4 crops of major importance (e.g. maize, sorghum, millet and sugarcane) and many important C_3 weeds. For example, if one considers the most troublesome weeds worldwide for nine principal crops, 11 of the 23 are C_3 (Table 15.2). For the USA, among the additional 15 worst weeds in these crops, nine are C_3 (Table 15.2). As is suggested from the relatively small overlap of weed species

Table 15.2. Troublesome weeds associated with the crops listed in Table 15.1. Troublesome weeds are those that are inadequately controlled and interfere with crop production and/or yield, crop quality or harvest efficiency. The list is not all inclusive, but includes those weeds most frequently associated with a given crop as obtained from farmer surveys. Troublesome weeds for the USA were tabulated from Bridges (1992). Troublesome weeds on a global basis were obtained from Holm *et al.* (1977). C_3/C_4 indicates photosynthetic pathway.

	Troublesome weeds	
Crop	US	World
Barley	*Avena fatua* L. (C_3)	*Avena fatua* L. (C_3)
	Cirsium arvense (L.) Scop. (C_3)	*Capsella bursa-pastoris* (L.) Medicus (C_3)
	Convolvulus arvensis L. (C_3)	*Chenopodium album* L. (C_3)
	Elytrigia repens (L.) Nevski (C_3)	*Galinsoga parviflora* Cav. (C_3)
	Kochia scoparia (L.) Schrad. (C_4)	*Lolium temulentum* L. (C_3)
		Sorghum halapense (L.) Pers. (C_4)
Maize	*Abutilon theophrasti* Medicus (C_3)	*Amaranthus retroflexus* L. (C_4)
	Chenopodium album L. (C_3)	*Chenopodium album* L. (C_3)
	Cirsium arvense (L.) Scop. (C_3)	*Cyperus rotundus* L. (C_4)
	Sorghum bicolor (L.) Moench (C_4)	*Digitaria sanguinalis* (L.) Scop. (C_4)
	Sorghum halapense (L.) Pers. (C_4)	*Echinochloa colona* (L.) Link (C_4)
		Eleusine indica (L.) Gaertn. (C_4)
Cotton	*Cynodon dactylon* (L.) Pers. (C_4)	*Cynodon dactylon* (L.) Pers. (C_4)
	Cyperus esculentus L. (C_4)	*Cyperus rotundus* L. (C_4)
	Cyperus rotundus L. (C_4)	*Digitaria sanguinalis* (L.) Scop. (C_4)
	Eleusine indica (L.) Gaertn. (C_4)	*Echinochloa crus-galli* (L.) Beauv. (C_4)
	Xanthium strumarium L. (C_3)	*Eleusine indica* (L.) Gaertn. (C_4)
		Portulaca oleracea L. (C_4)
		Sorghum halapense (L.) Pers. (C_4)
Potato	*Chenopodium album* L. (C_3)	*Chenopodium album* L. (C_3)
	Cirsium arvense (L.) Scop. (C_3)	*Echinochloa crus-galli* (L.) Beauv. (C_4)
	Cyperus esculentus L. (C_4)	*Galinsoga parviflora* Cav. (C_3)
	Echinochloa crus-galli (L.) Beauv. (C_4)	*Portulaca oleracea* L. (C_4)
	Elytrigia repens (L.) Nevski (C_3)	*Stellaria media* (L.) Vill. (C_3)

Table 15.2. *Continued*

Crop	Troublesome weeds	
	US	World
Rice (paddy)	*Brachiaria platyphylla* (Griseb.) Nash (C_4) *Cyperus esculentus* L. (C_4) *Echinochloa crus-galli* (L.) Beauv. (C_4) *Heteranthera limosa* (Sw.) Willd. (C_3) *Leptochloa fascicularis* (Lam.) Gray (?)	*Cyperus difformis* L. (C_4) *Echinochloa colona* (L.) Link (C_4) *Echinochloa crus-galli* (L.) Beauv. (C_4) *Fimbristylus miliacea* (L.) Vahl (?) *Monochoria vaginalis* (Burm. F.) Presl (C_3) *Scirpus maritimus* (C_3)
Sorghum	*Brachiaria platyphylla* (Griseb.) Nash (C_4) *Panicum dichotomiflorum* Michx. (C_4) *Sorghum bicolor* (L.) Moench (C_4) *Sorghum halapense* (L.) Pers. (C_4) *Xanthium strumarium* L. (C_3)	*Cyperus rotundus* L. (C_4) *Digitaria sanguinalis* (L.) Scop. (C_4) *Echinochloa colona* (L.) Link (C4) *Echinochloa crus-galli* (L.) Beauv. (C_4) *Eleusine indica* (L.) Gaertn. (C_4) *Sorghum halapense* (L.) Pers. (C_4)
Soybean	*Abutilon theophrasti* Medicus (C_3) *Cassia obtusifolia* L. (C_3) *Chenopodium album L.* (C_3) *Sorghum halapense* (L.) Pers. (C_4) *Xanthium strumarium* L. (C_3)	*Cyperus rotundus* L. (C_4) *Echinochloa colona* (L.) Link (C_4) *Echinochloa crus-galli* (L.) Beauv. (C_4) *Eleusine indica* (L.) Gaertn. (C_4) *Rottboellia exaltata* L.f. (C_4)
Sugarcane	*Cynodon dactylon* (L.) Pers. (C_4) *Cyperus rotundus* L. (C_4) *Panicum maximum* Jacq. (C_4) *Rottboelia cochinchinensis* (Lour.) (C_4) *Sorghum halapense* (L.) Pers. (C_4)	*Cynodon dactylon* (L.) Pers. (C_4) *Cyperus rotundus* L. (C_4) *Digitaria sanguinalis* (L.) Scop. (C_4) *Panicum maximum* Jacq. (C_4) *Rottboellia exaltata* L.f. (C_4) *Sorghum halapense* (L.) Pers. (C_4)
Wheat	*Allium vineale* L. (C_3) *Bromus secalinus* L. (C_3) *Cirsium arvense* (L.) Scop. (C_3) *Convolvulus arvensis* L. (C_3) *Lolium multiflorum* Lam. (C_3)	*Avena fatua* L. (C_3) *Chenopodium album* L. (C_3) *Convolvulus arvensis* L. (C_3) *Polygonum convolvulus* L. (C_3) *Stellaria media* (L.) Vill. (C_3)

listed as most troublesome for crops worldwide vs. those for the USA, there are regional differences in what weeds are important for a given crop species. Even within the USA, there is very little overlap in the weed species that are important in maize and soybean crops between the Lake states and the Gulf states, a north–south comparison within a humid region (Table 15.3). There are no large differences in the proportion of C_3 and C_4 weed species in these areas, nor in the proportion of perennial species (Table 15.3). However, it is interesting that the estimated crop losses due to weeds (Bridges, 1992) without the use of herbicides are substantially larger in the south than in the north in both maize (22 vs. 35%) and soybeans (22 vs. 64%). This may be due to the occurrence in the south of some very aggressive weeds whose presence is limited in the northern states by low temperatures. It is important to recognize that crop/weed interactions vary with locality, and how they may respond to

Table 15.3. List of most troublesome weeds for the Lake states (Michigan, Minnesota, Wisconsin) and the Gulf states (Alabama, Louisiana, Mississippi) for two principal crops: maize and soybean. Troublesome weeds are those that are inadequately controlled and interfere with crop production, crop quality or harvest efficiency. The list is not all inclusive, but includes all weeds associated with a given crop in each region (i.e. Gulf or Lake) as obtained from farmer surveys. Troublesome weeds for the USA were tabulated from Bridges (1992).

Crop	Weed	Common name	Pathway (C_3 or C_4)	Annual/ perennial
Lake states				
Soybean	*Abutilon theophrasti* (Medicus)	Velvetleaf	C_3	Annual
	Amaranthus retroflexus L.	Redroot pigweed	C_4	Annual
	Ambrosia artemisifolia L.	Common ragweed	C_3	Annual
	Ambrosia trifida L.	Giant ragweed	C_3	Annual
	Calystegia sepium (L.) R.Br.	Hedge bindweed	C_3	Perennial
	Chenopodium album L.	Lambsquarters	C_3	Annual
	Cirsium arvense (L.) Scop.	Canadian thistle	C_3	Perennial
	Echinochloa crus-galli (L.) Beauv.	Barnyardgrass	C_4	Annual
	Eriochloa villosa (Thunb.) Kunth	Woolly cupgrass	C_4?	Annual
	Solanum nigrum L.	Black nightshade	C_3	Annual
	Solanum ptycanthum Dur.	Eastern nightshade	C_3	Annual
	Solanum sarrachoides Sendtree	Hairy nightshade	C_3	Annual
	Sonchus arvensis L.	Perennial sawthistle	C_3	Perennial
Gulf states				
Soybean	*Acanthospermum hispidum* DC.	Bristly starbur	C_3	Annual
	Brachiara platyphylla (Griseb.) Nash	Broadleaf signalgrass	C_4	Annual
	Campsis radicans (L.) Seem. Ex Bur.	Trumpetcreeper	C_3	Perennial
	Caperonia palustris (L.) St. Hil.	Texasweed	C_3?	Annual
	Cardiospermum halicacabum L.	Balloonvine	C_3	Annual
	Cassia obtusifolia L.	Sicklepod	C_3	Annual
	Cassia occidentalis L.	Coffee sena	C_3	Annual
	Desmodium tortuosum (Sw.) DC.	Florida beggarweed	C_3	Annual
	Echinochloa crus-galli (L.) Beauv.	Barnyardgrass	C_4	Annual
	Euphorbia heterophylla L.	Wild poinsetta	C_3	Annual
	Euphorbia maculata L.	Spotted spurge	?	Annual
	Ipomoea hederacea Gray	Ivyleaf morning glory	C_3	Annual
	Ipomoea lacunosa L.	Pitted morning glory	C_3	Annual
	Panicum texanum Buckl.	Texas panicum	C_4	Annual
	Rottboellia cochinchinensis (Lour.)	Itchgrass	C_4	Annual
	Sesbania exaltata (Raf.) Rydb.	Hemp sesbania	C_3	Annual
	Sida spinosa L.	Prickly sida	C_3	Perennial
	Solanum carolinense L.	Horsenettle	C_3	Perennial
	Sorghum halapense (L.) Pers.	Johnsongrass	C_4	Perennial

Table 15.3. *Continued*

Crop	Weed	Common name	Pathway (C_3 or C_4)	Annual/ perennial
Lake states				
Maize	*Abutilon theophrasti* (Medicus)	Velvetleaf	C_3	Annual
	Amaranthus hybridus L.	Smooth pigweed	C_4	Annual
	Apocynum cannabinum L.	Hemp dogbane	C_3	Perennial
	Chenopodium album L.	Lambsquarters	C_3	Annual
	Cirsium arvense (L.) Scop.	Canadian thistle	C_3	Perennial
	Convolvulus arvensis L.	Field bindweed	C_3	Perennial
	Digitaria sanguinalis (L.) Scop	Large Crabgrass	C_4	Annual
	Elytrigia repens (L.) Nevski	Quackgrass	C_3	Perennial
	Eriochloa villosa (Thunb.) Kunth	Woolly cupgrass	C_4?	Annual
	Panicum dichotomiflorum Michx.	Fall panicum	C_4	Annual
	Panicum miliaceum L.	Wild millet	C_4	Annual
	Setaria faberi Herrm.	Giant foxtail	C_4	Annual
	Sorghum almum Parod.	Sorghum almum	C_4	Perennial
	Sorghum bicolor (L.) Moench	Shattercane	C_4	Annual
Gulf states				
Maize	*Brachiara platyphylla* (Griseb.) Nash	Broadleaf signalgrass	C_4	Annual
	Cassia obtusifolia L.	Sicklepod	C_3	Annual
	Cynodon dactylon (L.) Pers.	Bermudagrass	C_4	Perennial
	Cyperus esculentus L.	Yellow nutsedge	C_4	Perennial
	Cyperus rotundus L.	Purple nutsedge	C_4	Perennial
	Panicum dichotomiflorum Michx.	Fall panicum	C_4	Annual
	Panicum ramosum (L.) Stapf	Browntop millet	C_4	Annual
	Panicum texanum Buckl.	Texas panicum	C_4	Annual
	Solanum carolinensis L.	Horsenettle	C_3	Perennial
	Sorghum halapense (L.) Pers.	Johnsongrass	C_4	Perennial

environmental change depends on many ecophysiological characteristics which may or may not correlate with photosynthetic pathway.

Many of the world's most troublesome weeds, both C_3 and C_4, are confined to tropical or subtropical areas (Holm *et al.*, 1977, 1997) and poleward expansion seems to be limited by low temperatures. There is no evidence that C_4 metabolism does not function well at low temperatures (Leegood and Edwards, 1996), although the photosynthetic advantage of C_4 relative to C_3 plants in warm, low-light environments is reversed at cool temperatures (Ehleringer and Bjorkman, 1977). In fact, C_4 weeds such as *Amaranthus* spp., *Echinochloa crus-gali* and *Portulaca oleracea* occur in Canada and northern Europe. Rather, limitation of ranges of some weed species by low temperature seems to be related to their evolution in tropical or subtropical areas. Some weeds in which the temperature sensitivity of growth has been examined in relation to range extension are *Cassia obtusifolia*, *Crotolaria spectabilis*, *Imperata cylindrica*, *Panicum texanum*, *Rottboellia exaltata*, *Sesbania exaltata* and *Striga asiatica* (reviewed in Patterson, 1993), and *Lonicera japonica* and *Pueraria lobata* (Sasek and Strain, 1990). In addition to the possibility that global warming could result in a poleward extension of the ranges of some of these weeds, it should be noted that increased [CO_2] could, by itself, allow such range extension. This is because elevated [CO_2] has been found to increase the tolerance to low temperatures in several species (e.g. Sionit *et al.*, 1981, 1987; Potvin and Strain, 1985; Boese *et al.*, 1997). Possible effects of [CO_2] on range extension have been examined in *Lonicera japonica* and *Pueraria lobata* (Sasek and Strain, 1990).

15.2 Comparative Responses of Crops and Weeds

15.2.1 Climatic change and photosynthetic pathway

Almost all crop and weed species fall into two of the three major photosynthetic pathways: C_3 and C_4. Plants with the CAM pathway, such as pineapple and cacti, will not be discussed here (see Chapter 14, this volume). Numerous observations of the response of growth of C_3 and C_4 species to elevated [CO_2] support the expectation, discussed earlier, that C_3 species are more responsive than C_4 species. In a survey of responses of crop species, Kimball (1983) found that increases in biomass in response to [CO_2] doubling from 350 to 700 μmol mol^{-1} averaged 40% and 11% in C_3 and C_4 species, respectively. Very similar percentages resulted from the analysis of Cure and Acock (1986). However, studies that include non-crop C_4 species (reviewed in Poorter, 1993 and Patterson, 1995a) indicate substantial responses in some C_4 species. Ziska and Bunce (1997a) reported larger stimulation in both photosynthesis and biomass with [CO_2] doubling among weedy C_4 species as a group than among C_4 crop species. Reasons for this have not been identified. Among C_3 species, crop and weed species tend to have similar stimulation of growth by elevated [CO_2] (e.g. Poorter, 1993; Bunce, 1997), probably because they have similar, high relative growth rates, which enhance the stimulation of biomass compared with slower growing species (Hunt *et al.*, 1993; Poorter, 1993; Bunce, 1997). Not surprisingly, studies in which C_3 and C_4 species were grown in direct competition have almost invariably found that elevated [CO_2] favoured the C_3 species (reviewed in Patterson and Flint, 1990; see also Table 15.4). An exception was *Amaranthus retroflexus* competing with *Abutilon theophrasti* (Bazzaz *et al.*, 1989). *Amaranthus retroflexus* is one of the C_4 species most responsive to increased [CO_2] (Ziska and Bunce, 1997a). Because the range of responses of biomass stimulation for C_3 and C_4 species overlap (Poorter, 1993), such exceptions will occur.

It is generally thought that global warming would favour C_4 species relative to C_3 species. This is based on the observation that C_4 species are distributed in warmer environments and generally have higher optimum temperatures for photosynthesis and growth, and because their higher intrinsic water-use efficiency (the ratio of photosynthesis to transpiration) might better adapt them to the greater evaporative demand that would result from warming. However, because warmer temperatures are only likely to occur if

$[CO_2]$ is also high, the relevant question is whether the combination of increased temperature and elevated $[CO_2]$ would favour C_3 or C_4 plants.

At high $[CO_2]$, the optimum temperature for photosynthesis of many C_3 plants increases and becomes similar to that of C_4 plants (e.g. Osmond *et al.*, 1980; cf. Leegood and Edwards, 1996). This is expected since the main function of the C_4 cycle is to concentrate CO_2 at the same enzyme used by C_3 plants. Differences among species in the optimum temperature for photosynthesis at high $[CO_2]$ are probably related to membrane stability (Bjorkman *et al.*, 1980), although this may be manifest as enzyme deactivation (e.g. Kobza and Edwards, 1987). It might be expected, from the increase in the optimum temperature for photosynthesis at elevated $[CO_2]$, that the stimulation of growth of C_3 species by elevated $[CO_2]$ would be greater at warm than at cool temperatures (Long, 1991). This pattern is commonly observed (Imai and Murata, 1979; Idso *et al.*, 1987), although there are numerous exceptions, both for photosynthetic stimulation (Greer *et al.*, 1995) and for growth (e.g. Patterson *et al.*, 1988; Tremmel and Patterson, 1993; Ziska and Bunce, 1997b). Nevertheless, for the modest increases in temperature expected from global warming (i.e. 1.5–4.5°C), the stimulation of growth by the combination of elevated $[CO_2]$ and elevated temperature will probably be greater in C_3 than in C_4 species. This is consistent with some of the limited experimental data (Patterson *et al.*, 1988; Coleman and Bazzaz, 1992; Tremmel and Patterson, 1993). Exceptions have been observed when large temperature increases have been imposed (Read and Morgan, 1996) or critical threshold temperatures for reproductive damage in the C_3 species have been exceeded (Alberto *et al.*, 1996). High temperatures sometimes limit reproductive development and global warming may decrease reproductive output in such situations despite an increase in $[CO_2]$ (e.g. Wheeler *et al.*, 1996; Ziska *et al.*, 1997). It is unclear whether this is more likely to occur in C_3 than C_4 species, but if it were, it could alter weed community composition and affect crop/weed interactions. In the field, mean temperatures in temperate regions are usually substantially below the optimum temperature for vegetative growth even for C_3 species – especially early in the season, when competitive outcomes are often determined (e.g. Kropff and Spitters, 1991). This is probably true everywhere except in warm arid regions, since mean temperatures even in humid tropical areas are usually below 27°C. In temperate zones, it is likely that planting times would be adjusted such that crops and weeds would be exposed to the same temperatures during the establishment phase as they are now, even if global warming occurs.

Higher temperatures would also create increased evaporative demand. In some respects, with its high water-use efficiency and CO_2-saturated photosynthesis, C_4 metabolism is better adapted to high evaporative demand (e.g. Bunce, 1983). However, this is another situation where, if increased $[CO_2]$ occurs in combination with higher temperatures, it becomes unclear whether C_4 species would be favoured. This is because C_3 species show at least as large an increase in water-use efficiency at elevated $[CO_2]$ as do C_4 species. In addition to the reduced stomatal conductance that usually occurs in both photosynthetic types, C_3 species also have increased photosynthesis at elevated [CO_2]. This distinction is probably even more important in the field, where transpiration is much less sensitive to changes in stomatal conductance (McNaughton and Jarvis, 1991; Bunce *et al.*, 1997), and photosynthetic responses would dominate changes in water-use efficiency at elevated [CO_2].

15.2.2 Crop/weed competition

Many studies in which weed and crop species have been included in the same [CO_2] treatments have been interpreted in terms of crop/weed competition even though the plants did not compete with each other. For example, weed/crop biomass ratios were decreased at elevated [CO_2] in itchgrass (C_4)/soybean (C_3) comparisons (Patterson and Flint, 1980), and barnyardgrass, goosegrass and southern crabgrass (all C_4)/soybean (C_3) comparisons (Patterson, 1986). Weed/crop ratios were increased in itchgrass (C_4)/maize (C_4), velvetleaf (C_3)/maize (C_4), and velvetleaf (C_3)/soybean (C_3) comparisons (Patterson and Flint, 1980). Many additional such comparisons could be gleaned from the literature, but they are only suggestive of how crop/weed interactions might change with increased [CO_2], because the outcome of competition is notoriously difficult to predict from comparisons of the growth of species in isolation. This is illustrated by the work of Bazzaz and co-workers (Bazzaz and Carlson, 1984; Zangerl and Bazzaz, 1984; Bazzaz and Garbutt, 1988), who measured species responses to [CO_2] when plants were grown alone or in competition and found little relationship between the relative responses of isolated and competing plants.

The few studies in which crop and weed species have been grown in competition with each other at different [CO_2] can be divided into those in which the crops and weeds differed in photosynthetic pathway and those in which the pathway was the same. All studies in which the photosynthetic pathway differed have compared C_3 crops with C_4 weed species. Increasing [CO_2] increased the crop/weed (i.e. the C_3/C_4) ratio in all of these studies (Table 15.4). This pattern is consistent with the greater response of photosynthesis and growth to elevated [CO_2] in C_3 than in C_4 plants. Similarly, in rangelands of the southwestern USA, the increase in [CO_2] from preindustrial to the current values has strongly favoured C_3 over C_4 species (Johnson *et al.*, 1993; also see Chapter 14, this volume), although in that system the C_3 species are often considered weeds.

Only a very few studies have examined crops and weeds grown in competition with each other at different [CO_2] when both species had the same photosynthetic type (Table 15.4). Several studies, including a few field studies, have indicated that the species dominance in grassland or pasture mixtures may change with [CO_2] (e.g. Owensby *et al.*, 1993; Luscher *et al.*, 1998; see also Chapter 12, this volume). Two such studies of pasture mixtures in which the 'weed' species were identified are included in Table 15.4, and in both, the weed species were favoured by elevated [CO_2]. In the study involving *Chenopodium album* and sugarbeet (Table 15.4), the competitive advantage of sugarbeet at elevated [CO_2] was attributed to late emergence in the weed

Table 15.4. Studies in which crop and weed species were grown in competition as a function of $[CO_2]$.

Weed	Crop	High CO_2 favours	Environment	Reference
Differing photosynthetic pathway				
Sorghum halepense (C_4)	Meadow fescue	Crop	Glasshouse	Carter and Peterson, 1983
Sorghum halepense (C_4)	Soybean	Crop	Chamber	Patterson *et al.*, 1984
Echinochloa glabrescens (C_4)	Rice	Crop	Glasshouse	Alberto *et al.*, 1996
Paspalum dilatatum (C_4)	Various grasses	Crop	Chamber	Newton *et al.*, 1996
Various grasses (C_4)	Lucerne	Crop	Field	Bunce, 1993a
Same photosynthetic pathway				
Chenopodium album (C_3)	Sugarbeet	Crop	Chamber	Houghton and Thomas, 1996
Taraxacum officinale (C_3)	Lucerne	Weed	Field	Bunce, 1995
Plantago lanceolata (C_3)	Various grasses	Weed	Chamber	Newton *et al.*, 1996
Taraxacum and *Plantago* (C_3)	Various grasses	Weed	Field	Potvin and Vasseur, 1997

species in this particular experiment (Houghton and Thomas, 1996). The authors noted that competitive outcome in this system, as in many others, depends critically on the timing of emergence, which could be influenced by climatic change. In the lucerne study (Bunce, 1995), the competitive advantage of *Taraxacum officinale* probably occurred because lucerne growth was not stimulated by elevated $[CO_2]$ in the long term. The outcome of crop/weed competition at elevated $[CO_2]$ within a photosynthetic type will thus probably be determined more by factors affecting plant development than by differences in photosynthetic responses.

15.2.3 Elevated $[CO_2]$ and plant development

As has been recognized for many years by crop physiologists (e.g. Potter and Jones, 1977; Evans, 1993), differences in photosynthate partitioning or allocation (i.e. allometric relationships) are often more important in determining plant growth rates than are differences in single-leaf photosynthetic rates. This probably also applies to comparative responses to environmental change. For example, in a study of ten C_3 species, differences among species in leaf photosynthetic response to elevated $[CO_2]$ provided a good prediction of the stimulation of whole-plant photosynthetic rate, but the stimulation of photosynthesis was a poor predictor of the increase in biomass at elevated $[CO_2]$ (Bunce, 1997). The increase in biomass was more related to species differences in relative growth rate (Bunce, 1997), which is strongly influenced by patterns of allocation of photosynthate. Differences in allocation are probably even more crucial in determining competitive outcomes, where the difference between success and failure may sometimes simply depend on leaf height. Elevated [CO_2] sometimes increases height growth (e.g. Patterson and Flint, 1982), but it is unknown whether there is a differential response for crops and weeds grown in competition.

Many aspects of plant development are affected by [CO_2] – for example, rates of germination, leaf initiation, tillering, branching, flowering and senescence, all of which could affect crop/weed interactions. One potentially important effect which has been documented under field conditions is the more rapid emergence of weed seedlings at elevated [CO_2] (Ziska and Bunce, 1993). More rapid emergence at elevated [CO_2] occurred more often in species with small rather than large seeds, suggesting that weeds may be more affected than most crops.

Elevated [CO_2] can also alter the time of flower initiation or the rate of floral development. Flowering can be faster, slower or unchanged at elevated [CO_2], depending on species (reviewed in Patterson, 1995a), but no clear patterns have emerged relative to crop/weed interactions. Reekie *et al.* (1994) reported that elevated [CO_2] delayed flowering in four short-day species and hastened it in four long-day species, but these responses are not universal. Little is known concerning how increasing [CO_2] could affect seed quality or viability in weeds. In some crops (Conroy, 1992), there is a reduction in nitrogen or protein content in seeds developed at elevated [CO_2]. However, the impact of increasing [CO_2] on seed viability for either crops or weeds has not been investigated. None of the studies to date in which crop/weed competition was compared at different [CO_2] with annual crop species has been of long enough duration to include more than one generation of weeds. In one long-term field study with lucerne, elevated [CO_2] increased the dominance of one weed species relative to the other weeds (Bunce, 1995). Because weed species vary in their responses to global climatic change, the composition of weed communities will undoubtedly change (Zangerl and Bazzaz, 1984).

15.2.4 Interaction of [CO_2] with water and nutrients

Potential increases in global temperature may be accompanied by changes in precipitation patterns and increased frequency of drought. Even when water is limited, elevated [CO_2] can stimulate plant growth (e.g. Patterson, 1986; Choudhuri *et al.*, 1990). Plants with C_4 photosynthetic metabolism sometimes only have increased photosynthesis and growth at elevated [CO_2] under dry conditions (e.g. Patterson, 1986; Knapp *et al.*, 1993), when elevated [CO_2] slows the development of stress. Whether dry conditions would alter the competitive relationships between C_3 and C_4 species at elevated [CO_2] is not known. Within a photosynthetic pathway, crops and weeds have reasonably similar responses to drought (Patterson, 1995b), although the impact of weeds on crop production may decrease because of reduced growth of both weeds and crops (Patterson, 1995b). Studies which have examined the interactive effects of increasing [CO_2] and water stress on annual weeds and crops competing with each other are unavailable. In a pasture mixture (Newton *et al.*, 1996), the proportion of weed biomass increased with [CO_2] to a similar extent in wet and dry treatments.

Under extreme nutrient deficiencies, there may be no response of biomass to elevated [CO_2]; under moderate limitations more relevant to agricultural situations, the increase in biomass may be reduced but the relative stimulation by elevated [CO_2] is often similar (e.g. Wong, 1979; Rogers *et al.*, 1993; Seneweera *et al.*, 1994). In some species even moderate phosphate deficiencies eliminate biomass stimulation (e.g. Goudriaan and de Ruiter, 1983). As in the case of water stress, reductions in growth caused by nutrient deficiency may reduce the impact of weeds on crop production (Patterson, 1995b), since smaller plants interfere less with each other. In some C_4 species, the growth response to elevated [CO_2] may only occur under nutrient-deficient conditions (Wong and Osmond, 1991). However, as with the interactive effects of [CO_2] and water availability, we are not aware of any study that has directly examined the interaction between increasing [CO_2] and nutrient availability when crops and weeds were competing with each other. In a study of a pasture mixture, Schenk *et al.* (1997) presented data suggesting that elevated [CO_2] may affect how the outcome of competition between ryegrass and white clover varies with the nitrogen supply.

15.2.5 Uncertainties and limitations

To date, the majority of the data concerning crop/weed interactions with increasing [CO_2] and changing climate are based on studies in controlled environment chambers or glasshouses. This is a substantial limitation, since extrapolation of results from such environments to field conditions is highly uncertain. For example, Ghannoum *et al.* (1997) found that the biomass of a C_4 grass species increased substantially at elevated [CO_2] under field conditions, but did not respond in a controlled environment chamber. This was attributed to a photomorphogenic effect of the artificial light, which limited stem extension in the chamber. It is understandable that crop scientists doing field work on global change have tried to eliminate weeds as a variable in their experiments, but we urge that some field work be directed toward understanding the interactive responses of weeds as well as crops. Increasing [CO_2] could alter the production of secondary compounds, which could in turn affect allelopathy between crops and weeds, but this has not been examined to date. Most studies of crop/weed interactions and global change have dealt only with the separate responses of annual crops and their associated weeds. A largely different set of weed species are associated with perennial and woody crops (Holm *et al.*, 1977) and these have received almost no experimental attention with respect to global change. No information is available on how long-term exposure to increased [CO_2] by itself or in conjunction with other environmental changes (e.g. water, temperature and nutrients) will affect crop/weed interactions in the field. Yet, such information is necessary to predict how changes in [CO_2] and climate may alter weed populations and crop losses due to weeds.

15.3 Weed Management

It is clear that increasing atmospheric [CO_2] and associated changes in climate have the potential to directly affect weed physiology, geographical distribution and crop/weed interactions. Additionally, weed responses to environmental changes may also affect their response to control measures.

15.3.1 Herbicide control

It is well known that variables such as temperature, wind speed, soil moisture and atmospheric humidity influence the effectiveness of herbicide applications (reviewed in Muzik, 1976). In addition, these same environmental variables can affect crop injury due to herbicide application. These variables undoubtedly affect herbicide uptake by influencing stomatal opening (e.g. Dybing and Currier, 1961; Sargent and Blackman, 1962; Mansfield *et al.*, 1983), among other effects. One of the more ubiquitous effects of elevated [CO_2] in herbaceous plants is a reduction in stomatal aperture and conductance. The magnitude of the reduction in conductance varies with other environmental factors, but is often as large as 50% (e.g. Bunce, 1993b). Much as increased [CO_2] (e.g. Barnes and Pfirrmann, 1992; Mulchi *et al.*, 1992) and decreased humidity (McLaughlin and Taylor, 1981) protect plants from ozone damage, probably by reducing stomatal aperture and the amount of ozone reaching the interior of leaves, elevated [CO_2] may protect weeds from some post-emergence herbicides. Elevated [CO_2] might affect leaf thickness, cuticular thickness and stomatal density in ways that could reduce the effectiveness of herbicides. Decreased stomatal conductance with increasing [CO_2] concentration could reduce transpiration and the uptake of soil-applied herbicides. Stage of growth for effective weed control could also be affected by environmental changes. Post-emergence herbicide control of some weeds is primarily effective in the seedling stage. The time spent in the seedling stage could be reduced by elevated [CO_2] or warming. For perennial weeds, increasing [CO_2] could stimulate greater rhizome or tuber growth, making herbicidal control of such weeds more difficult and costly.

15.3.2 Biological control

Natural and manipulated biological control of weeds and other potential pests could be affected by increasing atmospheric [CO_2] and climatic change (Norris, 1982; Froud-Williams, 1996). Climatic changes could alter the efficacy of the biocontrol agent by changing the growth, development and reproduction of the selected weedy target. Direct effects of [CO_2] on increasing starch concentration in leaves and lowering nitrogen contents could also affect biocontrol by altering the behaviour and growth rate of herbivores (see Chapter 16, this volume). Conversely, as pointed out by Patterson (1995a), global warming could result in increased overwintering of insect populations, which could increase biological control of weeds in some cases. However, increasing insect populations could also include specific crop pests, which could increase the susceptibility of a given crop to weed competition.

15.3.3 Mechanical control

Tillage is a principal method of weed control in many agronomic systems. Roots and rhizomes in many weed species are organs of vegetative propagation and overwintering (Holm *et al.*, 1977). Elevated [CO_2] commonly stimulates the growth of roots and rhizomes more than that of shoots (reviewed in Rogers *et al.*, 1994). Increased root and rhizome growth in such species may make control more difficult as [CO_2] rises.

15.4 Summary and Future Research

The ongoing increase in the concentration of carbon dioxide in the atmosphere, as well as potential changes in temperature and precipitation, may have important consequences for crop losses due to weeds. The physiological plasticity of weeds and their greater intraspecific genetic variation compared with most crops could provide weeds with a competitive advantage in a changing environment. However, because so little experimental work on crop/weed interactions under global change conditions has been carried out under field conditions, it is premature to conclude the magnitude or direction of changes in the interactions. Despite the lack of direct experimental evidence, several effects seem probable. One is that C_3 species will be favoured relative to C_4 species as [CO_2] increases, although exceptions should be anticipated. The fact that many weeds are C_4 and most crops are C_3 may seem advantageous, but will be of little comfort to those trying to grow C_4 crops in competition with C_3 weeds. A second likely outcome, driven by increased [CO_2] itself and potentially exacerbated by warming, is that some tropical and subtropical weeds will extend their ranges poleward and become troublesome in areas where they are not currently a problem. Thirdly, and most speculative, because of the stimulation of seedling emergence and root and rhizome growth by increased [CO_2], and because of decreased stomatal conductance with increasing [CO_2], weed control could become more difficult both for mechanical and chemical control measures.

Although much has been learned about responses to global change factors from studies of crops and weeds in controlled environment chambers and glasshouses, field facilities to simulate future environments are now sufficiently available that we urge the development of long-term field studies of crop/weed interactions and weed control under global change conditions. Designing meaningful experiments will be challenging. Perhaps it is worth keeping in mind that crop/weed interactions are local events, and generalizations about how interactions may change with global climate are best built from specific examples. Information obtained from such studies could be of substantial benefit in the development of new weed control strategies in a changing climate.

References

Alberto, A.M., Ziska, L.H., Cervancia C.R. and Manalo, P.A. (1996) The influence of increasing carbon dioxide and temperature on competitive interactions between a C_3 crop, rice (*Oryza sativa*), and a C_4 weed (*Echinochloa glabrescens*). *Australian Journal of Plant Physiology* 23, 795–802.

Barnes, J.D. and Pfirrmann, T. (1992) The influence of CO_2 and O_3, singly and in combination, on gas exchange, growth and nutrient status of radish (*Raphanus sativus* L.). *New Phytologist* 121, 403–412.

Bazzaz, F.A. and Carlson, R.W. (1984) The response of plants to elevated CO_2. I. Competition among an assemblage of annuals at two levels of soil moisture. *Oecologia* 62, 196–198.

Bazzaz, F.A. and Garbutt, K. (1988) The response of annuals in competitive neighborhoods: effects of elevated CO_2. *Ecology* 69, 937–946.

Bazzaz, F.A., Garbutt, K., Reekie, E.G. and Williams W.E. (1989) Using growth analysis to interpret competition between a C_3 and a C_4 annual under ambient and elevated CO_2. *Oecologia* 79, 223–235.

Bjorkman, O., Badger, M.R. and Armond, P.A. (1980) Response and adaptation of photosynthesis to high temperatures. In: Turner, N.C. and Kramer, P.J. (eds) *Adaptation of Plants to Water and High Temperature Stress.* John Wiley & Sons, New York, pp. 233–250.

Boese, S.R., Wolfe, D.W. and Melkonian, J.J. (1997) Elevated CO_2 mitigates chilling-induced water stress and photosynthetic reduction during chilling. *Plant, Cell and Environment* 20, 625–632.

Bridges, D.C. (1992) *Crop Losses Due to Weeds in the USA.* Weed Science Society of America, Champaign, Illinois, 401 pp.

Bunce, J.A. (1983) Differential sensitivity to humidity of daily photosynthesis in the field in C_3 and C_4 species. *Oecologia* 54, 233–235.

Bunce, J.A. (1993a) Growth, survival, competition, and canopy carbon dioxide and water vapor exchange of first year alfalfa at an elevated CO_2 concentration. *Photosynthetica* 29, 557–565.

Bunce, J.A. (1993b) Effects of doubled atmospheric carbon dioxide concentration on the responses of assimilation and conductance to humidity. *Plant, Cell and Environment* 16, 189–197.

Bunce, J.A. (1995) Long-term growth of alfalfa and orchard grass plots at elevated carbon dioxide. *Journal of Biogeochemistry* 22, 341–348.

Bunce, J.A. (1997) Variation in growth stimulation by elevated carbon dioxide in seedlings of some C_3 crop and weed species. *Global Change Biology* 3, 61–66.

Bunce, J.A., Wilson, K.B. and Carlson, T.N. (1997) The effect of doubled CO_2 on water use by alfalfa and orchard grass: simulating evapotranspiration using canopy conductance measurements. *Global Change Biology* 3, 81–87.

Carter, D.R. and Peterson, K.M. (1983) Effects of a CO_2 enriched atmosphere on the growth and competitive interaction of a C_3 and C_4 grass. *Oecologia* 58, 188–193.

Choudhuri, U.N., Kirkham, M.B. and Kanemasu, E.T. (1990) Carbon dioxide and water level effects on yield and water use of winter wheat. *Agronomy Journal* 82, 637–641.

Coleman, J.S. and Bazzaz, F. (1992) Interacting effects of elevated CO_2 and temperature on growth and resource use of co-occurring C_3 and C_4 annuals. *Ecology* 73, 1244–1259.

Conroy, J.P. (1992) Influence of elevated atmospheric CO_2 concentrations on plant nutrition. *Australian Journal of Botany* 40, 445–456.

Cure, J.D. and Acock, B. (1986) Crop responses to carbon dioxide doubling: a literature survey. *Agricultural and Forest Meteorology* 38, 127–145.

Dippery, J.K., Tissue, D.T., Thomas, R.B. and Strain, B.R. (1995) Effects of low and elevated CO_2 on C_3 and C_4 annuals. 1. Growth and biomass allocation. *Oecologia* 101, 13–20.

Dybing, C.D. and Currier, H.B. (1961) Foliar penetration of chemicals. *Plant Physiology* 36, 169–174.

Ehleringer, J. and Bjorkman, O. (1977) Quantum yields for CO_2 uptake in C_3 and C_4 plants. *Plant Physiology* 59, 86–90.

Ehleringer, J.R., Cerling, T.E. and Helliker, B.R. (1997) C_4 photosynthesis, atmospheric CO_2 and climate. *Oecologia* 112, 285–299.

Evans, L.T. (1993) *Crop Evolution, Adaptation and Yield.* Cambridge University Press, Cambridge, UK, pp. 500.

Froud-Williams, R.J. (1996) Weeds and climate change: implications for their ecology and control. *Aspects of Applied Biology* 45, 187–196.

Ghannoum, O., von Caemmerer, S., Barlow, E.W.R. and Conroy, J.P. (1997) The effect of CO_2 enrichment and gas exchange of a C_3 (*Panicum laxum*) and a C_4 (*Panicum antidotale*) grass. *Australian Journal of Plant Physiology* 24, 227–237.

Goudriaan, J. and de Ruiter, H.E. (1983) Plant growth in response to CO_2 enrichment, at two levels of nitrogen and phosphorus supply. 1. Dry matter, leaf area and development. *Netherlands Journal of Agricultural Science* 31, 157–169.

Greer, D.H., Laing, W.A. and Campbell, B.D. (1995) Photosynthetic responses of thirteen pasture species to elevated CO_2 and temperature. *Australian Journal of Plant Physiology* 22, 713–722.

Grime, J.P. (1979) *Plant Strategies and Vegetation Processes.* John Wiley & Sons, New York, 222pp.

Holm, L.G., Plucknett, D.L., Pancho, J.V. and Herberger, J.P. (1977) *The World's Worst Weeds: Distribution and Biology.* University of Hawaii Press, Honolulu, 609 pp.

Holm, L., Doll, J., Holm, E., Pancho, J. and Herberger, J. (1997) *World Weeds: Natural Histories and Distribution.* John Wiley & Sons, New York, 1129 pp.

Houghton, J.T., Meira Filho, L.G., Callander, B.A., Harris, N., Kattenburg, A. and Maskell, K. (1996) *IPCC Climate Change Assessment 1995:. The Science of Climate Change.* Cambridge University Press, Cambridge, UK, 572 pp.

Houghton, S.K. and Thomas, T.H. (1996) Effects of elevated carbon dioxide concentration and temperature on the growth and competition between sugarbeet (*Beta vulgaris*) and fat-hen (*Chenopodium album*). *Aspects of Applied Biology* 45, 197–204.

Hunt, R., Hand, D.W., Hannah, M.A. and Neal, A.M. (1993) Further responses to CO_2 enrichment in British herbaceous species. *Functional Ecology* 7, 661–668.

Idso, S.B., Kimball, B.A., Anderson, M.G. and Mauney, J.R. (1987) Effects of atmospheric CO_2 enrichment on plant growth: the interactive role of air temperature. *Agriculture Ecosystems and Environment* 20, 1–10.

Imai, K. and Murata, Y. (1979) Effect of carbon dioxide concentration on growth and dry matter production on crop plants. VII. Influence of light intensity and temperature on the effect of carbon dioxide enrichment in some C_3 and C_4 species. *Japanese Journal of Crop Science* 48, 409–417.

Johnson, H.B., Polley, H.W. and Mayeux, H.S. (1993) Increasing CO_2 and plant–plant ineractions: effects on natural vegetation. *Vegetatio* 104/105, 157–170.

Kimball, B.A. (1983) Carbon dioxide and agricultural yield: an assemblage and analysis of 430 prior observations. *Agronomy Journal* 75, 779–788.

Knapp, A.K., Hamerlyn, C.K. and Owensby, C.E. (1993) Photosynthetic and water relations response to elevated CO_2 in the C_4 grass, *Andropogon gerardii*. *International Journal of Plant Science* 154, 459–466.

Kobza, J. and Edwards, G.E. (1987) Influences of leaf temperature on photosynthetic carbon metabolism in wheat. *Plant Physiology* 83, 69–74.

Kropff, M.J. and Spitters, C.J.T. (1991) A simple model of crop loss by weed competition from early observations on relative leaf area of the weeds. *Weed Research* 31, 97–105.

Leegood, R.C. and Edwards, G.E. (1996) Carbon metabolism and photorespiration: temperature dependence in relation to other environmental factors. In: Baker, N.R. (ed.) *Photosynthesis and the Environment*. Kluwer Academic Publishers, Dordrecht, pp. 191–221.

Long, S.P. (1991) Modification of the response of photosynthetic productivity to rising temperatures by atmospheric CO_2 concentrations: has its importance been underestimated? *Plant, Cell and Environment* 14, 729–739.

Luscher, A., Hendrey, G.R. and Nosberger, J. (1998) Long-term responsiveness to free air CO_2 enrichment of functional types, species and genotypes of plants from fertile permanent grassland. *Oecologia* 111, 37–45.

Mansfield, T.A., Pemadusa, M.A. and Smith, P.J. (1983) New possibilities for controlling foliar absorption via stomata. *Pesticide Science* 14, 294–298.

McLaughlin, S.B. and Taylor, G.E. (1981) Relative humidity: important modifier of pollutant uptake by plants. *Science* 211, 167–169.

McNaughton, K.G. and Jarvis, P.G. (1991) Effects of spatial scale on stomatal control of transpiration. *Agircultural and Forest Meteorology* 54, 279–302.

Mulchi, C.L., Slaughter, L., Saleem, M., Lee, E.H., Pausch, R. and Rowland, R. (1992) Growth and physiological characteristics of soybean in open-top chambers in response to ozone and increased atmospheric CO_2. *Agriculture Ecosystems and Environment* 38, 107–118.

Muzik, T.J. (1976) Influence of environmental factors on toxicity to plants. In: Audus, L.J. (ed.) *Herbicides. Physiology, Biochemistry, Ecology*. Academic Press, New York, pp 203–247.

Newton, P.C.D., Clark, H., Bell, C.C. and Glasglow, E.M. (1996) Interaction of soil moisture and elevated CO_2 on the above-ground growth rate, root length density and gas exchange of turves from temperate pasture. *Journal of Experimental Botany* 47, 771–779.

Norris, R.F. (1982) Interactions between weeds and other pests in the agro-ecosystem. In: Hatfield, J.L. and Thomason, I.J. (eds) *Biometeorology in Integrated Pest Management*. Academic Press, New York, pp. 343–406.

Osmond, C.B., Bjorkman, O. and Anderson, D.J. (1980) *Physiological Processes in Plant Ecology. Toward a Synthesis with Atriplex*. Springer-Verlag, Berlin, 486 pp.

Owensby, C.E., Coyne, P.I., Ham, J.A., Auen, L.M. and Knapp, A.K. (1993) Biomass production in a tallgrass prairie ecosystem exposed to ambient and elevated CO_2. *Ecological Applications* 3, 644–653.

Parker, C. and Fryer, J.D. (1975) Weed control problems causing major reduction in world food supplies. *FAO Plant Protection Bulletin* 23, 83–95.

Patterson, D.T. (1985) Comparative ecophysiology of weeds and crops. *Weed Physiology: Reproduction and Ecophysiology* 1, 102–129.

Patterson, D.T. (1986) Responses of soybean (*Glycine max*) and three C_4 grass weeds to CO_2 enrichment during drought. *Weed Science* 34, 203–210.

Patterson, D.T. (1993) Implications of global climate change for impact of weeds, insects and plant diseases. *International Crop Science* 1, 273–280.

Patterson, D.T. (1995a) Weeds in a changing climate. *Weed Science* 43, 685–701.

Patterson, D.T. (1995b) Effects of environmental stress on weed/crop interactions. *Weed Science* 43, 483–490.

Patterson, D.T. and Flint, E.P. (1980) Potential effects of global atmospheric CO_2 enrichment on the growth and competitiveness of C_3 and C_4 weed and crop plants. *Weed Science* 28, 71–75.

Patterson, D.T. and Flint, E.P. (1982) Interacting effects of CO_2 and nutrient concentration. *Weed Science* 30, 389–394.

Patterson, D.T. and Flint, E.P. (1990) Implications of increasing carbon dioxide and climate change for plant communities and competition in natural and managed ecosystems. In: Kimball, B.A., Rosenberg, N.J. and Allen, L.H. Jr (eds) *Impact of Carbon Dioxide, Trace Gases and Climate Change on Global Agriculture.* ASA Special Publication No. 53, American Society of Agronomy, Madison, Wisconsin, pp 83–110.

Patterson, D.T., Flint, E.P. and Beyers, J.L. (1984) Effects of CO_2 enrichment on competition between a C_4 weed and a C_3 crop. *Weed Science* 32, 101–105.

Patterson, D.T., Highsmith, M.T. and Flint, E.P. (1988) Effects of temperature and CO_2 concentration on the growth of cotton (*Gossypium hirsutum*), spurred anoda (*Anoda cristata*), and velvetleaf (*Abutilon theophrasti*). *Weed Science* 36, 751–757.

Poorter, H. (1993) Interspecific variation in the growth response of plants to an elevated ambient CO_2 concentration. *Vegetatio* 104/105, 77–97.

Potter, J.R. and Jones, J.W. (1977) Leaf area partitioning as an important factor in growth. *Plant Physiology* 59, 10–14.

Potvin, C. and Strain, B.R. (1985) Effects of CO_2 enrichment and temperature on growth in two C_4 weeds, *Echinochloa crus-galli*, and *Eleusine indica. Canadian Journal of Botany* 63, 1495–1499.

Potvin, C. and Vasseur, L. (1997) Long-term CO_2 enrichment of a pasture community. *Ecology* 78, 666–677.

Read, J.J. and Morgan, J.A. (1996) Growth and partitioning in *Pascopyrum smithii* (C_3) and *Bouteloua gracilis* (C_4) as influenced by carbon dioxide and temperature. *Annals of Botany* 77, 487–496.

Reekie, J.Y.C., Hicklenton, P.R. and Reekie, E.G. (1994) Effects of elevated carbon dioxide on time of flowering in four short-day and four long-day species. *Canadian Journal of Botany* 72, 533–538.

Rogers, G.S., Payne, L., Milham, P. and Conroy, J. (1993) Nitrogen and phosphorus requirements of cotton and wheat under changing atmospheric CO_2 concentrations. *Plant and Soil* 155/156, 231–234.

Rogers, H.H., Runion, G.B. and Krupa, S.V. (1994) Plant responses to atmospheric CO_2 enrichment with emphasis on roots and the rhizosphere. *Environmental Pollution* 83, 155–189.

Sargent, J.A. and Blackman, G.E. (1962) Studies on foliar penetration. I. Factors controlling the entry of 2,4-D. *Journal of Experimental Botany* 13, 348–368.

Sasek, T.W. and Strain, B.R. (1990) Implications of atmospheric CO_2 enrichment and climatic change for the geographical distribution of two introduced vines in the USA. *Climate Change* 16, 31–51.

Schenk, U., Jager, H.J. and Weigel, H.J. (1997) The response of perennial ryegrass/white clover swards to elevated atmospheric CO_2 concentrations. 1. Effects on competition and species composition and interaction with N supply. *New Phytologist* 135, 67–79.

Seneweera, S., Milham, P. and Conroy, J. (1994) Influence of elevated CO_2 and phosphorus nutrition on the growth and yield of a short-duration rice (*Oryza sativa* L. cv. Jarrah). *Australian Journal of Plant Physiology* 21, 281–292.

Sionit, N., Strain, B.R. and Beckford, H.A. (1981) Environmental controls on the growth and yield of okra. I. Effects of temperature and of CO_2 enrichment at cool temperatures. *Crop Science* 21, 885–888.

Sionit, N., Strain, B.R. and Flint, E.P. (1987) Interaction of temperature and CO_2 enrichment on soybean: growth and dry matter partitioning. *Canadian Journal of Plant Science* 67, 59–67.

Tremmel, D.C. and Patterson, D.T. (1993) Responses of soybean and five weeds to CO_2 enrichment under two temperature regimes. *Canadian Journal of Plant Science* 73, 1249–1260.

USDA (1988) *Agricultural Statistics 1988.* US Government Printing Office, Washington, DC, 544 pp.

Wheeler, T.R., Batts, G.R., Ellis, R.H., Hadley, P. and Morison, J.I.L. (1996) Growth and yield of winter wheat (*Triticum aestivum*) crops in response to CO_2 and temperature. *Journal of Agricultural Science* 127, 37–48.

Wong, S.C. (1979) Elevated atmospheric partial pressure of CO_2 and plant growth. I. Interactions of nitrogen nutrition and photosynthetic capacity in C_3 and C_4 plants. *Oecologia* 44, 68–74.

Wong, S.C. and Osmond, C.B. (1991) Elevated atmospheric partial pressure of CO_2 and plant growth. III. Interactions between *Triticum aestivum* (C_3) and *Echinochloa frumentacea* (C_4) during growth in mixed culture and different CO_2, N nutrition and irradiance treatments, with emphasis on below-ground responses estimated using the $\delta^{13}C$ value of root biomass. *Australian Journal of Plant Physiology* 18, 137–152.

Zangerl, A.R. and Bazzaz, F.A. (1984) The response of plants to elevated CO_2. II. Competitive interactions among annual plants under varying light and nutrients. *Oecologia* 62, 412–417.

Ziska, L.H. and Bunce, J.A. (1993) The influence of elevated CO_2 and temperature on seed germination and emergence from soil. *Field Crops Research* 34, 147–157.

Ziska, L.H. and Bunce, J.A. (1997a) Influence of increasing carbon dioxide concentration on the photosynthetic and growth stimulation of selected C_4 crops and weeds. *Photosynthesis Research* 54, 199–208.

Ziska, L.H. and Bunce, J.A. (1997b) The role of temperature in determining the stimulation of CO_2 assimilation at elevated carbon dioxide concentration in soybean seedlings. *Physiologia Plantarum* 100, 126–132.

Ziska, L.H., Namuco, O., Moya, T. and Quilang, J. (1997) Growth and yield response of field-grown tropical rice to increasing carbon dioxide and temperature. *Agronomy Journal* 89, 45–53.

16 Crop Ecosystem Responses to Climatic Change: Pests and Population Dynamics

Andrew Paul Gutierrez

Division of Ecosystem Science, University of California, 151 Hilgard Hall, Berkeley, CA 94720, USA

> Man stalks across the landscape, and desert follows his footsteps.
>
> Herodotus (5th century BC)

> And when it was morning, an east wind brought the locusts.
>
> *Exodus* X: 13

> Our problem is that we are too smart for our own good, and for that matter, the good of the biosphere. The basic problem is that our brains enable us to evaluate, plan, and execute. Thus, while all other creatures are programmed by nature and subject to her whims, we have our gray computer to motivate, for good or evil, our chemical engine. Indeed, matters have progressed to the point where we attempt to operate independently of nature, challenging her domination of the biosphere. This is a game we simply cannot win, and in trying we have set in train a series of events that has brought increasing chaos to the planet.
>
> On the species *Homo sapiens*
> Robert van den Bosch, 1978

16.1 Introduction

Climate change may disrupt not only pest dynamics in agriculture but also the dynamics of herbivores in stable ecosystems where regulation of their numbers has gone largely unnoticed. From an anthropocentric perspective, pests are organisms of any taxa (species of arthropods, fungi, plants, vertebrates, etc.) that cause annoyance, disease, discomfort or economic loss to humans. In an ecological context, pests are merely filling evolved roles in ecosystems that may have been simplified or disrupted through domestication, inputs of nutrients and toxic substances, or by sheer human mismanagement. Many pests are exotics, accidentally introduced to new environments and freed from natural constraints. Many of our crop, livestock and other domesticated species have wild progenitors in parts of the world other than where they are cultivated, and as a result of selection, domesticated species are often genetic shadows of the original genetic diversity of the species and may be more susceptible to pests. Pest genetics, on the other hand, may remain largely unchanged, or if change has occurred it may have been in response to changes in their environment wrought by humans (e.g. pesticide resistance). Pests are usually part of complex trophic interactions in linear food chains or complex food webs; disruption by pesticides often leads to the scourges of pest resurgence, secondary pest outbreaks and pesticide resistance as well as adverse ecosystem and human health effects (van den Bosch, 1978). Price (1984 and later editions) provides a very readable overview of the ecosystem concept that is entirely compatible with this chapter.

To examine the impact of climate change on these systems, we must take a holistic tritrophic view, often using models that are abstractions of reality. Some of the questions that must be answered include: can a pest survive in the environment in question, what is its potential geographical range, what might its phenology be, at what level will its numbers be regulated and how much damage will it cause? Unfortunately, the literature on these topics is sparse. Drake (1994) made a general statement about the possible impact of global warming on Australian pests. He stressed the need for pest-forecasting systems (modelling) and estimating the impact of global warming on pest activity. Heong (1995) edited a volume on the effects of global warming on rice and rice arthropods, and included some preliminary modelling results. Brasier and Scott (1994) made a theoretical assessment of the effects of the pathogen *Phytophthora cinnamoni* on *Quercus* spp. in the Mediterranean area using the geographical information system (GIS) based CLIMEX program (see Sutherst *et al*., 1991). The bases of CLIMEX are discussed below. The analysis suggests that a 1.5–3°C increase in temperature might increase the activity of the pest and extend its range. Morimoto *et al.* (1998), using temperature summation and day-length models, predicted that the northern boundary of *Plutella xylostella* would shift northward by 300 km in Japan. Ellis *et al.* (1997) used field data to show that peak flight in some Dutch microlepidoptera shifted to 11.6 days earlier during the period 1975–1994. Kobak *et al.* (1996) predicted that global warming effects would increase the area of steppes into larch forests as global warming and less moisture increased the frequency of fire and the severity of pathogens. Fleming (1996) and Fleming and Candau (1998) used a process-based approach to model the possible effects of global warming on spruce budworm in Canada, and discuss the possible changes in damage patterns. This latter approach is akin to some of the modelling methods suggested here.

16.2 The Use of Growth Indices in Climate Matching

In the earlier part of the 20th century, time series plots of daily, weekly or monthly temperature and other variables would be used to create graphs (e.g. thermohygographs) that characterized climate favourable to the distribution of species. This empirical approach did not include the physiological responses of species to the climatic factors. Fitzpatrick and Nix (1970) made an important innovation by characterizing the growth rates of temperate and tropical pastures in Australia as normalized hump-shaped physiological response indices to weather and edaphic factors (Fig. 16.1a). A low index value means that the factor was either below or above the optimum (i.e. the peak). The easiest effect to demonstrate is that of temperature on developmental rate (Fig. 16.1b). The derivative of the developmental rate function is also hump-shaped (e.g. Fig. 16.1a), suggesting that development occurs above a lower thermal threshold and stops at an upper one with the maximum at the peak. Temperature sets the baseline for the potential developmental rate of poikilotherms, with other factors altering growth, development and reproduction. The relationship of developmental rate to temperature is shown for three hypothetical species in a food chain (Fig. 16.1c). In this example, the

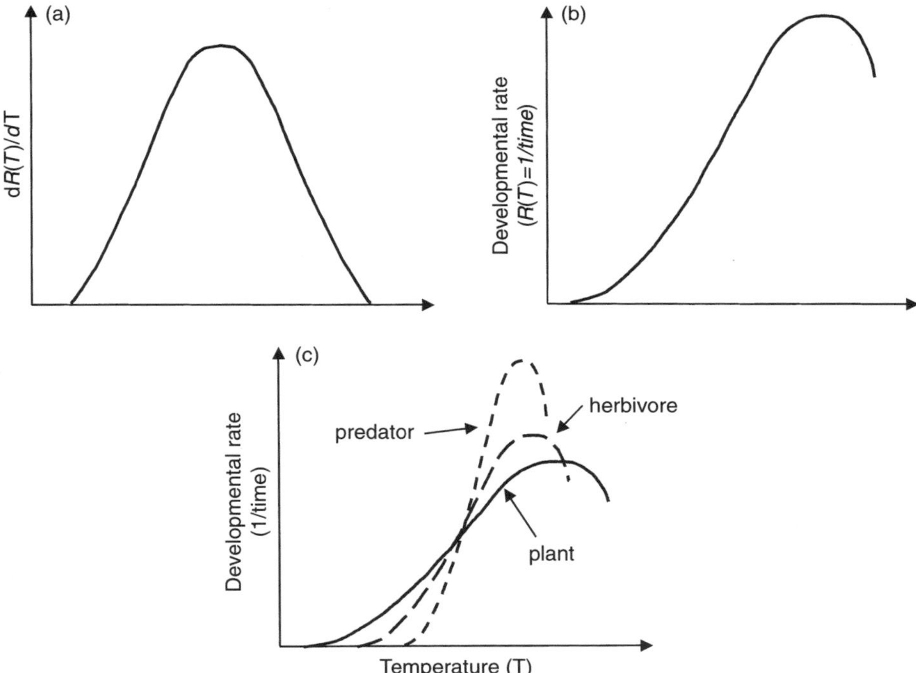

Fig. 16.1. Hypothetical developmental responses of a poikilotherm to temperature: (a) the developmental rate; (b) the temperature derivative of the developmental rate function in (a); (c) the hypothetical developmental rates on temperature of a plant, a herbivore and a predator.

plant has a broad response to temperature, the herbivore has an intermediate one, and that of the predator is quite narrow. This means that the plant may develop at lower and higher temperatures than both the herbivore and the predator, and that the herbivore has a broader range than the predator. The studies of Messenger and Force (1963) and Messenger (1964, 1968) on the spotted alfalfa aphid and its parasitoids are excellent examples of such thermal effects on trophic-level developmental and demographic rates, and the consequences on the geographical distribution of these species.

This exposé is a reasonable explanation for the formulation of the temperature index (*TI*) in the Fitzpatrick–Nix system. For factors such as soil nitrogen, a low index value might mean that either there is too little or that it is present in toxic quantities. In general, growth is assumed to be a function of several factors each assumed to have similarly shaped growth indices. In the Fitzpatrick–Nix approach, the final growth index is a product of all growth indices. Assuming indices for temperature (*TI*), nitrogen (*NI*), soil water (*WI*) and other factors and using the Fitzpatrick–Nix model, the growth index $GI(t)$ at time t at location ij is computed as:

$$GI_{ij}(t) = TI_{ij}(t) \times NI_{ij}(t) \times WI_{ij}(t) \ldots \qquad \text{(Eqn 16.1)}$$

A factor may be limiting (i.e. equal zero), making $GI(t) = 0$, and is a demonstration of von Liebig's Law of the Minimum. Normally, factors exhibit different levels of favourability, and the product of their indices reflect the compounding effects on $GI(t)$.

To demonstrate the utility of their approach, Fitzpatrick and Nix computed the weekly 30-year average *GI* for temperate and tropical pastures for all sites in Australia having the requisite meteorological data. Though not stated, the net flux of factors such as soil water, nitrogen and other essential nutrients have to be updated at each site to characterize their time-varying effects. The areas of favourability of pasture growth across Australia could then be mapped for the different periods of the year using equation 16.1.

Gutierrez *et al.* (1974) modified this approach and used it to explain the phenology and abundance of several species of aphid trapped weekly at 32 locations over a 2-year period (1969/70) in southeast Australia. Weather in this area is highly variable temporally and spatially, and many of the aphids do not survive year round in much of this area because of winter frost or summer drought. These aphids must recolonize areas when favourable conditions return. Small areas of microclimate may enable small numbers to persist but these are usually insufficient to repopulate the area. Their migration is undirected and winged aphids are blown on the winds to favourable and unfavourable areas alike. A good example of a species with this migratory habit is the cowpea aphid (*Aphis craccivora* Koch), which breeds on pasture legumes throughout Australia. In most areas, influxes of migrants are required to begin local population growth when conditions are favourable. It is not uncommon for areas with lush pasture to receive few or no aphid migrants and only small populations arise. Large influxes of migrants may arrive in areas with mostly bare ground during extended periods of drought (Gutierrez *et al.*, 1971, 1974). Populations of the cowpea aphid on occasion reach the very high densities reported by Johnson (1957). This may occur when local conditions are favourable for long periods and large influxes of migrants arrive. Normally the season is too short for such outbreaks to occur on a regular basis. Hence, soil moisture limits pasture growth during summer and frost limits the aphid during winter. Natural enemies play a minor role in regulating aphid numbers, hence cowpea dynamics in most areas of southeast Australia are weather-driven.

Temperature indices for the aphid and soil moisture indices for the plant were computed at the 32 trap locations for the period 1968–1970, using observed weather and 30-year average weather. Autumn is an important period for the build-up of cowpea aphid (Gutierrez *et al.*, 1971). Weekly values for autumn (roughly weeks 6 to 20) for 1969 and 1970 and 30-year average weekly values are shown in Fig. 16.2a, for Bathurst and Trangie, New South Wales. These data demonstrate the large variability of weather that regularly occurs in southeast Australia. The autumn of 1969 was wetter and cooler than the same period in 1970 which was a period of drought. Cowpea aphid populations developed in 1969 but not 1970 at these two locations.

During the years 1969 and 1970, cowpea populations developed at 18 of the 32 sites. The populations developed at different times of the year at the different locations. Average temperature and moisture indices for the periods of aphid activity at these sites are plotted in Fig. 16.2c. The data exhibit a remarkable degree of clustering, with the boundary around the data being the bivariate normal tolerance limits ($P < 0.05$). This figure may be viewed as a physiological thermohygrogram that defines the physiological limit of tolerance of the aphid to the two factors (i.e. an estimate of Shelford's Law of Tolerance for the two factors). This means that local conditions are favourable for cowpea aphid population growth when the indices are within this boundary. Price (1984) illustrated this concept quite nicely and Fig. 16.3 is redrawn from that work. Gutierrez *et al.* (1974) also examined the regional migration of cowpea aphid, explaining how the patterns of winds characteristic to southeast Australia enable it to bridge areas of favourability. The original papers should be consulted for details.

Other species may respond differently to the same two factors, and of course many other factors (more dimensions) may impinge on them. Similar growth indices computed for two grass aphids (*Rhopalosipum padi* and *R. maidis*) yielded similar clustering of the data, but in different areas of the physiological thermohygrogram (Fig.16.2d). The distribution of the growth index data reflects the fact that *R. maidis* population occurs late in the spring and early summer, when the Mediterranean grasses are drying. Populations of *R. padi* may develop during wet cool periods of autumn and early spring because they are protected at the base of bunch grasses. Gutierrez and Yaninek (1983) used different methods to examine the thermohygrograms for six other species, and included the size of the populations in computing the tolerance region.

The important point of these analyses is that site favourability determines when outbreaks can occur, but the arrival of migrants is a necessary prerequisite for the build-up of cowpea populations.

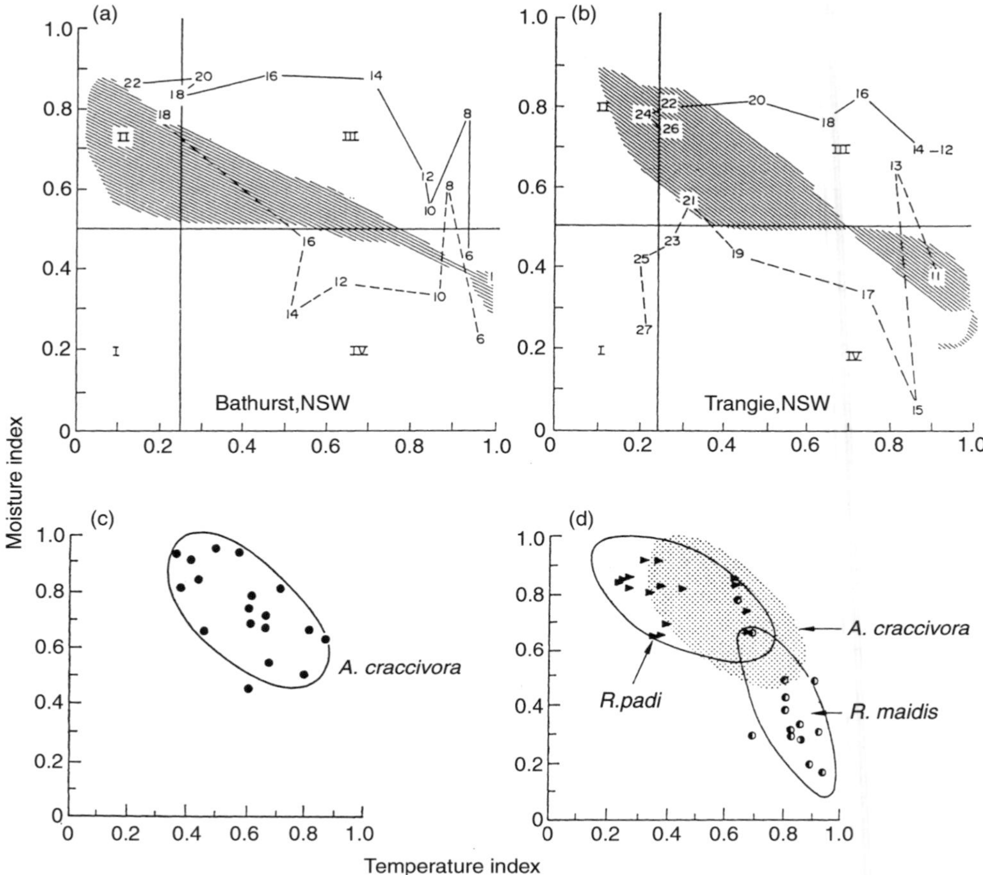

Fig. 16.2. Growth indices for cowpea aphid in southeast Australia (Gutierrez *et al.*, 1974) using the Fitzpatrick and Nix (1970) method: the historical 30 year average values (i.e. the margins of the shaded area) and the values for autumn 1969 (—) and 1970 (----) at (a) a highland site (Bathurst, New South Wales) and (b) on the interior plains (Trangie, New South Wales); (c) the average indices at 18 of the 32 locations for the periods when cowpea aphid populations developed (see Gutierrez *et al.*, 1974, and text); (d) the average indices for two grass aphids (the shaded area is that of cowpea aphid in (c).

The studies of Fitzpatrick and Nix (1970) and Gutierrez *et al.* (1974) predate the advent of modern GIS, which can now quickly capture and map regional data. The climate-matching GIS program CLIMEX developed by Sutherst *et al.* (1991 and citations therein) is also based on the Fitzpatrick–Nix approach; it uses 30-year weather averages and includes areas outside of Australia. This algorithm has been used effectively to map the potential range of the Russian grain aphid (*Diuraphis noxia*) (Hughes and Maywald, 1990), the pathogen *Phytophthora cinnamoni* on *Quercus* spp. (Brasier and Scott, 1994) and other pests. Its successful application has occurred because the Fitzpatrick–Nix indices capture the shape of some essential biology. A

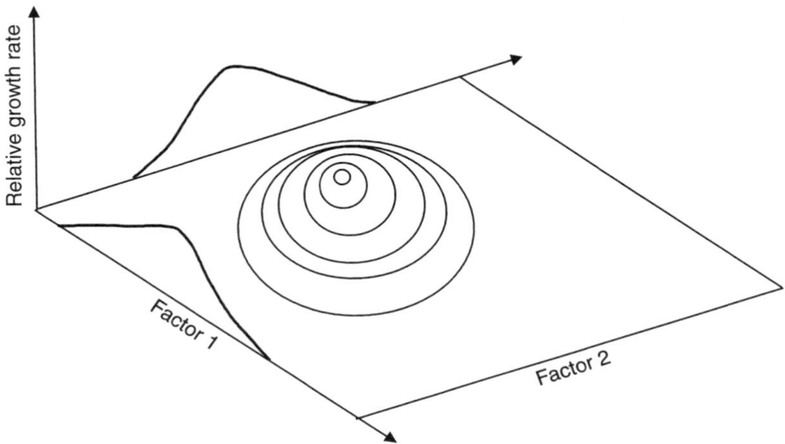

Fig. 16.3. Hypothetical physiological growth response of a poikilotherm to two environmental factors (cf. Price, 1984).

shortcoming of CLIMEX is that the physiological–mathematical bases have not been published. Below we develop a reasonable mathematical description of this biology.

16.3 Some Modelling Preliminaries

The question of how to model complex biological systems and how much detail to include in the model remains open. The debate is often couched in arguments about simple vs. complex models. Here the focus will be on physiologically based models of intermediate complexity (i.e. process models) that are driven by weather and influenced by various abiotic factors. For convenience, a simple food chain consisting of a plant (*P*), herbivore (*H*) and carnivore (*C*) is used to illustrate the issues. All of these species are assumed to be poikilotherms, though the effects of climate change on their interactions could also affect homeotherms that depend on them as resources. Two questions are addressed: how can this model be used to assess the distribution of species, and how do these physiological aspects affect their dynamics? The latter question is formulated first.

It is commonly accepted that populations may be affected by intraspecific competition for resources (bottom-up effects), by predation from higher trophic levels (top-down effects: Hairston *et al.*, 1960), and by lateral effects from interspecific competition for resources. Berryman and Gutierrez (1998) proposed that there are three basic ways of analysing population dynamics problems:

1. The first appeals to the laws of physics as a metaphor for population interactions – a mass action point of view (e.g. Lotka, 1925; Volterra, 1926, 1931).

2. The second approach models the effects of plant and carnivore densities on survivorship and is heavily rooted in classical life table analysis (e.g. Morris, 1963; Varley *et al.*, 1973).
3. The third models the general process of consumption (i.e. the flow of biomass or energy through the food chain – physiologically based models).

The physiological approach has its conceptual origins in older ecophysiological work (von Liebig, 1840; Candolle, 1855; Shelford, 1931; Hutchinson, 1959) and in ecological energetics (Phillipson, 1966; Odum, 1971). It is increasingly being used by plant physiologists (see de Wit and Goudriaan, 1978) and ecologists in weather-driven population dynamics models (Gutierrez and Wang, 1977; Gutierrez, 1992, 1996). Here, the physiological approach will be emphasised.

16.4 A Physiologically Based Approach

The basic premise of the physiological approach put forth here is that all organisms are consumers and have similar problems of resource acquisition (input) and allocation (output). The conditions that organisms face vary widely, as do the biological adaptations to solve these problems (Gutierrez and Wang, 1977). A model to analyse the effects of climate change on organisms must include the effects on its behaviour and physiology – on its ability to function and compete in the environment. It must include intraspecific competition for resources, prey availability and physiological processes of assimilation of resource to self in a tritrophic setting. Petrusewicz and McFayden (1970) and Batzli (1974) outlined the details of allocation physiology (Fig. 16.4). This well-known physiology is seldom included in population dynamics models, which is unfortunate because it is here that the effects of weather and climate change must enter.

16.4.1 Resource acquisition

Royama (1971, 1992) recognized two kinds of attack or resource acquisition functions: the 'instantaneous' and 'overall' hunting equations. These models are commonly called functional response models (Solomon, 1949; Holling 1966) and the instantaneous form is appropriate for continuous-time *differential* equation population dynamics models. If the time step of the model is large relative to the life span of the organism, then the integrated form of the instantaneous model or the overall hunting model is more appropriate. The dynamics of such systems should be written as discrete-time *difference* equations. The consumer may be a parasitoid or a predator, and these important differences must be taken into account in integrating the model. Readers are referred to Royama (1971, 1992) for complete details of this theory and Gutierrez (1996) for some applications.

There are many functional response models, but a variant of Watt's (1959) model is preferable because it incorporates all of the desiderata

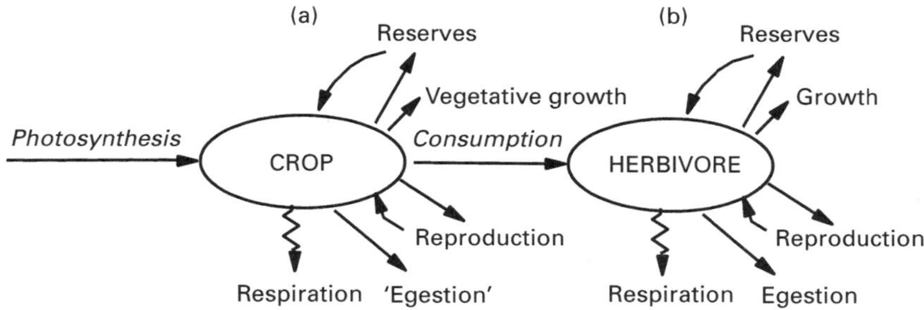

Fig. 16.4. The metabolic pool model of a plant and animal (cf. Gutierrez and Wang, 1977).

outlined above. Ivlev (1955) was the first to think about the functional responses from the consumer point of view (Berryman and Gutierrez, 1998), but unfortunately few ecologists followed his lead (e.g. Watt, 1959; Gilbert *et al.*, 1976; Gutierrez and Wang, 1977). Like Ivlev, Watt proposed that a consumer's attack rate depended on how hungry it was, and in addition on competition among consumers. The underlying logic of the models is outlined using the Berryman and Gutierrez notation. If δ_h is the food needed to satiate hunger per unit of herbivore mass (H) per dt and f_h is the amount consumed per unit per dt, then hunger of the consumers is the fraction of its demand not met, i.e. $(\delta_h - f_h)/\delta_h$). The relationship of consumption df_h to the change in resource (dP) must include H as a component of the total demand by all consumers ($\delta_h H$; see Gutierrez 1996).

$$\frac{df_h}{dP} = \frac{\alpha_h(\delta_h - f_h)}{\delta_h H} \qquad \text{(Eqn 16.2)}$$

The parameter α is the apparency of resource mass P to herbivore mass H. Integrating equation 16.2 gives the variant of Watt's functional response model used for estimating the quantity of resource obtained per unit of consumer as a function of resource and consumer abundance.

$$f_h = \delta_h(1 - e^{-\alpha_h P/\delta_h H}) \qquad \text{(Eqn 16.3)}$$

Equation 16.2 is an instantaneous hunting equation (see Gutierrez, 1996). In field applications, the amount of nutrients required by a poikilotherm organism per unit mass (δ_h) depends on temperature. The model saturates asymptotically to δ_h as the exponent becomes more negative (Fig. 16.5a). Royama (1971) integrated equation 16.3 to obtain the parasitoid and predator functional response models for use in discrete time dynamics models. The model proposed by de Wit and Goudriaan (1978) for modelling photosynthesis has the same form as equation 16.3, but its analytical origins are quite different.

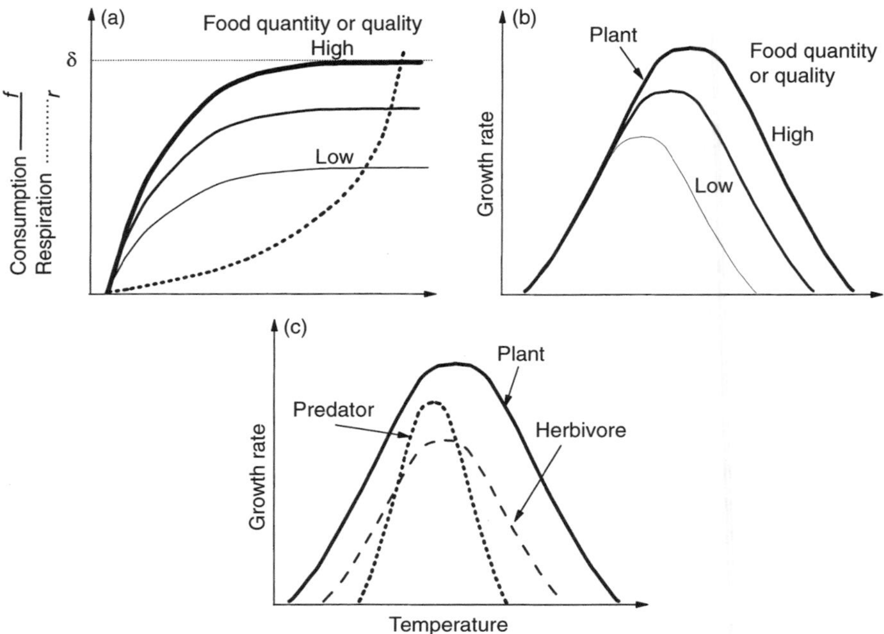

Fig. 16.5. (a) Hypothetical functional responses (consumption) under different levels of resources (solid lines of varying width) as well as the respiration rate (dashed line) both plotted on temperature. (b) Growth rate on temperature as the difference between consumption and respiration at different levels of resources, from (a). (c) Growth rates of the three species at constant resource across temperature.

16.5 Per Unit Mass Growth Dynamics

Suppose that each unit of herbivore mass (H) consumes f_h units of resource mass (P), and that r_h (respiration rate) units are required to sustain life, then $f_h - r_h$ units will be available for growth and reproduction. The effects of temperature and levels of food on f_h are illustrated in Figure 16.5a. Consumption (f) depends on temperature but also decreases with level of resource (the thinner lines). Also shown is the respiration rate r_h (i.e. the Q_{10} rule), which increases with temperature but is largely independent of resource level. The extra costs of searching at low levels of resource could affect r, but the effect is ignored here.

Hence, if θ_h units of offspring can be produced per unit of food in excess of respiration, then the *instantaneous* per unit mass rate of change of H is:

$$\frac{dH}{Hdt} = \theta_h (f_h - r_h) \qquad \text{(Eqn 16.4)}$$

in which dH/Hdt is a hump-shaped function on temperature that equals zero at the lower thermal threshold for feeding and when f_h equals r_h at a high temperature (Fig. 16.5b, Gutierrez, 1992). If both sides are divided by the maximum growth rate under non-limiting conditions, a humped growth index with respect to temperature, level of food and intraspecific competition arises. As the levels of resource decrease or competition increases, the maximum growth rate (i.e. the optimum) falls and the function shifts to the left (the thinner lines in Fig. 16.5b). Growth rate indices on temperature and high resource abundance for three hypothetical species in a food chain are illustrated in Fig. 16.5c.

Among the essential resources required by plants are light, minerals (e.g. N, K, P) and water, while animals may require, among other things, vitamins, micronutrients and water. There is a plethora of abiotic factors that may affect the growth rates of organisms. For example, soil pH, wind, saturation deficits or lack of mates, nesting sites or territories may affect growth rates of different species across trophic levels. A degree of imagination is needed to make some of these important analogies. Shortfalls of essential resources and levels of abiotic factor may affect f, and their acquisition is modelled using variants of equation 16.3 (Gutierrez *et al.*, 1988, 1993). The parameter α_j must reflect the apparency of the jth resource to the species. Rearranging the terms of equation 16.3 gives the supply/demand ratio f_j/δ_j for the jth factor. The right-hand side is the probability of acquiring δ_j units of resource j. The effect of shortfalls of all $j = 1,\ldots,J$ factors on a species may be viewed as compounded resource acquisition probabilities (ϕ^*):

$$0 < \phi_j^* = \frac{f_1}{\delta_1}\frac{f_2}{\delta_2}\cdots\frac{f_J}{\delta_J} < 1 \qquad \text{(Eqn 16.5)}$$

The effects of all resource shortfalls (equation 16.5) may be included in equation 16.4 as follows:

$$\frac{dH}{Hdt} = \theta_h\left[\phi_h^* f_h - r_h\right] \qquad \text{(Eqn 16.6)}$$

Substituting the full model in equation 16.2 for f_h in equation 16.5 yields:

$$\frac{dH}{Hdt} = \theta_h\left[\phi_h^*\delta_h\left(1 - e^{-\frac{\alpha_h P}{\delta_h H}}\right) - r_h\right] \qquad \text{(Eqn 16.7)}$$

This model may be normalizing by dividing both sides by the maximum growth rate, yielding a hump-shaped growth index ($0 \leq GI < 1$) on temperature, ranging from zero to unity. This model is akin to the Fitzpatrick–Nix model, but its origins are quite different. The shortfalls of all other factors are included in ϕ_h^* (Gutierrez *et al.*, 1994).

In a metapopulation context, the favourability of conditions for growth at different locations may be quite variable, but the same model (equation 16.7) may be used to estimate the growth index value at each site and the isolines of favourability over a large geographical area may be displayed using GIS.

To examine top-down effects, equation 16.6 may be cast as a tritrophic population model (equation 16.7) (Gutierrez and Baumgärtner, 1984; Gutierrez *et al.*, 1994) by including predator mass (C) in equation 16.7 and rearranging terms:

$$\frac{\mathrm{d}H}{\mathrm{d}t} = \theta_h \left[\phi_h^* \delta_h \left(1 - e^{-\frac{\alpha_h P}{\delta_h H}} \right) - r_h \right] H - \phi_c^* \delta_c \left(1 - e^{-\frac{\alpha_c H}{\alpha_c C}} \right) C \qquad \text{(Eqn 16.8)}$$

This model could be evaluated at equilibrium with temperature related parameters substituted to evaluate the effects of climate change on regulation (Gutierrez *et al.*, 1994). Gutierrez and colleagues have called equation 16.8 the 'metabolic pool' approach to modelling trophic interrelationships because the approach seeks to model behavioural (α) and weather-driven physiological processes (δ and r) that link and drive the trophic dynamics of species. These results could be plotted for a region using GIS. This model has been used extensively to model diverse tritrophic systems with and without age structure (Gutierrez and Wang, 1977, Gutierrez *et al.*, 1988, 1994; Gutierrez, 1992, 1996).

16.6 Non-resource Factors

Species with wide geographical distributions may have clines of biotypes adapted to diverse conditions. The alfalfa weevil is a well-studied example (Bosch *et al.*, 1981). Often the adaptation to survive either harsh cold or extremes of dryness and high temperatures is a period of dormancy (Nechols *et al.*, 1998). Dormancy is a very important factor in determining the range of species and must be included in a model to examine the effects of climate change on species distribution, phenology and dynamics. Some ubiquitous species appear to have undergone small levels of adaptation to diapause and the same biology may gives a plethora of responses to different weather over a large geographical area. Pink bollworm (Gutierrez *et al.*, 1981) and cabbage rootfly (Johnsen and Gutierrez, 1997; Johnsen *et al.*, 1997) are two well-worked examples.

In north temperate regions, winter dormancy in cabbage rootfly is strongly affected by the interplay between temperature and day-length. In Europe, this biology produces a panoply of phenologies across the varying environments from the Mediterranean in the south to the Baltic in the north. In the north, fly pupae enter diapause in the autumn as photoperiod and temperatures decline, and emerge as adults in response to increasing spring temperatures. In California, rootfly is widely distributed, but is economically important mainly along the fog shrouded coast. Its phenology is particularly interesting here because it is moulded by declining photoperiod in autumn, when temperatures are still moderately high, and in summer when temperatures may be cool but photoperiod is long. Under these conditions, active populations and diapause forms occur simultaneously all year long on wild mustard. This overwintering biology is complicated, and Johnsen and Gutierrez (1997) and Johnsen *et al.* (1997) modelled it as substages of physiological development. The same model was used to predict the emergence patterns in both Denmark and California. Global warming would likely alter its phenology, dynamics and economic importance.

Pink bollworm attacks cotton worldwide. In California's desert valleys, populations develop during spring and summer, and diapausing individuals are produced in the autumn. There is a strong nutritional component to the length of diapause, with the strongest diapause occurring in older cotton bolls (Ankersmit and Adkisson, 1968). The response surface of diapause on temperature × photoperiod was mapped by Gutierrez *et al.* (1981) for an Arizona population. Gutierrez *et al.* (1986) showed that different areas of this response surface also describe pink bollworm diapause in Brazil and California. However, in North America the range of this pest is limited by heavy frosts that kill diapausing larvae. This restricts it from invading the million acres of cotton in the San Joaquin Valley of California, where heavy winter frosts are common. In the frost-free areas of southern Brazil, the diapause response allows diapausing and non-diapausing pink bollworm to occur simultaneously nearly all year long (Gutierrez *et al.*, 1986).

The relationships of rootfly and pink bollworm to temperature and photoperiod produce unique site specific patterns of diapause induction and population dynamics in response to prevailing weather. This punctuate the need to understand the factors that determine the year-long phenology of a species before the effects of climate change on pest dynamics and impact can be estimated. Tritrophic effects must also be included.

16.7 Climate Effects on Tritrophic Systems

It has been shown that Australian aphid populations may develop when favourable weather conditions occur, and their tolerance to two factors in their environment can be estimated from field data. However, determining the level of regulation of a species by natural enemies is more difficult.

Assume, for the sake of simplicity, that the climatic optimum of each species occurs at the midpoint of their response limits in all dimensions. Hypothetical time series of the two indices at some mythical site are shown in Fig. 16.6 as the margins of the shaded area, with the physiological limits of each species in the tritrophic system superimposed. In the first example (Fig. 16.6a), the overlaps of the physiological limits of all three trophic levels and site-specific indices are quite good. In the second (Fig. 16.6b), the physiological limits of the plant and herbivore coincide well with each other and with the site's climatic indices, but the overlap with the physiological limits of the predator is marginal. Figure 16.6c illustrates a much poorer degrees of climate matching. In all cases, however, without specific details about the biology of the interacting species during adverse periods, nothing can be said about regulation. The case represented by Fig. 16.6c would appear to be similar to that of cowpea aphid in Australia, where brief periods of favourability allow population build-up and high migratory capacity permits regional persistence, with natural enemies being of little importance throughout much of the area studied. To evaluate the level of regulation of a population, it is necessary to assess the tritrophic dynamics as modified by weather and other factors at equilibrium. This is explored below.

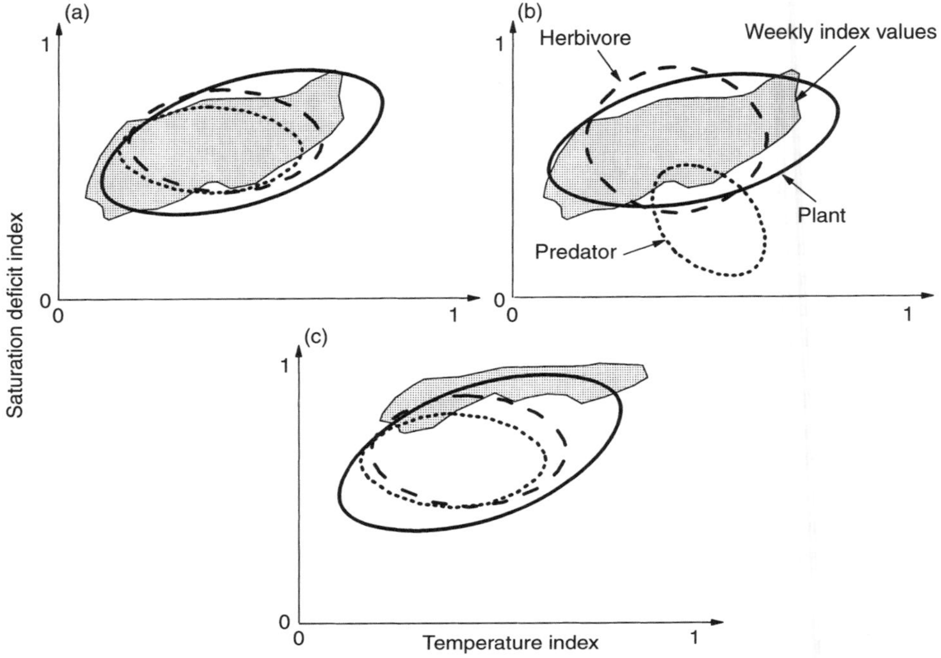

Fig. 16.6. The hypothetical physiological limits of poikilotherm species in a food chain to temperature and saturation deficit. The hypothetical indices for the location are the margins of the shaded area: (a) plant and herbivore and a well-adapted natural enemy; (b) the same as (a) but with a poorly adapted natural enemy; (c) the effects of shifts in weather at the site on relationships in (a).

16.8 The Effects of Climate Change on Biological Control

Assume a food chain wherein all species in a tritrophic system are well adapted to their environment (e.g. Fig. 16.6a), and that the natural enemy population regulates the herbivore population at low density. If climate change shifts site growth indices into unfavourable ranges (e.g. as in Fig. 16.6c), this would force all species to operate and survive near the margins of their physiological range or to perish. Extinction at the site might occur because of insufficient genetic diversity, but the species might survive in areas that remain favourable (refuges) or in new areas that become favourable as a result of climatic change. If climate change is slow, selection might enable the species to adapt. In another scenario, the species may become extinct locally as conditions become more favourable for their competitors. This may occur at any trophic level, but the lower down this occurs, the greater will be the number of species likely to be displaced (so-called keystone species effects).

There are numerous examples of the effects of climate on the establishment and efficacy of biological control agents. This is a well-known problem in the field of biological control and one of the major axioms has been to introduce climatically adapted natural enemies. Occasionally this axiom does not hold, and a good example is the control of cassava mealybug in Africa by the introduction of a parasitoid from a climatically different area of South America (Herren and Neuenschwander, 1991). However, the spotted alfalfa aphid in the great Central Valley of California was not controlled until a parasitoid biotype having the capacity to aestivate during the long dry summer was introduced (Flint, 1981). Similarly, the walnut aphid in California was not controlled until biotypes adapted to the different environments where walnut grew were introduced (Bosch *et al.*, 1970). The capacity to withstand high temperatures and to produce summer dormant individuals proved critical for the natural enemies of both aphid pests. The classical biological control of cottony cushion scale by the vedalia beetle (*Rodolia cardinalis*) has aspects of climatic limitations of natural enemy physiological activity (Quesada and DeBach, 1973). The parasitic fly *Cryptochaetum iceryae* was introduced but its range was restricted to cooler areas, while the vedalia beetle was able to control the pest in hotter areas (Quesada and DeBach, 1973). The classic study of DeBach and Sundby (1963) showed that successive introductions of parasitoids to control California red scale on citrus resulted in a sequence of one parasitoid displacing the former. This occurred until each found the subset of the total biotic and abiotic environment occupied by citrus favourable to its own development.

Such effects of climate on biological control interactions are recurring, and could be exacerbated by global warming if formerly effective natural enemies failed and new introductions were required for control. This problem could occur in agricultural and non-agricultural systems

16.9 Analysis of Species Persistence in Food Webs

To assess the regulation of a species in any trophic level by the interaction of top-down and bottom-up forces, a realistic model of the system must be formulated and evaluated at equilibrium (see Gutierrez, *et al.* 1994, for an analysis of a tritrophic system based on equation 16.8). Simplifications of model equation 16.8 by Schreiber and Gutierrez (1998) gave qualitative predictions about field systems that have proved useful in exploring species invasions into existing food webs and assembly rules for species persistence. The model may incorporate the effects of variable weather on each trophic level, making it a valuable tool for evaluating the effects of climate change over a large geographical area. These results could also be displayed using GIS.

An example of this kind of food web analysis is that of the pea aphid (*Acyrthosiphon pisum*)/blue alfalfa aphid (*A. kondoi*) system in California lucerne (cf. Schreiber and Gutierrez, 1998). All of the major actors in the food web, except the coccinellids, are exotic. Pea aphid is currently controlled at very low levels by the fungal pathogen *Pandora neoaphidis*, which likely entered the system when blue alfalfa aphid was introduced (Gutierrez *et al.*, 1990). This pathogen attacks blue alfalfa aphid at one-tenth of the rate of pea aphid and during a normal wet northern Californian winter it causes catastrophic mortality to pea aphid populations (Pickering and Gutierrez, 1991). In the absence of the pathogen, pea aphid out-competes blue alfalfa aphid during hot periods. Pea aphid is also attacked by two parasitoids, the host-specific *Aphidius smithi* and *A. ervi* . The latter also attacks the less preferred blue alfalfa aphid. Other aphidophagous predators (lady beetles, lacewings and others) attack both aphids in proportion to their occurrence, and are important in lowering aphid abundance but not in regulating their numbers. If rainfall decreased in California, a change in species dominance from blue alfalfa aphid to pea aphid would likely occur. Evidence for this was seen during years of drought when pea aphid numbers surged (Gutierrez *et al.*, 1990). Pea aphid has effective natural enemies (principally *A. smithi*) that regulate its numbers in the absence of the pathogen (Bosch *et al.*, 1966).

Using capital letters to represent the species just described, an assembly diagrams based on the Schreiber and Gutierrez (1998) model of this aphid system for wet and dry winter scenarios is shown in Fig. 16.7. The directional arrows indicate the sequence of when the different species might enter the system. In wet winters (Fig.16.7a), the tritrophic system A/B/E predominates starting from any sequence of introductions. However, if the winter is dry (Fig. 16.7b) and irrigation is available for plant growth, the system would be dominated by A/P/S. This outcome would be mediated by weather through the action of the pathogen and parasitoids on the dynamics of the aphids.

The key point here is that climate change in this and other systems may upset the balance between species in food webs, leading to new webs. It may lead to changes in the relative importance of species and hence regulation; it may cause the extinction of some species, and in the extreme result in an entirely new biota.

16.10 General Comments

In the face of climate change, one could envision the invasion of new areas by pests formerly limited by one or more constraints. For example, as explained in section 16.6, the range of pink bollworm in North American cotton is limited by winter frost; hence milder winters would increase its range northward. Boll weevil is limited by desiccation of fruit buds in hot dry areas (DeMichele *et al.*, 1976); hence increased summer rainfall might extend its geographical range in formerly dry areas. This occurred during the early 1980s, when a sequence of very wet years in Arizona and southern California, coupled with the cultivation of stub-cotton, temporarily increased the threat from boll weevil.

Some species may quickly reach outbreak proportions when conditions become favourable. In North Africa and the Middle East, desert locust numbers decline during periods of drought, but quickly explode from barely detectable numbers during prolonged region-wide rainy periods (e.g. Roffey and Popov 1968). This scenario is similar to that of the cowpea aphid in Australia described above (section 16.2). Under current weather, it might not be necessary to know the population dynamics of such species *per se*, but only whether conditions favour their increase and for how long. The growth index approach

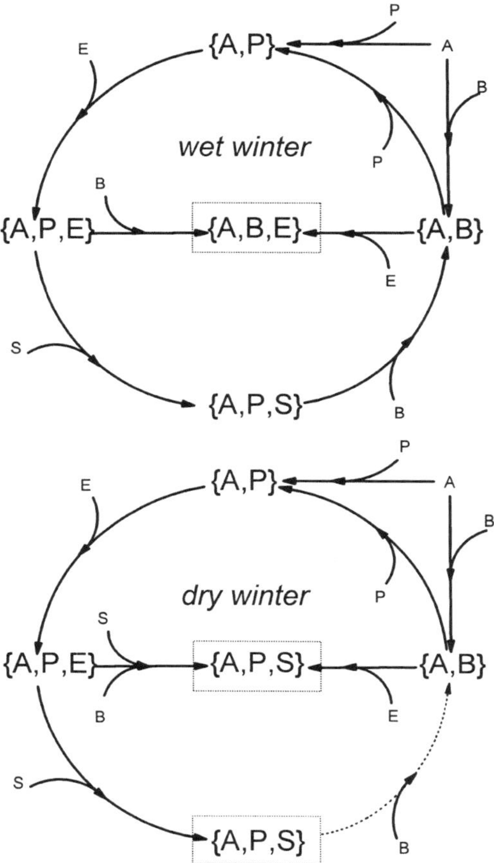

Fig. 16.7. Assembly diagrams for lucerne (A), pea aphid (P), blue alfalfa aphid (B) and the pea aphid specific parasitoids *A. smithi* (S) and stenophagous parasitoid *A. ervi* (E) during (a) wet and (b) dry seasons. The fungal pathogen *Pandora neoaphidis* is very active during wet years and less so during dry ones. The arrows indicate the entry of species, with the heavy arrows indicating the evolution of the system. The dashed line indicates an infeasible path.

worked well for the highly migratory cowpea aphid but would it work equally well for locust in the vast affected areas of North Africa and the Middle East, an area many times larger than southeast Australia? The current limitation to this approach is the lack of infrastructure for collecting the requisite weather data, and the unreliability of rainfall predictions using satellite remote sensing over this large landmass. If this data gap could be overcome, physiologically based models could provide a good way to evaluate the effects of weather and climate change on pest dynamics regionally using real-time weather.

What would happen to species such as desert locust in North Africa or cowpea aphid that are not regulated by natural enemies if the rainfall increased in response to climate change? Would cowpea aphid and desert locust population outbreaks be more frequent and prolonged? These are difficult questions to answer and require simplified but realistic models (e.g. equation 16.8). The models proposed by Gutierrez (1996), Schreiber and Gutierrez (1998) and in this chapter may be useful in answering such issues. However, field data and intuition suggest that the exotic cowpea aphid would become a more serious problem requiring new biological control agents, and possibly fungal pathogens might become more important in desert locust. In any case, the analysis would require a tritrophic perspective, and if feasible the results could easily be embedded in a GIS system (cf. Cherlet and Di Gregorio, 1991). In areas with greater infrastructure, linking of biologically rich models for pest/crop (e.g. cotton) systems in GIS would provide important information on their interactions as modified by weather, including the effects of climate change.

Lastly, from the point of view of global ecosystem sustainability, humankind remains the major pest of nature (Regev *et al.*, 1998), and appears to be the major cause of climate change. Global human activities based on greed require regulation that is not provided by the current the world economy.

References

Ankersmit, G.W. and Adkisson, P.L. (1968) Photoperiodic responses of certain geographic strains of *Pectinophora gossypiella* (Lepidoptera), *Journal of Insect Physiology* 13, 553–564.

Batzli, G.O. (1974) Production assimilation and accumulation of organic matter in ecosystems. *Journal of Theoretical Biology* 45, 205–217.

Berryman, A.A. and Gutierrez, A.P. (1998) Dynamics of insect predator–prey interactions. In: Huffaker, C.B. and Gutierrez, A.P. (eds) *Ecological Entomology*, 2nd edn. John Wiley & Sons, New York, pp. 389–423.

Bosch, R. van den (1978) *The Pesticide Conspiracy.* Doubleday, Garden City, New York.

Bosch, R. van den, Schlinger, E.I., Lagace, C.F. and Hall, J.C. (1966) Parasitism of *Acyrthosiphon pisum* by *Aphidius smithi*, a density dependent process in nature. *Ecology* 47, 1049–1055.

Bosch, R. van den, Frazer, B.D., Davis, C.S., Messenger, P.S. and Hom, R. (1970) An effective walnut aphid parasite from Iran. *California Agriculture* 24, 8–10.

Bosch, R. van den, Messenger, P.S. and Gutierrez, A.P. (1981) *An Introduction to Biological Control.* Plenum Publishing, New York.

Brasier, C.M. and Scott, J.K. (1994) European oak declines and global warming: a theoretical assessment with special reference to the activity of *Phytophthora cinnamomi. Bulletin OEPP* 24, 221–232.

Candolle, A.P. de. (1855) *Geographique Botanique.* Raisonée, Paris.

Cherlet, M. and Di Gregorio, A. (1991) Calibration and integrated modelling of remote sensing data for desert locust habitat monitoring. *Food and Agricultural Organization's Remote Sensing Center Report.* CLO/INT/004/BEL and GCP/INT/BEL, FAO, Rome.

DeBach, P. and Sundby, R.A. (1963) Competitive displacement between ecological homologues. *Hilgaardia* 34,105–166.

DeMichele, D.W., Curry, G.L., Sharpe, P.J.H. and Barfield, C.S. (1976) Cotton bud drying: a theoretical model. *Environmental Entomology* 5, 1011–1016.

Drake, V.A. (1994) The influence of weather and climate on agriculturally important insects: an Australian perspective. *Australian Journal of Agricultural Research* 45, 487–509.

Ellis, W.N., Donner, J.H. and Kuchlein, J.H. (1997) Recent shifts in phenology of Microlepidoptera, related to climatic change (Lepidoptera). *Entomologische Berichten (Amsterdam)* 57, 66–72.

Fitzpatrick, E.A. and Nix, H.A. (1970) The climatic factor in Australian grasslands ecology. In: Moore, R.M. (ed.) *Australian Grasslands.* Australian National University Press, Brisbane, pp 3–26.

Fleming, R.A. (1996) A mechanistic perspective of possible influences of climate change on defoliating insects in North America's boreal forests. *Silva Fennica* 30, 281–294.

Fleming, R.A and Candau, J.-N. (1998) Influences of climatic change on some ecological processes of an insect outbreak system in Canada's boreal forests and the implications for biodiversity. *Environmental Monitoring and Assessment* 49, 235–249.

Flint, M.L. (1981) Climatic ecotypes of *Trioxys complanatus*, a parasite of the spotted alfalfa aphid. *Environmental Entomology* 9, 501–507.

Frazer, B.D. and Gilbert, N. (1976) Coccinellids and aphids: a quantitative study of the impact of adult ladybirds (Coleoptera: Coccinellidae) preying on field populations of pea aphids (Homoptera: Aphididae). *Journal of the Entomological Society of British Columbia* 73, 33–56.

Gilbert, N., Gutierrez, A.P., Frazer, B.D. and Jones, R.E. (1976) *Ecological Relationships.* Freeman and Co., New York.

Gutierrez, A.P. (1992) The physiological basis of ratio dependent theory. *Ecology* 73, 1552–63.

Gutierrez, A.P. (1996) *Applied Population Ecology: a Supply–Demand Approach.* John Wiley & Sons, New York, 300 pp.

Gutierrez, A.P. and Baumgärtner, J.U. (1984) Multitrophic level models of predator–prey energetics. I. Age specific energetics models – pea aphid *Acyrthosiphon pisum* (Harris) (Homoptera: Aphididae) as an example. *Canadian Entomology* 116, 924–932.

Gutierrez, A.P. and Wang, Y.H. (1977) Applied population ecology: models for crop production and pest management. In: Norton, G.A. and Holling, C.S. (eds) *Pest Management.* International Institute for Applied Systems Analysis Proceeding Series, Pergamon Press, Oxford, UK.

Gutierrez, A.P. and Yaninek, J.S. (1983) Responses to weather of eight aphid species commonly found in pastures in southeastern Australia. *Canadian Entomology* 115, 1359–1364.

Gutierrez, A.P., Morgan, D.J. and Havenstein, D.E. (1971) The ecology of *Aphis craccivora* Koch and subterranean clover stunt virus. I. The phenology of aphid populations and the epidemiology of virus in pastures in Southeast Australia. *Journal of Applied Ecology* 8, 699–721.

Gutierrez, A.P., Havenstein, D.E., Nix, H.A. and Moore, P.A. (1974) The ecology of *Aphis craccivora* Koch and subterranean clover stunt virus. III. A regional perspective of the phenology and migration of the cowpea aphid. *Journal of Applied Ecology* 11, 21–35.

Gutierrez, A.P., Butler, G.D. Jr and Ellis, C.K. (1981) Pink bollworm: diapause induction and termination in relation to fluctuating temperatures and decreasing photophases. *Environmental Entomology* 10, 936–942.

Gutierrez, A.P., Pizzamiglio, M.A., Dos Santos, W.J., Villacorta, A. and Gallagher, K. D. (1986) Analysis of diapause induction and termination in field pink bollworm, *Pectinophora gossypilla* (Saunders 1843) in Brazil. *Environmental Entomology* 15, 494–500.

Gutierrez, A.P., Neuenschwander, P., Schulthess, F., Wermelinger, B., Herren, H.R., Baumgärtner, J.U. and Ellis, C.K. (1988) Analysis of the biological control of cassava pests in West Africa. II. The interaction of cassava and cassava mealybug. *Journal of Applied Ecology* 25, 921–940.

Gutierrez, A.P., Hagen, K.S. and Ellis, C.K. (1990) Evaluating the impact of natural enemies: a multitrophic level perspective. In: Mackauer, M.P., Ehler, L.E. and Rolands, J. (eds) *Critical Issues in Biological Control.* Intercept Press, Andover, UK.

Gutierrez, A.P., Neuenschwander, P. and van Alphen, J.J.M. (1993) Factors affecting the establishment of natural enemies: biological control of the cassava mealybug in West Africa by introduced parasitoids: a ratio dependent supply–demand driven model. *Journal of Applied Ecology* 30, 706–721.

Gutierrez, A.P., Mills, S.J., Schreiber, S.J. and Ellis, C.K. (1994) A physiologically based tritrophic perspective on bottom up–top down regulation of populations. *Ecology* 75, 2227–2242.

Hairston, N.G., Smith, F.E. and Slobodkin, L.B. (1960) Community structure, population control, and competition. *American Naturalist* 94, 421–425.

Herren, H.R. and Neuenschwander, P. (1991) Biological control of cassava pests in Africa. *Annual Review of Entomology* 36, 257–283.

Heong, K.L., Song, Y.H., Pimsamarn, S. Zhang, R. and Bae, S.D. (1995) Global warming and rice arthropod communities. In: Peng, S., Ingram, K.T., Neue, H.-U. and Ziska, L.H. (eds) *Climate Change and Rice.* Symposium, Manilla, Philippines, March 1994. Springer, Berlin, pp. 326–335.

Holling, C.S. (1966) The functional response of invertebrate predators to prey density. *Memoirs of the Entomology Society of Canada* 48, 3–86.

Hughes, R.D. and Maywald, G.W. (1990) Forecasting the favorableness of the Australian environment for the Russian wheat aphid, *Diuraphis noxia* (Homoptera: Aphididae), and its potential impact on Australian wheat yields. *Bulletin of Entomology Research* 80, 165–175.

Hutchinson, G.E. (1959) Homage to Santa Rosalia, or why there are so many kinds of animals? *American Naturalist* 93, 145–159.

Ivlev, V.S. (1955) *Experimental Ecology of the Feeding of Fishes.* English translation by D. Scott (1961), Yale University Press, New Haven, Connecticut.

Johnsen, S. and Gutierrez, A.P. (1997) Induction and termination of winter diapause in a Californian strain of the cabbage maggot (Diptera: Anthomyiidae). *Environmental Entomology* 26, 84–90.

Johnsen, S., Gutierrez, A.P. and Jørgensen, J. (1997) Overwintering in the cabbage root fly *Delia radicum*: a dynamic model of temperature-dependent dormancy and post-dormancy development. *Journal of Applied Ecology* 34, 21–28.

Johnson, B. (1957) Studies on the dispersal by upper winds of *Aphis craccivora* Koch in New South Wales. *Proceedings of the Linnean Society of New South Wales* 82, 191–208.

Kobak, K.I., Turchinovich, I.Y., Yu Kondrsheva, N., Schulze, E.D., Schulze, W., Koch, H. and Vygodskaya, N.N. (1996). Vulnerability and adaptation of the larch forest in Eastern Siberia in climate change. *Water, Air and Soil Pollution* 92, 119–127.

Liebig, J. von (1840) *Chemistry and its Applications to Agriculture and Physiology*, 4th edn 1847. Taylor and Walton, London.

Lotka, A.J. (1925) *Elements of Physical Biology.* Williams and Witkins, Baltimore, Maryland. Reissued (1956) as *Elements of Mathematical Biology.* Dover Publications, New York.

Messenger, P.S. (1964) Use of life-tables in a bioclimatic study of an experimental aphid–braconid wasp host–parasite system. *Ecology* 45, 119–131.

Messenger, P.S. (1968) Bioclimatic studies of the aphid parasite *Praon exsoletum.* 1. Effects of temperature on the functional response of females to varying host densities. *Canadian Entomology* 100, 728–741.

Messenger, P.S. and Force, D.C. (1963) An experimental host–parasite system: *Therioaphis maculata* (Buckton). *Praon palitans* Musebeck (Homoptera: Aphididae-Hymenoptera: Bracondidae). *Ecology* 44, 532–540.

Morimoto, N., Imura, O. and Kiura, T. (1998) Potential effects of global warming on the occurrence of Japanese pest insects. *Applied Entomology and Zoology* 33, 147–155.

Morris, R.F. (1963) Predictive population equations based on key factors. *Memoirs of the Entomology Society of Canada* 32, 16–21.

Nechols, J.R., Tauber, M.J., Tauber, C.A. and Masaki, S. (1998) Adaptations to hazardous seasonal conditions: dormancy, migration, and polyphenism . In: Gutierrez, A.P. and Huffaker, C.B. (eds) *Ecological Entomology.* John Wiley & Sons, New York, pp. 159–200.

Odum, E.P. (1971) *Fundamentals of Ecology,* 3rd edn. W.B. Saunders and Co., Toronto.

Petrusewicz, K. and MacFayden, A. (1970) *Productivity of Terrestial Animals: Principles and Methods.* IBP Handbook 13, Blackwell, Oxford, UK.

Phillipson, J. (1966) *Ecological Energetics.* Edward Arnold, London.

Pickering, J. and Gutierrez, A.P. (1991) Differential impact of the pathogen *Pandora neoaphidis* (R. & H.) Humber (Zygomycetes: Entomophthorales) on the species composition of *Acyrthosiphon* aphids in alfalfa. *Canadian Entomology* 123, 315–320.

Price, P.W. (1984) The Concept of the Ecosystem. In: Huffaker, C.B. and Rabb, R.L. (eds) *Ecological Entomology.* John Wiley & Sons, New York, pp. 19–52.

Quezada, J.R. and DeBach, P. (1973) Bioecological and population studies of the cottony scale, *Icerya purchasi* Mask., and its natural enemies, *Rodolia cardinalis* Mul. and *Cryptochaetum iceryae* Wil., in southern California. *Hilgardia* 41, 631–688.

Regev, U., Gutierrez, A.P., Schreiber, S. and Zilberman, D. (1998) Biological and economic foundation of renewable resource exploitation. *Ecological Economics* 26, 227–242.

Roffey, J. and Popov, G. (1968) Environmental and behavioural processes in desert locust outbreaks. *Nature* 219, 446–450.

Royama, T. (1971) A comparative study of models for predation and parasitism. *Res. Population Ecology, Kyoto University Suppl.* 1, 1–91.

Royama, T. (1992) *Analytical Population Dynamics.* Chapman & Hall, London.

Schreiber, S. and Gutierrez, A.P. (1998) A supply–demand perspective of species invasions and coexistence: applications to biological control. *Ecological Modelling* 106, 27–45.

Shelford, V.E. (1931) Some concepts of bioecology. *Ecology* 12, 455–467.

Solomon, M.E. (1949) The natural control of animal populations. *Journal of Animal Ecology* 18, 1–35.

Sutherst, R.W., Maywald, G.F. and Bottomly, W. (1991) From CLIMEX to PESKY, a generic expert system for risk assessment. *EPPO Bulletin* 21, 595–608.

Tucker, C.J., Townshend, R.G. and Goff, T.E. (1985) African landcover classification using satellite data. *Science* 227, 369–375.

Varley, C.G., Gradwell, G.R. and Hassell, M.P. (1973) *Insect Population Ecology: an Analytical Approach*. University of California Press, Berkeley, 212 pp.

Volterra, V. (1926) Variations and fluctuations of the number of individuals in animal species living together. *J. Cons. Perm. Int. Ent. Mer.* 3, 3–51. Reprinted in Chapman, R.N. (1931). *Animal Ecology,* New York.

Volterra, V. (1931) Principes de biologie mathematique. *Acta Biotheoretica* 3, 1–36.

Watt, K.E.F. (1959) A mathematical model for the effects of densities of attacked and attacking species on the number attacked. *Canadian Entomology* 91, 129–144.

Wit, C.T. de and Goudriaan, J. (1978) *Simulation of Ecological Processes,* 2nd edn. PUDOC Publishers, Wageningen, The Netherlands.

17 Crop Ecosystem Responses to Climatic Change: Soil Organic Matter Dynamics

GÖRAN I. ÅGREN[1] AND ANN-CHARLOTTE HANSSON[2]

[1]*Department of Ecology and Environmental Research, Swedish University of Agricultural Sciences, PO Box 7072, 07 Uppsala, Sweden SE-750;* [2]*Biology and Crop Protection Science, Swedish University of Agricultural Science, PO Box 7043, SE 75007 Uppsala, Sweden*

17.1 Introduction

On a global scale, soils contain 1500 Pg (Pg = 10^{15} g) C and 300 Pg N, compared with around 600 Pg C and 13 Pg N in terrestrial biomass, and 720 Pg C in the atmosphere (Chapter 2, this volume). Other important elements have similar relationships (Reeburgh, 1997). Another way to describe these relationships is that 1 m^2 of normal Swedish arable topsoil (0–30 cm) with 4% organic matter (OM) contains 400 kg mineral soil, 8 kg C in soil organic matter (SOM) and 0.1 kg C in roots, whereas the atmosphere above this 1 m^2 contains 1.4 kg C (Andrén *et al.*, 1990). Since the pool of C in the atmosphere is considerably smaller than the pool of C in the soil, a small relative change in the amount of C in the soil will have a substantial influence on the C content in the atmosphere. Thus, understanding the processes regulating the dynamics of SOM and how they respond to a changing environment is crucial for managing our natural resources, and for understanding how the state of SOM may feed back to the change in climate. Other recent relevant reviews of this topic can be found in Bouwman (1990), Anderson (1992) and Lal *et al.* (1995).

17.2 A Framework for Understanding

OM in the soil consists of non-living organic matter (NLOM; Zepp and Sonntag, 1995) and a living component composed of plant roots and soil microorganisms and animals. The NLOM is from either native plant material or remnants of microorganisms and soil animals, or NLOM that has been partially processed by soil organisms. OM coming from herbivores or their predators can for this discussion be regarded as plant material, since that is its source.

These different kinds of OM appear in a variety of structures and consist of a variety of chemical compounds (sugars, celluloses, lignin, etc.) in different proportions.

SOM dynamics can be discussed within the general framework shown in Fig. 17.1, which is divided into NLOM and decomposer biomass. The two, of course, are not physically separated as in the figure and therefore can be difficult to measure separately. However, most estimates of decomposer biomass suggest that it makes up around 4–5% of the total SOM (Wardle, 1992).

The soil is supplied with new NLOM primarily from plant litter input, a source that will change both quantitatively and qualitatively with climatic change. In the soil, the litter is mixed with old NLOM and provides the food for the primary decomposers (mainly fungi and bacteria; Berg *et al.*, 1998). The rate at which the primary decomposers consume NLOM depends both on its availability (chemical structure and association with the soil matrix) and on the physical (temperature, water) and chemical climate in the soil. Respiration associated with growth and sustenance of the decomposers leads to losses of C from the soil in the form of CO_2. When decomposers die, their biomass is included in NLOM, thus becoming a food source for other decomposers. An important aspect of this cycle between NLOM and decomposers is that the chemical composition of NLOM assimilated by the decomposers is different from that returned via the dead decomposers.

OM that has not been part of the decomposer biomass proper but that has been modified by decomposer activity should also be included in this cycle. The decomposer community consists of a complete food web: primary decomposers, fungivores and bacterivores and predators on them. Conceptually, this does not alter Fig. 17.1, but considerations of the soil community structure may be important for rates of transformation (e.g. Zheng *et al.*, 1997), although the food web aspect can usually be ignored (Andrén *et al.*, 1999).

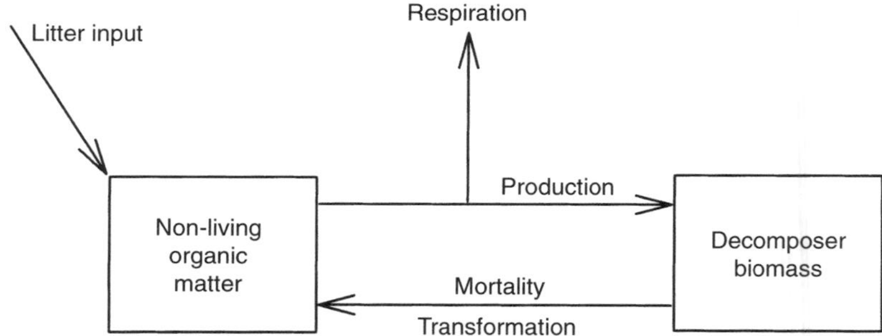

Fig. 17.1. A general framework for describing soil organic matter dynamics. (See text for further explanation.)

17.3 Measuring Soil Organic Matter

The spatial heterogeneity of SOM can be great; for example, two soil samples from the same field can differ greatly in the amount and quality of SOM and in its rate of decomposition, which depends on soil structure. It is normal for estimates of the amount of SOM to have a large variance. Direct determinations of flows are confronted with methodological difficulties concerning the estimation of the magnitude of root C inputs – all methods give results with high variance and low temporal resolution. Since the NLOM pool is large compared with the fluxes, this variance makes it difficult to determine them from changes in the NLOM pools (Buyanovsky and Wagner, 1995).

17.4 Sources of Soil Organic Matter

A major factor influencing SOM dynamics is the quantity and quality of OM entering the soil. Climatic effects on OM production are covered in several other chapters in this volume, but here we look at the implications for the soil. The quantity of litter input to the soil is rather unimportant because soil storage is approximately proportional to the litter input, although the slow turnover of SOM means that long periods are required for it to reach a new steady state. On a global scale, the stability of the CO_2 concentration ([CO_2]) during the last few thousand years before industrialization shows that litter input and respiration costs have been approximately balanced for a long time. The problems of balancing the global C budget (the non-identified terrestrial sink) show that this is currently not the case (Schimel, 1995). On a local level, it is normal that these flows do not balance, i.e. there is either an increase in SOM due to a higher input than output through mineralization, or vice versa.

Changes in the quality of the litter input can come mainly from two sources. Firstly, a climatic change can lead to changes in allocation in the plant; hence the relative contribution of plant components to litter input will change. Lawlor and Mitchell (1991) and Rogers *et al.* (1996) have reviewed the literature for effects of elevated [CO_2] on root-to-shoot ratios (R:S) and found a variety of responses. About 75% of the observations in the study by Rogers *et al.* (1996) showed changes in R:S of less than 30%, with a mean of +11%. The reason for this variability is not clear, but interactions with other factors, mainly nutrition (cf. Joffre and Ågren, 1999), can be suspected. Furthermore, a majority of the available root results only estimates the amount of structural root C and ignores the soluble C, thus underestimating the C input to the soil via the root system. Whether or not the elevated [CO_2] may result in a different ratio between structural and soluble root C compared with today needs to be studied further (Rogers *et al.*, 1994). If labile C in the soil increases under elevated [CO_2], it may influence soil N availability, because N mineralization may change (Davidson, 1994). However, there are divided opinions about whether this input significantly stimulates soil processes or influences soil communities (Pregitzer, 1993; Curtis *et al.*, 1994).

The second cause of changes in litter quality is a change in the chemical composition of a given litter type. Cotrufo *et al.* (1998) reviewed 75 published studies of effects of elevated [CO_2] on N concentration in plant tissues. They found an overall 14% decrease in N concentration, but with great variability between plant functional types. C_4 and N-fixing plants responded considerably less than C_3 plants. The implications of these results for litter quality are not immediate. Firstly, the observations are made on live tissue, and it is not known how the retranslocation of N during senescence will interact with this change in live-tissue N concentration. Secondly, although the C:N ratio has been widely used as an index of decomposibility (Melillo *et al.*, 1982), it is not unchallenged (Joffre and Ågren, 1999). Other more subtle changes in the chemical composition may play a more important role. For example, a small change in lignin concentration can be more important. The general conclusion today is that changes in [CO_2] *per se* will probably have minor effects on litter decomposition rates (Norby and Cotrufo, 1998).

17.5 Regulation of Rates of Soil Organic Matter Decomposition

In addition to the effects on changing the quality of litter, a climatic change will also include direct effects on the soil physical climate, with temperature being the most apparent variable affected. A number of studies have explored the relationship between temperature and SOM decomposition rate. Such temperature dependence is often expressed as a Q_{10} value (the relative change in rate with a 10°C increase in temperature) with an often-used value of around 2 for biological processes. Kirschbaum (1995) summarized a number of these studies and showed that the conventional assumption of an exponential temperature dependence is not valid. Instead, SOM decomposition shows a higher temperature sensitivity close to freezing, with Q_{10} values of 5–10, but values close to 2 at temperatures near 20–25°C. The exponential relation never goes to zero, and this causes problems in describing processes that are probably halted completely because of frozen soil.

Kätterer *et al.* (1998) investigated four different temperature response functions for OM decomposition: two Arrhenius-type functions, a Q_{10} function and a quadratic function. All four functions provided excellent fits to the observed data except at low temperatures, where the quadratic function was superior. This is not surprising, because all four functions are quite similar over limited temperature intervals. They also found a Q_{10} close to 2 when their response function was fitted over the entire range of observations. Since activities are low near 0°C, even large relative errors become small in absolute terms, thus making Q_{10} values estimated over large temperature ranges insensitive to errors in the fitting at the low temperature range. This explains the difference between the results of Kirschbaum (1995) and Kätterer *et al.* (1998). The technical difficulties of accurately measuring the low activities at low temperatures makes it uncertain which temperature response function best describes this region. High temperature where activity decreases is also an area that is not well studied, but could become important if soil temperatures rise. A number of studies around the world are in progress to provide a better understanding of the effects of soil warming on the rate of SOM decomposition.

The effects of soil moisture on SOM decomposition have been much less well studied than have temperature effects, but decomposition rates are limited by either too much or too little soil water. Howard and Howard (1993) showed that log(C mineralization rate of soils) could be expressed as a quadratic function of soil water content. Lomander *et al.* (1999a,b) expressed the mineralization rate directly as a quadratic function of soil water content. Because it is not the amount of water in the soil, but rather its availability, that matters, some authors have preferred to use soil water potential as an expression. Quemada and Cabrera (1997) and Seyferth and Persson (2000) found that C mineralization rates could be linearly related to log(–soil water potential).

Assessing the climatic effect on the mineralization rate is difficult because of the interaction between temperature and other factors. Increasing temperature is often associated with an increasing water deficit that neutralizes the rate of increase caused by increasing temperature. This effect can be large enough to make the temperature response appear practically linear (Andrén and Paustian, 1987). There might also be an important interaction between temperature and OM quality. O'Connell (1990) studied decomposition of low and high quality eucalyptus leaf litter at different temperatures and found that the low quality litter had a stronger temperature response than did the high quality litter. Such an interaction can have a disproportionately large effect on the stability of NLOM since the largest fraction of NLOM, is low quality materials.

17.6 Transformations of Soil Organic Matter

One of the least well understood processes in SOM dynamics is transformation among different organic compounds. One reason for this lack of understanding is the absence of a good method for isolating functional SOM fractions (Elliott *et al.*, 1994; Coleman and Crossley, 1996). There is a generally accepted view that, over time, OM is transformed towards more and more recalcitrant materials – the so-called humification process (Swift *et al.*, 1979). The rate at which the SOM is converted from easily degradable materials to recalcitrant ones is, however, of great importance for the size of the soil C pool. The models used to describe these transformations work with different conceptual fractions – fractions that are not necessarily the best ones and are also difficult to measure. The descriptions vary considerably, from those that represent NLOM as consisting of only two types of material with different turnover rates (Andrén and Kätterer, 1997) vs. those that describe NLOM with several discrete material types (e.g. Jenkinson *et al.*, 1987; Parton *et al.*, 1987) and those that prefer a continuous description (Ågren and Bosatta, 1998). It can be shown that the continuous description embraces the others (Bosatta and Ågren, 1995), but over short periods (a few decades) these approaches do not differ much (Paustian *et al.*, 1997a; Hyvönen *et al.*, 1998). The choice between representations becomes a matter of technical/mathematical convenience and the level of abstraction at which one prefers to work. The continuous description is more abstract and mathematically demanding.

Since the climatic effects on transformations of OM are mediated by the decomposers, it is most likely that any changes in this process should occur through changes among the categories of soil biota in the decomposer communities, i.e. shifts in the relative abundance of species (Whitford, 1992). Changes in soil microbial communities have been observed when soil samples were exposed to different temperatures (Zogg *et al.*, 1997). However, according to Smith *et al.* (1999), the effects of changed temperature and moisture on the existing community will be more important than any change in the community composition.

A key factor for SOM decomposability is the stability of organic C in soil. Carbon enclosed in aggregates is somewhat protected from decomposition. The aggregate structure and stability depend on physical properties of the mineral soil and on the soil biota. Thus, production of organic C compounds that bind aggregates, and hyphal entanglement of soil particles (microflora), production of faecal pellets and creation of biopores are all the result of soil biota activities (Coleman and Crossley, 1996). The extent to which this affects the selection of compounds that are degraded and the new compounds that are produced, as well as the long-term consequences of such changes, are not understood.

17.7 Experimental Results

In a tallgrass prairie, increases in soil microbial activity have been measured under chambers where the vegetation was exposed to increased [CO_2] (Rice *et al.*, 1994). The increases in this case were restricted to dry years. In wet years, the [CO_2] had no effect on the microbial activity. The effect of increased [CO_2], therefore, seems to depend on the increased soil moisture as a result of decreased transpiration of the grass when [CO_2] went up. However, Schortemeyer *et al.* (1996) observed a doubling of the population of *Rhizobium leguminosarum* bv. *trifolii* in the rhizosphere of white clover growing under 600 ppm [CO_2] vs. white clover growing under ambient [CO_2] (350 ppm). Heterotrophic and ammonium-oxidizing bacteria did not respond to the same treatment. The rhizosphere of perennial ryegrass growing under the same conditions showed no changes in bacterial populations. This indicates that N fixation might increase with increasing [CO_2] (Zanetti *et al.*, 1995). Responses of soil fauna to [CO_2] have also been observed (Yeates *et al.*, 1997), but results are conflicting (Runion *et al.*, 1994; Ross *et al.*, 1995). Again, indirect effects associated with soil moisture may be the most important factor (Harte *et al.*, 1996; Briones *et al.*, 1997).

Experimental determination of the effects of climatic change on SOM are difficult because of the long times required to reach conclusive results, the perturbations caused by the experimental set-up, and the costs of conducting the experiments. A number of techniques are currently being used (van de Geijn, 1998), but some of these experiments are so newly established that results are not yet available.

In an experiment with cotton under elevated [CO_2] in Arizona, increases in soil C were measured after 3 years of increased biomass production if the crop was irrigated (Wood *et al.*, 1994). In a parallel 2-year experiment with wheat, soil C storage also increased (Prior *et al.*, 1997). However, other studies showed no differences (Ross *et al.*, 1995; Nitschelm *et al.*, 1997; Torbert *et al.*, 1997). These conflicting results may be due to the short observation periods. Observed changes in C isotope signatures (Nitschelm *et al.*, 1997; Torbert *et al.*, 1997) have demonstrated clear changes in processes regulating soil C storage. A possible explanation for the absence of increases in soil C stocks is that the soil has become C-saturated (Hassink and Whitmore, 1997). The C-storage capacity of a soil depends on how much protection can be provided by soil mineral particles. If this storage capacity is exceeded, additional C will be immediately attacked and respired by decomposers.

17.8 Conclusions

How much soil C stocks will change as a result of a climatic change will depend on how much a particular factor (such as litter input) will change, and on how sensitive soil C storage is to that particular factor. If we disregard the possibility of a soil being C-saturated, an x% increase in litter input can be expected to increase soil C store by x% once the system has returned to steady state. Similarly, if the specific decomposition rate is changed by x%, the steady-state soil C store can be expected to change by about x% as long as x is not too large. These are the simple consequences of simple first-order kinetics. During the transition from one steady state to another, litter inputs will differ from respiration. However, at steady state the two must be equal, whatever the level of the C store.

Matters become less evident when the problem is considered in more detail. The simple first-order decomposition rate is an aggregate of conditions (such as temperature) that directly affect rates, and decomposer properties that determine the transformation of C compounds. It seems that many of the decomposer-related properties can have a much larger effect on the C store than the change in the property itself (Ågren *et al.*, 1996). The positive and negative feedbacks that are not fully understood or cannot be quantified further complicate estimation of the effects of climatic changes on SOM. An example of such a feedback is that decreased soil tilling leads to increased stability of aggregates. This leads to decreased decomposition, but the greater SOM leads to increased water-holding capacity, which in turn leads to increased decomposition rates. However, it seems that the first-order assumption works well over a wide range of conditions. Kätterer and Andrén (1999) showed that it was adequate to explain the influence of management in 99 long-term agricultural experiments in northern Europe.

Finally, land use and management have a large impact on whether soils function as sources or as sinks for C (Paustian *et al.*, 1997b). Rasmussen *et al.*

(1998) concluded, in a review of results of long-term agroecosystem experiments, that returning residues to soil over long periods has transformed many temperate soils from sources to sinks. Soil erosion leads to increased emission of CO_2 by exposing C locked within aggregates to decomposition. In general, soil management methods, such as no-till or reduced tilling, cover crops, mulching, etc., that are used to reduce soil erosion result in sequestration of C in the soil (see Chapter 3, this volume). All these methods aim at increased aggregate stability.

Another way to reduce the risk of C losses from arable land is to avoid cultivation of marginal lands, such as organic soils. The C content in the subsoil may be increased by cultivating deep-rooted crop species or varieties. Lal *et al.* (1998) have described in detail the potential of US cropland to sequester C and mitigate the greenhouse effect.

References

Ågren, G.I. and Bosatta, E. (1998) *Theoretical Ecosystem Ecology – Understanding Element Cycles*, 2nd edn. Cambridge University Press, Cambridge, UK, 233 pp.

Ågren, G.I., Johnson, D.W., Kirschbaum, M. and Bosatta, E. (1996) Ecosystem physiology – soil organic matter. In: Breymeyer, A., Hall, D.O., Melillo, J.M. and Ågren, G.I. (eds) *Global Change: Effects on Forests and Grasslands.* John Wiley & Sons, Chichester, pp. 207–228.

Anderson, J.M. (1992) Responses of soils to climate change. *Advances in Ecological Research* 22, 163–210.

Andrén, O. and Kätterer, T. (1997) ICBM – the introductory carbon balance model for exploration of soil carbon balances. *Ecological Applications* 7, 1226–1236.

Andrén, O. and Paustian, K. (1987) Barley straw decomposition in the field: a comparison of models. *Ecology* 68, 1190–1200.

Andrén, O., Lindberg, T, Paustian, K. and Rosswall, T. (eds) (1990) Ecology of arable land – organisms, carbon and nitrogen cycling. *Ecological Bulletins*, Munksgaard, Copenhagen, 40, 210 pp.

Andrén, O., Brussaard, L. and Clarholm, M. (1999) Soil organism influence on ecosystem-level processes – bypassing the hierarchy? *Applied Soil Ecology* 11, 177–188.

Berg, M.P., Kniese, J.P., Zoomer, R. and Verhoef, H.A. (1998) Long-term decomposition of successive strata in a nitrogen saturated Scots pine forest soil. *Forest Ecology and Management* 107, 159–172.

Bosatta, E. and Ågren, G.I. (1995) The power and reactive continuum models as particular cases of the q-theory of organic matter dynamics. *Geochimica et Cosmochimica Acta* 59, 3833–3835.

Bouwman, A.F. (ed.) (1990) *Soils and Greenhouse Effect.* John Wiley & Sons, Chichester, 575 pp.

Briones, M.J.I., Ineson, P. and Pearce, T.G. (1997) Effects of climate change on soil fauna; responses of enchytraeids, Diptera larvae and tardigrades in a transplant experiment. *Applied Soil Ecology* 6, 117–134.

Buyanovsky, G.A. and Wagner, G.H. (1995) Soil respiration and carbon dynamics in parallel native and cultivated ecosystems. In: Lal, R., Levine, E. and Stewart, B.A. (eds) *Soils and Global Change. Advances in Soil Science.* CRC Press, Boca Raton., Florida, pp. 209–217.

Coleman, D.C. and Crossley, D.A. Jr (1996) *Fundamentals of Soil Ecology.* Academic Press, San Diego, 205 pp.

Cotrufo, M.F., Ineson, P. and Scott, A. (1998) Elevated CO_2 reduces the nitrogen concentration of plant tissues. *Global Change Biology* 4, 43–54.

Curtis, P.S., O'Neill, E.G., Teeri, J.A., Zak, D.R. and Pregitzer, K.S. (1994) Belowground responses to rising atmospheric CO_2: implications for plants, soil biota and ecosystem processes. *Plant and Soil* 165, 1–6.

Davidson, E.A. (1994) Climate change and soil microbial processes: secondary effects are hypothesised from better known interacting primary effects. In: Rounsevell, M.D.A. and Loveland, P.J. (eds) *Soil Responses to Climate Change.* Springer-Verlag, Berlin, pp. 155–168.

Elliott, E.T., Janzen, H.H., Campbell, C.A., Cole, C.V. and Myers, R.J.K. (1994) Principles of ecosystem analysis and their application to integrated nutrient management and assessment of sustainability. In: *Proceedings Sustainable Land Management 21st Century,* Vol. 2, *Plenary Papers.* Canadian Society of Soil Science, Lethbridge, Alta, Canada.

Geijn, S.C. van de (1998) Experimental research facilities for the assessment of climate change impact on managed and natural ecosystems. In: Peters, D., Maracchi, G. and Ghazi, A. (eds) *Climate Change Impact on Agriculture and Forestry.* European Commission, Luxembourg, pp. 117–136.

Harte, J., Rawa, A. and Price, V. (1996) Effects of manipulated soil microclimate on mesofaunal biomass and diversity. *Soil Biology and Biochemistry* 28, 313–322.

Hassink, J. and Whitmore, A.P. (1997) A model of the physical protection of organic matter in soils. *Soil Science Society of America Journal* 61, 131–139.

Howard, D.M. and Howard, P.J.A (1993) Relationships between CO_2 evolution, moisture content and temperature for a range of soil types. *Soil Biology and Biochemistry* 25, 1537–1546.

Hyvönen, R., Ågren, G.I. and Bosatta, E. (1998) Predicting long-term soil carbon storage from short–term information. *Soil Science Society of America Journal* 62, 1000–1005.

Jenkinson, D.S., Hart, P.B.S., Rayner, J.H. and Parry, L.C. (1987) Modelling the turnover of organic matter in long-term experiments at Rothamsted. *INTECOL Bulletin* 15, 1–8.

Joffre, R. and Ågren, G.I. (1999) From plant to soil and back: litter fall and decomposition. In: Roy, J., Mooney, H.A. and Saugier, B. (eds) *Terrestrial Global Productivity. Past, Present, Future.* Academic Press, San Diego.

Kätterer, T. and Andrén, O. (1999) Long-term agricultural field experiments in Northern Europe: analysis of the influence of management on soil carbon stocks using the ICBM model. *Agriculture, Ecosystems and Environment* 72, 165–179.

Kätterer, T., Reichstein, M., Andrén, O. and Lomander, A. (1998) Temperature dependence of organic matter decomposition: a critical review using literature data analyzed with different models. *Biology and Fertility of Soils* 27, 258–262.

Kirschbaum, M.U.F. (1995) The temperature dependence of soil organic matter decomposition and the effect of global warming on soil organic C storage. *Soil Biology and Biochemistry* 27, 753–760.

Lal, R., Levine, E. and Stewart, B.A. (eds) (1995) *Worlds Soils and Greenhouse Effect: an Overview.* Advances in Soil Science, CRC Press Inc., Boca Raton, Florida, 440 pp.

Lal, R., Kimble, J.M., Follett, R.F. and Cole, C.V. (eds) (1998) *The Potential of US Cropland to Sequester Carbon and Mitigate the Greenhouse Effect.* Sleeping Bear Press, Ann Arbor, Michigan, 128 pp.

Lawlor, D.W. and Mitchell, R.A.C. (1991) The effects of increasing CO_2 on crop photosynthesis and productivity: a review of field studies. *Plant, Cell and Environment* 14, 807–818.

Lomander, A., Kätterer, T. and Andrén, O. (1999a) Carbon dioxide evolution from top- and subsoil as affected by moisture and constant and fluctuating temperature. *Soil Biology and Biochemistry* 30, 2017–2022.

Lomander, A., Kätterer, T. and Andrén, O. (1999b) Modelling the effects of temperature and moisture on CO_2 evolution from top- and subsoil using a multi-compartment approach. *Soil Biology and Biochemistry* 30, 2023–2030.

Melillo, J.M., Aber, J.D. and Muratore, J.F. (1982) Nitrogen and lignin control of hardwood leaf litter decomposition dynamics. *Ecology* 63, 621–626.

Nitschelm, J.J., Lüscher, A., Hartwig, U.A. and Kessel, C. van (1997) Using stable isotopes to determine soil carbon input differences under ambient and elevated atmospheric CO_2 conditions. *Global Change Biology* 3, 411–416.

Norby, R.J. and Cotrufo, M.F. (1998) A question of litter quality. *Nature* 396, 17–18.

O'Connell, A.M. (1990) Microbial decomposition (respiration) of litter in eucalypt forests of south-western Australia: an empirical model based on laboratory incubations. *Soil Biology and Biochemistry* 22, 155–160.

Parton, W.J., Schimel, D.S., Cole, C.V. and Ojima, D.S. (1987) Analysis of factors controlling soil organic matter levels in Great Plains Grasslands. *Soil Science Society of America Journal* 51, 1173–1179.

Paustian, K., Ågren, G.I. and Bosatta, E. (1997a) Modelling the role of litter quality on decomposition and nutrient cycling. In: Cadisch, G. and Giller, K.E. (eds) *Driven by Nature: Plant Litter Quality and Decomposition.* CAB International, Wallingford, UK, pp. 313–335.

Paustian, K., Levine, E., Post, W.M. and Ryzhova, I.M. (1997b) The use of models to integrate information and understanding of soil C at the regional scale. *Geoderma* 79, 227–260.

Pregitzer, K.S. (1993) Impact of climate change on soil processes and soil biological activity. In: Atkinson, D. (ed.) *Global Climatic Change: Its Implication for Crop Protection.* BCPC, Farnham, UK, pp. 71–82.

Prior, S.A., Torbert, H.A., Runion, G.B., Rogers, H.H., Wood, C.W., Kimball, B.A., LaMorte, R.L., Pinter, P.J. and Wall, G.W. (1997) Free-air carbon dioxide enrichment of wheat: soil carbon and nitrogen dynamics. *Journal of Environmental Quality* 26, 1161–1166.

Quemada, M. and Carbrera, M.L. (1997) Temperature and moisture effects on C and N mineralization from surface applied clover residue. *Plant and Soil* 189, 127–137.

Rasmussen, P.E., Goulding, K.W.T, Brown, J.R., Grace, P.R., Janzen, H.H. and Körschens, M. (1998) Long-term agroecosystem experiments: assessing agricultural sustainability and global change. *Science* 282, 893–896.

Reeburgh, W.S. (1997) Figures summarizing the global cycles of biogeochemically important elements. *Bulletin of the Ecological Society of America* 78, 260–267.

Rice, C.W., Garcia, F.O., Hampton, C.O. and Owensby, C.E. (1994) Soil microbial response in tallgrass praire to elevated CO_2. *Plant and Soil* 165, 67–74.

Rogers, H.H., Runion, G.B. and Krupa, S.V. (1994) Plant responses to atmospheric CO_2 enrichment with emphasis on roots and the rhizosphere. *Environmental Pollution* 83, 155–189

Rogers, H.H., Prior, S.A., Runion, G.B. and Mitchell, R.J. (1996) Root to shoot ratio of crops as influenced by CO_2. *Plant and Soil* 187, 229–248.

Ross, D.J., Tate, K.R. and Newton, P.C.D. (1995) Elevated CO_2 and temperature effects on soil carbon and nitrogen cycling in ryegrass/white clover turves of an Endoaquept soil. *Plant and Soil* 176, 37–49.

Runion, G.B., Curl, E.A., Rogres, H.H., Backman, P.A., Rodriguez-Kabana, R. and Helms, B.E. (1994) Effects of CO_2 enrichment on microbial populations in the rhizosphere and phyllosphere of cotton. *Agriculture and Forest Meteorology* 70, 117–130.

Schimel, D.S. (1995) Terrestrial ecosystems and the carbon cycle. *Global Change Biology* 1, 77–91.

Schortemeyer, M., Hartwig, U.A., Hendrey, G.R. and Sadowksy, M.J. (1996) Microbial community changes in the rhizospheres of white clover and perennial ryegrass exposed to free air carbon dioxide enrichment (FACE). *Soil Biology and Biochemistry* 28, 1717–1724.

Seyferth, U. and Persson, T. (2000) Effects of soil temperature and moisture on carbon mineralisation in a coniferous forest soil. *Soil Biology and Biochemistry* (in press).

Smith, P., Andrén, O., Brussaard, L., Dangerfield, M., Ekschmitt, K., Lavelle, P., Van Noordwijk, M. and Tate, K. (1999) Soil biota and global change at the ecosystem level: the role of soil biota in mathematical models. *Global Change Biology* 7, 773–784.

Swift, M.J., Heal, O.W. and Anderson, J.M. (1979) *Decomposition in Terrestrial Ecosystems.* Studies in Ecology, Vol. 5. Blackwell, Oxford, UK.

Torbert, H.A., Rogers, H.H., Prior, S.A., Schlesinger, W.H. and Runion, G.B. (1997) Effects of elevated atmospheric CO_2 in agro-ecosystems on soil carbon storage. *Global Change Biology* 3, 513–521.

Wardle, D.A. (1992) A comparative assessment of factors which influence microbial carbon and nitrogen levels in soil. *Biological Reviews* 67, 321–358.

Whitford, W.G. (1992) Effects of climate change on soil biotic communities and soil processes. In: Peters, R.L. and Lovejoy, T.E. (eds) *Global Warming and Biological Diversity.* Yale University Press, New Haven, Connecticut, pp. 124– 136.

Wood, C.W., Torbert, H.A., Rogers, H.H., Runion, G.B. and Prior, S.A. (1994) Free-air CO_2 enrichment effects on soil carbon and nitrogen. *Agricultural and Forest Meteorology* 70, 103–116.

Yeates, G.W., Tate, K.R. and Newton, P.C.D. (1997) Response of the fauna of a grassland soil to doubling of atmospheric carbon dioxide concentration. *Biology and Fertility of Soils* 25, 307–315.

Zanetti, S., Hartwig, U.A., Hebeisen, T., Lüscher, A., Hendrey, G.R., Blum, H. and Nösberger, J. (1995) Effect of elevated atmospheric CO_2 on performance of symbiotic nitrogen fixation in white clover in the field (Swiss FACE Experiment) and possible ecological implications. In: Tikhonovich, I.A., Romanow, V.I., Provorov, N.A. and Newton, W.E. (eds) *Nitrogen Fixation: Fundamentals and Applications.* Kluwer Academic, Dordrecht, p. 615.

Zepp, R.G. and Sonntag, C. (eds) (1995) *The Role of Nonliving Organic Matter in the Earth's Carbon Cycle.* John Wiley & Sons, Chichester, UK.

Zheng, D.W., Bengtsson, J. and Ågren, G.I. (1997) Soil food webs and ecosystem processes: decomposition in donor-control and Lotka-Volterra systems. *American Naturalist* 149, 125–148.

Zogg, G.P., Zak, D.R., Ringelberg, D.B., MacDonald, N.W., Pregitzer, K.S. and White, D.C. (1997) Compositional and functional shifts in microbial communities due to soil warming. *Soil Science Society of America Journal* 61, 475–481.

18 Crop Ecosystem Responses to Climatic Change: Interactive Effects of Ozone, Ultraviolet-B Radiation, Sulphur Dioxide and Carbon Dioxide on Crops

JAMES V. GROTH AND SAGAR V. KRUPA

Department of Plant Pathology, University of Minnesota, 495 Borlaug Hall, 1991 Upper Buford Circle, St Paul, MN 55108-6030, USA

18.1 Introduction

Any observed or predicted changes in the global climate are of utmost international concern (Houghton *et al.*, 1990, 1996). Such alterations in our climate are governed by a complex system of atmospheric processes and their products (Runeckles and Krupa, 1994). In the context of crop production, relevant atmospheric processes consist of losses in the beneficial stratospheric ozone ($[O_3]$) column and increases in the concentrations of tropospheric trace gases. Some of these, such as surface $[O_3]$ and carbon dioxide concentration ($[CO_2]$), have direct impacts on crops, while others such as methane ($[CH_4]$) and nitrous oxide ($[N_2O]$) are critical in altering the air temperature (Krupa, 1997). On the products side, increases in the surface level ultraviolet-B (UV-B, 280–315 or 320 nm) radiation and changes in air temperature and precipitation patterns are particularly important for agriculture (Bazzaz and Sombroek, 1996).

Currently there is a global network for measuring ambient $[CO_2]$ (World Wide Web Site 1). However, very few of these monitoring locations are in agricultural areas (Krupa and Groth, 1999). In comparison, measurements of surface $[O_3]$ have been the domain of federal and local regulatory agencies in many countries. Here too, at least in the USA, fewer than 10% of the monitoring sites are in rural agricultural or forested areas (US EPA, 1996). In contrast, in the USA and Canada, monitoring ambient sulphur dioxide concentration ($[SO_2]$) has been the responsibility of industries emitting the pollutant, but subject to the scrutiny of the local regulatory agencies. In other countries, such as Germany, provincial agencies operate the measurement networks.

Global measurements of UV-B have only begun recently (World Wide Web Site 2). On the other hand, the World Meteorological Organization (WMO) has operated a long-term global network for global radiation (total direct and diffuse radiation), air temperature and precipitation depth (World Wide Web Site 3). More recently WMO has added surface [O_3] to its database.

Independent of the existence of various measurement networks, global or local, large or small, there are significant temporal and spatial variabilities in the measured crop growth-regulating climatic variables (both chemical and physical). The corresponding stochastic relationships in cause and effects pose a major problem in conducting site-specific studies and scaling the results geographically by modelling (Krupa and Groth, 1999). Nevertheless, it is possible to derive first-order empirical conclusions, together with their associated uncertainties, and these form the basis for the following discussion.

18.2 The Concept of Interaction

18.2.1 General background

Some of the topics addressed by Krupa *et al.* (1998a) relate to the application of existing scientific knowledge of environment/crop interactions to the global spatial grid for revealing patterns of their spatial variability at the present time. Other topics are more fundamental for understanding the dynamic relationships that occur at any spatial level, from global to the local scale. The contents of this chapter are in the latter context, without considering the spatial term. The focus is on the combination of factors that impact production of a given crop, but with strong emphasis on climate-change factors.

The joint effects of elevated UV-B, [CO_2], surface [O_3], temperature, moisture, diseases and insect pests on crops are ideally dealt with in an integrated manner, if we are to understand or predict their potential limitations to crop production (Krupa *et al.*, 1998a). Such an effort requires a dynamic systems perspective. Due to the complexity of the task, it can only be implemented using advanced computer technologies and complex models yet to be developed. Experimental and observational findings of the joint effects, (even taken pairwise) of the aforementioned variables and their interactions on plant responses are fragmented and are only beginning to be studied in a quantitative manner.

Ever since cultivation of crops began, and indeed even before, plants have been exposed to increases in [CO_2], extremes of temperature, moisture, light and fluctuations in nutrients. All of these factors represent needed resources that can become growth limiting. In addition, crops have been and are being exposed to increases in tropospheric [O_3] and perhaps, in the future, increases in ground-level UV-B. Added to these factors are the effects of plant diseases, insect pests and competition from weeds. Therefore, agricultural production is in part a result of the ability of a crop to grow under the impacts of these multiple stress factors and within the management practices.

There is considerable interest in how plants respond to multiple environmental stresses in a changing environment. In this context, studies of plant stress have followed the normal course of examining each of the single main effects first, followed by studies of interactions of some of the effects (Forseth, 1997).

18.2.2 What is meant by an interaction?

The detection of interactions among effects (individual environmental stimuli such as some level of [CO_2] or UV-B) on measurements of plant growth is not a simple task, and we will begin by defining 'interaction' as it is used here, as well as whether these interactions detected are meaningful. Some confusion exists between strict statistical and biological definitions and methods of detecting interactions. Statistical interactions are precisely defined as quantitatively determined interaction sum of squares components in linear models that partition variance. Because statistical sophistication is not always present in otherwise reasonable papers that attempt to measure interactions, the following terminology is used here, although we are aware that other definitions of these terms can be found.

Two effects are *additive* only when they affect the measurement in a linear or arithmetic manner, and the combined effect is the sum of the individual effects. This additivity can involve the sign of the function, if the two individual effects influence the measurement in opposite directions.

Anything other than additive is considered as an *interaction*. Two kinds of interactions that can be identified are known by multiple terms:

- If the combined effect is less than the sum of the individual effects, the relative interaction is *ameliorative*, or *competitive*, or the effects are said to *cancel* or *nullify* one another.
- If the combined effect is greater than the sum of the individual effects, the interaction is *synergistic*, or the effects are said to *reinforce* or *amplify* one another.

Simple statistical tests can be used to show whether the combined effects are significantly greater or less than the additive expectation. The same ideas apply to two factors that act in an opposing manner, but terminology becomes confusing. Such cases show most clearly that statistical definitions of interactions are not always synonymous with biological definitions; many would consider cases of cancelling factors to be important biological interactions irrespective of whether the cancelling effects interact statistically. A consideration of such cancelling effects will determine the importance of studies of combinations of environmental influences on plant health.

18.2.3 Curve-fitting approaches

According to Krupa *et al.* (1998a), at their simplest, statistical methods of detecting interactions use one exposure dose of each of the independent variables. However, a more complete picture is desirable, and methods to describe the dose–response surface of two agents acting together on some measure of plant growth or productivity are outlined in Kreeb and Chen (1991). If the dose–response relationship of one or both agents is non-linear, the surface will not be planar. A non-planar surface does not automatically signify an interaction. The clearest indication of an interaction is probably the existence of a complex surface or landscape in which specific areas are significantly above or below the expected surface, based on mathematical or statistical curve fitting, and expected additivity. For suspected peaks or valleys, it should be possible to reduce the problem to a simple test of whether the actual effect on plant growth is significantly different from the expected effect from additivity of the two dose–response curves. In this sense, a response surface is valuable because it may reveal interactions of two agents at levels that might otherwise have been undetected if only one or a few levels were tested (Myers, 1971). Beyond the few mathematical relationships discussed by Kreeb and Chen (1991), there is desktop computer software (e.g. TableCurve 3D.19, by Jandel Scientific, San Rafael, California, USA) that enables the exploration and representation of response surfaces for data on three factors from among 243 polynomial equations, 260 rational equations and 172 non-linear models, fitting up to 36,582 equations, and then sorting the best equations according to goodness of fit (Krupa *et al.*, 1998a).

One of the most serious problems with showing interactions is the choice of response measurement and associated scale issues (Falconer, 1960). Reality of the level, duration, etc., of each effect is discussed in a later section of this chapter in relation to individual studies. Researchers have used many different measurements of plant responses to climatic change effects, and results are not always consistent; interactions detected with one type of plant measurement often fail to appear with another type of measurement in the same studies. There is no single best type of measurement to use, and the choice may be based on many things, including ease of measurement (e.g. pollen tube length), perceived importance (e.g. biomass), or economic significance (e.g. agricultural yield). The relationships between different measurements that might be used are often non-linear. Transformation of some measurements may be called for, because the data are not normally distributed or because means are not independent of variances. In reality, very few of the studies of possible interactions explore these statistical issues, and so it may be difficult to determine whether the interactions are genuine – and, if they are genuine, are the measurements that show them meaningful? Data transformation may make statistical sense, but does it make biological sense? At this stage of investigation any measurement can be defended as being informative. Eventually, it may be better to concentrate on measurements that are bound to translate into yield or quality loss in food and fibre crops.

18.2.4 Potential combinations of interactions

Even if stress factors or stimuli are taken singly, their effects on plant growth are complex. Some are inherently complex, such as plant diseases, which affect plants in a myriad of ways and can interact among themselves. Stress factors should not be taken singly, because they almost never occur singly, and when they occur together their interactions should be understood. For simplicity, assume that there are ten stress factors. The total number of combinations of n parameters taken as some number at a given time is $2^n - 1$. For a set of ten potential stress agents, this means 1023 potential interactions. Subtracting the ten single factorial relationships to a dependent response variable leaves a total of 1013 potential interactions of the combinations of two or more stress agents affecting a single crop response. Clearly, this is a daunting challenge for crop ecologists. It is easy to display the interactions in a graphic form for any three of the 11 dimensions (including the dependent response variable), but a larger number of interactions cannot be clearly and completely illustrated (Krupa *et al.*, 1998a).

18.2.5 Types of interactions

Synergistic change occurs when two or more environmental or ecological processes jointly interact, either simultaneously or sequentially, and the result is not a simple sum of the otherwise individual responses, but instead is multiplicative. This response appears, as an amplification of effects and a compounding of the impacts, to be consistent. Sometimes the tolerance of a species to one source of stress becomes reduced when other stresses are experienced simultaneously (Myers, 1992). Very little is known about such ecological or physiological synergistic interactions and the mechanisms responsible for them.

According to Krupa *et al.* (1998a), biologically, the value of harvestable crop parts for food, fuel, shelter or fibre is a result of stored energy (carbon, C), along with the nutrient content. For the purposes of national and international policy, initial considerations are probably given to the production of edible crop parts, since food is the most urgently needed commodity. A simple question is: 'What might happen to the crop productivity – namely, the amount of carbon stored?' Although there are elaborate models of crop growth processes, when viewed at the most fundamental and general level, the growth of any crop is a balance between carbon uptake and subsequent carbon storage, after the necessary carbon losses. This is similar to a financial budget: income is disbursed into savings after losses through expenses. The same line of reasoning can be used for nutrients such as nitrogen, but that aspect will be considered only in a limited way in the following discussion. Plants absorb carbon from the atmospheric carbon dioxide in the daytime (photosynthesis) and they normally lose varying portions of that carbon both in the day (photorespiration) and at night (dark respiration):

$$\text{C-uptake} = (\text{C-storage}) - (\text{C-losses}) \qquad \text{(Eqn 18.1)}$$

In addition to the normal maintenance costs, plant stress from unexpected increases in the environmental contaminants causes additional carbon losses within the crop as plants attempt to repair the stress effect.

If carbon uptake increases due to increased [CO_2] and possibly due to climatic warming, and if carbon losses also increase because of episodes of increased UV-B radiation interspersed between episodes of tropospheric [O_3], it is conceivable that the net effect on carbon storage might only be a negligible change (increase) in the balance:

C-uptake (increase) = [C-storage (increase)] – [C-losses (increase)] (Eqn 18.2)

Even with a direct fertilization effect of increased ambient [CO_2], if carbon losses increase substantially because of increases in atmospheric contaminants and from changes due to less optimal levels of meteorological resources needed for crop growth, then carbon storage in plants could decrease, i.e. there will be decreased crop production:

C-uptake (increase) = [C-storage (decrease)] – [C-losses (increase)] (Eqn 18.3)

This idealized concept would vary according to the differences in the sensitivities of crop species and cultivars to multiple, simultaneous or sequential exposures to different stress factors. Within an optimal range for crop growth, light, temperature, moisture and nutrients are necessary resources. The soil nutrient status is usually managed by fertilizer usage practices, but if decreases in seasonal light levels (e.g. increases in cloud cover), or changes in temperature and/or moisture occur in a given region, those changes coupled with any increases in UV-B radiation, tropospheric [O_3], or increases in 'biotic factors' such as weeds, diseases or insect pests might be environmentally undesirable. The questions that require attention are: 'How can increases in the various undesirable parameters *jointly affect* carbon losses from changes in carbon storage in plant parts?' and 'What is the nature of the experimental and/or observational evidence?'

There is a difference between the potential joint effects of two or more physicochemical factors (e.g. elevated UV-B and elevated [CO_2]) on individual species and such effects on species competition. Examples of the latter are discussed by Billick and Case (1994). Interactions between changes in the physicochemical processes and their resulting effects on plant host and pathogen and/or insect pests, and the interactions between the organismal species themselves in that system, comprise one of the most complex issues, because it can involve both types of interactions. However, relative to increased [CO_2], UV-B and [O_3] singly, current evidence suggests the general features presented in Table 18.1, although the emphasis of this chapter is on crop yield.

18.3 Interactive Effects on Crops

18.3.1 Ozone and UV-B

Tropospheric O_3 is all-pervasive and continues to be of environmental concern (National Research Council, 1992). There are numerous publications on the effects of ambient and elevated levels of O_3 on crops (US EPA, 1996; Krupa

Table 18.1. Overview of the effects of CO_2, UV-B and O_3 on plants in single-exposure mode. (Adapted from Krupa and Kickert, 1989).

	Plant response to environmental change		
Plant characteristic	Doubling of CO_2 only	Increased UV-B only	Increased tropospheric O_3 only
Photosynthesis	C_3 plants increase up to 100%, but C_4 plants show only a small increase	Decreases in many C_3 and C_4 plants	Decreases in many plants
Leaf conductance	Decreases in C_3 and C_4 plants	Not affected in many plants	Decreases in sensitive species and cultivars
Water-use efficiency	Increases in C_3 and C_4 plants	Decreases in most plants	Decreases in sensitive plants
Leaf area	C_3 plants increase more than C_4 plants	Decreases in many plants	Decreases in sensitive plants
Specific leaf weight	Increases	Increases in many plants	Increases in sensitive plants
Crop maturation rate	Increases	Not affected	Decreases
Flowering	Earlier flowering	Inhibits or stimulates flowering in some plants	Decreased floral number, delayed fruit setting and yield
Dry matter production and yield	C_3 plants nearly double, but C_4 plants show only small increases	Decreases in many plants	Decreases in many plants
Sensitivity between species	Major differences between C_3 and C_4 plants	Large variability between species	Large variability between species
Sensitivity within species (cultivars)	Can vary among cultivars	Differs between cultivars of a species	Differs between cultivars of a species
Drought-stress sensitivity	Less sensitive to drought	Less sensitive to UV-B, but sensitive to lack of water	Less sensitive to ozone but sensitive to lack of water
Mineral-stress sensitivity	Less responsive to elevated CO_2	Less, while others more sensitive to UV-B	More susceptible to ozone injury

et al., 1998b). Acute short-term effects of [O_3] on sensitive crops, under appropriate growth conditions, can result in visible foliar injury. Such injury on broadleaved plants consist of chlorosis, bleaching, bronzing, flecking, stippling and necrosis (Krupa *et al.*, 1998b). In a typical case, injury symptoms will occur on the upper leaf surface only, since the primary site of O_3 injury is the chloroplast (mesophyll layer). On grasses, injury occurs as chlorosis or flecking between the veins and severely so where the leaf blade is folded. Acute effects may or may not result in crop loss. In contrast, chronic long-term effects can result in reductions in crop growth and yield, with or without foliar injury symptoms (Krupa *et al.*, 1998b). Variations in crop response to [O_3] can be observed at the generic, species and cultivar levels. Much of our knowledge of the effects of O_3 on crops is from: (i) field surveys; (ii) artificial exposures in controlled environments, greenhouses and field fumigation chambers; (iii) use of protective chemicals; and (iv) comparisons of cultivars with known differences in sensitivity. Few studies have used chamberless, ambient field exposures, comparing crop responses at different study sites (Manning and Krupa, 1992). There is evidence from both the USA (US EPA, 1996) and Europe (Fuhrer *et al.*, 1998) that O_3 at current ambient concentrations may cause reductions in crop yield. A controversial issue relates to the artificial exposure methodologies used and the numerical definitions of O_3 exposures in the application of single-point models for establishing cause–effect relationships (Krupa and Kickert, 1997). This is an area that deserves significant improvement in the future. Nevertheless, the adverse effects (visible foliar injury) of surface [O_3] on crops has been shown worldwide (Krupa *et al.*, 1998b).

In contrast to [O_3], much less is known about the foliar injury symptoms of elevated UV-B exposures on crops, other than morphological deformations and necrosis (Krupa *et al.*, 1998a). Many plants accumulate melanin-type pigments in the leaf epidermal cell layer (as a defence mechanism) in response to elevated UV-B. There is only one field study showing crop yield reduction due to elevated UV-B (soybean, *Glycine max*: Teramura and Sullivan, 1988). Here, variability in response is known at the cultivar level. However, studies on crop response to elevated UV-B are fraught with many uncertainties (Krupa and Groth, 1999). These uncertainties are attributed to the types of artificial UV-B exposure methodologies and instrumentation used for quantifying the UV-B level, as well as the definition of exposure dosimetry. There is a significant need for research to address these deficiencies (Krupa *et al.*, 1998a).

There are very few studies on the joint effects of elevated [O_3] and UV-B on crops. High levels of O_3 and UV-B do not co-occur. In addition to the major role of O_3 destroying chemicals in the stratosphere, wavelength-dependent absorption of UV-B by tropospheric O_3 will result in the photolysis of that O_3 (Finlayson-Pitts and Pitts, 1986; Runeckles and Krupa, 1994). Consequently, reduced amounts of atmospheric O_3 will permit disproportionately large amounts of UV-B radiation to penetrate through the atmosphere. For example, with overhead sun and typical [O_3], a 10% decrease in the [O_3] column was predicted to result in a 20% increase in UV-B penetration at 305 nm, a 250% increase at 290 nm, and a 500% increase at 287 nm (Cutchis, 1974). Various plant processes have unique responses to radiation wavelengths (action spectra). In addition to this critical wavelength issue, any studies on the joint effects of elevated [O_3] and UV-B must include sequential exposures of the two factors. Krupa and Kickert (1989) have summarized the possible patterns of the occurrence of these two variables under different climatic scenarios. Independent of this, some scientists have examined the joint effects of simultaneously elevated [O_3] and UV-B on crops (Miller and Pursley, 1990; Booker *et al.*, 1992). Here the experimental methods and exposure protocols used make it difficult to interpret and extrapolate the results to the climatic change context. Nevertheless, Booker *et al.* (1992), in their studies with soybean, concluded that elevated UV-B neither promoted the effects of $[O_3]$ nor provided protection against those effects. Van de Staaij *et al.* (1997) examined the effects of sequential exposures to artificially elevated $[O_3]$ and UV-B on photosynthesis and growth of a native grass species in the Netherlands. Although they found no interactive effects, the exposure regimes of the two independent variables were closer to the observed patterns under ambient conditions (Albar, 1992; Runeckles and Krupa, 1994). In conclusion, future research on the joint effects of elevated $[O_3]$ and UV-B must use realistic experimental protocols.

18.3.2 Ozone and sulphur dioxide

Since the reviews of Ormrod (1982), Kohut (1985) and Mansfield and McCune (1988), there has been only one similar attempt (US EPA, 1996) to summarize the joint effects of $[O_3]$ and $[SO_2]$ on crops. In combination, the two pollutants can produce additive, less than additive or more than additive effects in the development of foliar injury symptoms (e.g. bifacial, interveinal necrosis: Legge *et al.*, 1998) and in the reduction of crop biomass and yield, although the underlying mechanisms are not well understood. Nevertheless, much of the information has been obtained from studies on radish (*Raphanus sativus*), soybean, bean (*Phaseolus* spp.) and petunia (*Petunia* sp.). Only isolated studies have been conducted with other crops. Furthermore, almost all of these studies were conducted under laboratory/greenhouse conditions or in field exposure chambers. Thus, very little is known about the influences of other crop growth-regulating environmental variables on the joint effects of these two pollutants (US EPA, 1996).

In addition to the considerations of the artificial conditions in the exposure chambers, most of these studies on the joint effects of $[O_3]$ and $[SO_2]$ have used pollutant concentrations and exposure regimes (exposure dynamics) that are dissimilar from the ambient environment. Frequently, $[SO_2]$ that are characteristic of the maxima in the vicinity of a point source are coupled to exposure durations and frequencies that are more characteristic of area sources. Similarly, in many cases the O_3 exposures have been with concentrations and durations greater than the ambient (Heck *et al.*, 1988; Jäger *et al.*, 1993). Therefore, it is difficult to draw conclusions regarding crop productivity under ambient conditions from these types of studies.

18.3.3 Carbon dioxide and ozone

Very few studies have been published on the joint effects of elevated $[CO_2]$ and $[O_3]$ on crop productivity. Scientists have begun to examine this issue only recently. In an early study, a cotton (*Gossypium hirsutum*) growth and yield model was applied to five cotton-growing areas in the USA from the early 1960s to the mid-1980s (Reddy *et al.*, 1989). The original model included plant growth as affected by weather variables, but neither [CO_2] nor [O_3]. Revised versions of the model incorporated, at first, the two decadal (1960s to 1980s), annual [CO_2], and subsequently the model included the annual [CO_2] plus the average summertime [O_3] and their effects on the growth functions. The most realistic results were obtained when the model was applied to data relevant for Fresno, California, USA. The inclusion of the 23-year increase in ambient [CO_2] showed an insignificantly small positive increase in the corresponding mean cotton (lint) yield, and this was attributed to several years of inadequate applications of nitrogen to the soil. The study concluded that, with adequate nitrogen, elevated [CO_2] probably would have increased lint yields by 10%. The inclusion of 23 years of summertime surface mean O_3 concentrations, along with the increased [CO_2], showed a 17% decrease in the corresponding simulated mean yield of lint. From these results it is not possible to determine the nature of the interaction between [CO_2] and [O_3], because no simulation was performed using only weather variables and elevated [O_3].

More recently, Mulchi *et al.* (1992) observed in an open-top chamber experiment with soybean that most effects caused by elevated [CO_2] and [O_3] appeared to be additive; the negative effects of elevated [O_3] were compensated by the positive effects of increased [CO_2], and vice versa. From another open-top chamber study, Rudorff *et al.* (1996) found no significant interactive effects of elevated [CO_2] and [O_3] on either wheat (C_3) or maize (*Zea mays* L.) (C_4). Carbon dioxide enrichment had a beneficial effect on wheat, but not on maize. Nevertheless, the authors concluded that O_3-induced stress will likely be diminished under increased [CO_2]. However, maximal benefits in wheat production in response to CO_2 enrichment will not materialize under concomitant increases in tropospheric [O_3]. According to Hertstein *et al.* (1995), 'future prospects of regional crop production in a high [CO_2] world would strongly depend on future trends in tropospheric [O_3], if the respective crop plants respond to both gases.'

18.3.4 Carbon dioxide and UV-B

There are very few studies on the joint effects of elevated [CO_2] and UV-B (Krupa *et al.*, 1998a). General plant response patterns show that increased [CO_2] stimulates plant growth, while increased UV-B suppresses it (Table 18.1). One study concluded that the combined effects appear to be additive (Rozema *et al.*, 1991). The growth rate of wheat (*Triticum aestivum*) was increased from elevated [CO_2] alone, but was reduced by a combination of both elevated [CO_2] and UV-B.

Another study examined pea (*Pisum sativum*), tomato (*Lycopersicon esculentum*) and aster (*Aster* spp.) shoot dry weight response to elevated [CO_2] and UV-B (Rozema *et al.*, 1990). In that report, the duration of the exposures was not stated; therefore, it is not possible to discern the total exposure dose. The two treatments were 2.8 W, and 0.0 W UV-B m^{-2}, where the first exposure level represented a cloudless summer midday value at Amsterdam, the Netherlands. Hence, the experiment was conducted to determine what happens when UV-B is eliminated, although it is not clear that the statistical approach used was designed to answer that question. If it were to have been an experiment on increased UV-B, then upper levels of 3.5–4.0 W m^{-2} should have been used, rather than 2.8 W. Also, light levels for the growth conditions were relatively low (175 μmol m^{-2} s^{-1}). Since only two treatment levels were used, this is insufficient to determine the quantitative nature of any interaction between [CO_2] and UV-B on plant response. In spite of this, both pea and tomato shoot dry weights showed statistically significant changes: decreases from increased UV-B alone and increases from elevated [CO_2] alone. When exposed simultaneously to increases in both factors over the same unspecified time period, there were no statistically significant ($P < 0.05$) interactions relative to the controls. The study concluded that the combined effect of UV-B with [CO_2] was additive, but critical inspection of the results appears to indicate that a positive effect from increased [CO_2] can be totally eliminated by a predominating negative effect of increased UV-B. A tentative observation might be that the combined effect was negative, and less than additive, because the [CO_2] effect did not appear to compensate for the foliar injury induced by UV-B.

In a phytotron experiment with rice (*Oryza sativa*), an antagonistic interaction was observed between a doubled ambient [CO_2] level and an increased UV-B radiation regime (approximating a 10% equatorial stratospheric [O_3] reduction) (Teramura *et al.*, 1990). By itself, a doubled [CO_2] significantly elevated photosynthesis, seed yield and biomass, but those increases disappeared under the combined effect.

In the early 1990s, an analysis of plant responses to elevated [CO_2] concluded that there was no observational or experimental evidence of a statistically significant joint interactive effect of elevated [CO_2] and increased UV-B on plant growth properties, even though there were reasons to expect that such interactions should exist (Rozema, 1993). However, no definition was given of what would comprise such statistically significant evidence.

A good introduction to modelling plant effects for establishing relationships under UV-B exposure was given by Allen (1994). In this context, although intended for modelling climatic change effects for longer time periods (several years), the need for evaluating model precision and realism are pertinent for shorter time spans as well. High precision means that the model predictions of the structure and functioning of the system closely agree with observations; high realism means that the model represents causal relationships in the system. Precision can be examined directly; realism cannot and must be inferred not only from high precision in predicting contemporary conditions of climate (e.g. UV-B), but also by rigorous testing of the model under simulated future conditions (Hänninen, 1995).

18.3.5 Carbon dioxide and sulphur dioxide

In the recent years, SO_2 emissions in western bloc countries have declined to some extent (GEMS/WHO, 1988). Since the mid 1970s, crop scientists have become more convinced that tropospheric [O_3] is a much more serious problem in the western bloc countries and worldwide than [SO_2] (Krupa and Manning, 1988). Furthermore, in the context of global climatic change, SO_2 is a radiatively cooling gas; with reference to adverse effects on crops, it is considered to be a local problem in the vicinity of its emission sources. Because of all these reasons, very few studies have been conducted on the joint effects of [CO_2] and [SO_2] on crops (Allen, 1990). Carlson and Bazzaz (1985) showed that exposures to elevated [CO_2] reduced the sensitivity of various plant species to SO_2 injury. Similarly Rao and De Kok (1994) found that combined exposures of wheat (a C_3 plant) to elevated CO_2 levels prevented the adverse effects of [SO_2]. Such a conclusion was also reached by Lee *et al.* (1997) in their studies on soybean (also a C_3 plant). It is not clear at this time how a C_4 crop would respond to the two trace gases combined.

18.3.6 Multiple factors

Only recently have studies begun to examine plant responses to some of the multiple growth-regulating factors. An analysis of the scientific literature on plant response to individual exposures of elevated [CO_2], UV-B or [O_3] indicated that among all crops, nine species have cultivars that are relatively sensitive to each of the factors (Table 18.2) (Krupa and Kickert, 1989, 1993). Five of these crops are important in some developing countries of Asia (Krupa *et al.*, 1998a). Together with their relative ranking (in parentheses) in sensitivity to increases in surface [CO_2], UV-B and [O_3], they are: pea (2), bean (3), lettuce (*Lactuca sativa*) (6), cucumber (*Cucumis sativus*) (7) and tomato (9). Potato (*Solanum tuberosum*), an important crop in the transition market of the countries of Eastern Europe, ranked fourth in its sensitivity. Rice, ranking eighth in sensitivity to each of the three factors, is important in the developing Asian countries. The transition market countries of Eastern Europe lead the world in oat (*Avena sativa*) production, which ranked fifth among all the crops examined. The highest ranking crop, sorghum (*Sorghum vulgare*), is important in the developing African countries.

An evaluative review of published experimental data covers what is known about the joint effects of surface UV-B, [O_3], [CO_2], other contaminant gases, light, moisture, plant diseases and insect herbivores (Runeckles and Krupa, 1994). The review concluded that: (i) many growth chamber studies are flawed, since light levels used are far less than the ambient; this mitigates against photorepair processes in the plants and can show a higher adverse effect to UV-B exposure than would occur in an ambient environment; (ii) soybean vegetative growth can show a less-than-additive biomass response to combined exposures of UV-B and [O_3] (Miller and Pursley, 1990); (iii) in some annual crop plants, sequential exposure to UV-B followed by [O_3] can show a negative but less-than-additive response in pollen tube growth (Feder and Shrier, 1990), which is important for seed production; (iv) under water stress, some plants can become less sensitive to elevated levels of UV-B and more sensitive to the water stress; (v) little is known about any possible interaction

Table 18.2. Comparison of the sensitivities of agricultural crops to enhanced CO_2 (mean relative yield increases of CO_2-enriched to control) (after Kimball, 1983a,b, 1986; Cure, 1985; Cure and Acock, 1986), for CO_2 concentrations of 1200 μl l^{-1} or less (Kimball, 1983a,b), or 680 ppm (Cure and Acock, 1986); to enhanced UV-B radiation; and to ground-level O_3. Species considered to be sensitive to all three factors are indicated in bold. (Reprinted from Krupa, S.V. and Kickert, R.N. (1989) The Greenhouse Effect: impacts of ultraviolet-B (UV-B) radiation, carbon dioxide (CO_2), and ozone (O_3) on vegetation, *Environmental Pollution* 61, 263–393 with permission from Elsevier Science.)

Crop type	Crop[a]	Enhanced CO_2; mean relative yield increase[b]	Sensitivity to enhanced UV-B	Sensitivity to O_3
Fibre crops	Cotton[1] (*Gossypium hirsutum*)	3.09	Tolerant	Sensitive
C_4 grain crops	**Sorghum (*Sorghum vulgare*)**	**2.98**	**Sensitive**	**Intermediate**
Fibre crops	Cotton[1]	2.59–1.95	–	–
Fruit crops	Aubergine (*Solanum melongena*)	2.54–1.88	Tolerant	Unknown
Legume seeds	**Peas (*Pisum sativum*)**	**1.89–1.84**	**Sensitive**	**Sensitive**
Roots and tubers	Sweet potato (*Ipomoea batatas*)	1.83	Unknown	Unknown
Legume seeds	**Beans (*Phaseolus* spp.)**	**1.82–1.61**	**Sensitive**	**Sens./Intermed.**
C_3 grain crops	Barley[2] (*Hordeum vulgare*)	1.70	Sensitive	Tolerant
Leaf crops	Swiss chards (*Beta vulgaris* var. *cicla*)	1.67	Sensitive	Unknown
Roots and tubers	**Potato[3] (*Solanum tuberosum*)**	**1.64–1.44**	**Sens./Toler.**	**Sensitive**
Legume crops	Lucerne (*Medicago sativa*)	1.57[c,d]	Tolerant	Sensitive
Legume seeds	Soybean[4] (*Glycine max*)	1.55[e]	Sensitive	Tolerant
C_4 grain crops	Maize[5] (*Zea mays*)	1.55	Tolerant	Sensitive
Roots and tubers	Potato[3]	1.51	–	–
C_3 grain crops	**Oats (*Avena sativa*)**	**1.42**	**Sensitive**	**Sensitive**
C_4 grain crops	Maize[5]	1.40[e]	–	–
C_3 grain crops	Wheat[6] (*Triticum aestivum*)	1.37–1.26	Tolerant	Intermediate
Leaf crops	**Lettuce (*Lactuca sativa*)**	**1.35**	**Sensitive**	**Sensitive**
C_3 grain crops	Wheat[6]	1.35	–	–
Fruit crops	**Cucumber (*Cucumis sativus*)**	**1.43–1.30**	**Sensitive**	**Intermediate**
Legume seeds	Soybean[4]	1.29	–	–
C_4 grain crops	Maize[5]	1.29	–	–
Roots and tubers	Radish (*Raphanus sativus*)	1.28	Tolerant	Intermediate
Legume seeds	Soybean[4]	1.27–1.20	–	–
C_3 grain crops	Barley[2] (*Hordeum vulgare*)	1.25	–	–
C_3 grain crops	**Rice[7] (*Oryza sativa*)**	**1.25**	**Sensitive**	**Intermediate**
Fruit crops	Strawberry (*Fragaria* spp.)	1.22–1.17	Unknown	Tolerant
Fruit crops	Sweet pepper (*Capsicum frutescens*)	1.60–1.20	Sens./Toler.	Unknown
Fruit crops	**Tomato (*Lycopersicon esculentum*)**	**1.20–1.17**	**Sensitive**	**Sens./Intermed.**
C_3 grain crops	Rice[7]	1.15	–	–
Leaf crops	Endive (*Cichorium endivia*)	1.15	Unknown	Intermediate
Fruit crops	Muskmelon (*Cucumis melo*)	1.13	Sensitive	Unknown

Table 18.2. *Continued*

Crop type	Crop[a]	Enhanced CO_2; mean relative yield increase[b]	Sensitivity to enhanced UV-B	Sensitivity to O_3
Leaf crops	Clover (*Trifolium* spp.)	1.12	Tolerant	Sensitive
Leaf crops	Cabbage (*Brassica oleracea* var. *capitata*)	1.05	Tolerant	Intermediate
Flower crops	Nasturtium (*Tropaeolum* spp.)	1.86	–	–
Flower crops	Cyclamen (*Cyclamen* spp.)	1.35	–	–
Flower crops	Rose (*Rosa* spp.)	1.22	Tolerant	–
Flower crops	Carnation (*Dianthus caryophyllus*)	1.09	–	Intermediate
Flower crops	Chrysanthemum (*Chrysanthemum morifolium*)	1.06	Tolerant	Intermediate
Flower crops	Snapdragon (*Antirrhinum majus*)	1.03	–	–

[a]Crops with superscript numbers have more than one ranking.
[b]From Kimball (1983a,b). If shown, the second value is from Kimball (1986).
[c]Mean relative yield increase of CO_2-enriched (680 ppm) to control crop (300–350 ppm), after Cure and Acock (1986).
[d]Based on biomass accumulation; yield not available.
[e]Field-based result from Rogers *et al.* (1983a,b).

between plant nutrient deficiencies and response to UV-B exposure; (vi) a 43% increase in ambient [CO_2] can lessen the growth and physiological effects of O_3 exposure to soybean (Mulchi *et al.*, 1992); (vii) in general, plant injury from O_3 tends to increase under nitrogen stress, but plant O_3 susceptibility often decreases under drought; (viii) very little is known about actual plant responses to the interaction of biotic pathogens and UV-B; (ix) even less is known about actual plant responses to the interaction of insect herbivores, UV-B and other physical, chemical and biotic stress factors; and (x) there have been numerous reports of reduced plant susceptibility to elevated [O_3], conferred by prior infection of viruses, bacteria or fungi (Runeckles and Krupa, 1994).

18.4 Uncertainties Associated with Current Understanding

Current knowledge of atmospheric processes and their products in global climatic change and the ability to predict their impacts on crop growth and productivity in the future are fraught with significant uncertainties (Krupa and Kickert, 1993). There are several sources of these uncertainties. Atmospheric measurements of several crop growth-regulating parameters are inadequate to capture satisfactorily their spatial and temporal variabilities (Krupa and Groth, 1999). Such information is critical in developing realistic experimental designs to establish cause–effect relationships. Many experimental exposure systems and the corresponding definitions of the exposure dosimetry do not capture the stochasticity of the relevant relationships. A much closer working relationship is required between the atmospheric scientists and crop biologists.

Most of the knowledge of the effects of global climatic change variables on crop growth and productivity is based on univariate experimental studies (Krupa and Kickert, 1989). In these situations, crop growth-regulating variables by themselves have produced opposing effects on the biology of the same crop. The few bivariate artificial exposure studies to date, have shown that the effect of one independent variable can be negated by another, the net result being essentially no change in the crop biomass.

There is a significant increase in the use of computer simulation models to predict the impacts of global climatic change on future agriculture (Bazzaz and Sombroek, 1996). While such efforts are very educative, their use of input parameters does not satisfactorily address many of the concerns in initializing the models. Furthermore, such model outputs need validation.

Many published growth-simulation computer models are available for the important crops of the world, especially for those in the temperate latitudes. A few models have been published for crop competition with weeds. Almost all of these are designed to simulate crop growth and production in response to solar radiation, temperature, moisture and, in some cases, nutrient levels, with only a few such models incorporating a direct [CO_2] response (Krupa *et al.*, 1998a). We know of no crop growth and production simulation models that are currently designed to address changes in [O_3] and UV-B, even though one plant response model was designed to study the effects of increased UV-B and light on photosynthesis in wheat and wild oats (Ryel *et al.*, 1990). Modifications of the current crop simulation models to respond to changes in [CO_2], [O_3] and UV-B should be undertaken to explore the level of uncertainty in the assessments of these and other climatic changes on crops.

A conceptual dynamic, compartmental, plant carbon-growth model has been described that could integrate the combined interactions of day-to-day changes in photosynthetically active solar radiation, ambient [CO_2], moisture, nutrient availability and stress effects from air pollutants on the mass balance of carbohydrate production and consumption by plant organs (roots, stems, leaves and fruits) (Jäger *et al.*, 1991). Known and hypothetical interactions of UV-B with these factors, and the consequential joint effects of such interactions, should be incorporated into the structure of such a model (Krupa *et al.*, 1998a).

18.5 Concluding Remarks

Many currently grown crops are considered to be genetically depauperate (Hoyt, 1988). Elite lines and hybrids derived from such germplasms are designed to yield well under relatively narrow and well-defined growing conditions. However, superior genotypes are used heavily in various crop breeding programmes, often resulting in considerable relatedness among cultivars grown across large geographical areas. Nevertheless, some crops are more narrow in their geographical distributions than others. In general, the narrow genetic base and specific goals used in breeding virtually all modern crops make it unlikely that crop breeders will be able to accommodate climatic changes, if such changes are large and rapid. Confounding this overall discussion are the issues of changes in the epidemiology of crop diseases, insect pest incidences and crop/weed competition.

In the final analysis, if climatic change occurs gradually, production agriculture will be able to adapt to such changes. This is more likely if continued improvements are made in modifying the crop biology through the use of molecular tools. However, it is difficult to predict the future with any confidence, given the uncertainties associated with current knowledge of the overall subject.

Acknowledgements

We thank the University of Minnesota Agricultural Experiment Station for its support in kind and Leslie Johnson for her critical assistance in the preparation of this manuscript.

References

Albar, O.F. (1992) The spectral distribution of solar ultraviolet radiation at the ground. PhD thesis, University of Nottingham, UK.

Allen, L.H. Jr (1990) Plant responses to rising carbon dioxide and potential interactions with air pollutants. *Journal of Environmental Quality* 19, 15–34.

Allen, L.H. Jr (1994) Plant effects and modeling responses to UV-B radiation. In: Biggs, R.H. and Joyner, M.E.B. (eds) *Stratospheric Ozone Depletion/UV-B Radiation in the Biosphere.* Springer-Verlag, Berlin, pp. 303–310.

Bazzaz, F. and Sombroek, W. (eds) (1996) *Global Climate Change and Agricultural Production: Direct and Indirect Effects of Changing Hydrological, Pedological and Plant Physiological Processes.* John Wiley & Sons, Chichester, UK, 345 pp.

Billick, I. and Case, T.J. (1994) Higher order interactions in ecological communities: what are they and how can they be detected? *Ecology* 75, 1529–1543.

Booker, F.L., Miller, J.E. and Fiscus, E.L. (1992) Effects of ozone and UV-B radiation on pigments, biomass and peroxidase activity in soybean. In *Tropospheric Ozone and the Environment. II. Effects, Modeling and Control, Proceedings of the Specialty Conference, Atlanta, Georgia.* Air and Waste Management Association, Pittsburgh, Pennsylvania, 955 pp.

Carlson, R.W. and Bazzaz, F.A. (1985) Plant response to SO_2 and CO_2. In: Winner, W.E., Mooney, H.A. and Goldstein, R.A. (eds) *Sulfur Dioxide and Vegetation.* Stanford University Press, Stanford, California, pp. 313–331.

Cure, J.D. (1985) Carbon dioxide doubling response: a crop survey. In: Strain, B.R. and Cure, J.D. (eds) *Direct Effects of Increasing Carbon Dioxide on Vegetation.* DOE/ER-0238, US Department of Energy, Washington, DC, pp. 99–116.

Cure, J.D. and Acock, B. (1986) Crop responses to carbon dioxide doubling: a literature survey. *Agricultural and Forest Meteorology* 38, 127–145.

Cutchis, P. (1974) Stratospheric ozone depletion and solar ultraviolet radiation on earth. *Science* 184, 13–19.

Falconer, D.S. (1960) *Introduction to Quantitative Genetics.* Oliver and Boyd, London, 365 pp.

Feder, W.A. and Shrier, R. (1990) Combination of UV-B and ozone reduces pollen tube growth more than either stress alone. *Environmental and Experimental Botany* 30, 451–454.

Finlayson-Pitts, B.J. and Pitts, J.N. Jr (1986) *Atmospheric Chemistry: Fundamentals and Experimental Techniques.* John Wiley & Sons, New York, 1098 pp.

Forseth, I.N. (1997) Plant response to multiple environmental stresses: implications for climatic change and biodiversity. In: Reaka-Kudla, M.L., Wilson, D.E. and Wilson, E.O. (eds) *Biodiversity II – Understanding and Protecting Our Biological Resources.* Joseph Henry Press, Washington, DC, pp. 187–196.

Fuhrer, J., Skärby, L. and Ashmore, M.R. (1998) Critical levels for ozone effects on vegetation in Europe. *Environmental Pollution* 97, 91–106.

GEMS/WHO (1988) *Assessment of Urban Air Quality Worldwide.* WHO, Geneva, 100 pp.

Hänninen, H. (1995) Assessing ecological implications of climatic change: can we rely on our simulation models? *Climatic Change* 31, 1–4.

Heck, W.W., Taylor, O.C. and Tingey, D.T. (eds) (1988) *Assessment of Crop Loss from Air Pollutants.* Elsevier Applied Science, London, 552 pp.

Hertstein, U., Grünhage, L. and Jäger, H.-J. (1995) Assessment of past, present and future impacts of ozone and carbon dioxide on crop yields. *Atmospheric Environment* 29, 2031–2040.

Houghton, J.T., Jenkins, G.J. and Ephraums, J.J. (eds) (1990) *Climate Change: The IPCC Scientific Assessment.* Cambridge University Press, Cambridge, UK, 364 pp.

Houghton, J.T., Meira Filho, L.G., Callander, B.A., Harris, N., Kattenberg, A. and Maskell, K. (eds) (1996) *Climate Change 1995: Summary for Policy Makers.* Cambridge University Press, Cambridge, UK.

Hoyt, E. (1988) *Conserving the Wild Relatives of Crops.* IBPGR Headquarters, c/o FAO of the United Nations, Rome, 45 pp.

Jäger, H.-J., Grünhage, L., Dämmgen, U., Richter, O. and Krupa, S. (1991) Future research directions and data requirements for developing ambient ozone guidelines or standards for agroecosystems. *Environmental Pollution* 70, 131–141.

Jäger, H.-J., Unsworth, M., De Temmerman, L. and Mathy, P. (eds) (1993) *Effects of Air Pollutants on Agricultural Crops in Europe: Results of the European Open-Top Chambers Project.* Report no. 46, Air Pollution Series of the Environmental Research Programme of the Commission of the European Communities, Directorate-General for Science, Research and Development, Brussels, Belgium.

Kimball, B.A. (1983a) *Carbon Dioxide and Agricultural Yield: an Assemblage and Analysis of 770 Prior Observations.* Report 14, USDA, US Water Conservation Laboratory, Phoenix, Arizona.

Kimball, B.A. (1983b) Carbon dioxide and agricultural yield: an assemblage and analysis of 430 prior observations. *Agronomy Journal* 75, 779–788.

Kimball, B.A. (1986) Influence of elevated CO_2 on crop yield. In: Enoch, H.Z. and Kimball, B.A. (eds) *Carbon Dioxide Enrichment of Greenhouse Crops*, Vol. II – *Physiology, Yield, and Economics.* CRC Press, Boca Raton, Florida, pp. 105–115.

Kohut, R. (1985) The effects of SO_2 and O_3 on plants. In: Winner, W.E., Mooney, H.A. and Goldstein, R.A. (eds) *Sulfur Dioxide and Vegetation: Physiology, Ecology and Policy Issues.* Stanford University Press, Stanford, California, pp. 296–312.

Kreeb, K.H. and Chen, T. (1991) Combination effects of water and salt stress on growth, hydration and pigment composition in wheat (*Triticum aestivum* L.): a mathematical modeling approach. In: Esser, G. and Overdieck, D. (eds) *Modern Ecology – Basic and Applied Aspects.* Elsevier, New York, pp. 215–231.

Krupa, S.V. (1997) Global climate change: processes and products – an overview. *Environmental Monitoring and Assessment* 46, 73–88.

Krupa, S.V. and Groth, J.V. (1999) Global climate change and crop responses: uncertainties associated with the current methodologies. In: Agrawal, S.B. and Agrawal, M. (eds) *Environmental Pollution and Plant Responses.* Lewis Publishers, Boca Raton, Florida, pp. 1–18.

Krupa, S.V. and Kickert, R.N. (1989) The Greenhouse Effect: impacts of ultraviolet-B (UV-B) radiation, carbon dioxide (CO_2), and ozone (O_3) on vegetation. *Environmental Pollution* 61, 263–393.

Krupa, S.V. and Kickert, R.N. (1993) The Greenhouse Effect: the impacts of carbon dioxide (CO_2), ultraviolet-B (UV-B) radiation and ozone (O_3) on vegetation. *Vegetatio* 104/105, 223–238.

Krupa, S.V. and Kickert, R.N. (1997) Ambient ozone (O_3) and adverse crop response. *Environmental Reviews* 5, 55–77.

Krupa, S.V. and Manning, W.J. (1988) Atmospheric ozone: formation and effects on vegetation. *Environmental Pollution* 50, 101–137.

Krupa, S.V., Kickert, R.N. and Jäger, H.-J. (1998a) *Elevated Ultraviolet (UV)-B Radiation and Agriculture.* Springer-Verlag, Heidelberg/Berlin, and Landes Bioscience, Georgetown, Texas, 296 pp.

Krupa, S.V., Tonneijck, A.E.G. and Manning, W.J. (1998b) Ozone. In: Flagler, R.B., Chappelka, A.H., Manning, W.J., McCool, P.M. and Shafer, S.R. (eds) *Recognition of Air Pollution Injury to Vegetation: a Pictorial Atlas.* Air and Waste Management Association, Pittsburgh, Pennsylvania, pp. 2.1–2.28.

Lee, E.H., Pausch, R.C., Rowland, R.A., Mulchi, C.L. and Rudorff, B.F.T. (1997) Responses of field-grown soybean (cv. Essex) to elevated SO_2 under two atmospheric CO_2 concentrations. *Environmental and Experimental Botany* 37, 85–93.

Legge, A.H., Jäger, H.-J. and Krupa, S.V. (1998) Sulfur dioxide. In: Flagler, R.B., Chappelka, A.H., Manning, W.J., McCool, P.M. and Shafer, S.R. (eds) *Recognition of Air Pollution Injury to Vegetation: a Pictorial Atlas.* Air and Waste Management Association, Pittsburgh, Pennsylvania, pp. 3.1–3.42.

Manning, W.J. and Krupa, S.V. (1992) Experimental methodology for studying the effects of ozone on crops and trees. In: Lefohn, A.S. (ed.) *Surface Level Ozone Exposures and Their Effects on Vegetation.* Lewis Publishers, Chelsea, Michigan, pp. 93–156.

Mansfield, T.A. and McCune, D.C. (1988) Problems of crop loss assessment when there is exposure to two or more gaseous pollutants. In: Heck, W.W., Taylor, O.C. and Tingey, D.T. (eds) *Assessment of Crop Loss From Air Pollutants.* Elsevier Applied Science, London, pp. 317–344.

Miller, J.E. and Pursley, W.A. (1990) Effects of ozone and UV-B radiation on growth, UV-B absorbing pigments and antioxidants in soybean. *Plant Physiology* 93 (Suppl.), 101.

Mulchi, C.L., Slaughter, L., Saleem, M., Lee, E.H., Pausch, R. and Rowland, R. (1992) Growth and physiological characteristics of soybean in open-top chambers in response to ozone and increased atmospheric CO_2. *Agriculture, Ecosystems and Environment* 38, 107–118.

Myers, N. (1992) Synergisms: joint effects of climate change and other forms of habitat destruction. In: Peters, R.L. and Lovejoy, T.E. (eds) *Global Warming and Biological Diversity.* Yale University Press, New Haven, Connecticut, pp. 344–354.

Myers, R.H. (1971) *Response Surface Methodology.* Allyn and Bacon, Boston, Massachusetts, 246 pp.

National Research Council (1992) *Rethinking the Ozone Problem in Urban and Regional Air Pollution.* National Academy Press, Washington, DC, 500 pp.

Ormrod, D.P. (1982) Air pollutant interactions in mixtures. In: Unsworth, M.H. and Ormrod, D.P. (eds) *Effects of Gaseous Air Pollution in Agriculture and Horticulture.* Butterworth Scientific, London, pp. 307–331.

Rao, M.V. and De Kok, L.J. (1994) Interactive effects of high CO_2 and SO_2 on growth and antioxidant levels in wheat. *Phyton (Horn)* 34, 279–290.

Reddy, V.R., Baker, D.N. and McKinion, J.M. (1989) Analysis of effects of atmospheric carbon dioxide and ozone on cotton yield trends. *Journal of Environmental Quality* 18, 427–432.

Rogers, H.H., Bingham, G.E., Cure, J.D., Smith, J.M. and Surano, K.A. (1983a) Responses of selected plant species to elevated carbon dioxide in the field. *Journal of Environmental Quality* 12, 569–574.

Rogers, H.H., Thomas, J.F. and Bingham, G.E. (1983b) Response of agronomic and forest species to elevated atmospheric carbon dioxide. *Science* 220, 428–429.

Rozema, J. (1993) Plant responses to atmospheric carbon dioxide enrichment: interactions with some soil and atmospheric conditions. *Vegetatio* 104/105, 173–190.

Rozema, J., Lenssen, G.M. and Staaij, J.W.M. van de (1990) The combined effect of increased atmospheric CO_2 and UV-B radiation on some agricultural and salt marsh species. In: Goudriaan, J., Keulen, H. van and Laar, H.H. van (eds) *The Greenhouse Effect and Primary Productivity in European Agro-Ecosystems.* Pudoc, Wageningen, the Netherlands, pp. 68–71.

Rozema, J., Staay, J. van de, Costa, V., Pereira, J.T., Broekman, R., Lenssen, G. and Stroetenga, M. (1991) A comparison of the growth, photosynthesis and transpiration of wheat and maize in response to enhanced ultraviolet-B radiation. In: Abrol, Y.P., Wattal, P.N., Govindjee, Ort, D.R., Gnanam, A. and Teramura, A.H. (eds) *Impact of Global Climatic Changes on Photosynthesis and Plant Productivity.* Oxford and IBH Publishing, New Delhi, India, pp. 163–174.

Rudorff, B.F.T., Mulchi, C.L., Lee, E.H., Rowland, R. and Pausch, R. (1996) Effects of enhanced O_3 and CO_2 enrichment on plant characteristics in wheat and corn. *Environmental Pollution* 94, 53–60.

Runeckles, V.C. and Krupa, S.V. (1994) The impact of UV-B and ozone on terrestrial vegetation. *Environmental Pollution* 83, 191–213.

Ryel, R.J., Barnes, P.W., Beyschlag, W., Caldwell, M.M. and Flint, S.D. (1990) Plant competition for light analysed with a multispecies canopy model. I. Model development and influence of enhanced UV-B conditions on photosynthesis in mixed wheat and wild oat canopies. *Oecologia* 82, 304–310.

Staaij, J.W.M. van de, Tonneijck, A.E.G. and Rozema, J. (1997) The effect of reciprocal treatments with ozone and ultraviolet-B radiation on photosynthesis and growth of perennial grass *Elymus athericus. Environmental Pollution* 97, 281–286.

Teramura, A.H. and Sullivan, J.H. (1988) Effects of ultraviolet-B radiation on soybean yield and seed quality: a six-year field study. *Environmental Pollution* 53, 466–468.

Teramura, A.H., Sullivan, J.H. and Ziska, L.H. (1990) Interaction of elevated ultraviolet-B radiation and CO_2 on productivity and photosynthesis characteristics in wheat, rice and soybean. *Plant Physiology* 94, 470–475.

US EPA (1996) *Air Quality Criteria for Ozone and Other Photochemical Oxidants, Vol. II.* EPA-600/P-93/00bF, US Environmental Protection Agency, National Center for Environmental Assessment, Research Triangle Park, North Carolina.

World Wide Web Site 1. http://cdiac.esd.ornl.gov/trends/co2/contents.htm.

World Wide Web Site 2. http://www.wdc.rl.ac.uk/wdcmain/appendix/gdappenb2.html.

World Wide Web Site 3: http://www.wmo.ch/web/ddbs/clmdata.html.

19 Crop Breeding Strategies for the 21st Century

ANTHONY E. HALL[1] AND LEWIS H. ZISKA[2]

[1]Department of Botany and Plant Sciences, University of California, Riverside, CA 92521-0124, USA; [2]USDA-ARS, Climate Stress Laboratory, Bldg 0461, 10300 Baltimore Avenuce, Beltsville, MD 20705-2350, USA

19.1 Introduction

This chapter focuses on crop breeding strategies needed to solve some of the problems and exploit some of the opportunities that may result from global climate change. We will emphasize the choice of traits and methods for selecting these traits. A unique feature of global climate change is the projected increase in atmospheric carbon dioxide concentration ([CO_2]). This should result in increases in productivity for many crop species. Grain yields of some cereal crops may have increased by 7–25% over the last 100 years solely as a result of the increase in [CO_2] during this period (Goudriaan and Unsworth, 1990; Mayeaux *et al.*, 1997). But, the major increases in crop yield that have occurred during the last 50–100 years were mainly due to factors other than increased [CO_2] including various technological advances (Amthor, 1998). However, as will be shown in this review, potential crop productivity may be increased by breeding or selecting cultivars that are more responsive to elevated [CO_2]. Climatic changes, such as increases in temperature, also may occur in the 21st century (Houghton *et al.*, 1996) and in some cases decrease crop productivity. Plant breeding has the potential to overcome some of the problems caused by heat stress. Plant breeding programmes operate within time frames of decades and it is likely that this will continue into the near future, with genetic engineering complementing rather than replacing traditional breeding methods, especially for annual crop species. The changes in atmospheric [CO_2] and climate are likely to be rapid; consequently, it is appropriate to begin designing new breeding strategies and cultivars for the 21st century. This can be facilitated by considering past, present and possible future climates. Our analysis will emphasize the separate and interactive effects of increases in atmospheric [CO_2] and global warming on the productivity, grain quality and water relations of crops, and potential effects on

(eds K.R. Reddy and H.F. Hodges)

problems caused by pests. Earlier reviews of part of this topic were provided by Badger (1992) and Hall and Allen (1993).

19.2 Plant Responses to Increases in Atmospheric [CO_2]

Substantial changes occurred in atmospheric [CO_2] in the geological past and in the 20th century (Allen, 1994). Millions of years ago, atmospheric [CO_2] was about 1200–4000 μmol mol^{-1} but it decreased substantially over centuries (Sundquist, 1986). Analysis of air trapped in polar ice indicates that prior to 1800, atmospheric [CO_2] fluctuated between 180–290 μmol mol^{-1} for at least 220,000 years (Barnola *et al.*, 1987; Jouzel *et al.*, 1993). Since 1800, ice core data indicate accelerating increases in atmospheric [CO_2] from 280 to 300 μmol mol^{-1} by 1900 and 315 μmol mol^{-1} by 1958. Direct measurements indicate substantial increases since 1958 from 315 to 360 μmol mol^{-1} by the 1990s (Hall and Allen, 1993). Global reliance on fossil fuels could result in further increases in atmospheric [CO_2]. The moderate scenarios in the report of the Intergovernmental Panel on Climate Change (IPCC; Houghton *et al.*, 1996) predict that atmospheric [CO_2] will exceed 600 μmol mol^{-1} by the end of the 21st century. Even with full implementation of the agreements to reduce the use of fossil fuels, made at the meeting in Kyoto, Japan, in December 1997, atmospheric [CO_2] is expected to increase substantially in the 21st century (Bolin, 1998).

Plants with the C_4 photosynthetic system evolved during an early period after the atmospheric [CO_2] became low and this system represents a specific adaptation to the low [CO_2] environments of the last 200,000 years (Bowes, 1993). The growth responses of C_4 and crassulacean acid metabolism (CAM) plants to elevated [CO_2] are smaller than those of C_3 species (Poorter, 1993) and their responses to [CO_2] will not be considered in this review.

The extent and nature of the evolution of plants with the C_3 photosynthetic system, with respect to low [CO_2], are not known. Prior to 1900, these plants were subjected to low [CO_2] for thousands of years. The characteristics of the enzyme involved in the initial fixation of CO_2 in C_3 plants, ribulose bisphosphate carboxylase (Rubisco), may not have changed very much (Morell *et al.*, 1992). However, it is likely that low [CO_2] resulted in evolutionary modifications to whole plant processes, such as increases in the ratio of photosynthetic source to carbohydrate sink tissues, and that some C_3 plants may not be well adapted to either future or even present day levels of atmospheric [CO_2].

Empirical selection for yield under field conditions may indirectly select plants that are responsive to the continually increasing levels of CO_2 (Kimball, 1985). However, it is likely that indirect selection will be inefficient because the atmospheric [CO_2] is increasing rapidly, and yield is dependent upon many abiotic and biotic factors. Direct selection in breeding nurseries that are subjected to elevated [CO_2] would not appear feasible in commercial programmes due to the large numbers of plants that must be screened and the high costs of nursery environments that have elevated [CO_2] compared with conventional field nurseries. A unique experiment was conducted by Maxon Smith (1977) over 5 years in special glasshouses. Segregating populations of lettuce (*Lactuca sativa* L.) were selected for yield and agronomic traits in two environments: day-time CO_2 enrichment with higher temperatures, and ambient [CO_2] with lower temperatures. In both environments, selection was effective in improving agronomic traits but did not enhance yield, and the selection in the elevated CO_2 and high-temperature environment did not increase lettuce responsiveness to elevated CO_2. As was pointed out by Maxon Smith (1977), the lack of success in increasing yield and responsiveness to elevated [CO_2] could have been due to the small number of plants that were screened compared with the numbers that can be screened in field nurseries. However, if the biochemical, physiological and morphological traits that contribute to responsiveness to elevated [CO_2] could be identified, they could be used to supplement empirical selection based on yield under field conditions and thereby enhance breeding programmes.

Possible beneficial modifications that could be exploited by plant breeders may be discovered by examining plant responses to elevated [CO_2]. At intermediate temperatures, doubling atmospheric [CO_2] increased grain yield of various small grain cereals by 32% and grain legumes by 54% (Kimball, 1983), but the increases often were less than the increases in photosynthesis that occur with short-term doubling of [CO_2] at the same temperatures (Poorter, 1993; Allen, 1994). There are several possible explanations for the smaller yield responses to long-term CO_2 enrichment compared with the short-term photosynthetic responses. A major factor is the down-regulation of photosynthetic capacity under long-term exposure to elevated [CO_2] that occurred in some experiments. This down-regulation could either be a consequence of artificial growth conditions or it may indicate that current cultivars of C_3 plants are not well-adapted to elevated [CO_2]. Down-regulation has been attributed to feedback mechanisms that operate when the supply of carbohydrates from photosynthesis exceeds sink demands for carbohydrates (Allen, 1994). Arp (1991) pointed out that strong down-regulation was observed when plants were grown in small pots that would have resulted in a much smaller root sink for carbohydrates than occurs in nature. In addition, CO_2-induced growth enhancement can depend on nutrient supplies in the root zone (McConnaughay *et al.*, 1993). Conroy (1992) proposed, based on a review of the literature, that the greatest absolute increases in productivity with CO_2 enrichment will occur when soil N and P availabilities are high. More recently, Ziska *et al.* (1996b) showed for rice (*Oryza sativa* L.) that growth response to elevated [CO_2] was negligible when soil N was very deficient and increased with greater supplies of soil N. Increases in growth responses to high [CO_2] with increases in nitrogen supply also were observed for wheat (*Triticum aestivum* L.) and cotton (*Gossypium hirsutum* L.). These data were consistent with the hypothesis that increases in N supply increase the ratio of the carbohydrate sinks to the photosynthetic source which then result in increased growth responses to elevated [CO_2] (Rogers *et al.*, 1996a,b).

Large differences in the responses of growth or yield to elevated [CO_2] have been observed among C_3 species that could have been due to differences in either environmental conditions or genotypic effects. The genotypic effects are of particular importance to this review. In studies comparing C_3 species, Poorter (1993) and Poorter *et al.* (1996) reported a positive correlation between responsiveness to elevated [CO_2] (in terms of the absolute value of the increase in relative growth rate) and the relative growth rate in atmospheric [CO_2]. More simply, faster-growing species were more responsive, and the authors commented that the more responsive plants may have had larger sink strengths. Comparisons of contrasting cultivars within species, however, would permit more rigorous tests of hypotheses concerning traits influencing responsiveness to elevated [CO_2], because different species have substantial variation in genetic background that could obscure responses and hinder interpretations.

Progress during the 20th century in increasing the productivity of several annual C_3 crops through plant breeding was estimated as mainly (77%) resulting from increases in harvest index (typically harvest index (HI) = grain yield/total shoot biomass) with only 23% due to increases in total shoot biomass (Gifford, 1986). A general explanation for the increases in grain yield associated with increases in HI is increased efficiency of carbohydrate partitioning, although evidence for traits that were consistently associated with the rise in HI is not available (Evans, 1993). An alternative explanation is that the increases in both grain yield and HI resulted from inadvertent selection of plant types that were better adapted to the higher [CO_2]s of the last half of the 20th century compared with the low [CO_2]s in earlier centuries. As hypothesized by Hall and Allen (1993), plants with higher HI would be expected to have greater sinks for carbohydrates in relation to their photosynthetic capacity and thus be more responsive to elevated [CO_2].

Responses to elevated [CO_2] of contrasting genetic lines of cowpea (*Vigna unguiculata* L. Walp.) support this hypothesis. In growth chamber studies, a genotype with enhanced pod set due to heat-tolerance genes (No. 518) produced more shoot biomass and had greater pod yield with elevated [CO_2], under either high night temperatures or more optimal temperatures, than a genetically similar cultivar (CB5) that does not have the heat-tolerance genes (Ahmed *et al.*, 1993). Subsequent field studies with six pairs of cowpea lines either having or not having the heat-tolerance genes demonstrated that the heat-tolerance genes cause dwarfing and enhance HI under both hot and more optimal temperatures (Ismail and Hall, 1998). Taken together, these studies indicate that selection for higher HI through enhancing pod set might produce cowpea cultivars whose grain yields are more responsive to elevated [CO_2] under both hot and more optimal temperatures. This hypothesis has not been rigorously evaluated such as by testing the responses of these genetic lines to elevated [CO_2] under field conditions.

Responses to elevated [CO_2] of six contrasting cultivars of spring wheat were studied in open-top chambers in field conditions but with plants in pots (Manderscheid and Weigel, 1997). These cultivars had been introduced in Germany between 1890 and 1988. The more recent cultivars had less tillering, shorter stems, and higher HI, but they were less responsive to elevated [CO_2] with respect to shoot biomass production, number of tillers, stem weight and stem height than the older cultivars. Grain yield responses varied inconsistently with the year of introduction. The authors argued that the greater shoot biomass responses of the older cultivars to elevated [CO_2] were due to the increased sink strength resulting from their greater tendency to tiller. These results may not, however, be relevant to optimal field conditions in that the newest cultivars had similar grain yields as the oldest cultivars under ambient [CO_2]. One would expect a cultivar released in 1988 to have higher grain yields under optimal present-day environmental conditions than a cultivar released in 1890. Generally, newer small grain cultivars with greater HI have greater grain yields under optimal field conditions than older cultivars. The results of Manderscheid and Weigel (1997) may be explained by their use of abnormally wide spacing that would have benefited the older cultivars due to their greater tendency for tillering and preventing the newer cultivars from achieving their potential grain yields per unit area. However, the number of spikes per plant (four to five) under ambient [CO_2] were similar to the values expected to occur under the high plant densities of optimal field conditions and do not support this argument. If their results are relevant to optimal field conditions, they indicate the need to select wheat plants with a greater tillering tendency and smaller HI, which is the opposite of current breeding practice. The cultivar effects reported by Manderscheid and Weigel (1997) are consistent with the conclusion of Rogers *et al.* (1996a) that the greater responsiveness to elevated [CO_2] of wheat under high soil N was due to stimulation of tillering by the high N. Neither of these studies gave a clear indication of whether cultivars of wheat with greater tillering tendency would be more responsive than modern cultivars to elevated [CO_2] in terms of grain yield under optimal field conditions and this is a critical issue for plant breeding programmes.

Contrasting cultivars of rice were studied by Ziska *et al.* (1996a) with plants in pots at wide spacing in glasshouses. At intermediate temperatures (day/night, 29/21°C), there were substantial differences among 17 cultivars in responses of both total plant biomass and grain yield to [CO_2] of 664 μmol mol^{-1}. As with the studies of spring wheat by Manderscheid and Weigel (1997), responsiveness of rice to elevated [CO_2] was associated with increased numbers of fertile tillers, there was no association between responsiveness and HI (measured as grain yield/total plant biomass), and the cultivars differed in many traits. Rice cultivar differences in response to elevated [CO_2] also were observed in field conditions (Moya *et al.*, 1998). Among three cultivars tested, one (N-22) exhibited a large increase in grain yield (+57%), another (IR72) had a 20% increase, and an experimental line had a non-significant 8% decrease in grain yield. Additional field studies with IR72 (Ziska *et al.*, 1997) confirmed that grain yield responses to elevated [CO_2] were associated with increases in numbers of fertile tillers; but the grain yield responses (+15% and +27%) were much smaller than the responses of total plant biomass (+31% and +40%). This indicates that grain yield may have been sink limited and that responsiveness to elevated [CO_2] in rice might be enhanced by breeding to enhance reproductive sink strength.

To date, there have been few reports of studies aimed at determining the specific traits that enhance responsiveness of grain yield to elevated [CO_2].

The few studies reported were constrained by the use of cultivars that differed in many traits and the use of growth chambers or greenhouses or potted plants, which makes it difficult to predict responses under optimal field conditions. More definitive tests of various hypotheses would be obtained by comparative studies of responsiveness to elevated [CO_2] under optimal field conditions with genetic lines that differ mainly in HI (or some other trait such as tillering) and also exhibit positive correlations between grain yield and HI under ambient [CO_2]. There is a critical need for field studies of this type, because current selection strategies may be appropriate or in the opposite direction from those needed for breeding cultivars for future environments that have elevated [CO_2].

19.3 Plant Responses to Increases in Air Temperature

Global warming of a few degrees Celsius has been predicted to occur as a consequence of the increases in [CO_2] and other gases that absorb infrared radiation in the earth's atmosphere (Kerr, 1986). There may be greater increases at night rather than day temperatures (Kukla and Karl, 1993) and at extreme latitudes rather than at the tropics. The extent of global warming that may occur in the 21st century is poorly understood, but it is appropriate to discuss the effects of heat stress on cultivars because major agricultural regions presently experience weather that is sufficiently hot to reduce crop yields.

Reproductive development of many crop species is particularly sensitive to high temperature, especially high night temperature, such that fruit or seed yield is reduced more than overall biomass production (Hall, 1992). Sorghum (*Sorghum bicolor* L. Moench) exhibited a 28% reduction in grain yield when subjected to a 5°C increase in night temperature for 1 week during floret differentiation under field conditions (Eastin *et al.*, 1983). Cowpea subjected to different night-time temperatures under field conditions using enclosures (Nielsen and Hall, 1985a) exhibited a 4.4% decrease in grain yield per degree Celsius increase in night-time temperature above a threshold daily minimum night temperature of 15°C (Nielsen and Hall, 1985b; Hall and Allen, 1993). The lower yields were associated with reductions in the proportion of flowers producing pods. Pollen development of sensitive cowpea cultivars is damaged by high night temperature during a specific period occurring 9–7 days before anthesis (Ahmed *et al.*, 1992). In a more recent study using contrasting field environments, cowpea exhibited a 13.6% decrease in grain yield per degree Celsius increase in average daily minimum night temperature above 16°C during early flowering, mainly due to decreases in pod set (Ismail and Hall, 1998). High temperatures damaged the reproductive sink far more than the photosynthetic source. Total shoot biomass production was reduced by only 5.6% $°C^{-1}$ compared with the 13.6% decrease in grain yield. For wheat, phytotron studies indicated reductions in grain yield of 3–4% $°C^{-1}$ increase in temperature above a mean daily temperature of 15°C (Wardlaw *et al.*, 1989). From studies with wheat across contrasting field environments, Fischer (1985) estimated that there were 4% fewer kernels per unit area per degree Celsius increase in temperature during the 30 days preceding anthesis. In field studies with rice, plots were subjected to a 4°C increase in air temperature compared with controls using open-top chambers (Moya *et al.*, 1998). In three different growing seasons the elevated temperature treatment reduced grain yields of rice cultivar IR72 by 26, 16 and 13% (Moya *et al.*, 1998) due to reductions in spikelet and pollen fertility (Matsui *et al.*, 1997). Studies in sun-lit growth chambers demonstrated that both high day and high night temperature can reduce spikelet fertility in rice (Ziska and Manalo, 1996). Apparently, increases in either day or night temperature due to global warming could cause reductions in grain yield in several small grain cereals and grain legumes due to impairment of reproductive development.

Problems caused by heat stress have been partially solved by breeding cultivars with the ability to set seeds and fruits under hot conditions, especially for tomato (*Lycopersicon esculentum* Mill.), Pima cotton (*Gossypium barbadense* L.) and cowpea (reviewed by Hall, 1992, 1993). The positive and potentially negative effects of the heat-tolerance genes incorporated into cowpea have been investigated (Ismail and Hall, 1998). Six pairs of cowpea lines that either have or do not have heat tolerance, with each pair having similar genetic background, were compared in eight field environments with contrasting temperatures (average daily minimum night temperatures during early flowering of 15–25°C). These field environments mainly differed in temperature. They had similar sunny conditions, no significant plant disease problems, and optimal management of irrigation, soil fertility and pests. Comparisons of the pairs of lines showed that the heat-tolerance genes had progressively greater effects at higher night temperatures, increasing pod set and grain yield and causing dwarfing. Correlation analyses predicted that, at 21°C minimum night temperature, heat-tolerant plants would have 50% greater pod set and grain yield but also 55% shorter internodes and 30% less vegetative shoot biomass than heat-susceptible plants. The heat-tolerance genes clearly have positive effects in hot environments by increasing pod set and grain yield but do they also have negative effects? The dwarfing that occurred may be due to either pleiotropic effects of the heat-tolerance genes or effects of other closely linked genes. The potentially negative effects of the dwarfing include incomplete ground cover that reduces both interception of solar radiation and competitiveness with weeds. Both of these potentially negative effects could be overcome by growing cowpea in narrower rows than the 76–102 cm currently used by many farmers. Indeed, under narrow-row conditions the dwarf cultivars may have greater yields than current cultivars under either narrow or conventional wider row widths. Heat-tolerant cowpeas have been developed for the subtropical production zone in California (Hall, 1993). Glasshouse studies demonstrated that the heat-tolerance genes which were effective in subtropical conditions also confer high pod set and grain yield under the short-day, high night temperature conditions typical of tropical zones where cowpea is grown (Ehlers and Hall, 1998). Pedigree breeding procedures have been developed for incorporating heat tolerance during reproductive development into cowpea (Hall, 1992, 1993) that should be effective with some other crop species. These procedures include choosing a field or glasshouse nursery where segregating populations can be subjected to stressfully high night temperature during critical periods of reproductive development. Major genes conferring heat tolerance during the early floral, anther development and embryo development stages are incorporated by visual selection. Individual plants are chosen in the F_2 generation that have no heat-induced floral bud suppression and produce abundant flowers. In more advanced generations, individual plants are chosen from families with abundant pod set and sufficient seeds per pod of adequate quality. Also, a field screening procedure has been developed for grain sorghum with the objective of combining tolerance to heat during reproductive development with drought resistance (Eastin *et al.*, 1988).

For some crop species and environmental conditions the photosynthetic source may be damaged more by heat than the reproductive sink. The water-splitting component of photosystem II appears to be particularly sensitive to high temperatures (Berry and Björkman, 1980). Recently, it has been suggested that a small heat-shock protein in the chloroplast may provide some protection to photosystem II electron transport during heat stress (Heckathorn *et al.*, 1998). Grain yield of wheat can become limited by the photosynthetic capacity of the flag leaf when heat stress occurs during heading and grain filling. Comparisons of spring wheat cultivars with contrasting heat tolerance growing in hot, irrigated environments demonstrated that grain yield was positively correlated with photosynthetic CO_2 uptake and leaf conductance of flag leaves and canopy temperature depression (Reynolds *et al.*, 1994). For environments with sunny conditions and a large vapour pressure deficit, canopy temperature depression was proposed to be a useful trait for selecting for heat tolerance in wheat grown under irrigated conditions (Amani *et al.*, 1996). Cultivars with a greater canopy temperature depression had greater stomatal conductance and faster rates of CO_2 assimilation. Progress has been made in using measurements of canopy temperature depression in breeding programmes to enhance heat tolerance in spring wheat (Reynolds *et al.*, 1998). In environments with moderate temperatures, progress in increasing yields of spring wheat also has been associated with cooler canopies, greater stomatal conductance and faster rates of CO_2 assimilation (Fischer *et al.*, 1998). Pima cotton cultivars that had increased heat tolerance due to selection for ability to set bolls in hot environments (Hall, 1992) also had greater stomatal conductance (Lu and Zeiger, 1994). Greater reproductive sink strength may be linked with greater photosynthetic capacity that in turn is associated with greater stomatal conductance (Schulze and Hall, 1982). For sorghum lines, negative correlations were observed between grain yield and canopy minus air temperature, indicating positive correlations between grain yield and stomatal conductance, and the correlations were stronger under dryland than irrigated conditions (Eastin *et al.*, 1988). Condon and Hall (1997) discussed the positive genetic associations between biomass production and stomatal conductance that have been observed with several crop species in well-watered and dry environments. They proposed that the evolution of these annual crop plants resulted in conservative stomatal performance, that is, a tendency for stomata to be at least partially closed on many occasions. This could have occurred if plant performance during very dry years, when conservative stomatal performance may be adaptive, had disproportionate influences on seed production and long-term evolutionary success due to soil 'seed banks' being much less effective after 1 year. Further selection for greater stomatal conductance, or more efficient surrogates such as canopy temperature depression, may be beneficial since the increases in atmospheric [CO_2] that occur will have a tendency to cause decreases in stomatal conductance (Allen, 1994). Blum (1988) has developed procedures that can screen large numbers of genotypes of wheat for differences in canopy temperature under field conditions.

19.4 Interactive Effects of Increases in Atmospheric [CO_2] and Global Warming on Plants

The interactive effects of elevated [CO_2] and higher temperatures on plants are complex (Conroy *et al.*, 1994) but can be simplified if one separates cases where heat stress limits the reproductive sink from cases in which heat stress limits the photosynthetic source. In many cases, reproductive development is more sensitive to high temperatures than overall plant biomass production. Several studies showed the detrimental effects of high temperatures on reproductive development are not ameliorated by elevated [CO_2]. For soybean (*Glycine max* [L.] Merr.) grown under controlled-environment field conditions, HI progressively decreased with increasing temperature under either 660 or 330 μmol mol^{-1} [CO_2] (Baker *et al.*, 1989). Reproductive development of Pima cotton is also sensitive to high temperatures, such that the plants may not produce either fruiting branches or bolls (Reddy *et al.*, 1992). Studies in naturally sunlit, controlled-environment chambers demonstrated that elevated [CO_2] of 700 μmol mol^{-1} did not ameliorate this problem (Reddy *et al.*, 1995, 1997). Controlled-environment field studies with rice demonstrated that grain yield decreased 10% °C^{-1} increase in average temperature above 26°C at [CO_2]s of either 330 or 660 μmol mol^{-1} (Baker and Allen, 1993). The decreases in rice grain yield were due mainly to fewer grains per panicle. High day and high night temperatures can cause decreases in viability of pollen grains at anthesis, increases in floret sterility and decreases in seed set in rice (Ziska and Manalo, 1996). Elevated [CO_2] aggravated the effect on pollen, causing a 1°C decrease in the threshold maximum canopy surface temperature after which the percentage of spikelets having ten or more germinated pollen grains exhibited a precipitous decline (Matsui *et al.*, 1997). Heat-induced increases in floral sterility may have been responsible for the down-regulation of photosynthesis observed in rice under high temperatures and elevated [CO_2] through indirect effects associated with reductions in reproductive sink strength (Lin *et al.*, 1997).

With high night temperatures, many cowpea genotypes do not produce flowers, while others produce flowers but not pods (Ehlers and Hall, 1996). Growth chamber studies with contrasting cowpea genotypes grown in pots demonstrated that these heat-induced detrimental effects are present under either 350 or 700 μmol mol^{-1} [CO2] (Ahmed *et al.*, 1993). A heat-tolerant cowpea genotype had greater pod production under elevated [CO_2] at both high and more optimal night temperatures than a genetically similar cultivar that does not have the heat-tolerance genes. This suggests that the heat-tolerance genes may confer responsiveness to elevated [CO_2] in this species. The possibility that genes which enhance heat tolerance during reproductive development also enhance grain yield responses to elevated [CO_2] under a range of temperatures should be evaluated with other species that exhibit sensitivity to heat during reproductive development.

For cases where the photosynthetic source is particularly sensitive to heat stress, the interactive effects of elevated [CO_2] is less clear. The review of Allen (1994) indicates that for C_3 plants, photosynthetic responses to elevated [CO_2] of individual leaves often increased with increasing temperature, up to some maximum temperature, and in some cases biomass responded in the same way. Simulations based on a model of leaf photosynthesis, photorespiration and respiration predicted that the response of net canopy CO_2 uptake to elevated [CO_2] by C_3 plants could be greater at higher temperatures (Long, 1991). However, studies of Ziska and Bunce (1997) with soybean demonstrated that even though photosynthetic responses to elevated [CO_2] of individual leaves increased with increasing temperature, photosynthetic rates of whole plants did not. Ziska and Bunce (1995) also found that cultivars of soybean differed in their responsiveness to elevated [CO_2] and higher temperature with respect to photosynthesis and biomass production. Criteria that breeders could use to select plants that have enhanced photosynthetic responses to elevated [CO_2] have not been established but the study of Ziska and Bunce (1995) indicated that genetic variability for this trait may be present within species.

19.5 Crop Water Relations and Global Climate Change

The major effects of global climate change on crop water relations that are relevant to plant breeding are likely to occur due to changes in rainfall and soil drought for crops grown without irrigation. Where rainfall is reduced, breeding may have to give greater emphasis to enhancing adaptation to drought and this can be a difficult task (Blum, 1988; Eastin *et al.*, 1988). In regions where global climate change substantially reduces rainfall, it may be necessary for farmers to grow alternative crop species that have enhanced adaptation to dry environments.

Crop water relations are also influenced by transpiration rates under both well-watered and dry soil conditions (Schulze and Hall, 1982). Plants that transpire faster develop lower leaf water potentials in the short term and more rapidly deplete the water in the root zone on a long-term basis. The effects of elevated atmospheric [CO_2] on transpiration are difficult to predict, because increased leaf growth rate results in greater interception of solar radiation and could result in faster transpiration, whereas the partial stomatal closure and earlier leaf senescence that can result from elevated [CO_2] could reduce seasonal transpiration. Due to different magnitudes for these effects, elevated [CO_2] has resulted in greater seasonal water use for cotton (Samarakoon and Gifford, 1996) but less seasonal water use for spring wheat (Kimball *et al.*, 1995). The reduced water use of spring wheat could have been responsible for their greater grain yield response to elevated [CO_2] under water-limited compared with well-watered soil conditions. Any increases in air temperature would cause plants to transpire faster and experience lower leaf water potentials. Effects on crops of changes in transpiration rate, due to global climate change, do not appear to warrant developing special plant breeding procedures.

19.6 Crop Pest Relations and Global Climate Change

A major potential problem confronting farmers would be where global climate change results in the occurrence of much larger populations of troublesome insect pests in specific regions. In temperate and subtropical climates, the extent to which insect pest populations build up during the growing season can depend on how well the insect pests survive the winter and how fast their populations increase during the growing season. Relatively small increases in minimum temperatures during the winter may result in substantial increases in the survival of specific insect pests that could cause catastrophic increases in their populations during the growing season. Higher temperatures during the growing season would also aggravate this effect by increasing the development rates of insect pests. Plant breeding can contribute to solving this problem by developing cultivars with resistance to various insect pests, but this is not an easy task, in many cases, and may become more complicated due to effects on plants of elevated [CO_2].

Elevated [CO_2] has often caused increases in the C/N ratio of plant tissue (Conroy, 1992) and this reduces the nutritive value of leaves for insects. Also, elevated [CO_2] may cause changes in the concentrations of proteins that are toxic to insects and other allelochemicals (Lincoln *et al.*, 1993). These changes in tissue composition can influence the feeding behaviour and performance of insect pest herbivores. For example, in different experiments, insect herbivores ate 20–80% faster when fed foliage from plants grown under elevated compared with normal [CO_2], presumably due to lower and limiting nitrogen content of foliage from plants grown under elevated [CO_2] (Lincoln *et al.*, 1993). For optimal crop production, higher soil N levels will be needed for environments where potential productivity is enhanced by increases in atmospheric [CO_2] and this will act to decrease the C/N ratio of plant tissue. However, the critical nitrogen concentration in leaves that is required for optimal grain production may be lower under elevated atmospheric [CO_2] than under present-day conditions (Conroy, 1992), so even with higher soil N levels the C/N ratios in plant tissues still may be lower than current levels. Consequently, it is possible that under the elevated [CO_2] of the next century the extent of crop damage by insect pest herbivores may change and the extent of varietal resistance to these pests also may change.

19.7 Grain Quality and Global Climate Change

Where leaf nitrogen content is smaller for plants grown under enriched [CO_2] there may be some changes in grain quality. Reductions in N and protein content have been reported for flour produced from grain of wheat plants grown under elevated [CO_2] (Conroy *et al.*, 1994). For field-grown rice, percentage grain protein decreased with either elevated [CO_2] or increasing temperature (Ziska *et al.*, 1997). Plant breeders may have to impose selection pressure for higher protein concentration in grain to offset any decreases accompanying the increases in atmospheric [CO_2].

19.8 Conclusions

Increased grain yields may be achieved by breeding for coordinated increases in both the reproductive sink and the photosynthetic source (Evans, 1993). In applying this strategy to plant breeding, projected changes in climate must be considered. Increases in atmospheric [CO_2] will tend to make photosynthetic sources more effective per unit leaf area. Consequently, maintaining a balance between photosynthetic sources and reproductive sinks may require selecting plants with much greater reproductive sinks. Breeding to maintain an appropriate balance will be particularly important for cases where high temperatures result in greater damage to reproductive development than to the photosynthetic source. Breeding for heat tolerance during reproductive development has been effective (Hall, 1992, 1993). Genes for heat tolerance during reproductive development enhance sink strength and HI (Ismail and Hall, 1998), and there are indications that they may enhance responsiveness to elevated [CO_2] under optimal as well as high temperatures in the indeterminate crop, cowpea (Hall and Allen, 1993). For determinate cereal crops, such as wheat and rice, it is not yet known whether responsiveness to elevated [CO_2] will be influenced by selection for increases or decreases in traits such as HI or the extent of tillering. Selection for greater sink strength could indirectly enhance photosynthetic capacity and activity through minimizing the feedback effects that down-regulate the photosynthetic system. Making full use of elevated [CO_2] may also require selection to enhance those components of the photosynthetic system, other than Rubisco, that become limiting if the CO_2 assimilation per unit leaf area of C_3 species is to reach much higher rates. Selection for more open stomata may be useful, except for environments with extreme drought, since elevated [CO_2] will tend to cause partial stomatal closure. Indirect selection procedures based on the measurement of canopy temperatures using remote sensing could be more effective than direct selection based on the measurement of the stomatal conductance of individual leaves and promising progress has been made in applying this technique to spring wheat (Fischer *et al.*, 1998; Reynolds *et al.*, 1998). Adjustments in particular aspects of programmes may be needed, such as in breeding for resistance to insect pests. The damage caused by insect pests and the extent of varietal resistance may be influenced by effects of elevated [CO_2] on plant tissue composition. Plant breeders also may have to impose selection pressure for higher protein concentration in grain to offset decreases that may accompany the increases in atmospheric [CO_2]. Overall, elevated atmospheric [CO_2] and global climate change will provide both opportunities that may be exploited by plant breeding to increase productivity and some additional problems for plant breeders and other scientists to solve in the 21st century.

References

Ahmed, F.E., Hall, A.E. and DeMason, D.A. (1992) Heat injury during floral development in cowpea (*Vigna unguiculata,* Fabaceae). *American Journal of Botany* 79, 784–791.

Ahmed, F.E., Hall, A.E. and Madore, M.A. (1993) Interactive effects of high temperature and elevated carbon dioxide concentration on cowpea [*Vigna unguiculata* (L.) Walp.]. *Plant, Cell and Environment* 16, 835–842.

Allen, L.H. Jr (1994) Carbon dioxide increase: direct impacts on crops and indirect effects mediated through anticipated climatic changes. In: Boote, K.J., Bennett, J.M., Sinclair, T.R. and Paulsen, G.M. (eds) *Physiology and Determination of Crop Yield.* Crop Science Society of America, Madison, Wisconsin, pp. 425–459.

Amani, I., Fischer, R.A. and Reynolds, M.P. (1996) Canopy temperature depression associated with yield of irrigated spring wheat cultivars in a hot climate. *Journal of Agronomy and Crop Science* 176, 119–129.

Amthor, J.S. (1998) Perspective on the relative insignificance of increasing atmospheric CO_2 concentration to crop yield. *Field Crops Research* 58, 109–127.

Arp, W.J. (1991) Effects of source-sink relations on photosynthetic acclimation to elevated CO_2. *Plant, Cell and Environment* 14, 869–875.

Badger, M. (1992) Manipulating agricultural plants for a future high CO_2 environment. *Australian Journal of Botany* 40, 421–429.

Baker, J.T. and Allen, L.H. Jr (1993) Contrasting crop species responses to CO_2 and temperature: rice, soybean and citrus. *Vegetatio* 104/105, 239–260.

Baker, J.T., Allen, L.H. Jr, Boote, K.J., Jones, P. and Jones, J.W. (1989) Responses of soybean to air temperature and carbon dioxide concentration. *Crop Science* 29, 98–105.

Barnola, J.M., Raynaud, D., Korotkevich, Y.S. and Lorius, C.D. (1987) Vostok ice core provides 160,000 year record of atmospheric CO_2. *Nature* 329, 408–414.

Berry, J.A. and Björkman, O. (1980) Photosynthetic responses and adaptation to temperature in higher plants. *Annual Review of Plant Physiology* 31, 491–543.

Blum, A. (1988) *Plant Breeding for Stress Environments,* CRC Press, Boca Raton, Florida.

Bolin, B. (1998) The Kyoto negotiations on climate change: a science perspective. *Science* 279, 330–331.

Bowes, G. (1993) Facing the inevitable: plants and increasing atmospheric CO_2. *Annual Review of Plant Physiology and Plant Molecular Biology* 44, 309–332.

Condon, A.G. and Hall, A.E. (1997) Adaptation to diverse environments: variation in water-use efficiency within crop species. In: Jackson, L.E. (ed.) *Ecology in Agriculture.* Academic Press, San Diego, pp. 79–116.

Conroy, J.P. (1992) Influence of elevated atmospheric CO_2 concentrations on plant nutrition. *Australian Journal of Botany* 40, 445–456.

Conroy, J.P., Seneweera, S., Basra, A.S., Rogers, G. and Nissen-Wooler, B. (1994) Influence of rising atmospheric CO_2 concentrations and temperature on growth, yield and grain quality of cereal crops. *Australian Journal of Plant Physiology* 21, 741–758.

Eastin, J.D., Castleberry, R.M., Gerik, T.J., Hultquist, J.H., Mahalakshmi, V., Ogunlela, V.B. and Rice, J.R. (1983) Physiological aspects of high temperature and water stress. In: Raper, C.D. Jr and Kramer, P.J. (eds) *Crop Reactions to Water and Temperature Stresses in Humid Temperate Climates*, Westview Press, Boulder, Colorado, pp. 91–112.

Eastin, J.D., Verma, P.K., Dhopte, A., Witt, M.D., Krieg, D.R. and Hatfield, J.L. (1988) Stress screening and sorghum improvement. In: Unger, P.W., Jordan, W.R., Sneed, T.V. and Jensen, R.W. (eds) *Challenges in Dryland Agriculture – a Global Perspective*, Proceedings of an International Conference on Dryland Farming, 15–19 August 1988, Amarillo/Bushland, Texas, pp. 716–723.

Ehlers, J.D. and Hall, A.E. (1996) Genotypic classification of cowpea based on responses to heat and photoperiod. *Crop Science* 36, 673–679.

Ehlers, J.D. and Hall, A.E. (1998) Heat tolerance of contrasting cowpea lines in short and long days. *Field Crops Research* 55, 11–21.

Evans, L.T. (1993) *Crop Evolution, Adaptation and Yield.* Cambridge University Press, Cambridge, UK.

Fischer, R.A. (1985) Number of kernels in wheat crops and the influence of solar radiation and temperature. *Journal of Agricultural Science, Cambridge* 105, 447–461.

Fischer, R.A., Rees, D., Sayre, K.D., Lu, Z., Condon, A.G., Larque-Saavedra, A. and Zeiger, E. (1998) Wheat yield progress associated with higher stomatal conductance and photosynthetic rate, and cooler canopies. *Crop Science* 38, 1467–1475.

Gifford, R.M. (1986) Partitioning of photoassimilate in the development of crop yield. In: Lucas, W.J. and Cronshaw, J. (eds) *Phloem Transport.* Alan R. Liss, New York, pp. 535–549.

Goudriaan, J. and Unsworth, M.H. (1990) Implications of increasing carbon dioxide and climate change for agricultural productivity and water resources. In: Kimball, B.A., Rosenberg, N.R. and Allen, L.H. Jr (eds) *Impact of Carbon Dioxide, Trace Gases and Climate Change on Global Agriculture*, Special Publication Number 53, American Society of Agronomy, Madison, Wisconsin, pp 111–130.

Hall, A.E. (1992) Breeding for heat tolerance. *Plant Breeding Reviews* 10, 129–168.

Hall, A.E. (1993) Physiology and breeding for heat tolerance in cowpea, and comparisons with other crops. In: Kuo, C.G. (ed.) *Adaptation of Food Crops to Temperature and Water Stress.* Asian Vegetable Research and Development Center, Taipei, Taiwan, pp. 271–284.

Hall, A.E. and Allen, L.H. Jr (1993) Designing cultivars for the climatic conditions of the next century. In: Buxton, D.R., Shibles, R., Forsberg, R.A., Blad, B.L., Asay, K.H., Paulsen, G.M. and Wilson, R.F. (eds) *International Crop Science I.* Crop Science Society of America, Madison, Wisconsin, pp. 291–297.

Heckathorn, S.A., Downs, C.A., Sharkey, T.D. and Coleman J.S. (1998) The small, methionine-rich chloroplast heat-shock protein protects photosystem II electron transport during heat stress. *Plant Physiology* 116, 439–444.

Houghton, J.T., Meira-Filho, L.G., Callander, B.A., Harris, N., Kattenberg, A. and Maskll, K. (1996) *IPCC Climate Change Assessment 1995. The Science of Climate Change,* Cambridge University Press, Cambridge, UK, 572 pp.

Ismail, A.M. and Hall, A.E. (1998) Positive and potential negative effects of heat-tolerance genes in cowpea. *Crop Science* 38, 381–390.

Jouzel, J., Barkov, N.I. and Barnola, J.M. (1993) Extending the Vostok ice-core record of the paleoclimate to the penultimate glacial period. *Nature* 364, 407–412.

Kerr, R.A. (1986) Greenhouse warming still coming. *Science* 232, 573–575.

Kimball, B.A. (1983) Carbon dioxide and agricultural yield: an assemblage and analysis of 430 prior observations. *Agronomy Journal* 75, 779–788.

Kimball, B.A. (1985) Adaptation of vegetation and management practices to a higher carbon dioxide world. In: Strain, B.R. and Cure, J.D. (eds) *Direct Effects of Increasing Carbon Dioxide on Vegetation.* US Department of Energy, Carbon Dioxide Research Division, DOE/ER-0238, Washington, DC, pp. 185–204.

Kimball, B.A., Pinter, P.J. Jr, Garcia, R.L., LaMorte, R.L., Wall, G.W., Hunsaker, D.J., Wechsung, G., Wechsung, F. and Kartschall, T. (1995) Productivity and water use of wheat under free-air CO_2 enrichment. *Global Change Biology* 1, 429–442.

Kukla, G. and Karl, T.R. (1993) Nighttime warming and the greenhouse effect. *Environmental Science and Technology* 27, 1468–1474.

Lin, W., Ziska, L.H., Namuco, O.S. and Bai, K. (1997) The interaction of high temperature and elevated CO_2 on photosynthetic acclimation of single leaves of rice *in situ. Physiologia Plantarum,* 99, 178–184.

Lincoln, D.E., Fajer, E.D. and Johnson, R.H. (1993) Plant–insect herbivore interactions in elevated CO_2 environments. *Trends in Ecology and Evolution* 8, 64–68.

Long, S.P. (1991) Modification of the response of photosynthetic productivity to rising temperature by atmospheric CO_2 concentrations: has its importance been underestimated? *Plant, Cell and Environment* 14, 729–739.

Lu, Z.-M. and Zeiger, E. (1994) Selection for higher yields and heat resistance in Pima cotton has caused genetically determined changes in stomatal conductance. *Physiologia Plantarum* 92, 273–278.

Manderscheid, R. and Wiegel, H.J. (1997) Photosynthetic and growth responses of old and modern spring wheat cultivars to atmospheric CO_2 enrichment. *Agriculture, Ecosystems and Environment* 64, 65–73.

Matsui, T., Namuco, O.S., Ziska, L.H. and Horie, T. (1997) Effects of high temperature and CO_2 concentration on spikelet sterility in indica rice. *Field Crops Research* 51, 213–219.

Maxon Smith, J.W. (1977) Selections for response to CO_2-enrichment in glasshouse lettuce. *Horticultural Research* 17, 15–22.

Mayeaux, H.S., Johnson, H.B., Polley, H.W. and Malone, S.R. (1997) Yield of wheat across a subambient carbon dioxide gradient. *Global Change Biology* 3, 269–278.

McConnaughy, K.D.M., Bernston, G.M. and Bazzaz, F.A. (1993) Limitations to CO_2-induced growth enhancement in pot studies. *Oecologia* 94, 50–557.

Morell, M.K., Paul, K., Kane, H.J. and Andrews, T.J. (1992) Rubisco: maladapted or misunderstood? *Australian Journal of Botany* 40, 431–441.

Moya, T.B., Ziska, L.H., Namuco, O.S. and Olsyk, D. (1998) Growth dynamics and genotypic variation in tropical, field-grown paddy rice (*Oryza sativa* L.) in response to increasing carbon dioxide and temperature. *Global Change Biology* 4, 645–656.

Nielsen, C.L. and Hall, A.E. (1985a) Responses of cowpea (*Vigna unguiculata* (L.) Walp.) in the field to high night temperatures during flowering. I. Thermal regimes of production regions and field experimental system. *Field Crops Research* 10, 167–179.

Nielsen, C.L. and Hall, A.E. (1985b) Responses of cowpea (*Vigna unguiculata* (L.) Walp.) in the field to high night temperatures during flowering. II. Plant responses. *Field Crops Research* 10, 181–196.

Poorter, H. (1993) Interspecific variation in the growth responses of plants to an elevated ambient CO_2 concentration. *Vegetatio* 104/105, 77–97.

Poorter, H., Roumet, C. and Campbell, B.D. (1996) Interspecific variation in the growth responses of plants to elevated CO_2: a search for functional types. In: Körner, C. and Bazzaz, F.A. (eds) *Carbon Dioxide, Populations and Communities.* Academic Press, San Diego, pp. 375–412.

Reddy, K.R., Hodges, H.F., McKinion, J.M. and Wall, G.W. (1992) Temperature effects on Pima cotton growth and development. *Agronomy Journal* 84, 237–243.

Reddy, K.R., Hodges, H.F. and McKinion, J.M. (1995) Carbon dioxide and temperature effects on Pima cotton development. *Agronomy Journal* 87, 820–826.

Reddy, K.R., Hodges, H.F. and McKinion, J.M. (1997) A comparison of scenarios for the effect of global climate change on cotton growth and yield. *Australian Journal of Plant Physiology* 24, 707–713.

Reynolds, M.P., Balota, M., Delgado, M.I.B., Amani, I. and Fischer, R.A. (1994) Physiological and morphological traits associated with spring wheat yield under hot, irrigated conditions. *Australian Journal of Plant Physiology* 21, 717–730.

Reynolds, M.P., Singh, R.P., Ibrahim, A., Ageeb, O.A.A., Larqué-Saavedra, A. and Quick, J.S. (1998) Evaluating physiological traits to complement empirical selection for wheat in warm environments. *Euphytica* 100, 85–94.

Rogers, G.S., Milham, P.J., Gillings, M. and Conroy, J.P. (1996a) Sink strength may be the key to growth and nitrogen responses in N-deficient wheat at elevated CO_2. *Australian Journal of Plant Physiology* 23, 253–264.

Rogers, G.S., Milham, P.J., Thibaud, M.-C. and Conroy, J.P. (1996b) Interactions between rising CO_2 concentration and nitrogen supply in cotton. I. Growth and leaf nitrogen concentration. *Australian Journal of Plant Physiology* 23, 119–125.

Samarakoon, A.B. and Gifford, R.M. (1996) Water use and growth of cotton to elevated CO_2 in wet and drying soil. *Australian Journal of Plant Physiology* 23, 63–74.

Schulze, E.-D. and Hall, A.E. (1982) Stomatal responses, water loss and CO_2 assimilation rates of plants in contrasting environments. In: Lange, O.L., Nobel, P.S., Osmond, C.B. and Zeigler, H. (eds) *Encyclopedia of Plant Physiology, Physiological Plant Ecology,* Vol. 12B, Springer-Verlag, New York, pp. 180–230.

Sundquist, E.T. (1986) Geological analogs: their value and limitations in carbon dioxide research. In: Trabalka, J.R. and Reichle, D.E. (eds) *The Changing Carbon Cycle.* Springer-Verlag, New York, pp. 371–402.

Wardlaw, I.F., Dawson, I.A., Munibi, B. and Fewster, R. (1989) The tolerance of wheat to high temperatures during reproductive growth. I. Survey procedures and general response patterns. *Australian Journal of Agricultural Research* 40, 1–13.

Ziska, L.H. and Bunce, J.A. (1995) Growth and photosynthetic response of three soybean cultivars to simultaneous increases in growth temperature and CO_2. *Physiologia Plantarum* 94, 575–584.

Ziska, L.H. and Bunce, J.A. (1997) The role of temperature in determining the stimulation of CO_2 assimilation at elevated carbon dioxide concentration in soybean seedlings. *Physiologia Plantarum* 100, 126–132.

Ziska, L.H. and Manalo, P.A. (1996) Increasing night temperature can reduce seed set and potential yield of tropical rice. *Australian Journal of Plant Physiology* 23, 791–794.

Ziska, L.H., Manalo, P.A. and Ordonez, R.A. (1996a) Intraspecific variation in the response of rice (*Oryza sativa* L.) to increased CO_2 and temperature: growth and yield response of 17 cultivars. *Journal of Experimental Botany* 47, 1353–1359.

Ziska, L.H., Weerakoon, W., Namuco, O.S. and Pamplona, R. (1996b) The influence of nitrogen on the elevated CO_2 response in field-grown rice. *Australian Journal of Plant Physiology* 23, 45–52.

Ziska, L.H., Namuco, O., Moya, T. and Quilang, J. (1997) Growth and yield response of field-grown tropical rice to increasing carbon dioxide and air temperature. *Agronomy Journal* 89, 45–53.

20 Role of Biotechnology in Crop Productivity in a Changing Environment

NORDINE CHEIKH[1], PHILIP W. MILLER[2] AND GANESH KISHORE[2]

Monsanto Company, [1]1920 Fifth Street, Davis, CA 95616, USA; [2]700 Chesterfield Parkway North, Chesterfield, St Louis, MO 63198, USA

20.1 Introduction

The world today is faced with great challenges to produce adequate food, feed and industrial products for the globe's 6 billion people. Nearly 200 new residents are added to our planet every minute and approximately 80 million every year. Furthermore, the area of arable land available to feed this population is shrinking and is subject to significant loss of topsoil and groundwater. Future population growth will continue, most noticeably in developing nations. Closely linked with this population explosion and subsequent increasing poverty and demand for food is the potential decline in quality of the environment in which food is grown and produced.

The complex triangle of poverty has been described as the assembly of environmental decline, rapid population growth and related problems that hinder development. In developed countries, urban sprawl has covered some of the best arable land, and poor land management has slowly degraded the quality of the agricultural land. All of these negative trends will be compounded by the predicted climatic change resulting from global warming.

Crop growth and yield potential and distribution of marketable products are limited in many regions of the world by a variety of environmental abiotic and biotic stress factors and challenges. Consequently, there is a substantial deficit between potential and realized crop yields.

There are about 14 Gha ($\times 10^9$) of land in the world, but only about 1.4 Gha are arable, non-stressed crop land. Of the remaining area, 2.9 Gha are under mineral stress (1.0 Gha salt stress), 3.7 Gha endure drought stress, 1.6 Gha are limited by excess water, 3.2 Gha have very shallow soil profiles, and 2.0 Gha are subject to permanent freezing. The total arable and potentially arable land, which could produce reasonable crop harvest, is estimated at about 3.2 Gha (Christiansen, 1982).

 Climate Change and Global Crop Productivity (eds K.R. Reddy and H.F. Hodges)

Farmers, people in agribusiness and consumers in recent history have looked to improvements and new discoveries in cultural practices (irrigation, fertilizers, and pesticides), breeding and other agricultural technologies to increase yield potential of crops, maintain yield stability in the face of environmental stresses and provide other desirable characteristics to improve food and feed. The history of agricultural development and improvement in crop production can be characterized as an evolution through four distinct eras: the labour era, the mechanical era, the chemical era, and the present biotechnological era. Advances in plant breeding have served as a backbone throughout this evolution. The first improvements in resistance or tolerance to stress involved determining the genetics of inheritance of desirable traits. These efforts have been generally successful in providing resistance to pathogens and insects and in developing many desirable agronomic characteristics. However, many of the successes have been overshadowed by the 8- to 12-year process of conventional cross-breeding and the likelihood of these traits being linked to inheritance of non-desirable traits.

Molecular genetic mapping of the plant genome has provided RFLP (restriction fragment length polymorphism) and RAPD (rapid amplified polymorphic DNA) markers which are sufficiently closely linked to known resistance genes to make their isolation, sequencing and transfer likely in the near future. Biotechnology, or production of transgenic plants with novel genes imparting stress tolerance or resistance, offers possibilities of a quantum leap in the ability to confer or improve stress tolerance in crop plants and, in combination with conventional breeding techniques, provide new solutions to some of these 'old problems'. The development and delivery of these technologies give considerable optimism for the future. Agricultural biotechnology will allow increases in crop yields and maintenance of yield stability without increasing land usage. Furthermore, farmers can exert some degree of control over weeds, insects and disease-causing pathogens that threaten crop production and distribution. Most importantly, farmers may be provided with the flexibility to plant crops that have been genetically altered to tolerate and grow in 'less than ideal' soils or weather conditions and in varied geographical locations.

There are several obvious opportunities for improving crop performance and productivity through optimization of cultural practices, plant breeding, and new developments in biotechnology. Certainly, it is quite clear that one important contribution of plant genetic engineering will be to improve stress-tolerance traits in crops of commercial value. The opportunities for enhancing crop performance under stress conditions lie in identifying key traits that require enhancement, stress-relieving candidate genes and the appropriate plant stage of development where enhancement should occur. This process of gene identification and gene expression patterns associated with quantitative genetic traits in crops is becoming more attractive and promising as the scientific community develops and gains access to large-scale genomics resources such as EST (expressed sequence tag) databases, high-throughput gene expression profiling technologies and genomic mapping information. This chapter focuses on key examples from recent publications in which stress tolerance of plants has been improved through gene transfer, or at least identifies cases in which genetic engineering of plants has shed some light on the mechanisms by which stress tolerance or resistance is conferred.

To improve the ability of crops to cope with existing and new stresses, it is imperative to develop a basic understanding of the mechanisms and processes by which plants respond to stress. Mechanisms of plant response or adaptation and their key processes that are affected by environmental stresses are summarized in Fig. 20.1. This diagram illustrates that many components that are part of the initial stress perception and response could be similar. If multiple stress perception and response mechanisms are similar, this makes certain transgenic approaches very powerful – namely, those in which manipulations of one component could confer tolerance or resistance to several environmental stresses. This chapter focuses mainly on advances in biotechnology to improve plant response to abiotic stresses.

20.2 Biotechnology and Abiotic Stresses

Recent advances in plant and microbial genomics have provided new and exciting molecular tools that allow scientists to dissect the biochemical and molecular mechanisms of stress tolerance or resistance. This information can be utilized to develop new crop plants that are protected from the impact of abiotic stresses. Since many of these major abiotic stresses arise as a result of a common biochemical phenomenon, efforts to improve tolerance to one

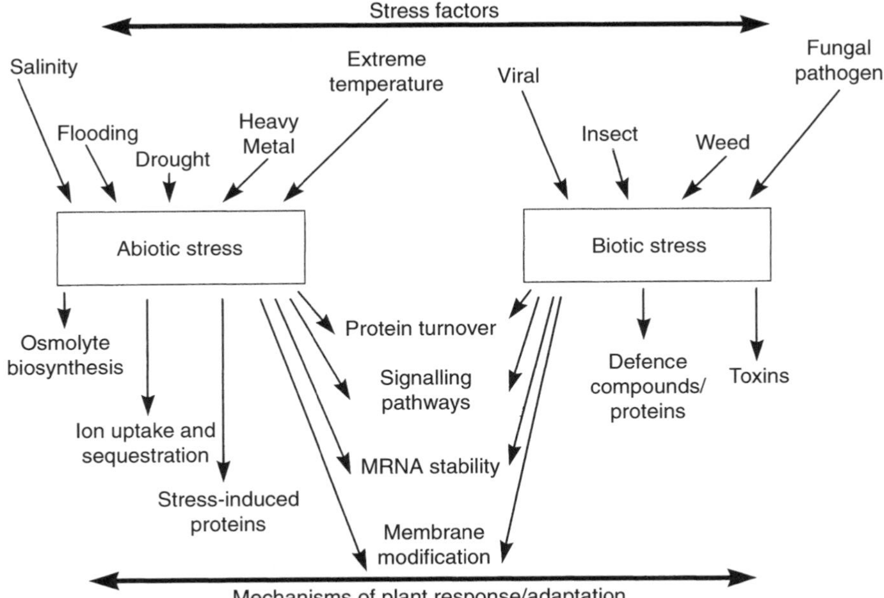

Fig. 20.1. Mechanisms of plant response or adaptation and key processes that are affected by different environmental stresses

abiotic stress have potential to confer tolerance to other abiotic stresses. Recently, new scientific efforts have focused on modifying higher plants with stress tolerance-conferring genes. These efforts have demonstrated enhanced abiotic stress tolerance to drought, salinity, temperature, oxidative, pH, heavy metal and flooding stress, and have also provided validation to the concept of enhancing abiotic stress tolerance in crop plants via biotechnology.

20.2.1 Salinity and drought tolerance

Drought and salinity are the major and most widespread environmental stresses that substantially constrain crop productivity in both non-irrigated and irrigated agriculture (Epstein *et al.*, 1980). The detrimental impact of salinity in agriculture is exacerbated by irrigation management practices used to increase crop outputs. Cellular water deficits due to salinity, drought or temperature can result in changes in cell volume, including loss of turgor, solute concentration, ion displacement, membrane structure and integrity and protein denaturation, as well as alterations of various cellular processes (Bray, 1997). Due to the large economic impact of these abiotic stresses, much of the initial biotechnological research has centred on drought and salinity tolerance. Recent metabolic engineering reports have demonstrated that the accumulation of low molecular weight osmoprotectants in the form of sugars (fructan), polyhydroxylated sugar alcohols (polyols), amino acids such as proline, and quartenary amines such as glycinebetaine provide a mechanism for osmoprotection and osmotic adjustment in plants.

Pilon-Smits *et al.* (1995) demonstrated that the drought tolerance of tobacco plants could be improved by engineering the plants to produce increased fructan sugar levels. Tobacco (*Nicotiana tabacum*) plants engineered with the bacterial *sacB* gene, which encoded a fructosyl transferase enzyme targeted to the plant vacuole, exhibited a 55% increase in growth rate and a 59% increase in biomass production relative to wild-type control plants under water-deficit conditions. The researchers concluded that the introduction of fructans into non-fructan-producing species will potentially enhance resistance to drought stress.

Trehalose, a non-reducing disaccharide, has been implicated as a component in the protective role in plants that can withstand complete dehydration (Drennan *et al.*, 1993). Overproduction of trehalose in transgenic tobacco was achieved by introducing the trehalose-6-phosphate synthase gene of yeast under the control of the Rubisco small subunit promoter (Holstrom *et al.*, 1996). Trehalose-producing plants exhibited both increased water retention and drought tolerance at various stages of development, though the cellular levels of trehalose were small (0.5 mM in cytosol).

Many photosynthetic organisms accumulate proline naturally in response to osmotic stress. The accumulation of proline is believed to serve as a redox sink, radical scavenger, pH stabilizer and protective solute that provides drought and salinity tolerance to the cells (Kishor *et al.*, 1995). The mothbean structural gene for Δ^1-pyrroline-5-carboxylate synthase (a bifunctional enzyme that catalyses the conversion of glutamate to Δ^1-pyrroline-5-carboxylate, the immediate precursor to proline) has been introduced into tobacco (Kishor *et al.*, 1995). The transgenic plants produced 10- to 18-fold more proline and exhibited a relatively higher osmotic potential in the leaf sap than controls under drought stress. In addition, the proline-overproducing plants had enhanced biomass and flower development under salinity-stress conditions.

The ability to engineer production of the osmoprotectant mannitol in higher plants has been demonstrated in both *Arabidopsis* (Thomas *et al.*, 1995) and tobacco (Tarczynski *et al.*, 1993). Seeds from *Arabidopsis* plants overexpressing the bacterial *mltd* gene encoding mannitol 1-phosphate dehydrogenase exhibited enhanced germination rates in the presence of elevated salt concentrations. Similarly, tobacco plants expressing the *mltd* gene produced a maximum concentration of mannitol of 100 mM, which led to increased biomass production under high saline conditions relative to the control.

Overproduction of cyclic sugar alcohols (cyclitols, such as ononitol) by expression of the *lmt1* gene that encodes a myo-inositol *o*-methyl transferase in tobacco was used by Vernon *et al.* (1993) as a model system to test the role of cyclitol overproduction in conferring tolerance to osmotic stress. However, no conclusive evidence was presented to correlate overproduction of cyclitols and stress tolerance in these transgenic plants.

Saneko *et al.* (1995) demonstrated that glycinebetaine-containing maize (*Zea mays* L.) lines were less sensitive to salinity stress than were mutants deficient in glycinebetaine production, thus providing evidence of the protective potential of this osmolyte. By introducing the *Escherichia coli betA* gene that encodes choline dehydrogenase into tobacco, Lullis *et al.* (1996) demonstrated increased salinity tolerance in the transgenic plants. Tobacco plants expressing the *betA* gene exhibited increased glycinebetaine production that resulted in an 80% increase in salt tolerance over wild-type plants, as measured by dry weight accumulation. Similar results have been achieved in *Arabidopsis* by overexpressing the bacterial *codA* gene for choline oxidase (Hayashi *et al.*, 1997). *Arabidopsis* plants overexpressing *codA* have also shown tolerance to cold stress and exhibited stability of photosystem II and normal germination under high salt.

In addition to small molecules playing a role in plant stress tolerance, a series of proteins known as late embryogenesis abundant (LEA) proteins, first identified in plant embryos undergoing desiccation, have been implicated as having a role in stress tolerance. An LEA protein gene, *HVA1*, from barley (*Hordeum vulgare* L.) was introduced into rice (*Oryza sativa* L.) plants under the regulation of the rice actin promoter (Xu *et al.*, 1996). Transgenic *HVA1* rice plants showed significant increases in tolerance to water deficits and salinity stress. Increased tolerance was exhibited by increased growth rates under stress, delayed development of damage symptoms, and improved recovery upon removal of the stress.

Although the ability of the aforementioned effectors to increase water-deficit tolerance in plants has been demonstrated, it is possible that some of these single effectors may not be fully adequate to confer substantial stress tolerance to a wide range of plants. The ability to control several effectors in a coordinated manner may be required to achieve consistent stress tolerance. To this end, recent engineering efforts have focused on modulating signal transduction cascades and transcriptional regulators as a means of altering plant endogenous stress-coping mechanisms. Pardo *et al.* (1998) demonstrated that overexpression of a truncated form of calcineurin (CaN, a Ca^{2+}-calmodulin-dependent protein phosphatase involved in salt-stress signal transduction in yeast) mediated salt adaptation in transgenic tobacco plants. The modified plants exhibited less perturbation of root growth and increased shoot survival. The increased tolerance of plants with the *CaN* transgene in the segregating population provides evidence that a common signal pathway exists in yeast and plants, and that heterologous protein is capable of modulating stress tolerance.

Genetic engineering studies with melon cultivars (*Cucumis melo*) to express the yeast salt-tolerance gene *HAL1* (a gene encoding a water-soluble protein that modulates monovalent ion channels) showed that placing transgenic melon plants under high salt conditions reduced root and vegetative growth. Plants containing this trait consistently had a higher level of tolerance than the control plants without this trait (Bordas *et al.*, 1997).

20.2.2 Temperature stress

Plants vary tremendously in their ability to survive both high and low temperature extremes. What determines the ability of plants to survive or adapt to temperature extremes is only partially understood, but the need to alter crop plants to withstand temperature extremes is evident. As weather patterns change due to human impact, it will be increasingly important to develop crop species that function in broader geographical and temperature ranges.

Low temperature is one of the major environment stresses that limits plant growth. Significant advances in understanding cold tolerance and modifying cold susceptibility in plants have been made using biotechnology. The increase of plastid membrane fatty acid desaturation has been correlated with increased cold tolerance in plants (Steponkus *et al.*, 1993). Several attempts have been made to modify cold tolerance via altering the fatty acid composition of plants. Transgenic tobacco overexpressing the *Arabidopsis* ω-3 fatty acid desaturase (*Fad7*) gene were analysed for altered cold tolerance (Kodama *et al.*, 1994). When exposed to 1°C temperatures for 7 days and then returned to optimal temperatures, transgenic *Fad7* plants did not exhibit the growth suppression and chlorosis that was displayed by the wild-type controls. Thus, the transgenic alteration appeared to have protected the tobacco from cold injury. The cold tolerance was increased due to the overexpressing of the *Fad7* gene.

A broad specificity Δ9-desaturase gene (*Des9*) from *Anacystis nidulans* was fused to a plant chloroplast-targeting sequence, and was introduced into tobacco by Ishizaki-Nishizawa *et al.* (1996). They demonstrated a 17-fold increase of Δ9-monosaturated fatty acids in transgenic plants. When these plants were exposed to 1°C for 11 days, they showed no signs of chlorosis. This suggests that plants with the increased Δ9-monosaturated fatty acids were protected from low-temperature injury.

Since freezing temperatures modify the cellular water balance, plants that normally show drought or salinity tolerance also have tolerance to subfreezing temperatures. Transgenically modified plants that have increased drought and salinity tolerance, apparently because of increased amounts of low molecular weight osmolytes, also have increased low-temperature tolerance. *Arabidopsis* plants overexpressing the *codA* gene exhibited elevated levels of glycine-betaine and proved to have both improved salinity tolerance and enhanced cold tolerance (Hayashi *et al.*, 1997). When the bacterial *betA* gene for choline dehydrogenase was introduced into potato (*Solanum tuberosum* L.), a cold-tolerant phenotype was observed (Holmberg and Bulow, 1998).

Many of the genes induced in cold-acclimated plants encode hydrophilic proteins (COR proteins) of undetermined function and these are believed to play a key role in cold tolerance (Thomashow, 1998). Transgenic *Arabidopsis* plants expressing one of the COR proteins (COR15am) constitutively showed that chloroplasts were 1–2°C more tolerant than non-transgenic control plastids (Artus *et al.*, 1996).

Formation of ice crystals in plant cells exposed to transient freezing temperatures leads to a mechanical disruption that kills the cells and can impact both crop productivity and harvested produce quality. Many freezing-tolerant plants contain antifreeze proteins that lower the temperature at which cellular ice forms (Thomashow, 1998). Antifreeze proteins in the blood of polar fishes has been shown to inhibit ice nucleation. Hightower *et al.* (1991) demonstrated that introduction of a gene encoding a fish antifreeze protein (AFP) into tobacco inhibited ice recrystallization in the transgenic tissue. Although no clear demonstration of freezing tolerance was demonstrated in these transgenic plants, these experiments provide some validation for using AFPs to confer tolerance to rapid freezing in plants.

Less effort has been put into engineering high-temperature tolerance in plants than low-temperature tolerance. Often, breeders and physiologists have associated heat stress with drought stress when assessing yield stability and crop performance in the field. It is believed that high temperature stress on plant growth and development occurs in photosynthetic functions and reproductive processes such as flowering, pollination and seed set. Plant cells respond to heat stress by rapidly accumulating heat shock proteins (HSPs). Although there is only correlative evidence for HSPs protecting cells from high-temperature stress, attempts have been made with some success to modify plant thermotolerance by overexpressing HSP–protein fusions. Lee *et al.* (1995) and Hinderhofer *et al.* (1998) have demonstrated that the basal thermotolerance of *Arabidopsis* can be increased when HSP-reporter fusions genes are introduced into plants.

Jaglo-Ottosen *et al.* (1998) demonstrated the ability to engineer freezing tolerance in *Arabidopsis* by constitutively expressing the *Arabidopsis* low-temperature transcription activator protein CBF-1. By modifying the expression of this protein, identified as one of the master switches of cold-tolerance inducing genes, the researchers demonstrated that it is possible to redirect the genetic resources of a plant to be cold tolerant in the absence of an acclimation period. This lends credence to the ability to modify thermotolerance and other stress responses via engineering of plant signal and regulatory pathways.

20.2.3 Heavy metal stress

Over the past decade, modern agricultural practices such as the use of sewage sludge have resulted in contamination of arable soils with heavy metals. In addition, runoff from mining and smelting operations has produced large areas of land contaminated with copper (Cu) and zinc (Zn) (Petalino and Collins, 1984). Aluminium (Al) is the most common metal in the earth's crust (predominantly in the form of insoluble aluminosilicates). When present in the solubilized form (primarily as Al^{3+}) in acid soils with pH of 5.5 or less, Al becomes toxic to many crop plants and presently is believed to represent a major obstacle to crop yield and productivity on 40% of the world's arable soils (Ryan *et al.*, 1995). Soil acidification is accelerated by certain farming practices and by occurrence of acid rain. In order to maintain productivity in soils that are adversely affected by increased levels of these pollutants, consistent applications of lime are required. This remedial practice may itself create runoff pollution, thus generating additional water quality issues. The increasing levels and areas of contaminated soils require the development of crop plants that are capable of growth in the presence of heavy metals and the sequestration of heavy metals in non-edible plant parts. This must be done through either conventional breeding or genetic engineering.

The primary effect of Al toxicity on plants is the inhibition of root growth. This results in nutrient deficiency and reduction of growth, and thus results in a negative impact on crop yield (Ryan *et al.*, 1995). One mechanism that some Al-tolerant plant species are postulated to utilize is the release of organic acids such as citric acid, which chelates Al outside the cell plasma membrane, thus preventing its uptake (Miyasaka *et al.*, 1991). De la Fuente *et al.* (1997) demonstrated that transgenic tobacco and papaya (*Carica papaya* L.) plants, engineered to overproduce citrate, resulted in tolerance to significant levels of Al. Tobacco and papaya plants were engineered with the bacterial gene that encodes citrate synthase (*CSb*) under the regulation of a constitutive promoter. Plants expressing the *CSb* gene exhibited a tenfold higher level of citrate in their roots and exhibited root growth at concentrations of 300 μM Al, whereas the controls showed severe inhibition at 50 μM.

An alternative solution to heavy metal tolerance is to bind and sequester the heavy metals within the plant to prevent their toxic effects, thus imparting heavy-metal resistance. Several researchers have utilized this approach by engineering plants with low molecular weight metal-binding proteins known as metallothioneins. The metallothioneins (MTs) are found in mammals and fungi and function to detoxify heavy metals in these systems. Transgenic brassica and tobacco plants have been created that constitutively express a mammalian *MT-II* gene (Misra and Gedamu, 1989). Plants bearing these constructs were shown to grow on toxic concentrations of cadmium (Cd) up to 100 μM, whereas controls exhibited severe root/shoot stunting and chlorosis. This approach was also used to engineer tobacco plants with the mouse (*MT-I*) gene under the control of a constitutive promoter (Pan *et al.*, 1994). Transgenic *MT-1* tobacco plants were unaffected by concentrations of cadmium up to 200 μM, whereas control plants exhibited chlorosis at 10 μM. These early experiments provide promising evidence that crop plants can be engineered to grow in these heavy-metal-laden soils.

20.2.4 Oxidative stress

A frequent consequence of environmental stress is the phenomenon of oxidative stress. These oxidative events arise from the disruption of various cellular responses that generate reactive oxygen species which cause oxidative damage in the cell. These oxidative stresses can also arise directly from exposure to ozone and pesticides. Although the plant system for protection against secondary oxidative stress damage is complex, the key enzymatic defences are superoxide dismutases (SODs) (Bowler *et al.*, 1992). Three different classes of SODs have been identified in plants and are separated on the basis of their cofactor metal: (i) copper/zinc SOD in cytosol and chloroplast; (ii) iron SOD in chloroplast; and (iii) manganese SOD in the mitochondria (Bowler *et al.*, 1994). Several researchers have reported improved performance in tobacco and lucerne (*Medicago sativa* L.) plants under oxidative stress by overexpressing distinct *SOD* genes targeted to plant organelles (Van Camp *et al.*, 1994; Aono *et al.*, 1995; Slooten *et al.*, 1995; Mckersie *et al.*, 1996; Van Camp *et al.*, 1996). In addition, Aono *et al.* (1995) demonstrated that a combination of *SOD* and the gene for glutathione reductase exhibited greater protection against oxidative stress than either gene alone.

20.3 Conclusions

Increasingly, biotechnology is providing new solutions to old agronomic problems. We have learned in recent years to improve crops by introducing herbicide- or insect-resistance. We are discovering and understanding how to optimize metabolic pathways and physiological processes in plants to increase crop yield potential and crop production under challenging agricultural environments. Efforts are also under way to improve crop quality and nutritional value for human consumption as well as animal feed.

Molecular biology tools will never replace the input and role of crop breeders in improving agronomic traits, but these tools will enable them to be more responsive in both time and breadth of environmentally sensitive traits to meet agricultural market needs and opportunities. Biotechnology tools combined with conventional breeding should position us to be able to take greater care of the production environment and allow us to achieve adequate food production and security for the growing world population.

Recent advancements in biotechnology are providing the research communities with new tools such as genomics and proteomics that will allow researchers to discover new genes and understand their function in higher numbers, and with greater speed and more precision. The knowledge gained from these technologies has mushroomed in recent years and is bringing more opportunities and tools to solve agricultural problems that once were hard to approach or understand.

References

Aono, M., Saji, H., Sakamoto, A., Tanaka, K., Kondo, N. and Tanaka, K. (1995) Paraquat tolerance of transgenic *Nicotiana tabacum* with enhanced activities of glutathione reductase and superoxide dismutase. *Plant Cell Physiology* 36, 1687–1691.

Artus, N.N., Uemura, M., Steponkus, P.L., Gilmour, S.J., Lin, C. and Thomasow, M.F. (1996) Constitutive expression of the cold-regulated *Arabidopsis thaliana COR15a* gene affects both chloroplast and protoplast freezing tolerance. *Proceedings of the National Academy of Sciences USA* 93, 13404–13409.

Bordas, M., Montesinos, C., Dabanza, M., Salvador, A., Roig, L.A., Serrano, R. and Moreno, V. (1997) Transfer of the yeast salt tolerance gene *HAL1* to *Cucumus melo* L. cultivars and *in vitro* evaluation of salt tolerance. *Transgenic Research* 6, 41–50.

Bowler, C., Van Montague, M. and Inz'e, D. (1992) Superoxide dismutase and stress tolerance. *Annual Review Plant Physiology Plant Molecular Biology* 43, 83–116.

Bowler, C., Van Camp, W., Van Montague, M. and Inz'e, D. (1994) Superoxide dismutase in plants. *Critical Reviews in Plant Sciences* 13, 199–218.

Boyer, J.S. (1982) Plant productivity and environment. *Science* 318, 443-448.

Bray, E.A. (1997) Plant responses to water deficit. *Trends in Plant Science* 2, 48–54.

Christiansen, M.N. (1982) World environmental limitations to food and fiber culture. In: Christiansen, M.N. and Lewis C.F. (eds) *Breeding Plants for Less Favorable Environments.* John Wiley & Sons, New York, pp. 1–11.

de la Fuente, J.M., Ramírez-Rodríguez, V., Cabera-Ponce, J.L. and Herrera-Estrella, L. (1997) Aluminum tolerance in transgenic plants by alteration of citrate synthesis. *Science* 276, 1566–1568.

Drennan, P.M., Smith, M.T., Goldsworthy, D. and van Staten, J. (1993) The occurrence of trehalose in the leaves of the desiccation-tolerant angiosperm *Myrothamnus flabellifoliius* Welw. *Journal of Plant Physiology* 142, 493–496.

Epstein, E., Norlyn, J.D., Rush, D.W., Kingsbury, R.W., Kelly, D.B., Cunningham G.A. and Wrona, A.F. (1980) Saline culture of crops: a genetic approach. *Science* 210, 399.

Hayashi, H., Mustardy, L., Deshnium, P., Ida, M. and Mursata, N. (1997) Transformation of *Arabidopsis thaliana* with the codA gene for choline oxidase; accumulation of glycinebetaine and enhanced tolerance to salt and cold stress. *The Plant Journal* 12, 133–142.

Hightower, R., Baden, C., Penzes, E., Lund, P. and Dunsmuir, P. (1991) Expression of antifreeze proteins in transgenic plants. *Plant Molecular Biology* 17, 1013–1021.

Hinderhofer, K., Eggers-Schumacher, G. and Schoffl, F. (1998) HSF3, a new heat shock factor from *Arabidopsis thaliana*, derepresses the heat shock response and confers thermotolerance when overexpressed in transgenic plants. *Molecular General Genetics* 258, 269–278.

Holmberg, N. and Bulow, L. (1998) Improving stress tolerance in plants by gene transfer. *Trends in Plant Science* 3, 61–66.

Holstrom, K.O., Welln, E.M.B., Talapo Palva, A.M., Tunnela, O.E. and Londesborough, J. (1996) Drought tolerance in tobacco. *Nature* 379, 683–684.

Ishizaki-Nishizawa, O., Fujii, T., Azuma, M., Sekiguchi, K., Murata, N., Ohtani, T. and Toguri, T. (1996) Low-temperature resistance of higher plants is significantly enhanced by a nonspecific cyanobacterial desaturase. *Nature Biotechnology* 14, 1003–1006.

Jaglo-Ottosen, K.R., Gilmour, S.J., Zarka, D.G., Schabenberger, O. and Thomashow, M.F. (1998) *Arabidopsis CBF-1* overexpression induces *COR* genes and enhances freezing tolerance. *Science* 280, 104–106.

Kishor, P.B.K., Hong, Z., Miao, G., Hu, C.A. and Verma, D.P.S. (1995) Overexpression of Δ^1-pyrroline-5-carboxylase synthase increases proline production and confers osmotolerance in transgenic plants. *Plant Physiology* 108, 1387–1394.

Kodama, H., Hamada, T., Horiguchi, G., Nishimura, M. and Iba, K. (1994) Genetic enhancement of cold tolerance by expression of a gene for chloroplast ω-3 fatty acid desaturase in transgenic tobacco. *Plant Physiology* 105, 601–605.

Lee, J.H., Hubel, A. and Schoffl, F. (1995) Derepression of the activity of genetically engineered heat shock factors causes constitutive synthesis of heat shock proteins and increased thermotolerance in transgenic *Arabidopsis*. *The Plant Journal* 8, 603–612.

Lullis, G., Holmberg, N. and Bulow, L. (1996) Enhanced NaCl stress tolerance in transgenic tobacco expressing bacterial choline dehydrogenase. *Bio/Technology* 14, 177–180.

Mckersie, B.D., Bowley, S.R., Harjanto, E. and Leprince, O. (1996) Water-deficit tolerance and field performance of transgenic alfalfa overexpressing superoxide dismutase. *Plant Physiology* 111, 1177–1181.

Misra, S. and Gedamu, L. (1989) Heavy metal tolerant transgenic *Brassica napus* L. and *Nicotiana tabacum* L. plants. *Theoretical Applied Genetics* 78, 161–168.

Miyasaka, S.C., Buta, J.G., Howell, R.K. and Foy, C.D. (1991) Mechanism of aluminum tolerance in snapbeans. *Plant Physiology* 96, 737–43.

Pan, A., Yang, M., Tie, F., Li, L., Chen, Z. and Ru, B. (1994) Expression of mouse metallothionein-I gene confers cadmium resistance in transgenic tobacco plants. *Plant Molecular Biology* 24, 341–351.

Pardo, J.M., Reddy, M.P., Yang, S., Maggio, A., Huh, G., Matsumoto, T., Coca, M.A., Paino-D'Urzo, M., Koiwa, H., Yun, D., Watad, A.A., Bressen, R.A. and Hasegawa, P.M. (1998) Stress signaling through Ca^{2+}/calmodulin-dependent protein phosphatase calcineurin mediates salt adaptation in plants. *Proceedings of the National Academy of Sciences USA* 95, 9681–9686.

Petalino, J.G. and Collins, G.B. (1984) Cellular approaches to environmental stress resistance. In: Collins, G.B. and Petalino J.G. (eds) *Applications of Genetic Engineering to Crop Improvement.* Nijhoff/W. Junk, Boston, Dordrecht, pp. 341–499.

Pilon-Smits, E.A.H., Ebskamp, M.J.M., Paul, M.J., Jeuken, M.J.W., Weisbeek, P.J. and Smeekens, S.C.M. (1995) Improved performance of transgenic fructan-accumulating tobacco under drought stress. *Plant Physiology* 107, 125–130.

Ryan, P.R., Delhaize, E. and Randall, P.J. (1995) Characterization of Al-stimulated efflux of malate from the apices of Al-tolerant wheat roots. *Planta* 196, 103–110.

Saneko, H., Nagasaka, C., Hahn, D.T., Yang, W., Premachandra, G.S., Joly, R.J. and Rhodes, D. (1995) Salt tolerance of glyucinebetaine-deficient and -containing maize lines. *Plant Physiology* 107, 631–638.

Slooten, L., Capiau, K., Van Camp, W., Van Montagu, M., Sybesma, C. and Inz'e, D. (1995) Factors affecting the enhancement of oxidative stress tolerance in transgenic tobacco overexpressing manganese superoxide dismutase in the chloroplasts. *Plant Physiology* 107, 737–750.

Steponkus, P.L., Uemura, M. and Webb, M.S. (1993) A contrast of the cryostability of the plasma membrane of winter rye and spring oat. Two species that differ in their freezing tolerance and plasma membrane lipid composition. In: Steponkus, P.L. (ed.) *Advances in Low Temperature Biology*, Vol. 2. JAI Press, London, pp. 338–362.

Tarczynski, M.C., Jensen, R.G. and Bohnert, H.J. (1993) Stress protection of transgenic tobacco by production of the osmolyte mannitol. *Science* 259, 508–510.

Thomas, J.C., Sepahi, M., Arendall, B. and Bohnert, H.J. (1995) Enhancement of seed germination in high salinity by engineering mannitol expression in *Arabidopsis thaliana. Plant, Cell and Environment* 18, 801–806.

Thomashow, M.F. (1998) Role of cold-response genes in plant freezing tolerance. *Plant Physiology* 118, 1–7.

Van Camp, W., Willekens, H., Bowler, C., Van Montagu, M., Inz'e, D., Reupold-Popp, R., Sanderman, H. and Langebartels, C. (1994) Elevated levels of superoxide dismutase protect transgenic plants against ozone damage. *Bio/Technology* 12, 165–168.

Van Camp, W., Capiau, K., Van Camp, W., Van Montagu, Inz'e, D. and Slooten, L. (1996) Enhancement of oxidative stress tolerance in transgenic tobacco plants overproducing Fe-superoxide dismutase in chloroplasts. *Plant Physiology* 112, 1703–1714.

Vernon, D.M., Tarczynsky, M.C., Jensen, R.G. and Bohnert, H.J. (1993) Cyclitol production in transgenic tobacco. *The Plant Journal* 4, 199–205.

Xu, D., Duan, X., Baiyang, W., Hong, B., Ho, H.T. and Wu, R. (1996) Expression of a late embryogenesis abundant protein gene, *HVA1*, from barley confers tolerance to water deficit and salt stress in transgenic rice. *Plant Physiology* 110, 249–257.

21 Global, Regional and Local Food Production and Trade in a Changing Environment

JOHN REILLY[1], DAVID SCHIMMELPFENNIG[2] AND JAN LEWANDROWSKI[2]

[1]*MIT Joint Program on the Science and Policy of Global Change, Building E40-263, 77 Massachusetts Avenue, Cambridge, MA 02139, USA;* [2]*US Department of Agriculture, Economic Research Service, 1800 M Street, Washington, DC 20036, USA*

21.1 Introduction

Serious attempts to investigate the impacts of potential changes in climate due to increased greenhouse gases date to the mid to late 1980s. Climatic change studies date to 1970 but few at that time considered economic impacts (Reilly and Thomas, 1993). An exception was a National Defense University study, which considered warming and cooling impacts on US agriculture using a Delphi approach (Gard, 1980).

Some of the advances in impact analysis since the first studies include: (i) explicit modelling of adaptation within the farm sector; and (ii) consideration of agriculture as a global system that interacts through trade in agricultural products and input flows. Some work has been done but more is needed on: (i) how government policies and programmes – ranging from crop insurance and disaster assistance to acreage reduction programmes, tariffs and quotas, water pricing, and the level of agricultural research – will affect the response of the farm sector; (ii) consideration of climatic change as only one of many forces, along with changes in population, economic activity and technology, that will shape the agricultural economy over the coming decades; (iii) explicit consideration of the high levels of uncertainty in climatic change prediction and the implications for adaptation of high levels of natural variability in weather; and (iv) modelling of competition for land, water, labour and capital as other sectors of the economy are affected by climatic change.

Before addressing the state of current research results with regard to the above issues, this chapter provides an overview of the methods used to estimate climatic impacts on agriculture, summarizes the scenarios used in existing studies and reviews some of the critical uncertainties in climatic projections.

21.2 Methods for Estimating Climatic Impacts on Agriculture

To grapple with the considerable complexity in impact analysis, two basic methods have been used. These are: (i) structural modelling of crop and farmer response, combining the agronomic response of plants with economic/management decisions of farmers; and (ii) spatial analogue models that exploit observed differences in agricultural production and climate among regions.

For the first approach, sufficient structural detail is needed to represent specific crops and crop varieties whose responses to different conditions are known through detailed experiments. This structural detail is represented in crop response models, which require details about weather on a daily or hourly basis and specific information on soil properties. Farm management models at a similarly detailed level can be linked to such crop response models and include details about the timing of field operations throughout the growing season, choice of which crop and crop variety to plant, how much fertilizer to use and whether and when to irrigate. They are solved to maximize the profits of farmers given expectations about what the weather, yields and prices will be over the season. An advantage of this approach is that the results can provide detailed understanding of the physical, biological and economic responses and adjustments. A disadvantage is that, for aggregate studies, inferences must be made from relatively few sites and crops to large areas and diverse production systems.

The spatial analogue approach uses cross-section data. It relates an agricultural performance measure, such as observed yields, production, revenue, profits or the value of agricultural land, to climatic conditions that vary regionally. The simplest application of this approach is to map shifts in cropping boundaries as a function of temperature or precipitation changes. More elaborate versions of this approach use statistical analyses of data across geographical areas. Such statistical analyses allow researchers to separate the effects of various climatic factors (temperature, precipitation, temperature extremes, variation in precipitation, etc.) from factors other than climate, such as soil type, that may also explain production differences across regions. An advantage of the spatial analogue approach is that it provides direct evidence on how farmers operating under commercial conditions have responded to different climatic conditions. An inherent assumption is that the adjustments that have been made over long periods of time across regions could be made easily as climate changes in a region.

21.3 Potential Changes in Climate Due to Greenhouse Gases

The impacts of climatic change on agriculture will depend on the ultimate form of climatic change. The geographical pattern of temperature and precipitation changes and any change in the variability of weather could be quite important, but there is little confidence in such details of climatic change scenarios produced by general circulation models (GCMs) because of the coarse resolution of these models. The analyses presented in subsequent sections of this chapter generally rely on climatic projections generated by GCMs based on equilibrium $2 \times CO_2$ simulations. The typical approach is to compute the average difference between 5–10 years of simulated $1 \times CO_2$ and $2 \times CO_2$ data of monthly means for temperature and precipitation. These calculated differences are then applied to actual climatic data for the specific location or regions being investigated. In the structural modelling approach, this involves applying the monthly changes to a detailed weather record. As a result, the pattern of temperature and precipitation over the month as displayed in the historical record is preserved in the climatic change scenario. Impact models are typically run over a period of 10–30 years, with the average impact reported, to avoid generating results that are due to choice of a single year that may not be representative of the site's climate.

In some of the studies reported, other climatic scenarios have been used. A unique approach used by Kaiser *et al.* (1993) is to construct a simple statistical weather generator. This allows construction of many different weather scenarios that show gradual warming over time consistent with a predetermined final temperature. While this approach is limited to the sites for which it was developed, it provides a way to generate time paths of climatic change in the absence of such data directly from GCM runs.

While equilibrium $2 \times CO_2$ scenarios have been standard model experiments reported from GCMs, these experiments do not provide information on when these changes will occur. The timing of climatic change depends on the specific path of CO_2 concentration increase and climatic system interactions with the ocean. Figure 21.1 indicates how the global mean temperature changes in these scenarios compare with the time path presented by IPCC (Houghton *et al.*, 1996). These scenarios generally represent global temperature increases beyond what is expected by 2100. New transient climatic scenarios (GCM scenarios that trace climatic change over time) have been developed recently but have generally not yet been used in impact analysis.

Regional changes in mean surface temperature and precipitation differ from the global means and there are large differences in the pattern of change among the different GCM scenarios used in these impact analyses. There is some agreement that higher-latitude areas will warm more than the global average, that high-latitude areas will receive proportionately more precipitation and that mid-continental mid-latitude areas will become drier, in terms of reduced soil moisture (Houghton *et al.*, 1996).

The four GCM scenarios presented above are representative of possible climatic changes under a $2 \times CO_2$ climate, but should not be considered 'predictions' of what will happen under climatic change. In general, a number of limitations of these and most available GCM climatic scenarios exist. These include:

- The global time path and local rate of global change. Localized changes may be more rapid, slower, or in a different direction than the global average because geographical patterns can change while the global mean

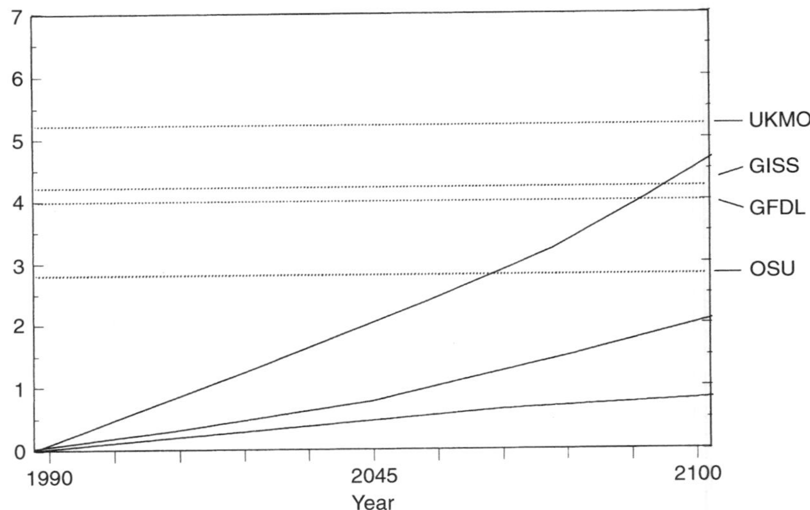

Fig. 21.1. Global mean temperature rise. Projections of global mean temperature from 1990 to 2100 for three climate sensitivities (4.5, 2.5 and 1.5°C) and a median emissions scenario including uncertainty in future aerosol concentrations. Increasing aerosols (e.g. sulphur, also from coal-burning) are estimated to have a cooling effect (after Houghton *et al.*, 1996). See Table 21.1 for identification of GCMs.

is changing. Changes in regular storm tracks could, over the course of a few years, lead to greatly reduced rainfall in one area and increased rainfall in a new area.

- Changes in the daily and seasonal pattern of climatic change. Daily, monthly and seasonal patterns of temperature and precipitation will likely be affected. Recent history shows an upward trend in night-time low temperatures in the northern hemisphere but little or no change in daytime high temperature (Kukla and Karl, 1993). Results from GCMs at this level are generally considered to provide little real information, because of coarse resolution of the models and inability to simulate phenomena such as ENSO (El Niño Southern Oscillation). Schimmelpfennig and Yohe (1994) estimate an index of crop vulnerability that provides a first step in understanding how changes in variability of climate affect production.
- Changes in the intensity of weather events. Heavy rain and high winds damage crops and cause soil erosion. Some scientific findings suggest that rainfall could become more intense (Pittock *et al.*, 1991). The frequency and strength of regular weather cycles such as ENSO and the strength of the jet stream may alter and thus change weather patterns. These factors and others leading to the occurrence of hurricanes, tornadoes, hail and wind storms are not adequately modelled by coarse-resolution GCM simulations.
- Sulphate aerosols. Sulphur, emitted as a result of burning coal, is thought to have a significant cooling effect but is only now being introduced into

climatic models (Houghton *et al.*, 1995). Sulphur emissions remain in the atmosphere only a few days and hence their cooling effect is confined to areas downwind from where significant amounts are released.

Given these and other limitations of GCM scenarios, impact analyses derived from them are illustrative of what might happen rather than a prediction for which confidence bounds can be specified. It may be possible to improve the realism of climatic scenarios by addressing the limitations identified above but firm 'predictions' of climate and its impact on agriculture decades into the future may never be possible. Hence, agricultural impact methods need to be improved to take advantage of the improving realism of climatic scenarios. It is important to understand what farmers, agricultural policy-makers, agricultural researchers and others can do to prepare for climatic changes that cannot be precisely predicted.

21.4 Climatic Change Impacts

Possible impacts for the world from one study are illustrated in Table 21.1 (Darwin *et al.*, 1995). Without adaptation, cereal production was estimated to fall globally by 19–29%. With farm-level, market and land use adaptation, there is potential to reduce these losses substantially so that impact on cereal production ranges between ± 1%. This study is based on an analogous region method. The initial impact on cereal production compares very closely to work by Rosenzweig and Parry (1994). The Darwin study differs in that adaptation is estimated as able to offset completely the yield losses at a global level. Issues remain with regard to whether the full potential for adaptation can be realized. These results do not include an adjustment for the possible beneficial effect of CO_2 fertilization. Both of these efforts relate agricultural economic impacts to a global model of the entire world economy. The Darwin *et al.* approach also

Table 21.1. Percentage changes in the supply[a] and production of cereals for the world by climate change scenario. (Source: Darwin *et al.*, 1994.)

General circulation model (GCM)	Supply		Production	
	No adaptation	Land use fixed	Land use fixed	No restrictions
GISS[b]	−22.6	−2.4	0.2	0.9
GFDL[c]	−23.5	−4.4	−0.6	0.3
UKMO[d]	−29.3	−6.4	−0.2	1.2
OSU[e]	−18.6	−3.9	−0.5	0.2

[a]Changes in supply represent the additional quantities that firms would be willing to sell *at 1990 prices* under the alternative climate. Changes in production are changes in equilibrium quantities.
[b]GISS = Goddard Institute for Space Studies.
[c]GFDL = Geophysical Fluid Dynamics Laboratory.
[d]UKMO = UK Meteorological Office.
[e]OSU = Oregon State University.

Table 21.2. Regional crop yield for 2 × CO_2 GCM equilibrium climates. (Source: Reilly *et al.*, 1996.)

Region	Crop	Yield impact (%)	Countries studied/comments
Latin America	Maize	−61 to increase	Argentina, Brazil, Chile, Mexico. Range is across GCM scenarios, with and without the CO_2 effect
	Wheat	−50 to −5	Argentina, Uruguay, Brazil. Range is across GCM scenarios, with and without the CO_2 effect
	Soybean	−10 to +40	Brazil. Range is across GCM scenarios, with CO_2 effect
Former Soviet Union	Wheat Grain	−19 to +41 −14 to +13	Range is across GCM scenarios and region, with CO_2 effect
Europe	Maize	−30 to increase	France, Spain, N. Europe. With adaptation, CO_2 effect. Longer growing season; irrigation efficiency loss; northward shift
	Wheat	Increase or decrease	France, UK, N. Europe. With adaptation, CO_2 effect. Longer season: northward shift; greater pest damage; lower risk of crop failure
	Vegetables	Increase	
North America	Maize Wheat	−55 to +62 −100 to +234	USA and Canada. Range across GCM scenarios and sites; with/without CO_2 effect
	Soybean	−96 to +58	USA. Less severe or increase in yield when CO_2 effect and adaptation considered
Africa	Maize	−65 to +6	Egypt, Kenya, South Africa, Zimbabwe. With CO_2 effect, range across sites and climate scenarios
	Millet	−79 to −63	Senegal. Carrying capacity fell by 11–38%
	Biomass	Decrease	South Africa; agrozone shifts
South Asia	Rice Maize Wheat	−22 to +28 −65 to −10 −61 to +67	Bangladesh, India, Philippines, Thailand, Indonesia, Malaysia, Myanmar. Range over GCM scenarios, and sites; with CO_2 effect; some studies also consider adaptation
Mainland China and Taiwan	Rice	−78 to +28	Includes rain-fed and irrigated rice. Positive effects in NE and NW China, negative in most of the country. Genetic variation provides scope for adaptation
Other Asia and Pacific Rim	Rice	−45 to +30	Japan and South Korea. Range is across GCM scenarios. Generally positive in northern Japan; negative in south
	Pasture	−1 to +35	Australia and New Zealand. Regional variation
	Wheat	−41 to +65	Australia and Japan. Wide variation, depending on cultivar

includes, in a crude way, water supply change impacts and competition for water and land resources from other sectors, but the modelling effort imposes climatic change on the world economy as it exists today. Rosenzweig and Parry (1994) estimate climatic impacts under a $2 \times CO_2$ world and linearly impose these changes through time (assuming the full effect occurs in 2060) in an economic model where economies grow and change through time.

Yield impact studies using crop response models have been conducted for many countries. Table 21.2 summarizes the basic findings. Results vary widely, depending on the climatic scenario, methods and particular site examined. All studies tend to show negative impacts for equatorial and subtropical regions, while areas nearer the poles often show improved cropping conditions. Beyond these broad regional patterns, which appear to be observed in most studies, greater details on regional impacts generally do not hold up across different climatic scenarios.

21.5 Issues in Impact Assessment

21.5.1 Adaptation and adjustment

Economic analysis of adaptation generally follows a path of investigation that considers those options that could be used to ameliorate losses without increasing costs. Once this baseline impact is established, further analysis of changing markets identifies whether commodity prices would increase or decrease as a result of changes in yields around the world. Then the analysis considers technological options and assumes that only those that would improve profitability are adopted. The following general classes of technological options show promise in ameliorating agricultural losses: changed sowing dates; multiple crops per growing season; planting of different crops or different crop varieties; development of new crop varieties; use of irrigation; changes in fertilizer use and tillage practices; and improved short-term weather forecasting (Reilly *et al.*, 1996).

A separate issue from adaptation is the cost of adjustment. Adjustment costs arise when firms are forced to retire equipment or capital investment before it would normally wear out, or only slowly learn about and adopt new technology and management options. Whether farmers would face significant adjustment costs in adapting to climate change depends on the lifetime of capital equipment, adoption rates and the rate at which climate changes. Some evidence exists on the lifetimes of important agricultural investments and times required for adoption of new technologies. This evidence suggests that 3–14 years are required for new variety adoption, 8–15 years for new variety development, 10–12 years for adoption of new tillage systems, 15–30 years to adopt a new crop, 3–10 years to open new agricultural lands, 20–25 years for the lifetime of irrigation equipment, and 50–100 years for the lifetime of dams and irrigation systems (Reilly, 1995). Except for longer-lived investments such as dams and irrigation equipment, adjustment would not appear to impose greater costs if local climate changes gradually at the rate indicated by global transient runs. Changes in precipitation patterns, however, may result in abrupt local changes in climate and thus impose adjustment costs. Very little solid evidence on local transient rates of climatic change exists; thus, it is difficult to establish whether adjustment costs will be significant.

21.5.2 Trade

Effects of climatic change on a region, national economy, and consumers of food and fibre depend on both how global production potential changes and how climate affects a region's agriculture directly. Trade affects the distribution of costs and benefits of climatic change; regions that are negatively affected by climatic change will seek to import food and fibre from areas with surplus production. If global production declines and agricultural prices rise, then producers in a region may gain from the higher prices even if yields fall somewhat. Consumers worldwide would be adversely affected by higher food and fibre prices. If world prices fall, agricultural producers may lose even if their yields improve slightly, because revenue (price × quantity) declines if the decrease in price is greater than the increase in production. Reilly *et al.* (1994) illustrated these effects for a range of climatic scenarios. Fischer *et al.* (1994) demonstrated that, under some circumstances, differential capacity to adapt can lead to further shifts of cereal production from tropical regions to temperate regions as a result of trade.

21.6 Future Trends and Issues Facing World Agriculture

An important aspect of the impact of climatic change on agriculture is the social, economic and agricultural system context in which climatic change will occur. The conditions of the social, economic and agricultural system in a country can affect the capacity to adapt, the vulnerability of society, and the anticipatory response of society to climatic change. People in isolated areas, who are highly dependent on local agriculture for income and food and without social support systems to assure food for the hungry in times of need, are likely to see more severe impacts than people in a wealthier country. In such a situation they may lack financial resources to change production practices, will see the consequences of production shortfalls reflected as more severe human consequences such as hunger and malnutrition, and may be less able to take effective anticipatory actions. The above considerations mean that the social and economic conditions across the world and future changes in these conditions will affect how people will be affected.

Some commentators (e.g. Erlich and Erlich, 1990) worry that rising demand for food over the next century due to population growth will lead to increasing global food scarcity and a worsening of hunger, malnutrition and associated problems in developing countries. Climatic change could thus exacerbate a bad situation. Concern about the ability to meet food demand stems from the observations that most of the potentially arable land is already under cultivation and much is suffering degradation, freshwater supplies are dwindling, and only a small amount of less fertile land remains to meet additional future needs.

More conventional agricultural economic modelling (e.g. Mitchell and Ingco, 1995) suggests that the world food situation is likely to improve, given assumptions about agricultural productivity growth. The world's food system might thus be less vulnerable to climatic change. This view is supported by past evidence of declining global food scarcity and real prices for food commodities, and recognition that while population continues to grow, the growth rate has slowed. For example, an index of food commodity prices by the World Bank (1992) shows an overall decline of 78% between 1950 and 1992. Increasing yields and improving agricultural productivity appear to be largely responsible for this trend; although farm subsidy increases in Japan, the EEC and the USA through the 1970s and into the 1980s may also have contributed.

Behind these different views of future world food security are different views about three broad forces that affect the world's food situation. These are: (i) demand growth due to rising populations and incomes; (ii) natural resource (land, water) availability and degradation of these resources; and (iii) the continuing ability of agricultural research to boost productivity and yields.

21.6.1 Future demand growth

For the future, there is a general consensus among forecasters that world population will be 8–9 billion by 2025 (McCalla, 1994), implying a growth rate of approximately 1.3% year^{-1} from the 6 billion population in 1998 – less than the 1.5% year^{-1} growth since 1970. Parikh (1994) estimates that world population will be 10.3–12.8 billion by 2050, based on extensions of UN (1994) estimates, implying a further slowing of the growth rate to around 1.1%. Studies projecting the impacts of climatic change around the middle of the 21st century (Chen and Kates, 1994; Fischer *et al.*, 1994; Rosenzweig and Parry, 1994) have assumed world population levels of approximately 10 billion. Nearly all (95%) of the predicted population increase is in the developing world, and the rate of growth is greatest in Africa, where population is projected to increase three- to fivefold over the 1990 level.

Demand for food will also increase with income. Table 21.3 shows projections for future real income growth by Fischer *et al.* (1994) for 1980–2060. Over this period, growth is expected to be a little higher in the developing countries (2.4% year^{-1}) than in the developed world (1.6%). However, in absolute changes, in per capita terms, the picture is not quite so optimistic for the developing world. The World Bank (1992) predicts that by 2030 real per capita income will increase by US$160 in sub-Saharan Africa to $500, by $1500 in Asia-Pacific to $2000, by $3200 in the Middle East/North Africa to $4000, and by $3300 in Latin America to $5500. Even though growth rates are more rapid than in the developed world, per capita income levels remain below current levels in the major developed countries.

Income and population growth rates are the basis of demand growth projections of models such as those of Fischer *et al.* (1994) (Table 21.4). Parikh (1994) finds that caloric demands increase between 1.5 and threefold by 2050 over the 2000 levels. The projected demands for cereals in Fischer *et al.* (1994) are broadly consistent with these estimates. The estimates display no particular bias toward growth in meat and dairy products, as might be expected with rising incomes, although Parikh's estimates show a wide range of possible growth in dairy products and meat.

The extent of shifts in the aggregate consumption among agricultural products depends on how income increases are distributed among income classes. Growth in income of the poorest will mean greater demand for basic foods such as grains, pulses and potatoes as they increase their basic calorie intake. Income growth concentrated among the wealthier segments of the populations in developing countries could generate more of a shift toward

Table 21.3. Projections of GDP growth. (Source: Fischer *et al.*, 1994.)

	GDP growth (%)	
Region	1980–2000	1980–2060
World	2.9	1.8
Developed	2.6	1.6
Developing	4.3	2.4
Africa	4.6	3.0
Latin America	3.9	2.1
Southeast Asia	4.7	2.4
West Asia	4.4	2.8

Table 21.4. Projected global production of food commodities in 2060. (Source: Fischer *et al.*, 1994.)

Commodity[a]	1980	2000	2020	2040	2060
Wheat	441	603	742	861	958
Rice	249	367	480	586	659
Coarse grains	741	1022	1289	1506	1669
Bovine and ovine meat	65	83	105	123	136
Dairy	470	613	750	877	997
Other animal produce	17	25	33	41	48
Protein feed	36	52	64	76	85
Other food	225	326	433	538	629
Non-food	26	34	41	47	52
Agriculture	310	438	572	700	810

[a]Wheat, rice, course grain in million tonnes; bovine and ovine meat in million tonnes carcass weight; dairy products in million tonnes whole milk equivalent; other animal produce, and protein feed in million tonnes protein equivalent; other food in 1970 US$ billion.

meat and dairy products if these consumers adopt consumption patterns more like those displayed by consumers in wealthier countries. The extent to which consumption patterns of different societies reflect cultural differences or income differences, and hence the extent to which patterns will change as income increases, remains a subject of research.

The implications of demand growth resulting from income growth vs. population growth are also quite different. Income-generated growth means that at least some people are eating more, while population-generated growth means that per capita consumption will fall if production does not keep pace. Income-generated growth, when income growth occurs mainly among the wealthier segments of society, could worsen the food situation for the poor. Rising demand by wealthier food consumers would increase prices and, by placing additional demand on the world's resources, make it more difficult for the poor to afford food. Whatever the trend, there is general agreement that the bulk of the future demand increase will come from the developing world, since in developed countries the average person is already reasonably well fed.

Can these demands be met? The projected food demands of the developed countries (USA, Canada, Europe, Japan, Australia and New Zealand) only increase modestly above their current levels in the study by Parikh (1994). Given that these countries as a group are essentially self-sufficient in agriculture at the moment, Parikh (1994), Crosson and Anderson (1994) and others predict that they can probably satisfy their future food demands relatively easily. Indeed, Mitchell and Ingco (1995) predict that net grain exports of the developed countries will increase from their current level of 117 Mt to 194 Mt in 2010; thus they see the developed countries as growing in importance in feeding people in developing countries.

One limit of these studies is that they do not directly consider resource availability and quality. Instead they extrapolate yield growth rates, assuming that productivity enhancements can more than overcome any effects of resource degradation. Typically, however, yield growth rates are assumed to slow in the future, but, with population growth rates slowing as well, growth in agricultural supply can still outpace growth in demand.

21.6.2 Resource availability and quality

While existing global food demand and supply models generally do not explicitly consider resource degradation, there is a variety of analyses that have considered resource issues that could constrain increases in agricultural production. One approach to considering resource adequacy is carrying capacity. Table 21.5 shows estimates from a United Nations Food and Agriculture Organization (FAO) study by Higgins *et al.* (1982) of the maximum population that could be supported by the available quantity of land and other resources in developing countries (China was not included in the analysis). This approach tends to show that the capacity supports much higher populations than were projected at that time for the year 2000, and indeed through

Table 21.5. Population carrying capacities of developing countries. (Source: Parikh, 1994.)

Location	Total land area (million ha)	Population in 2000 (millions)	Persons ha^{-1} in 2000	Potential population-supporting capacity in 2000 (persons ha^{-1})	
				Low inputs	High inputs
Total	6495	3590	0.55	0.86	5.11
Africa	2878	780	0.27	0.44	4.47
S.W. Asia	677	265	0.39	0.27	0.48
S. America	1770	393	0.22	0.78	6.99
C. America	272	215	0.79	1.07	4.76
S.E. Asia	898	1937	2.16	2.74	7.06

2050 in all regions of the world, if high levels of inputs are used. At an average of 5.11 people per hectare with a high level of input use and total land availability of 6.495 Gha, these estimates suggest that a total population of 33.2 billion could be supported. The projected demands in Parikh (1994) imply a 50% higher calorie intake than in the FAO study; adjusting for this, the population carrying capacity of developing countries is still 22.1 billion.

A number of caveats are associated with carrying capacity calculations. Obviously, key to the calculation is the level of inputs applied at low-input levels. Carrying capacity would be exceeded in some regions over the period to 2050. A well-functioning economic system should provide the incentive to intensify production as needed to meet demand. However, impoverished people suffering from hunger lack the income to have their food needs realized as demands that are recognized in an economic system. Hence, the agricultural production system will not respond to produce this food unless these demands are realized in the market, either by solving problems of poverty or by developing food assistance programmes that provide food for the poor.

Other assumptions of carrying-capacity calculations are also open to question. Typically the estimates assume that many areas that currently achieve low yields and production rates could achieve yield and production rates like those in other high-yield areas. Existing data do not support a full analysis of whether the resource conditions actually exist to support these production levels. The open debate is whether yields are low in some areas because resource conditions do not support higher production or because high yielding farming practices have not been adopted and, if so, under what economic conditions they would be adopted. Expanding agricultural land area and intensifying production may also impose significant environmental costs, including threats to biodiversity, water quality and natural ecosystems. Limits on input use and land expansion to reduce such impacts are generally not taken into account in these carrying-capacity calculations. On the other hand, these calculations assume existing technologies and do not factor in possible improvements in yields and productivity. A number of these issues have been addressed in separate analyses.

Land resources

Crosson and Anderson (1994) argue that a potentially important constraint on increasing the supply of cropland is that much of the uncultivated land is in areas that are poorly connected by roads, rail and air to existing domestic and foreign markets. Moreover, efforts to preserve natural forestlands could exclude much of the uncultivated land from conversion. As a result, they see little new land being converted.

The other land resource concern is degradation of existing agricultural land, with consequent effects on productivity. Soil erosion, compaction, nutrient depletion and desertification are principal dryland degradation concerns. Table 21.6 reports estimates of percentage of dry land suffering from degradation or desertification, based on estimates of FAO and reported by Norse (1994). These estimates show 70% of agricultural land and more for most continental regions as degraded. Unfortunately, these data do not indicate the extent to which degradation is affecting yields. Soil erosion, while degrading water quality in streams, rivers and estuaries, can have modest effects on yields and may be correctable by improved farming. For example, one study for the USA by Crosson (1992) suggested that in the long run these losses are typically small relative to the gains from technological progress.

These results may not be transferable to many developing countries. Soils in tropical areas are in general thinner, more vulnerable to erosion and less easily restored. Sustainable agricultural production has often been difficult to achieve on nutrient-poor soils. For example, in Brazil's Amazon region, government policies in the 1980s encouraged the conversion of tropical forests to crop and livestock production. Much of this land has since been abandoned (Repetto, 1988; Binswanger, 1989; Mahar, 1989).

Of the 3569 Mha that has so far been degraded in dryland areas of the world, 29% is in Africa and 37% in Asia, where desertification has been thought to be a significant problem. There is continuing debate about how to define desertification as well as its causes, extent and consequences. With respect to cause, the issue is whether it results from non-sustainable production practices or climatic variation on the order of a decade or so and how these factors interact. Nelson (1988) and Bie (1990) concluded that the incidence of desertification is a lot less than originally thought and that the contribution of desert margins to food production is small. For example, low-rainfall areas accounted

Table 21.6. Desertification and dry land degradation. (Source: Norse, 1994.)

Region	Total dry land area in agriculture (Mha)	Area degraded (Mha)	Dry land area degraded (%)
Africa	1430	1050	73
Asia	1880	1310	70
Australia	700	380	54
Europe	146	94	65
N. America	578	429	74
S. America	420	306	73
Total	5154	3569	69

for only 12% of domestic cereal production in sub-Saharan Africa and 1% in Asia in the early 1980s (Norse, 1994). There is also considerable controversy as to whether and under what conditions desertification is reversible.

Another potential threat is salinization of soils and water. Mainly a problem for irrigated areas, salinization also occurs in hot dry climates, where evaporation can increase salt concentration in soils. In irrigated areas, salinization is usually the result of poor construction, inadequate maintenance of canals, or excessive use of water. The end result is waterlogging, salinization, reduced crop yields and ultimately the loss of agricultural land. The UN (1992) estimated that 1–1.5 Mha $year^{-1}$ are lost because of this process, with another 30 Mha being at risk.

Water resources

Water resources are important for agriculture. While only 17% of global cropland is irrigated, this 17% of land accounts for more than one-third of world food production. An estimated additional 137 Mha have potential to be irrigated, compared with the 253 Mha currently irrigated (World Bank, 1994). As with land resources, the cost of developing these irrigation systems may be prohibitive. Current water systems in many developing countries achieve low efficiencies of water distribution and average crop yields are well below potential (Yudelman, 1993; Crosson, 1995; Agcaoili and Rosegrant, 1995).

As already mentioned, soil salinization from poorly designed and managed irrigation systems is a problem. Another concern with irrigation is the spread of waterborne diseases. Unpriced and heavily subsidized water resources, inadequate planning and maintenance of water systems, unassigned water rights and conflicts among goals are seen as sources of these problems (Repetto, 1988). Solutions to these problems are available in most cases, and a recent study found that investments in irrigation have been at least as profitable as investments in other agricultural enterprises (World Bank, 1994). Perhaps the most important solution is to move to full cost pricing of water. Full cost pricing would encourage adoption of water-conserving practices while providing funds to maintain and improve management of irrigation and drainage systems.

Technological innovation and adoption

The invention and diffusion of yield-enhancing technologies and farming practices have been more important factors in raising global agricultural output over the last 50 years than the expansion in quantities of land and water (Crosson and Anderson,1994). There is debate about whether the trend toward ever higher yields can be maintained. Some see evidence in recent history that yields may have reached a plateau; Reilly and Fuglie (1998), however, see no evidence of an absolute plateau in yields for major field crops in the USA.

Reilly and Fuglie (1998) also review evidence beyond simple time-trend statistics. As mentioned with regard to carrying capacity, a gap in yields among sites is often taken as evidence of an opportunity to increase production. Related research tasks include: (i) sustaining present yields, recognizing that yields will erode without maintenance research; (ii) closing yield gaps between low- and high-yielding areas; and (iii) increasing ideal condition yield ceilings. Agronomic potential for increases in the future may come from: (i) more attention to soil, water and other environmental factors, as most attention to date has been on genetic factors; (ii) management of the production process; and (iii) aspects of plant growth and yield, such as increasing the photosynthetic rate, observing that much of the past gains in yield have come from increasing the harvest index (grain-to-straw ratio). One study made what was considered reasonable adjustments in the harvest index, photosynthetic efficiency, pest management, leaf area and harvest efficiency for several crops to generate yield increases of 70–80%. They assumed that these could be achieved in 30–40 years. The increase was equivalent to a 1.3–1.6% per year rate of exponential growth, not unlike the rates assumed in global agricultural studies.

21.7 Future World Food Security

With prices for food declining over the longer term, some would describe the world as one of increasing global food abundance. Despite this abundance, many countries suffer from disrupted agricultural production and distribution systems such that famine is a distinct threat. Continuing poverty means that many people suffer from hunger. The number of people suffering from chronic hunger worldwide has declined from an estimated 844 million in 1979 to 786 million in 1990 (Bongaarts, 1994) but, as Chen and Kates (1994) show, hunger, malnutrition and famine cannot be defined in simple terms. While those at risk of famine are estimated to be less than 1% of the world's population, as many as 34% of the world's children suffer from food poverty and large percentages of the world's population suffer from micronutrient deficiencies or parasitic infestation that contribute to malnutrition.

Three major studies of the future world food situation suggest that, in the absence of climatic change, the food supply will continue to expand faster than demand over the next 20–30 years, with world prices projected to fall (Alexandratos, 1995; Mitchell and Ingco, 1995; Agcaoili and Rosegrant, 1995). Table 21.7 is an attempt by one forecast modelling exercise to convert food supply, demand and economic growth assumptions into estimates of hunger (Fischer *et al.*, 1994). While less optimistic than some, these estimates show that the percentage of people at risk of hunger declines significantly by 2060 but absolute numbers rise.

21.8 Conclusions

There has been much improvement in agricultural climatic change impact assessment in recent years. Uncertainties and limits in climatic scenarios are a severe limit on agricultural impact assessments. Results should be seen as illustrative of the potential magnitude of impacts rather than as predictive. That said, analyses of data suggest that global food production as a whole may not

Table 21.7. Projected number of people at risk of hunger (millions). Numbers in parentheses show percentages of population. (Source: Fischer *et al.*, 1994).

Region	1980	2000	2020	2040	2060
Developing	501	596	717	696	641
	(23)	(17)	(14)	(11)	(9)
Africa	120	185	292	367	415
	(26)	(22)	(21)	(19)	(18)
Latin America	36	40	39	33	24
	(10)	(8)	(6)	(4)	(3)
S. and S.E. Asia	321	330	330	232	130
	(25)	(17)	(13)	(8)	(4)
West Asia	27	41	55	64	72
	(18)	(16)	(14)	(12)	(11)

suffer significantly as a result of climatic change, at least until 2100, given current predictions of climate. Analyses do show significant effects at the regional level. Some key assumptions and caveats are necessary. Firstly, these studies have not considered changes in variability, extreme events, storm intensity and the like – such events have large and devastating effects on agriculture. If climatic change means significantly more variable and extreme conditions, impacts could be more severe than predicted in the current generation of impact assessments. Secondly, more complex interactions of climate and pests, soils and hydrology have not been incorporated in assessment models. Thirdly, the possibly beneficial effects of climatic change on world agriculture are based on estimates that adaptation potential is substantial. If farmers are unable to adapt, or adjustment costs are substantial, then impacts could be more severe.

A substantial interest at this point is to integrate climatic change impacts along with other changes in society, the economy and the agricultural system. Recognizing that regional effects may be severe even if the global impacts on the food system are small, and that the more severe effects may be in the poorer countries of the world, has focused more attention on what the human consequences of climatic changes may be for these areas. These areas are more vulnerable to severe human consequences than the developed world. While all major projections show a world where food availability is increasing relative to population and per capita incomes increasing worldwide, the broad regions of the world considered as developing remain poorer than the developed world of today. Hence, they will remain vulnerable to adverse consequences of climatic change and may have difficulty adapting.

Climatic change will also affect other resource and environmental problems such as land and water degradation and competition for land and water resources. If economic incentives do not encourage production practices that conserve soil and water resources, then climatic change could make these problems more severe. Problems of poverty and hunger may lessen over the next several decades but many will still suffer from hunger and risk famine. Impoverished people will be highly vulnerable if climatic change disrupts local agricultural production, unless programmes are in place to assure adequate health and nutrition. Improving nutrition, reducing the risks of famine and correcting economic incentives to use resources more wisely are actions that make sense today, whether or not climatic changes occur. These same actions can also reduce vulnerability to climatic change.

References

Agcaoili, M. and Rosegrant, M.W. (1995) Global and regional food demand, supply and trade prospects to 2010. In: Islam, N. (ed.) *Population and Food in the Early 21st Century: Meeting Future Food Demand of an Increasing World Population.* Occasional Paper, International Food Policy Research Institute (IFPRI), Washingon, DC, pp. 61–90.

Alexandratos, N. (1995) The outlook for world food production and agriculture to the year 2010. In: Islam, N. (ed.) *Population and Food in the Early 21st Century: Meeting Future Food Demand of an Increasing World Population.* Occasional Paper, International Food Policy Research Institute (IFPRI), Washington, DC, pp. 25–48.

Bie, S.W. (1990) *Dryland Degradation Measurement Techniques.* Working Paper 26, World Bank Environment Department, Washington, DC.

Binswanger, H. (1989) *Brazilian Policies that Encourage Deforestation in the Amazon.* Working Paper 16, World Bank Environment Department, Washington DC.

Bongaarts, J. (1994) Can the growing human population feed itself? *Scientific American* 270, 36–42.

Chen, R.S. and Kates, R.W. (1994) World food security: prospects and trends. *Food Policy* 4, 192–208.

Crosson, P. (1992) Temperate region soil erosion. In: Ruttan, V.W. (ed.) *Sustainable Agriculture and the Environment: Perspectives on Growth and Constraints.* Westview Press, Boulder, Colorado.

Crosson, P. (1995) Future supplies of land and water for world agriculture. In: Islam, N. (ed.) *Population and Food in the Early 21st Century: Meeting Future Food Demand of an Increasing World Population.* Occasional Paper, International Food Policy Reserach Institute (IFPRI), Washingon, DC, pp. 143–160.

Crosson, P. and Anderson, J.R. (1994) Demand and supply trends in global agriculture. *Food Policy* 19, 105–119.

Darwin, R., Tsigas, M., Lewandrowski, J. and Raneses, A. (1995) *World Agriculture and Climate Change: Economic Adaptations.* AER-703, USA Department of Agriculture, Washington, DC.

Erlich, P. and Erlich, A. (1990) *The Population Explosion.* Simon & Schuster, New York.

Fischer, G., Frohberg, K., Parry, M.L. and Rosenzweig, C. (1994) Climate change and world food supply, demand and trade: who benefits, who loses? *Global Environmental Change* 4, 7–23.

Gard, R.C. Jr (1980) *Crop Yields and Climate Change to the Year 2000,* Vol. 1, Research Directorate of the National Defense University, Washington, DC.

Higgins, G.M., Kassam, A.H., Naiken, L., Fischer, G. and Shah, M.M. (1982) *Potential Population Supporting Capacities of Lands in the Developing World.* Technical Report INT/75/P13, Food and Agriculture Organization, Rome.

Houghton, J.T, Meira Filho, L.G., Bruce, J., Lee, H., Callander, B.A., Haites, E., Harris, N., and Maskell, K., (eds) (1995) *Climate Change 1994: Radiative Forcing of*

Climate Change and an Evaluation of the IPCC IS92 Emission Scenarios. Cambridge University Press, Cambridge, UK.

Houghton, J.T., Meira Filho, L.G., Callander, B.A., Harris, N., Kattenburg, A. and Maskell, K. (eds) (1996) *IPCC Climate Change 1995: the Science of Climate Change.* Cambridge University Press, Cambridge, UK, 572 pp.

Kaiser, H.M., Riha, S.J., Wilks, D.S. and Sampath, R. (1993) Adaptation to global climate change at the farm level. In: Kaiser, H.M. and Drennen, T.E. (eds) *Agricultural Dimensions of Global Climate Change.* St Lucie Press, Delray Beach, Florida, pp. 136–152.

Kukla, G. and Karl, T.R. (1993) A nighttime warming and the greenhouse effect. *Environmental Science Technology* 27, 1468–1474.

Mahar, D.J. (1989) *Government Policies and Deforestation in Brazil's Amazon Region.* World Bank, Washington, DC.

McCalla, A.F. (1994) *Agriculture and Food Needs to 2025: Why We Should Be Concerned.* Consultative Group on International Agricultural Research, Washington, DC.

Mitchell, D.O. and Ingco, M.D. (1995) *Global and Regional Food Demand and Supply Prospects.* Working Paper, World Bank, Washington, DC.

Nelson, R. (1988) *A Dryland Management: the Desertification Problem.* Working Paper 8, World Bank Environment Department, World Bank, Washington, DC.

Norse, D. (1994) Multiple threats to regional food production: environment, economy, population. *Food Policy* 4, 133–148.

Parikh, K.S. (1994) Agriculture and food system scenarios for the 21st century. In: Ruttan, V.W. (ed.) *Agriculture, Environment and Health: Sustainable Development in the 21st Century.* University of Minnesota Press, St Paul, Minnesota.

Pittock, A.B., Fowler, A.M. and Whetton, P.H. (1991) Probable changes in rainfall regimes due to the enhanced greenhouse effect. *International Hydrology and Water Resources Symposium*, pp. 182–186.

Reilly, J. (1995) Climate change and global agriculture: recent findings and issues. *American Journal of Agricultural Economics* 77, 727–733.

Reilly, J. and Fuglie, K. (1998) Future yield growth in field crops: what evidence exists? *Soil and Tillage Research* 47, 275–290.

Reilly, J. and Thomas, C. (1993) *Toward Economic Evaluation of Climate Change Impacts: a Review and Evaluation of Studies of the Impact of Climate Change.* MIT CEEPR 93-009WP, Center for Energy and Environmental Policy Research, MIT.

Reilly, J., Hohmann, N. and Kane, S. (1994) Climate change and agricultural trade: who benefits, who loses? *Global Environmental Change* 4, 24–36.

Reilly, J., Baethgen, W., Chege, F.E., van de Geijn, S.C., Erda, L., Iglesias, A., Kenny, G., Patterson, D., Rogasik, J., Rutter, R., Rosenzweig, C., Sombroek, W. and Westbrook, J. (1996) Agriculture in a changing climate: impacts and adaptation. In: Watson, R.T., Zinyowera, M.C. and Moss, R.H. (eds) *Climate Change 1995: Impacts, Adaptations, and Mitigation of Climate Change: Scientific and Technical Analyses*, Cambridge University Press, Cambridge, UK, pp. 427–467.

Repetto, R. (1988) *The Forest for the Trees? Government Policy and the Misuse of Forest Resources.* World Resources Institute, Washington, DC.

Rosenzweig, C. and Iglesias, A. (eds) (1994) *Implications of Climate Change for International Agriculture: Crop Modelling Study.* US Environmental Protection Agency, Washington, DC, 312 pp.

Rosenzweig, C. and Parry, M.L. (1994) Potential impact of climate change on world food supply. *Nature* 367, 133–138.

Schimmelpfennig, D.E. and Yohe, G.W. (1994) Vulnerability of agricultural crops to climate change: a practical method of indexing. Paper presented at *Global Environmental Change and Agriculture: Assessing the Impacts*, Economic Research Service and Farm Foundation Conference, Rosslyn, Virginia.

UN (1992) International Conference on Water and the Environment: Development Issues for the 21st Century, Dublin, Ireland.

UN (1994) *World Population Prospects: The 1994 Revision.* Annex Tables, United Nations, New York.

World Bank (1992) *Price Prospects for Major Primary Commodities.* International Economics Department, Washington, DC.

World Bank (1994) A review of world bank experience in irrigation. Report 13676, World Bank Operations Department, Washington, DC.

Yudelman, M. (1993) *Demand and Supply of Foodstuffs up to 2050 with Special Reference to Irrigation.* International Irrigation Institute, Colombo, Sri Lanka, 100 pp.

Index